Renewable Energy Engineering

Fully revised and updated, the second edition of *Renewable Energy Engineering* provides students with a quantitative and accessible introduction to the renewable technologies at the heart of efforts to build a sustainable future. Key features include new chapters on essential topics in energy storage, off-grid systems, microgrids and community energy; revised chapters on energy and grid fundamentals, wind energy, hydro power, photovoltaic and solar thermal energy, marine energy and bioenergy; appendices on foundational topics in electrical engineering, heat transfer and fluid dynamics, for students from a wide range of STEM backgrounds; discussion of how real-world projects are developed, constructed and operated; over 60 worked examples linking theory to real-world engineering applications; and over 150 end-of-chapter homework problems, with solutions for instructors.

Accompanied online at www.cambridge.org/jenkins2e by extended exercises and datasets, enabling instructors to create unique projects and coursework, this new edition remains the ideal multi-disciplinary introduction to renewable energy, for senior undergraduate and graduate students in engineering and the physical sciences.

Nick Jenkins is Professor of Renewable Energy at Cardiff University. He is a Fellow of the IEEE, the IET and the Royal Academy of Engineering.

Janaka Ekanayake is Senior Professor and the Chair of Electrical and Electronic Engineering of the University of Peradeniya, Sri Lanka. He is a Fellow of the IEEE, IET, IESL and the Sri Lanka National Academy of Sciences.

'Jenkins and Ekanayake's comprehensive description of the main sources of renewable energy provides an excellent support for an introductory course on that topic. The addition of chapters on storage, off-grid, microgrid and community energy systems is particularly valuable.'

Daniel Kirschen, University of Washington

'An expertly crafted introduction to the engineering which underpins the key renewable energy technologies so vital to our future in a net-zero world. With its clear and informative presentation of both theory and practice, supported by many helpful worked examples and well formulated problems to aid understanding, this engineering textbook will prove invaluable for students and teachers of renewables.'

David Quarton, University of Bristol

'The 2nd edition of *Renewable Energy Engineering* by Professors Jenkins and Ekanayake is a concise and very "user friendly" book on renewable energy. It is an excellent book for undergraduate and postgraduate students. The science of different sources of renewable energy is explained with the aid of worked examples. Chapters on energy storage and electrical energy systems makes this a very well-balanced book.'

Maria Vahdati, University of Reading

Renewable Energy Engineering

SECOND EDITION

Nick Jenkins
Cardiff University

Janaka Ekanayake
University of Peradeniya

Shaftesbury Road, Cambridge CB2 8EA, United Kingdom

One Liberty Plaza, 20th Floor, New York, NY 10006, USA

477 Williamstown Road, Port Melbourne, VIC 3207, Australia

314–321, 3rd Floor, Plot 3, Splendor Forum, Jasola District Centre, New Delhi – 110025, India

103 Penang Road, #05–06/07, Visioncrest Commercial, Singapore 238467

Cambridge University Press is part of Cambridge University Press & Assessment, a department of the University of Cambridge.

We share the University's mission to contribute to society through the pursuit of education, learning and research at the highest international levels of excellence.

www.cambridge.org
Information on this title: www.cambridge.org/highereducation/isbn/9781009295789

DOI: 10.1017/9781009295734

First published 2017
Second edition 2024

Printed in the United Kingdom by TJ Books Limited, Padstow Cornwall

A catalogue record for this publication is available from the British Library

A Cataloging-in-Publication data record for this book is available from the Library of Congress

ISBN 978-1-009-29578-9 Hardback
ISBN 978-1-009-29576-5 Paperback

Additional resources for this publication at www.cambridge.org/jenkins2e

Contents

The plate section is to be found between pages 272 and 273

Preface to the Second Edition

The years since the preparation of the manuscript of the first edition of this book have seen a continuing rapid increase in the importance of renewable energy as a subject of study at undergraduate and MSc/MS level. The consequences of burning fossil fuels are now obvious with extreme weather events, prolonged droughts and catastrophic floods occurring in many parts of the world. There is near universal recognition, including among students, that ways must be found to provide the power, light, heat and transport that society needs while minimizing damage to the environment.

The use of renewable sources of energy has become accepted practice, and an increasing number of countries generate a significant fraction of their electricity from renewable technologies. This increases the need to store energy, and the new Chapter 11 of this second edition, 'Storage of Renewable Energy', discusses this important topic. Off-grid renewable energy systems now have access to rapidly maturing technologies and are becoming more cost-effective and reliable. Microgrids, both grid-connected and off-grid, are seen as increasingly attractive while community energy systems are being demonstrated to reduce dependence on large central generation. These topics are addressed in a new Chapter 12, 'Off-Grid Systems, Microgrids and Community Energy Systems'. All other chapters have been carefully reviewed and brought up to date. All comments received on the first edition of the book from students and colleagues have been carefully considered, and this second edition has benefited greatly from this input.

There is a critical shortage of engineers and technologists to develop, construct and operate renewable energy schemes. This book provides material for an initial course in renewable energy that is suitable for undergraduate engineering and science students as well as those following MSc/MS conversion courses who have previously studied a range of numerate subjects. Students from a wide variety of backgrounds wish to study the engineering aspects of renewable energy, and this textbook is intended to be accessible to all of them. A general level of high school physics and mathematics is assumed, and examples throughout the text demonstrate the various calculation techniques. Problems are provided at the end of each chapter with their numerical answers. The problems are graded in terms of their difficulty and the early questions of each chapter can be used to quickly check understanding of the subject matter. The full solutions of the problems, as well as extended exercises for coursework, are on the companion website www.cambridge.org/jenkins2e that is intended for instructors/teachers.

The book provides ample material to support the teaching of a one-semester course, giving an introduction to the commonly used renewable energy technologies. It describes the various renewable energy resources, how they can be quantified and the fundamental principles of their conversion to useful energy. The material to be studied can be chosen based on the particular interests and background of the students. After Chapter 1, 'Energy in the Modern World', the chapters can be studied in almost any order, with the exceptions that Chapter 4, 'The Solar Energy Resource', is a pre-requisite for Chapter 5, 'Photovoltaic Systems', and Chapter 6, 'Solar Thermal Systems'. Chapter 7, 'Marine Energy', uses concepts from Chapter 2, 'Wind Energy', and Chapter 3, 'Hydro Power', and so should be read after them. Chapters 10, 11 and 12 are likely to be particularly interesting for those interested in the generation of electricity from renewable sources of energy. The three tutorial chapters on electricity, heat transfer and fluid flow provide an introduction to these subjects for those who have not previously studied them and are intended for private study.

In addition to their academic careers, the authors have had direct experience of developing and installing renewable energy schemes. Chapter 9, 'Development and Appraisal of Renewable Energy Projects', describes the role that engineers play in the development of projects, while Chapter 10, 'Electrical Energy Systems', addresses the increasingly important question of how to integrate high penetrations of renewable energy into electrical power systems.

As this is an engineering textbook, a large number of drawings are used throughout the book to demonstrate the principles of the renewable energy technologies, as well as photographs to show examples of plant and equipment. Dave Thompson created all the illustrations for the book and his assistance and patience are gratefully acknowledged. Photographs were kindly provided by: Ceylon Electricity Board, Drax Power, Gilkes, Lanka Transformers Ltd, New Mills Engineering, Renewable Energy Systems, RenewableUK, Richard Jones, Rhico and RWE Innogy UK. The authors would also like to acknowledge the generous assistance given by their present and former colleagues, particularly Muditha Abeysekera, Meysam Qadrdan, Richard Marsh, Agustin Valera Medina and Tracy Sweet of Cardiff University, and Tony Burton, Paul Gardner and David Quarton for their expert reviews of particular sections.

Acknowledgement of Sources

Chapter 1

Figures 1.1, 1.2 and 1.3 used data from the *Energy Institute Statistical Review of World Energy*, 2023.

Figures 1.6 and 1.7 were kindly supplied by Richard Jones. RichardJones/BusinessVisual. Rights Managed.

Chapter 2

Figure 2.2 was kindly supplied by RWE Innogy UK.

Figures 2.11 and 2.26 were kindly supplied by RES, Renewable Energy Systems.

Figure 2.23: Alamy Stock Photo: Rob Arnold/Alamy Stock Photo.

Figure 2.27 and Table 2.2: A. Burton, N. Jenkins, D. Sharpe and E. Bossanyi, *Wind Energy Handbook*, 3rd ed., 2021 © John Wiley and Sons.

Chapter 3

Figures 3.3, 3.9, 3.11, 3.17 and 3.21 were kindly supplied by Gilkes, www.gilkes.com.

Figure 3.4: © E.M. Wilson, *Engineering Hydrology*, 1990, Red Globe Press, an imprint of Bloomsbury Publishing Plc.

Figures 3.5, 3.6 and 3.31: Ian David Jones, 'Assessment and design of small-scale hydro-electric power plants', 1988, PhD thesis, University of Salford: http://usir.salford.ac.uk/2212/1/234664.pdf.

Figure 3.7: Godfrey Boyle, *Renewable Energy: Power for a Sustainable Future*, 1996. By permission of Oxford University Press.

Figures 3.12 and 3.25 were kindly supplied by New Mills Engineering Ltd.

Figures 3.20 and 3.22: B.S. Massey and J. Ward-Smith, *Mechanics of Fluids*, 9th ed., © 2012, Spon Press. Reproduced by permission of Taylor & Francis Group.

Figure 3.23: Shutterstock Stock Photo: sulaco229/Shutterstock.com.

Figure 3.27: Godfrey Boyle, *Renewable Energy: Power for a Sustainable Future*, 1996. By permission of Oxford University Press, USA.

Figures 3.28 and 3.29: E. Mosanyi, *Water Power Development*, Volumes I and II, 1957, Hungarian Academy of Sciences.

Figure 3.30: Figure 3 of JWG 11/14.09 'Adjustable speed operation of hydroelectric turbine generators', *Electra* No. 167, August 1996 © CIGRE.

Figure 3.32 was developed from an image kindly provided by Spaans Babcock Ltd.

Figure 3.33: Alamy Stock Photo: Stephen Fleming/Alamy. Rights Managed.

Figure 3.34 was kindly supplied by Renewables First Ltd, www.renewablesfirst.co.uk.

Chapter 4

The data for *Figures 4.3 and 4.4* are from the PVGIS program of the European Union Joint Research Centre: http://re.jrc.ec.europa.eu/pvgis/apps4/pvest.php.

Figure 4.6: M.A. Green, *Solar Cells: Operating Principles, Technology and System Applications*, 1998, University of New South Wales. Reproduced by permission of the author.

Figures 4.8, 4.9 and 4.10: G.M. Masters, *Renewable and Efficient Electric Power Systems*, 2004, Wiley.

Chapter 5

Figures 5.1 and 5.2 were kindly supplied by RES, Renewable Energy Systems.

Figures 5.18 and 5.19 were adapted from F. Lasnier, T. Gan Ang and K.S. Lwin, 'Solar photovoltaic handbook', 1988. Unpublished manuscript.

Section 5.12 was written with Dr Tracy Sweet of Cardiff University.

Chapter 6

Figure 6.1: data from www.gov.uk, *Energy Consumption in the UK (ECUK)*, 2021, Ch. 3, Primary energy consumption.

Figure 6.2: data from www.gov.uk, *Energy Trends*, ET 71-APR 22.

Figure 6.14: Godfrey Boyle, *Renewable Energy: Power for a Sustainable Future*, 1996. By permission of Oxford University Press, USA.

Figures 6.16, 6.17 and 6.20: Volker Quaschning, *Understanding Renewable Energy Systems*, 2005, Earthscan.

Figure 6.18: Shutterstock Stock Photo: Roy Pedersen/Shutterstock.com.

Figure 6.19 was kindly supplied by Rhico Ltd: www.rhico.co.uk.

Figure 6.24: Shutterstock Stock Photo: Tom Grundy/Shutterstock.com.

Figure 6.25: adapted from *The Parabolic Trough Power Plants Andasol 1 to 3*, 2008, Solar Millennium.

Figure 6.27: Shutterstock Stock Photo: Eunika Sopotnicka /Shutterstock.com.

Figure 6.28: Shutterstock Stock Photo: raulbaenacasado/Shutterstock.com.

Table 6.5: data from H. Müller-Steinhagan, 'Concentrating solar thermal power', *Proceedings of the Royal Society Philosophical Transactions* A, **371** (2013).

Chapter 7

Table 7.2: data from *Energy Paper Number 57,* 1989, The Stationery Office. Contains public sector information licensed under the Open Government Licence v3.0.

Figure 7.3: R.H. Clarke, *Elements of Tidal-Electric Engineering,* 2007, IEEE Press–Wiley.

Table 7.4: data from G.A. Alcock and D.T. Pugh, 'Observations of tides in the Severn Estuary and Bristol Channel', 1980, Report No. 112 to the UK Department of Energy. Unpublished manuscript.

Figures 7.11, 7.12, 7.13 and *7.16*: data kindly supplied by Tidal Energy Ltd.

Figures 7.15 and 7.17 were kindly supplied by RenewableUK.

Table 7.7: data from ETSU R120. *A Brief Review of Wave Energy: A Report to the UK Department of Energy*, 1999. Contains public sector information licensed under the Open Government Licence v3.0.

Figures 7.19 and 7.20: *Wind, Waves and Shallow Water Phenomena,* 1999, Open University, with permission from Elsevier.

Figure 7.22: reprinted from Godfrey Boyle (ed.), *Renewable Energy: Power for a Sustainable Future*, 1996, by permission of Oxford University Press, USA.

Figures 7.25 and 7.28: data from ETSU R120. *A Brief Review of Wave Energy: A Report to the UK Department of Energy*, 1999. Contains public sector information licensed under the Open Government Licence v3.0.

Figure 7.26: data from ETSU V/06/00181/REP, *Pelamis – Conclusions of Primary R&D*, Ocean Power Delivery Ltd. Contains public sector information licensed under the Open Government Licence v3.0.

Figure 7.31: Alamy Stock Photo: Ashley Cooper pics/Alamy. Rights Managed.

Chapter 8

Table 8.2: data from P. Quaak, H. Knoef and H. Stassen, 'Energy from Biomass', 1999, World Bank Technology Paper No. 422.

Equation 8.2: from S. Gaur and T.B. Reed, *An Atlas of Thermal Data for Biomass and Other Fuels*, NREL/TP-433-7965, 1995, National Renewable Energy Laboratory, Golden, CO.

Figures 8.5–8.7, 8.12 and 8.13 were kindly provided by Lanka Transformers Ltd.

Sections 8.4.1–8.4.4 are based on the lecture notes of Professor Richard Marsh and Professor Agustin Valera, Cardiff University, and used with permission.

Tables 8.6 and 8.7: all values taken from B.M. Jenkins *et al.*, Combustion properties of biomass, *Fuel Processing Technology*, **54** (1998) 17–46 with permission of Elsevier.

Figures 8.8 and 8.9, and the information for *Tables 8.8 and 8.9*, were kindly supplied by Drax Power Ltd.

Chapter 9

Figures 9.1 and 9.2 were kindly supplied by RES, Renewable Energy Systems.

Chapter 10

Figure 10.5 was kindly provided by the Ceylon Electricity Board.

Figure 10.14: reprinted from *IEEE 1547: Standard for Interconnection and Interoperability of Distributed Energy Resources with Associated Electric Interfaces*, with permission from IEEE PES.

Figure 10.24 was kindly provided by Alstom.

Chapter 11

Figure 11.4: data from IEA Special Reports on Hydro Power, July 2021.

Figure 11.9 was kindly provided by Siemens Energy.

Figure 11.15: data from M.S. Imamura, P. Helm and W. Palz, *Photovoltaic System Technology,* 1992. Published by H.S. Stephens and Associates on behalf of the European Commission.

Figure 11.22 was kindly supplied by RES, Renewable Energy Systems.

Chapter 12

Figures 12.3 and 12.4 are from a Cardiff University student-led project that was supported by the Mothers of Africa charity and Cardiff University. System design and photographs provided by D.J. Rogers, L.J. Thomas and J.M. Stevens.

Figure 12.9 was kindly provided by the Ceylon Electricity Board.

1 Energy in the Modern World

INTRODUCTION

An adequate supply of energy is essential for the working of any modern society and at present most energy comes from burning fossil fuels – coal, oil and gas. Energy is needed for cooking food, heating or cooling, lighting, materials processing and for transport. Primary energy from fossil fuels is either used directly to provide heat and power transport or converted into electricity. An increasing quantity of electrical energy is consumed in the operation of computers and communication equipment.

Over the past 20 years, the world demand for energy has grown continuously at a rate of increase of around 2% per year. This level of consumption cannot be sustained because of the emission of greenhouse gases from burning fossil fuels and the subsequent impact on our environment. The reserves of coal, oil and gas that are now being used were laid down over millions of years and have been exploited for less than three centuries. The carbon that was captured through photosynthesis as the reserves of fossil fuel were formed is now being released rapidly into the atmosphere as CO_2.

There is clear agreement among climate scientists and policy makers that burning fossil fuels and the consequent emission of CO_2 (and other gases) into the atmosphere is leading to dangerous climate change. In response to this threat, many countries are committed to reducing emissions from their energy system, and renewable energy has a central role in this transition. However, renewable energy brings its own challenges. The initial capital cost of renewable energy schemes remains high, although it is decreasing rapidly, and their output depends on the resource and so varies with the strength of the sun and wind.

This book examines the various sources of renewable energy and how they can be used effectively. It focuses on those technologies that can make a significant contribution to energy supply over the next 20 years or so and pays particular attention to the renewable energy resource. Without a good energy resource, it is impossible to develop a cost-effective renewable energy scheme.

This introductory chapter starts with a review of energy use in the modern world and demonstrates the scale of the challenge that the world faces as it moves to a sustainable energy system. The implications of the simple equation describing exponential growth are described. Most governments are making strenuous efforts to improve the efficiency with which energy is used and to control energy demand. However, as the world population increases and societies become richer, the worldwide use of energy continues to grow. Approaches to controlling demand for energy are described and the difficulties of reducing energy use discussed. Discounted cash flow analysis is commonly used for the financial evaluation of projects to improve energy efficiency and for renewable energy supply. This simple technique is explained, together with the consequences of using it to evaluate projects that are initially expensive but have benefits in the future. The final section of the chapter re-emphasizes the need for renewable energy as the environmental impact of burning fossil fuel increases.

1.1 Energy in the Modern World

Figure 1.1 shows the worldwide consumption of primary energy from 2000 to 2022 and Figure 1.2 shows the percentage share of each source of energy. World energy consumption has increased steadily over this period with occasional variations in the rate of increase due to changes in the level of economic activity. There was a pause in the rate of increase of energy consumption in 2009–10 caused by the economic crisis in western countries and a larger reduction in 2020–21 due to the worldwide pandemic. However, these reductions were temporary and only delayed

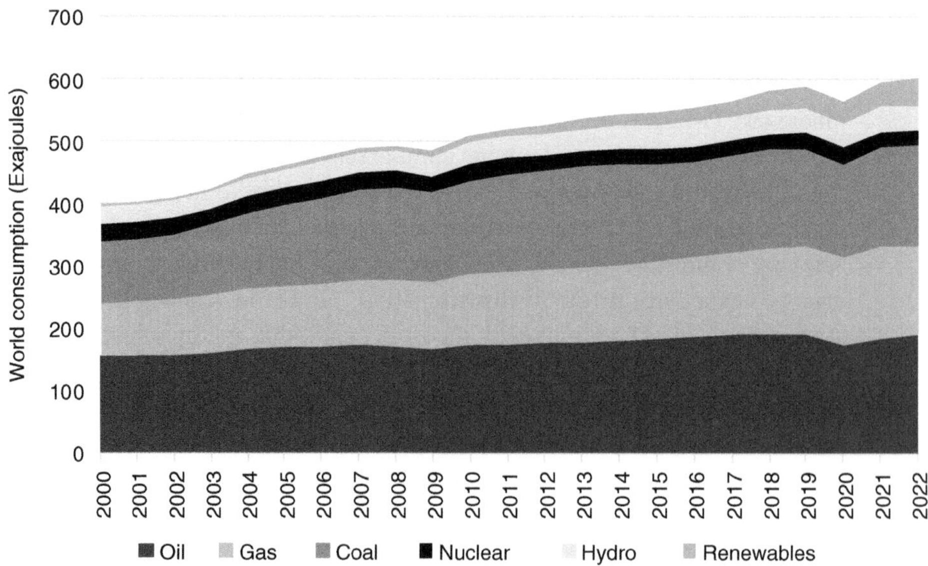

Figure 1.1 World consumption of primary energy.
Source: *Energy Institute Statistical Review of World Energy*, 2023.
(A black and white version of this figure will appear in some formats. For the colour version, refer to the plate section.)

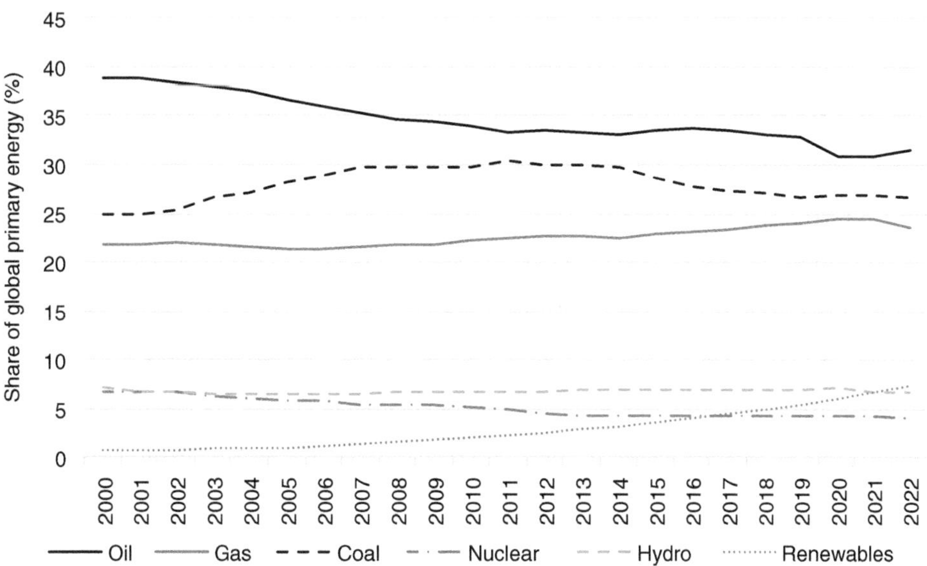

Figure 1.2 World share of primary energy.
Source: *Energy Institute Statistical Review of World Energy*, 2023.
(A black and white version of this figure will appear in some formats. For the colour version, refer to the plate section.)

the underlying increase. It can be seen that the use of all fossil fuels continues to increase.

Oil remains the most widely used fuel but with a considerable fraction of world energy coming from coal. The share of both fuels is decreasing but the fraction of natural gas is increasing. Natural gas (methane) is used for heating, electricity generation and as a chemical feedstock. The exploitation of shale gas, particularly in the USA, has increased supplies of natural gas considerably. The share of electricity generated from nuclear and hydro plants has remained substantially constant in recent times while the share from renewables is growing rapidly.

Figure 1.1 shows only those energy sources that are traded commercially and so does not include traditional biomass. Wood and other biomass fuels are particularly important in some areas of developing countries where they are used for cooking. Shortage of wood often leads to over-exploitation of the resource, environmental damage and high prices. These are major problems for some poor rural communities.

Electricity generation from renewables is growing rapidly. Figure 1.3 shows that the generation of electrical energy from renewable sources had an average annual growth rate between 2010 and 2020 of 16%. During this time the worldwide generating capacity from solar photovoltaics (PV) increased at an annual average rate of 38% and the capacity of wind generation increased at an annual average rate of 15%. These are extremely high rates of growth and imply that the installed capacity of solar generating doubled every two years, and the capacity of wind energy generation doubled every five years.

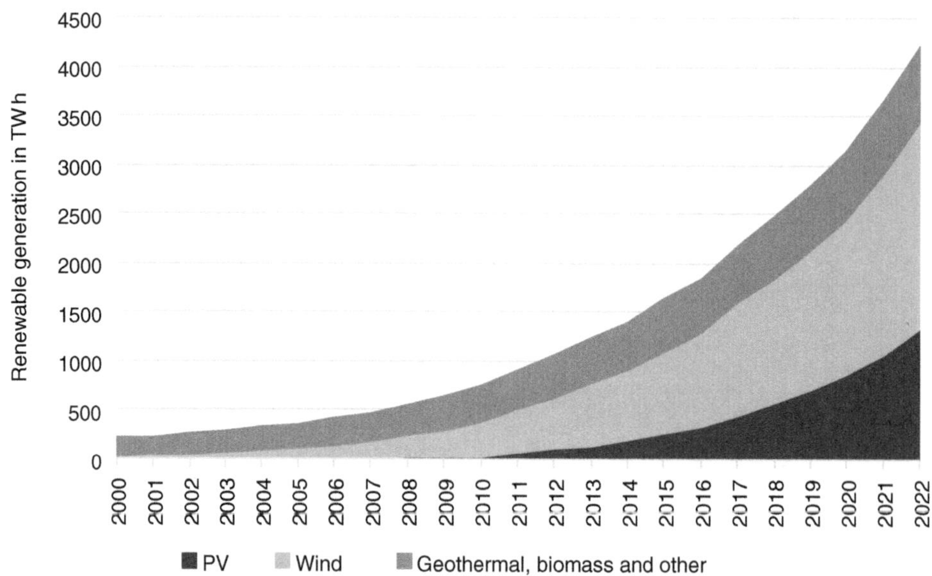

Figure 1.3 World annual generation of electricity from renewable sources.
Source: *Energy Institute Statistical Review of World Energy*, 2023.
(A black and white version of this figure will appear in some formats. For the colour version, refer to the plate section.)

Units Used to Quantify Energy

joule (J): this is the SI unit of energy. An exajoule (EJ) is 10^{18} joules.

kilowatt-hour (kWh): this is the usual unit of electricity production or consumption. 1 kWh is equal to 3.6×10^6 J. However, large generators produce many thousands of kWh expressed in MWh, GWh and TWh. The relationships between these are:

$$1\,\text{MWh} = 1000\,\text{kWh}$$

$$1\,\text{GWh} = 1000\,\text{MWh}$$

$$1\,\text{TWh} = 1000\,\text{GWh, or } 10^9\,\text{kWh}$$

Energy from technologies that generate electricity directly, e.g. nuclear, hydro and renewables, is sometimes expressed as the electrical output (Wh) or as the equivalent energy (J) that would be required by a thermal power station to produce the same amount of electricity. In this chapter, an efficiency of generation of between 36% and 40% is assumed, depending on the date of the data and the efficiency of generating stations at that time.

tonne of oil equivalent: when discussing energy, it is often convenient to use a single unit. With this unit, energy from different fuels is converted into energy produced by burning one tonne of oil. The heat content of one tonne of crude oil depends on the origin of the oil but is generally taken as 41.87 GJ. The following conversions are usually used:

$$\text{Heat value of 1 tonne of coal} = 29.3\ \text{GJ}$$
$$= 0.7\ \text{tonnes of oil equivalent}$$
$$1\ \text{MWh of electrical energy} = 3.6 \times 10^6 \times 1000\ \text{J}$$
$$= 3.6\ \text{GJ}$$
$$= 0.086\ \text{tonnes of oil equivalent}$$

1.1.1 Exponential Growth

Since 2000, world energy consumption has grown continuously at around 2% per year and continues to do so. This apparently rather small rate of increase does not at first sight give cause for great concern. However, this is an exponential rate of increase described by the equation

$$E_n = E(1 + r_e)^n \tag{1.1}$$

where

E_n is the energy use in year n,
E is the energy use at present,
r_e is the rate of increase of use of energy,
n is the number of years.

Equation 1.1 shows that over 25 years a rate of growth of 2% per year results in an increase in energy use by a factor of 1.64. If this rate of growth of energy use is maintained, then world energy consumption will increase from its value in 2022 of around 600 EJ to around 984 EJ over the next 25 years. An alternative representation of exponential growth is the doubling time. This is the time in years for an exponentially increasing quantity to double; it is shown in Table 1.1. The doubling time may be estimated approximately by dividing 72 by the percentage rate of increase. A worldwide primary energy use increasing at a long-term rate of 2% per year implies a doubling every 36 years.

Table 1.1 Doubling time of exponential increase

Rate of increase (% per year)	Approximate doubling time (years)
0.1	720
0.5	144
1	72
2	36
3	24
5	14

Example 1.1 – Exponential Growth

Use of natural gas is increasing at a rate of 3% per year and is expected to maintain at least this rate of growth. At present the worldwide use of gas as a primary energy source is 140 exajoules. Construct a table showing the use of gas over the next 30 years if the rate of increase varies between 1% and 4% per year. How much gas will be used in 30 years' time?

Solution

Use of gas over 30 years, using Equation 1.1 with a present gas consumption of 140 EJ

	Rate of increase per year			
Number of years	1%	2%	3%	4%
5	147	155	162	170
10	155	171	188	207
20	171	208	253	307
30	189	254	340	454

It may be seen that if an annual rate of increase of 3% is maintained over 30 years this would lead to a more than doubling of the use of gas from 140 to 340 EJ with a similar increase in emissions of CO_2.

1.2 Limiting Energy Use

The increase in worldwide energy use that can be anticipated simply from continuation of the historical rate of increase will place great strain on the ecosystems and energy resources of the planet. The obvious first response is to reduce energy use or at least limit its rate of increase, and the governments of most countries that consume significant quantities of energy have active programmes to improve the effectiveness with which energy is used. These can be considered in two aspects: energy efficiency and energy conservation.

1.2.1 Energy Efficiency

Energy efficiency is achieving the same goals and maintaining the same levels of services and comfort but with the use of less energy. An example of a successful energy efficiency measure is the switch from incandescent to low-energy light bulbs. This change in the way energy is used was made possible by the development of lighting technology, compact fluorescent (CF) and light-emitting diode (LED) lamps, together with new product standards that effectively stopped the sale of incandescent bulbs. Low-energy lights produce the same level of light as incandescent bulbs while using only 15–30% of the electrical energy; see Table 1.2. The same outcome, i.e. the same level and quality of light, is achieved with greater energy efficiency.

Table 1.2 Light output and power consumption of different lamps

	Power consumption		
Light output (lumens)	Incandescent lamp (W)	Compact fluorescent lamp (W)	Light-emitting diode lamp (W)
400–500	40	7–11	4–5
650–900	60	13–18	6–8
1100–1750	75	18–22	9–13
>1800	100	23–30	16–20
>2700	150	30–55	25–28

Another good example of a successful energy efficiency measure was the regulation made in the UK that all new domestic central heating gas boilers had to be high-efficiency condensing units. These produce the same level of heat as the older non-condensing boilers but consume less gas.

Around 40% of the primary energy used in the USA is within buildings. The energy is used for space heating, cooling and ventilation together with lighting and electrical appliances. Studies show that this energy use could be reduced by up to 30% with no reduction in the performance of the building if suitable improvements in energy efficiency were made using technology that is currently available. The cost of improving the insulation of buildings and installing advanced building controls would be balanced by savings in the use of energy over the coming years. The same level of comfort and services within the buildings would be maintained.

Improving energy efficiency is, at first sight, a technical question of using higher-efficiency equipment and limiting energy losses. It might be thought that if an economic case can be made that shows a benefit greater than the cost, then the appropriate action would be taken. The energy and financial savings would then follow. However, experience has shown that improving energy efficiency is, in practice, much more complex than a simple engineering and economic appraisal would indicate. There is a considerable body of literature that attempts to explain why people and organizations are reluctant to implement energy efficiency measures that, at first sight, give obvious financial benefit. Examples of the well-documented difficulties of implementing energy efficiency programmes include the following.

- For many individuals, the decision when buying certain goods, e.g. a car, does not include any assessment of its future energy use. Many modern cars have engines much more powerful than are necessary for their simple transportation function and hence have lower fuel efficiency than is necessary to transport the driver and passengers.
- The landlord–tenant problem refers to the situation where the cost of implementing energy efficiency measures in a building falls on the landlord but the reduced energy bills would benefit the tenant. The landlord is then reluctant to make the initial investment. Similarly, developers and builders of new dwellings or offices may be more concerned with reducing initial capital cost (which they bear) rather than achieving long-term low-energy performance (which benefits the building user).

- Individuals are more concerned with immediate expenditure than with benefits that accrue in the future. This is sometimes described as individuals having a high personal discount rate.
- Organizations and individuals may be limited in the amount of capital they can raise. Hence even if an investment in energy efficiency shows an attractive payback in, say, three years, it may not be possible to fund the initial changes that are required. This is a particular problem in public buildings that often operate with maintenance budgets over only one year.
- Improving the efficiency of existing buildings or industrial processes may involve disruption and this can delay an improvement that is economically attractive.

1.2.2 Economic Appraisal of Energy Efficiency Measures

Energy efficiency measures typically require an initial investment that is recovered by savings in energy costs that are made over a number of years.

The simplest way to evaluate an energy efficiency project is to calculate the payback period:

$$\text{Payback period} = \frac{\text{Capital cost}}{\text{Annual cost saving}}$$

However, most people and organizations have a time preference for money and they would rather receive money today not next year and would prefer to pay out money next year in preference to today. This makes it more difficult to justify an energy efficiency measure that has an initial expenditure but in which the benefits accrue only in later years. Financial analysis that recognizes the time value of money is known as discounted cash flow (DCF) appraisal; it is summarized here and discussed in more detail in Section 9.2.1.

The mechanics of DCF analysis are simple and the calculation may be implemented easily on a spreadsheet; the main functions are often included in commercially available spreadsheet packages.

The present value of a sum received or paid n years in the future is:

$$V_p = \frac{V_n}{(1+r_d)^n}$$

where

V_p is the present value of the sum,
V_n is the value of the sum in year n,
n is the number of years,
r_d is the discount rate.

The discount rate chosen reflects the value that the decision maker places on money and the risks that are anticipated in the project. The effect of a high discount rate is to make any expenditure or income in the future appear less significant. A discount rate of 10–15% would not be unusual for industrial energy efficiency projects.

Example 1.2 – Economic Appraisal of an Energy Efficiency Measure

Consider the installation of a building management system that controls the heating and cooling systems of a large building. The cost of installing the management system is £30 000 and it will result in savings of £10 000/year. Evaluate the project by calculating the payback period and the Net Present Value (NPV) at a discount rate of 15%.

Solution

The simple (non-discounted) appraisal is

$$\text{Payback period} = \frac{\text{Capital cost}}{\text{Annual cost saving}} = \frac{30\,000}{10\,000} = 3\,\text{years}$$

However, the simple payback period does not reflect the time value of money. Assume that a discount rate of 15% is chosen.

Year n	Expenditure (£)	Saving (£)	Discount factor $\frac{1}{(1+r_d)^n}$	Net present value at year n
0	–30 000		1	–30 000
1		10 000	0.869	–21 304
2		10 000	0.756	–13 743
3		10 000	0.657	–7168
4		10 000	0.571	–1450
5		10 000	0.497	3522

The NPV after three years is still negative at −£7168 when the payback period calculation indicates the project is recovering its costs. A negative NPV indicates the project should not proceed. The project only starts to have a positive NPV part way through year 5 or if a lower discount rate is chosen.

This calculation illustrates how the use of DCF analysis can make energy efficiency measures (where an initial expenditure is recovered only over time) appear unattractive.

1.2.3 Energy Conservation

Energy conservation[1] goes further than energy efficiency by changing the behaviour of individuals and the way they use energy. Achieving improvements in energy efficiency, where standards of comfort and service are maintained and technical improvements lead to reduced use of energy, is difficult due to the non-technical barriers discussed in Section 1.2.1. Implementing energy conservation, where energy demand is reduced through individuals modifying their behaviour and so reducing their demand for energy, is even more difficult. An example of a successful energy conservation measure is raising the set point of thermostats in air-conditioned office buildings and encouraging staff to dress less formally (e.g. men without jackets and

[1] Sometimes referred to as 'energy sufficiency'.

ties) so that they are comfortable in a higher temperature indoor environment. However, unless it is introduced carefully such an initiative involving changes in office dress can be threatening to some staff as it can alter the apparent social order of the office. In a similar way, reducing the set point of the heating thermostat in a home so that sweaters are required indoors during the winter is an obvious way to reduce energy use but may be perceived as being less comfortable and lead to disagreements within the family.

1.2.4 Management of Energy Demand Only Through Price

An over-simple economic view might be that increasing the price of energy will result in reduced energy consumption. It might be argued that, provided the price of energy reflects all costs, including those to the environment, known as external costs, then by allowing prices to rise the use of energy would be limited to an economically optimal level. However, relying only on increases in price to reduce energy demand is unsatisfactory. The energy demand of wealthy individuals and some industries where energy is a small fraction of overall costs does not reduce significantly as price increases. This can be stated more formally as that the demand for energy is inelastic with respect to its price. Poorer individuals and families spend a disproportionally large fraction of their income on energy and often live in dwellings that are badly insulated. Increasing energy prices leads to some reduction in their energy demand but can cause significant hardship to lower-income groups living in dwellings that cannot maintain reasonable levels of comfort with reduced energy use. It has been shown that lack of energy in a poorly insulated dwelling has serious consequences for human health.

1.2.5 Smart Meters

Traditionally, electricity and gas consumption was measured using mechanical meters that were read on a regular basis (e.g. three-monthly) and a paper account prepared and sent to the consumer. Many countries are now introducing smart meters that will give much more accurate and timely information on gas and electricity energy use to domestic consumers. The meters may be supplemented by an in-home display that is placed in a convenient place such as the kitchen and shows the energy being used in the dwelling in real time. The in-home display shows how much electricity and gas is being used throughout the day and helps identify where energy is wasted or used at a time when its cost is high. Smart meters and in-home displays provide better information on energy use to help customers to reduce their energy consumption or change the time of its use.

Smart meters can have two distinct objectives in support of demand reduction. They can be used to encourage consumers to:

- reduce overall energy demand,
- switch power consumption away from times of high system load and price.

Table 1.3 Reduction in energy use due to the use of smart meters

Customer type	Gas (%)	Electricity (%)
Domestic	2.2	3.0
Non-domestic	4.5	2.8

A reduction in the overall (annual) energy demand leads to reduced use of fuel and lower emissions. This is always desirable but improved information is only one factor affecting energy use. The results of trials of electricity smart meters are conflicting but they indicate that reductions in annual energy use of around 3% are possible. However, in the successful trials constant engagement by the energy supplier with the customers was necessary if the initial reductions were to be maintained.

Table 1.3 shows the percentage reduction in energy assumed by the UK Government when evaluating the performance of smart meters.

Switching demand away from times of peak load reduces the need for electricity generators and transmission capacity but may not affect annual energy use. Reducing peak demand is particularly important if the capital that is available to build new energy supply capacity is limited. By measuring the energy consumed every half an hour throughout the day, time-of-use tariffs can be offered to customers. These apply different charges during the day reflecting the value of electricity and are particularly important in electric power systems with a large fraction of renewable generation.

1.2.6 Demand-Side Response and the Variable Value of Electricity

The value of electricity varies throughout the year and peaks at times of maximum load or when there is a shortage of low-cost generating capacity, while at times of surplus generation the value of electricity can become very low. In an alternating current (ac) power system it is necessary to balance supply and demand on a second-by-second basis to maintain the frequency at 50 or 60 Hz. Rather than rely only on varying the output of generators to match demand it is becoming increasingly common to incentivize consumers to reduce their load at times of shortage of supply. In this way the need for generators to meet the peaks in system demand is reduced. This is common practice for large industrial loads and is becoming increasingly common for smaller commercial and domestic loads.

Demand-side response is particularly important in those electricity systems with a significant quantity of renewable generation that depends on the renewable resource (wind or sun). At times of plentiful renewable resource, consumers are encouraged to use electricity at a low price, while at times of shortage a high price is used to limit demand. This is illustrated in Figure 1.4, which shows how some

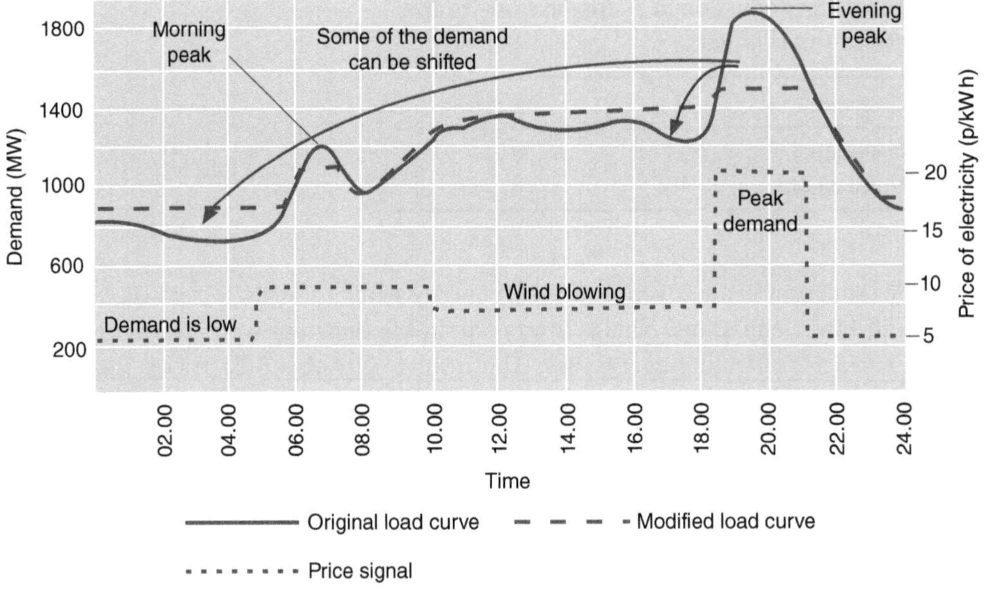

Figure 1.4 Using the price of electricity to shift demand.

of the demand of a region can be moved from the evening peak to the night with a time-varying price signal.

In summary, although limiting energy use is the obvious first step when addressing the world's energy problem, experience has shown that it is not at all easy to achieve. There are a number of rather simple and well-demonstrated approaches to increase the efficiency with which energy is used and also to conserve energy by encouraging individuals to modify their behaviour. These measures would appear to be obviously sensible and economically attractive for individuals and companies to undertake and would lower the impact of energy use on the environment. There is no doubt that energy efficiency and energy conservation measures should be implemented vigorously. However, for a variety of administrative and human reasons, experience shows that initiatives to improve energy efficiency and conserve energy are difficult to implement and the results of schemes to encourage energy demand reduction are often disappointing.

1.3 The Need for Renewable Energy

Worldwide energy use is increasing continuously and although every effort should be made to manage demand through energy efficiency and conservation this is not at all easy to achieve. Continuing to rely on fossil fuels to provide the world's energy has many problems, the most pressing of which is the emissions of the greenhouse gases and it is now recognized that some fossil fuel deposits, particularly coal, will need to be left in the ground unused if catastrophic climate change is to be avoided.

1.3.1 Environmental Impact of Burning Fossil Fuels

The consequences of burning fossil fuel can be listed as:

- local effects – particulate emissions and effect on air quality,
- regional effects – acid rain,
- global effects – climate change.

Thermal power stations, internal combustion engines and building heating systems all produce gaseous emissions and very small particles that are damaging to human health. Examples of the local consequences of such emissions are the photo-chemical smogs that occur in some large cities, often from vehicle exhausts. The famous London smog of the 1950s, which was caused primarily by the burning of coal for industry and domestic heating, resulted in legislation to require the use of smokeless fuel in UK cities and the electricity-generating stations to be moved out of the cities. Poor air quality in a number of large cities, particularly where atmospheric conditions concentrate the pollutants, is a significant cause of poor human health. Poor air quality disproportionally affects poorer members of society living close to the sources of the pollution.

Burning coal in power stations produces sulphur dioxide (SO_2) and other pollutants that when emitted to the atmosphere cause acid rain and considerable environmental damage, particularly to lakes and forests. Acid rain tends to be a regional effect. Historically British coal-fired power stations caused environmental damage in Germany and Scandinavia; the prevailing winds blow from the west. Acid rain has been recognized as an important environmental problem; regulations in parts of the USA limit the sulphur content of coal that can be burnt in power stations while in Europe large coal-fired power stations that are not fitted with flue gas de-sulphurization equipment have been phased out. Flue gas de-sulphurization equipment extracts the SO_2 from the flue gases of the power station boilers, but the equipment adds cost and takes significant power to operate, thus reducing the efficiency of the generating unit.

There is a now clear agreement among scientists and policymakers that the earth's climate is being changed by human activity through the emission of greenhouse gases. The main greenhouse gases are carbon dioxide (CO_2), methane (CH_4), nitrous oxides (NO_x) and fluorocarbons. Water vapour also plays a major role in the greenhouse effect. The greenhouse effect is a most complicated phenomenon with important impacts of gases and particles in the atmosphere that can either increase or lower the earth's temperature. However, it can be understood simply as the effect of gases in the upper atmosphere absorbing the long-wavelength radiation that is emitted from the earth's surface (Figure 1.5). The sun is a high-temperature source of energy with an effective temperature at its outer surface of around 6000 K. It emits short-wavelength (high-frequency) radiation that passes through the earth's atmosphere. This radiation strikes the earth, warming it, and the earth then re-radiates long-wavelength (low-frequency) radiation from its lower surface temperature. The high-frequency radiation from the

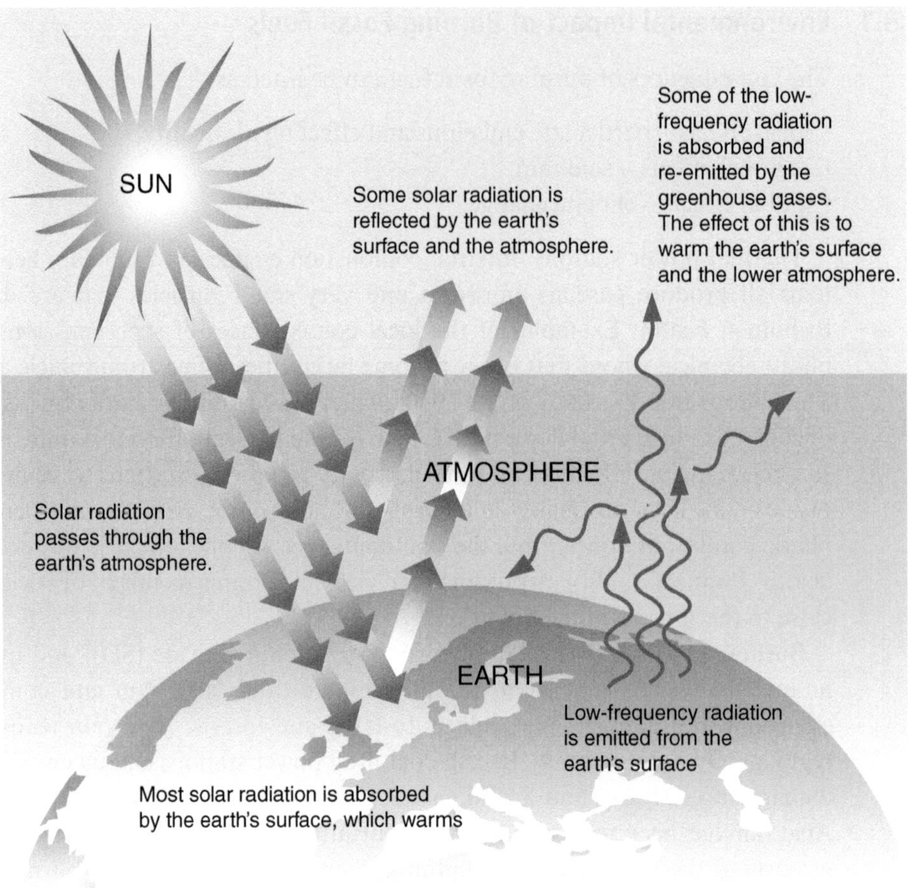

Figure 1.5 Simple representation of the greenhouse effect.

sun passes through the earth's atmosphere largely unaffected, while the concentration of gases in the upper atmosphere absorbs and reflects the lower-frequency (longer-wavelength) radiation.

The temperature of the earth depends on the balance between the incoming high-frequency radiation from the sun and the lower-frequency radiation re-radiated from the earth's surface. By increasing the concentration of greenhouse gases in the atmosphere, more of the low-frequency radiation is trapped and so the temperature of the earth increases. The concentration of existing greenhouse gases in the earth's atmosphere causes the temperature of the earth to be maintained at a level suitable for life; without it the earth would be colder by some 30 °C. By increasing the concentration of greenhouse gases, as we are currently doing, the earth's temperature increases, and the climate is changed. Although an increase in average temperature has significant implications, the consequent effects such as increasing sea levels due to melting of the ice in the polar regions and increases in the frequency of extreme weather events are potentially even more serious. Greenhouse gases disperse throughout the earth's atmosphere and so the effect is global.

Carbon dioxide is an inevitable product of burning fossil fuels and once emitted it remains in the atmosphere for more than 100 years. It is one of the most important greenhouse gases and many countries have policies to reduce emissions of CO_2. Countries responsible for 90% of the world's emissions of greenhouse gases have agreed to reduce emissions to net-zero by 2050. Net-zero is the state when the greenhouse gases emitted into the atmosphere, particularly CO_2, are balanced by actions to remove them. By restricting the emission of CO_2 and other greenhouse gases it is hoped to limit the rise in global mean surface temperature rise to around 1.5 °C above preindustrial levels. A rise of this magnitude with the associated increase in extreme weather events will still have important consequences for agriculture and biodiversity. Immediate action to reduce emissions has been shown to be essential as well as more cost-effective than delay.

1.3.2 Low-Carbon Electricity Generation

Electricity is only one form of energy, but its generation is easier to decarbonize with renewable energy than other uses of energy such as domestic or industrial heating, steel making or cement manufacture. A number of countries are making good progress in reducing the CO_2 emitted by their electricity systems and the UK has set a goal of decarbonizing its electrical sector by 2035.

Table 1.4 shows the carbon intensities of the fossil fuel generation technologies that have been used in the interconnected national power system of Great Britain (GB). Carbon intensity of generation is the amount of CO_2 emitted for every unit of electricity generated. Coal generation is the most carbon intensive, but generation from coal has now largely ceased in GB. Heavy fuel oil is relatively expensive and hardly used in GB for electricity generation. There is a considerable amount of electricity generated from natural gas usually in combined cycle units of a gas turbine followed by a steam turbine. The carbon intensity of the GB generating portfolio has more than halved since 2008 mainly due to the move away from coal and the increase in renewable energy generation (solar PV, wind and biomass). Nuclear

Table 1.4 Carbon dioxide emissions from electricity generation in Great Britain

Fuel	Tonnes of CO_2/GWh of electrical output
Coal	915
Oil	633
Gas (combined cycle)	405
Carbon intensity of GB generation portfolio in 2008	535
Carbon intensity of GB generation portfolio in 2021	185

Source: *Digest of UK Energy Statistics* (DUKES).

generation has remained roughly constant although some nuclear generating stations are reaching the end of their lives.

The options to generate electricity without emitting CO_2 are limited to:

- renewable energy,
- nuclear energy,
- fossil fuel generators equipped with carbon capture and storage.

There are those who consider that nuclear fission (i.e. splitting a heavy atomic nucleus into two lighter nuclei to release a large quantity of energy) is an attractive technology, and that nuclear generation should be expanded. However, generation of electricity using nuclear fission has some important difficulties including high capital costs and continuing uncertainty over the disposal of nuclear waste. Also, the close links between civil and military nuclear programmes during its historical development has led to concerns over its use in a number of countries. Even if nuclear fission maintains or expands its role in electricity generation in countries where the population supports its use, it is unlikely to be able to provide sufficient energy to substitute for fossil fuels in a world of increasing energy demand. Nuclear fusion (i.e. the combination of two light nuclei to form a heavy nucleus and release an even larger quantity of energy) remains a technology under development.

Removing carbon either before or after fossil fuels are burnt and storing CO_2 underground has the attraction that modified conventional fossil-generating units can continue to be used. This is known as Carbon Capture and Storage (CCS) or Carbon Capture Utilization and Storage (CCUS) where some of the CO_2 is used in chemical processes. CCS is an important future technology, particularly in combination with biomass generation, which results in the CO_2 in the atmosphere being reduced. However, neither the technology for extracting the carbon from fossil fuel generation nor the technology for storing the CO_2 have yet been implemented at commercial scale.

Example 1.3 – Carbon Intensity of Generation

A nation's annual electricity demand of 420 TWh is met by 280 TWh of wind generation, 60 TWh of nuclear, 45 TWh of solar, 25 TWh of biomass and 10 TWh generation from gas. The estimated carbon dioxide emission from natural gas generation is 405 tonnes/GWh, biomass 190 tonnes/GWh, solar and wind 25 tonnes/GWh and from nuclear generators is 15 tonnes/GWh.

(a) Calculate the carbon intensity of generation (CO_2 emission per GWh of the electricity generated).

(b) If the combined cycle gas turbine generators are equipped with Carbon Capture and Storage (CCS) their carbon dioxide emissions are reduced to 25 tonnes/GWh. What is now the carbon intensity of generation of the portfolio?

(c) If the biomass generators are fitted with CCS, the effect is to reduce the amount of CO_2 in the atmosphere by −50 tonnes/GWh. What is now the carbon intensity of generation?

Solution

(a) This calculation is conveniently done with a spreadsheet. The carbon intensity of the original portfolio is 42.4 tonnes of CO_2/GWh.

Technology	Carbon intensity of generation (tonnes CO_2/GWh)	Electricity generated (TWh)	Carbon emissions (Mtonnes CO_2)	Carbon intensity of portfolio (tonnes CO_2/GWh)
Wind	25	280	7.0	
Nuclear	15	60	0.9	
Solar	25	45	1.1	
Biomass	190	25	4.8	
Gas turbine	405	10	4.1	
Total portfolio		420	17.8	42.4

(b) If carbon capture and storage is fitted to the gas turbine generators, then the carbon emissions from gas generation reduce to 0.3 Mtonnes and the carbon intensity of generation of the portfolio reduces to 33.4 tonnes of CO_2/GWh.

Technology	Carbon intensity of generation (tonnes CO_2/GWh)	Electricity generated (TWh)	Carbon emissions (Mtonnes CO_2)	Carbon intensity of portfolio (tonnes CO_2/GWh)
Wind	25	280	7.0	
Nuclear	15	60	0.9	
Solar	25	45	1.1	
Biomass	190	25	4.8	
Gas turbine	405	10	0.3	
Total portfolio		420	14.0	33.4

(c) If the biomass generation is also fitted with CCS, then it captures and stores 1.3 Mtonnes of CO_2. The carbon intensity of generation of the portfolio then further reduces to 19.1 tonnes of CO_2/GWh.

Technology	Carbon intensity of generation (tonnes CO_2/GWh)	Electricity generated (TWh)	Carbon emissions (Mtonnes CO_2)	Carbon intensity of portfolio (tonnes CO_2/GWh)
Wind	25	280	7.0	
Nuclear	15	60	0.9	
Solar	25	45	1.1	
Biomass	–50	25	–1.3	
Gas turbine	25	10	0.3	
Total portfolio		420	8.0	19.1

1.4 Challenges of Renewable Energy

As with all forms of energy, renewable energy has its own disadvantages, and the transition away from fossil fuels will be challenging. Fossil fuels have the important advantages of being dense forms of energy, with respect to both weight and volume, and can be stored for months or even years in simple containers without degradation or excessive losses. The high-energy densities of fossil fuels and their low cost of storage is in contrast to the characteristics of renewable energy. Most renewable energy is derived ultimately from the sun, which is a diffuse energy source with a maximum irradiance power at the earth's surface of around 1 kW/m^2 and provides energy at a maximum rate of 6–8 kWh/m^2 per day. Although some forms of renewable energy, e.g. wind, biomass and hydro energy, concentrate the solar energy, the energy density of almost all renewable sources is much lower than that of fossil fuels. Many forms of renewables are converted into electrical energy. Electricity is the most flexible form of energy; it is quick and easy to control and at its point of use has very little environmental impact. However, there is no generally applicable, practical way to store large quantities of electrical energy cheaply, and electrical systems have to be operated by balancing their input and output power on a second-by-second basis.

1.4.1 Capacity Factors of Renewable Generation

No electricity generator operates at its full rated output all the time. Some types of generator are expensive to build but relatively cheap to run and are designed to operate at full output almost continuously. These are often known as baseload generators and are typically nuclear or large thermal units. Other generators are cheaper to build but more expensive to operate and are intended to operate only during times of peak demand or when other generators fail. These are known as peaking plant and are typically open-cycle gas turbines or reciprocating engines, or older, less efficient fossil generators.

The measure of how effectively a generator is used over the year is given by its Annual Capacity Factor (ACF) defined as:

$$\text{ACF} = \frac{\text{Electrical energy generated over a year } [\text{Wh}]}{\text{Generator rating } [\text{W}] \times \text{hours in a year}(8760)} \times 100\%$$

The ACF of a baseload thermal or nuclear generator can be up to 90%; there will always be some time needed for maintenance. Peaking plant, such as reciprocating engines or small gas turbines, may have a capacity factor as low as 20% and operate only when there is a shortage of other generators, and the price of electricity is high.

Renewable generators are expensive to build but have no fuel cost and, once built, are operated to maximize use of the renewable energy resources, and hence generation and revenue. The ACF of renewable generation is determined by:

- the variability of the renewable energy resource,
- the availability of transmission capacity to connect the generator to the load,
- breakdowns of the generating equipment.

Table 1.5 Typical Annual Capacity Factor of renewable energy generators in GB (%)

Generating technology	Annual Capacity Factor (%)
Solar PV	10
Onshore wind	25
Offshore wind	40
Hydro	40
Biomass (with storage)	80

The ACF of a well-built renewable energy generator that is connected to a strong power system will be determined by the variability of the renewable energy resource. Typical capacity factors for renewable generation in GB are shown in Table 1.5. Depending on the solar resource, the capacity factor of PV generation will be low, while that of biomass can be high if a good supply of biomass is available and can be stored. Wind speeds offshore are significantly higher than onshore, giving a higher capacity factor for wind turbines sited offshore.

The effect of low capacity factors is that an electric power system with a high fraction of renewable generation will have much more generating capacity than if fossil generation is used to meet the load. This is expressed as the Capacity Margin (CM) and is the ratio of generating capacity to the peak load.

$$\text{Capacity Margin}(\text{CM}) = \frac{\text{Total generating capacity} \quad [\text{W}]}{\text{Peak load} \quad [\text{W}]}$$

A typical CM of a fossil-based generating system would be 1.2 to allow for breakdowns and maintenance. An electric power system with a high proportion of renewable generation might have a CM as high as 3. This obviously has implications for the cost and amount of material used in manufacturing the renewable energy generators (e.g. the relatively scarce material used for some PV panels and for the permanent magnets of generators).

Example 1.4 – Capacity Margin of a Renewable Power System

A low carbon electric power system has a peak load of 50 GW. It generates 60% of its electrical energy from variable renewables with 15% of electricity from each of onshore and offshore wind, solar PV and hydro. The remaining 40% of electrical energy comes from biomass generation. The capacity factors of the generating technologies are shown in Table 1.5. Estimate the total capacity of all the generators needed and the capacity margin.

Solution

Example of onshore wind

Onshore wind turbines provide 15% of the load of 50 GW, i.e. 7.5 GW.

Generating 7.5 GW of electrical energy from onshore wind turbines that have a capacity factor of 25% requires a generating capacity of

$$\text{Capacity} = \frac{\text{Energy generated}}{\text{Capacity Factor}} = \frac{7.5}{0.25} = 30\ \text{GW}$$

Summing the capacity needed of each generating technology (given in the table below) shows that a total of 167.5 GW of generating capacity is needed. With a peak load of 50 GW this is a capacity margin of

$$\text{CM} = \frac{\text{Total generating capacity}}{\text{Peak load}} = \frac{167.5}{50} = 3.35$$

The peak generating capacity needed is 167.5 GW and the capacity margin is 3.35.

Technology	ACF of generating technology (%)	Percentage of electrical energy generated by technology (%)	Generating capacity (GW)
Onshore wind	25	15	30.00
Offshore wind	40	15	18.75
Solar	10	15	75.00
Hydro	40	15	18.75
Biomass	80	40	25.00
		Total generating capacity	167.50
		Capacity margin	3.35

1.4.2 Environmental and Social Impacts of Renewable Energy

Renewable energy schemes have their own environmental and social impacts that need to be managed. Wind farms can have significant visual impact and can create noise nuisance unless sited carefully; their impact on flora and fauna must also be considered. Large solar farms occupy a significant area of land that cannot be cultivated or used for grazing large animals. Some photovoltaic panels contain toxic material and have to be disposed of carefully at the end of their useful life. All hydro schemes change the flow of water and large reservoir schemes may displace significant numbers of rural people while the electricity generated may benefit an urban population.

The environmental consequences of a scheme are considered through the Environmental Impact Assessment that is undertaken before any project can proceed. With an appropriate choice of location the environmental effects of smaller renewable energy schemes tend to be limited and local but large schemes can be controversial. Considering the impacts of the transition to renewable energy more widely, some of the minerals used in batteries, permanent magnet generators and PV modules are only found in a few parts of the world and the mining practices used to extract them are socially and environmentally damaging.

The transition away from fossil fuels to renewables has led to considerable social change, which is likely to continue. Figures 1.6 and 1.7 show a wind and solar energy plant sited on the hills above a former coal mining area of South Wales. The local

Figure 1.6 Wind farm in South Wales.
Photo: RichardJones/BusinessVisual. Rights Managed.
(A black and white version of this figure will appear in some formats. For the colour version, please refer to the plate section.)

Figure 1.7 Solar photovoltaic farm in South Wales.
Photo: RichardJones/BusinessVisual. Rights Managed.
(A black and white version of this figure will appear in some formats. For the colour version, please refer to the plate section.)

coal mines closed some years ago and this led to a considerable loss of employment and social change. The environmental impact of these renewable energy schemes is limited and the land continues to be used for grazing sheep. However, the two schemes shown would not provide sufficient energy to compensate for the loss of a major source of fossil fuel such as a coal mine, and many more similar renewable energy projects will be needed in the future.

PROBLEMS

1. Distinguish clearly between 'energy' and 'power'.
2. Explain the difference between capacity factor and capacity margin.
3. What is meant by controlling demand through energy efficiency and energy conservation? Give an example of each approach from:
   (a) the home,
   (b) industry/commerce,
   (c) personal transport.
4. Write brief notes on the technologies that might be used to decarbonize the electric power sector.
5. What is meant by the time preference for money? Why does choosing a high discount rate make it difficult to justify the installation of energy efficiency measures?
6. What is smart metering and how can it contribute to reducing
   (a) greenhouse gas emissions,
   (b) the capital cost of the power system?
7. List and discuss the environmental impacts of generating electricity from fossil fuels. Distinguish clearly between local, regional and global impacts. Discuss the future use of fossil fuels in electrical power generation.
8. Calculate the percentage increase in energy consumption of a country if its population increases by 25% and the per capita energy consumption increases from 2000 kg of oil equivalent to 4000 kg of oil equivalent.

   [**Answer:** 150%.]

9. Household A uses electric heaters and the electricity consumption per year is 3000 kWh. Household B uses gas for heating and the annual gas consumption is 1000 m^3.
   Convert the energy use in both houses into kWh and kg of oil equivalent.
   Compare the annual CO_2 emissions in the two houses.

   Data:
   Average CO_2 emission factor for electricity is 0.527 kg/kWh,
   CO_2 generated by burning natural gas is 0.185 kg/kWh,
   With 100% conversion efficiency, 1 tonne of oil equivalent corresponds to 11.6 MWh of electrical energy output.
   Assume that the conversion efficiency of the steam turbine generator unit is 35%.
   Burning 1 m^3 of natural gas provides 11.2 kWh of useful heat energy.

   [**Answer:** 739 kg oe, 965 kg oe, 1581 kg CO_2, 2072 kg CO_2.]

10. A commercial building has 200 15 W compact fluorescent lamps. It is proposed to replace them with LED lamps providing an equivalent light output.

    Data:
    Cost of a 7 W LED lamp = £6.
    Average electricity price = 16p/kWh,
    Lamps operate an average 5 hours per day.
    The building is occupied 250 days per year.
    Calculate the simple payback period.

    [**Answer:** 3.75 years.]

11. The tenant of the building described in Problem 10 has a lease for five years and applies a discount factor of 15% when evaluating energy efficiency projects. What is the NPV of replacing the lamps and will the lamp replacement project proceed?

    [**Answer:** −£127. The lamp replacement will not proceed.]

12. Show that with exponential growth the doubling time in years may be estimated by dividing 72 by the percentage annual rate of increase.

13. If the carbon intensity of electricity generation in the UK is to decrease from 450 tonnes of CO_2/GWh to 50 tonnes of CO_2/GWh over 20 years what is the annual rate of decrease required, assuming an exponential reduction?

    [**Answer:** 10.4%.]

FURTHER READING

A number of classic books and reports have discussed the challenges society faces over its increasing use of fossil fuels. These books have made important contributions to the framing of energy and environmental policy as well as emphasizing the importance of renewable energy.

Brundtland G.O., *Our Common Future: Report of the World Commission on Environment and Development*, 1987, reprinted 2009, Oxford University Press.

The influential report that led to the concept of sustainable development, i.e. 'development that meets the needs of the present without compromising the ability of future generations to meet their own needs', becoming widely accepted.

Meadows D., Randers J. and Meadows D., *Limits to Growth: The 30-Year Update*, 2005, Earthscan.

An update to the Report to the Club of Rome published in 1972 that addressed how global ecological constraints of resource use and emissions from fossil fuels would influence developments in the twenty-first century.

Schumacher E.F., *Small is Beautiful: A Study of Economics as if People Mattered*, 1973, Blond and Briggs.

A classic book that questioned the assumptions of conventional economics with particular reference to energy. 'Fritz' Schumacher was Chief Economist to the British National Coal Board and the founder of the Intermediate Technology Development Group.

Gore A., *An Inconvenient Truth*, 2006, Bloomsbury.

A classic book that accompanied an important film describing the consequences of global warming.

Websites

Some data were taken from:

Energy Institute Statistical Review of World Energy, 2022. www.energyinst.org/statistical-review

ONLINE RESOURCES

Exercise 1.1 Carbon intensity of generation: online resources are used to investigate the carbon intensity of GB generation.

Exercise 1.2 Optimization of generation: Excel and Solver are used to determine the optimal mix of generators of an island power system.

2 Wind Energy

INTRODUCTION

Generating electricity from the wind is one of the most effective and rapidly growing ways of harnessing renewable energy, and increasing numbers of wind turbines are being installed in many countries. Modern wind turbines can be very large with rotor diameters greater than 200 m and similar tower heights. Figure 2.1 shows the rapid growth in the maximum size of commercially available turbines over 25 years. Advanced composite materials are used in the blades to resist the forces of the wind as well as those created by gravity acting on these large rotating structures. The most important loads are those that are caused by extreme wind speeds and the fatigue loads created by variations in the wind acting over the very large number of rotations of the rotor. All very large modern wind turbines operate at variable rotational speed and the generator output is connected to the electrical grid through power electronic converters. As wind turbines become larger their structures are increasingly dynamic, using active control to manage the loads through sophisticated control systems.

As with all renewable energy technologies, the resource determines if a scheme will be profitable, and this chapter includes a description of the wind resource and how the energy produced by a wind farm can be estimated. In many countries the environmental impact of a wind farm, particularly visual impact and noise, will determine if permission for its construction can be obtained from the civil authorities. The visual impact is reduced, and the average wind speed increased, if the turbines are located offshore. In shallow seas with water depths of up to around 50 m, offshore turbines are mounted on foundation structures placed on the seabed, while in deeper water floating wind turbines are more cost-effective. Locating turbines offshore leads to higher costs but the turbines are more productive in the higher wind speeds, and very large wind farms can be built over a wide area.

This chapter describes the wind energy resource, wind turbines and wind farms. Sections 2.1–2.3 present the history, advantages and disadvantages of wind energy, the power developed by a wind turbine and its power curve and how the electrical energy from a wind turbine can be estimated. Section 2.4 introduces

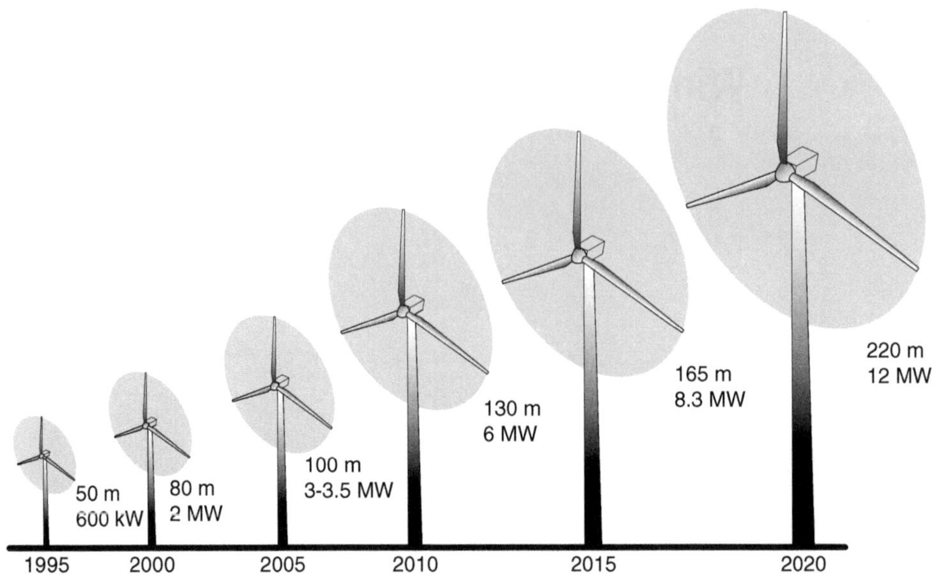

Figure 2.1 Growth in size of wind turbines showing increase in diameter of rotor and rated power output.

a simple theory of operation of a wind turbine based on linear momentum theory. Sections 2.5 and 2.6 deal with fixed-speed wind turbines and the techniques to control power above the rated wind speed. Section 2.7 extends this discussion to variable-speed turbines, and Section 2.8 discusses offshore wind energy. Section 2.9 presents the characteristics of the wind resource and its statistical representation, and finally Section 2.10 reviews the factors that are considered when developing wind farm projects.

2.1 Wind Turbines

2.1.1 History

The wind has been an important source of motive power for hundreds of years. Horizontal-axis windmills were used throughout Europe for grinding corn from the twelfth century until mills driven by steam engines superseded them in the middle of the nineteenth century. Vertical-axis windmills using aerodynamic drag, rather than lift forces, were in use even earlier in Iran and China. In the USA, small multi-bladed wind turbines became widespread for pumping water from around 1850 and remained important for agriculture until rural electrification in the 1930s allowed electric pumps to be used more cheaply and conveniently. Throughout the late nineteenth and early twentieth centuries a number of experimental wind turbines were developed to generate electricity including the 53 m diameter, 1250 kW Smith–Putnam turbine in Vermont in 1941. However, during the period of low and stable prices of oil that lasted from 1945 until 1973 there

Figure 2.2 Wind turbines on a site in the USA. Photo: RWE Innogy UK.
(A black and white version of this figure will appear in some formats. For the colour version, refer to the plate section.)

was limited interest in wind power for electricity generation. This changed dramatically with the first oil shock of 1973 when a number of national governments initiated programmes to develop large wind turbines to supply power to the electricity grid. Now wind energy is one of the most effective ways of decarbonizing the electricity supply system.

2.1.2 Advantages and Disadvantages of Wind Energy

The advantages of wind energy generation are that:

- each wind turbine can be large (up to 200 m rotor diameter, 12 MW rating),
- once planning (permitting) permission is obtained, a wind farm can be constructed quickly, within six months,
- in high-wind-speed onshore sites, it is low cost, e.g. 5p/kWh, 7c/kWh.

The main disadvantages are:

- the intermittent nature of the wind and hence the output power,
- the visual impact of the turbines.

Figure 2.2 shows a row of wind turbines on flat terrain in the USA. Note how they are positioned in a line across the prevailing wind direction.

2.2 Operation of a Wind Turbine

All wind turbines operate by extracting energy from the air that flows across the rotor (Figure 2.3).

The mass flow rate of the air that would pass through an area of the rotor disk is

$$\dot{m} = \rho A_D U_\infty \quad [\text{kg/s}]$$

Thus the kinetic energy per unit time is

$$\frac{1}{2}\dot{m}U_\infty^2 = \frac{1}{2}\rho A_D U_\infty . U_\infty^2 \quad [\text{W}]$$

and the power in the air flow is

$$P_{AIR} = \frac{1}{2}\rho A_D U_\infty^3 \quad [\text{W}]$$

where

$\dot{m}$ is the mass flow rate in kg/s,
ρ is the air density, 1.225 kg/m^3,
A_D is the swept area of the rotor (πR^2) in m^2,
U_∞ is the free-stream wind speed in m/s.

A wind turbine cannot extract all the power available in the wind flow and so a power coefficient (C_P) is defined. The power coefficient is simply the ratio of power captured by the wind turbine rotor (P_{WT}) to the power available in a free-stream air flow of the area of the rotor disk (P_{AIR}):

$$C_P = \frac{P_{WT}}{P_{AIR}}$$

Thus

$$P_{WT} = C_P P_{AIR} = C_P \frac{1}{2}\rho A_D U_\infty^3 \quad [\text{W}] \qquad (2.1)$$

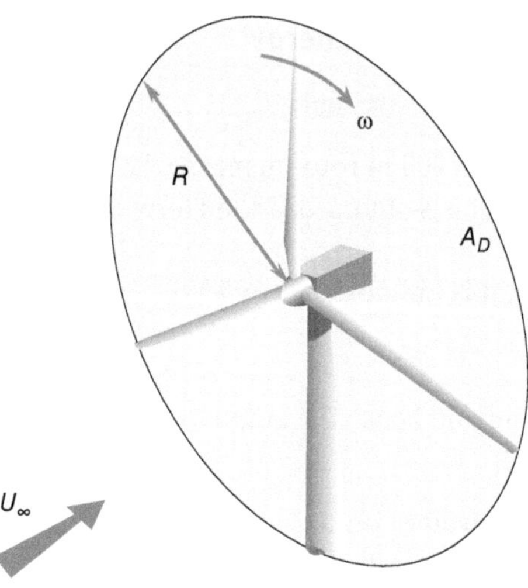

Figure 2.3 Wind turbine rotor: U_∞ is the free-stream wind speed, A_D is the area of the rotor disk, R is the radius of the rotor and ω is the rotational speed.

As a wind turbine operates by extracting energy from the air flowing through the area swept by the rotor, the rotor may rotate about a horizontal or vertical axis. Vertical-axis wind turbines, using either straight vertical blades or forming an egg-beater shape (Darrieus type), were developed and sold in the 1980s, but were not commercially successful and are now not produced in large numbers. In a vertical-axis wind turbine, there is a large cyclic torque on the rotor caused by the blades experiencing different apparent wind speeds as they move upwind and downwind. These cyclic forces have to be resisted by a strong, and hence expensive, rotor and support structure and so vertical-axis wind turbines were found not to be commercially competitive. Another disadvantage is that vertical-axis wind turbines usually do not self-start and have to be run up to speed with the generator operating as a motor.

The rotor of a wind turbine can be constructed with any number of blades. A rotor with only a few blades runs at a high rotational speed to extract the maximum power from the air passing through the rotor disk. The noise generated by the rotor increases with the fifth or sixth power of the tip speed. Hence the speed of the rotor tip is usually limited to 60–80 m/s for wind turbines installed on land, where noise has to be controlled. Water-pumping wind turbines (which require a high starting torque at low wind speeds) have large numbers of blades (e.g. 24 or more) and rotate slowly.

It is now accepted practice to use three-bladed, horizontal-axis, upwind rotors although two-bladed designs and even single-bladed horizontal-axis turbines were offered commercially in the 1980s. Three-bladed wind turbines are visually more pleasing than two- or single-bladed designs whose rotation appears to the eye not to be smooth. Although this apparently uneven speed of rotation is an optical illusion, it is crucially important how wind turbines are perceived by the public as this determines which wind farms receive permission to be built. The importance of public perception when trying to develop a wind farm project (or any other renewable energy scheme) cannot be over-stated. In the future, lower-cost, two-bladed designs may be used for very large offshore wind turbines where visual impact and noise are less important than on land.

The remainder of this chapter is devoted to three-bladed, horizontal-axis wind turbines.

2.2.1 Power Curve of a Wind Turbine

Figure 2.4 shows the power that can be captured by a wind turbine rotor 50 m in diameter assuming a constant power coefficient (C_P) of 0.4. With a wind speed below about 4 m/s there is very little power available in the wind and the turbine rotor is kept stationary by a shaft brake or allowed to idle. From 5 m/s the power increases rapidly following the cubic relationship with wind speed of Equation 2.1. A turbine rated wind speed is chosen above which all the power available in the wind is not transferred to the rotor shaft and so the loads on the turbine and the power rating of the components (e.g. gearbox and generator) are limited. The choice of rated wind speed is made

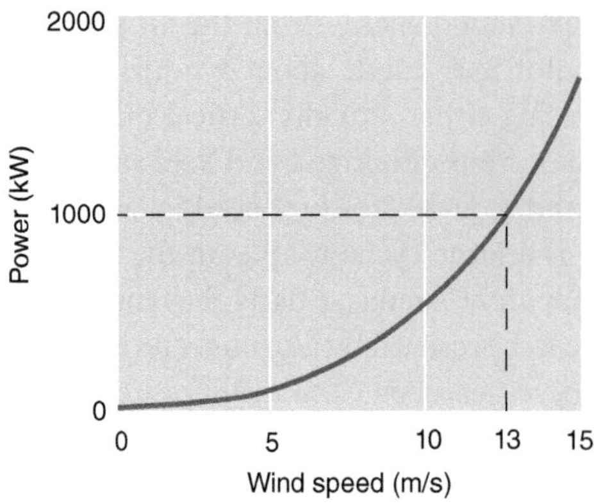

Figure 2.4 Power that could be extracted by a 50 m diameter rotor with a C_P of 0.4.

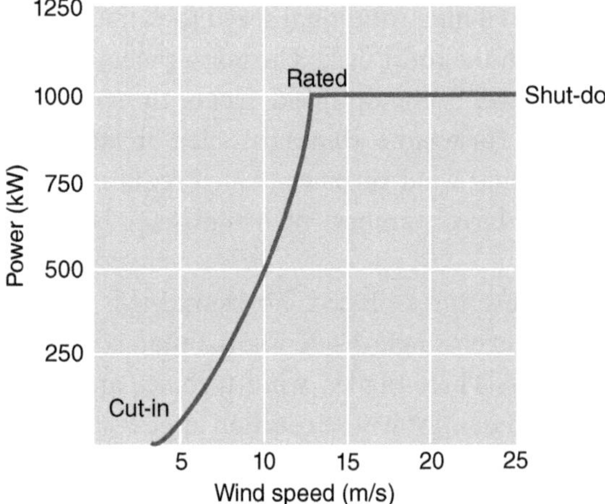

Figure 2.5 Power curve of a 50 m diameter wind turbine rated at 1 MW.

by the turbine designer considering the anticipated wind speeds of the sites where the turbines will be installed. The choice of the rated wind speed and power of the turbine is made so that site winds above the rated wind speed do not occur frequently and the additional energy that could be generated in these higher winds does not justify the extra expense of a wind turbine with a higher power rating.

In this example, a rated wind speed of 13 m/s was chosen and a power rating of 1000 kW. At wind speeds higher than 13 m/s the power captured by the rotor is limited by controlling the lift developed by the wind turbine blades. Control of the power is achieved either by rotating the blades about their longitudinal axis (pitching) or by ensuring the blades stall (see Section 2.4). Very high wind speeds (above 25 m/s) produce very high loads on the wind turbine structure but occur very infrequently, so generate little energy over the year. At these wind speeds the turbine is shut down with the rotor either parked or allowed to idle.

The choices of rated and shut-down wind speeds made by the designer result in the power curve of the wind turbine, shown in Figure 2.5. The power curve is the overall operating characteristic of a wind turbine, describing the relationship between the free-stream wind speed and the electrical power produced, usually measured at the base of the tower. It includes all the mechanical and electrical losses of the wind turbine (e.g. in the gearbox and generator) as well as variations of C_P. The power curve is the key operating characteristic of a wind turbine and is guaranteed by the manufacturer. It is used to predict the energy that would be developed by a wind turbine in the wind speeds of a particular site.

The detailed shape of the power curve of a wind turbine varies slightly with the control techniques used: fixed or variable speed and stall or pitch power regulation. However, all wind turbines have a power curve broadly as shown in Figure 2.5. The main function of the power curve is to predict energy output, and the wind speeds are 30 minute or hourly average values. The maximum speeds of the gusts that the turbine must withstand safely for short periods are considerably higher.

2.3 Energy Output of a Wind Turbine

The energy produced by a wind turbine is determined by the power curve and the wind speeds of the site where it is located. Figure 2.6a shows a set of measured hourly mean wind speeds for a site in the UK. These wind speeds were measured with an anemometer at the hub height of the turbines to be used at the site and one-hour mean average values taken. The dataset consists of hourly mean measurements taken over a year and so there are 8760 measurements (24 × 365). The measured hourly mean values are then allocated to 1 m/s bins centred on integer values of wind speed (e.g. the 7 m/s bin is incremented for each hourly mean wind speed that is between 6.5 and 7.5 m/s).

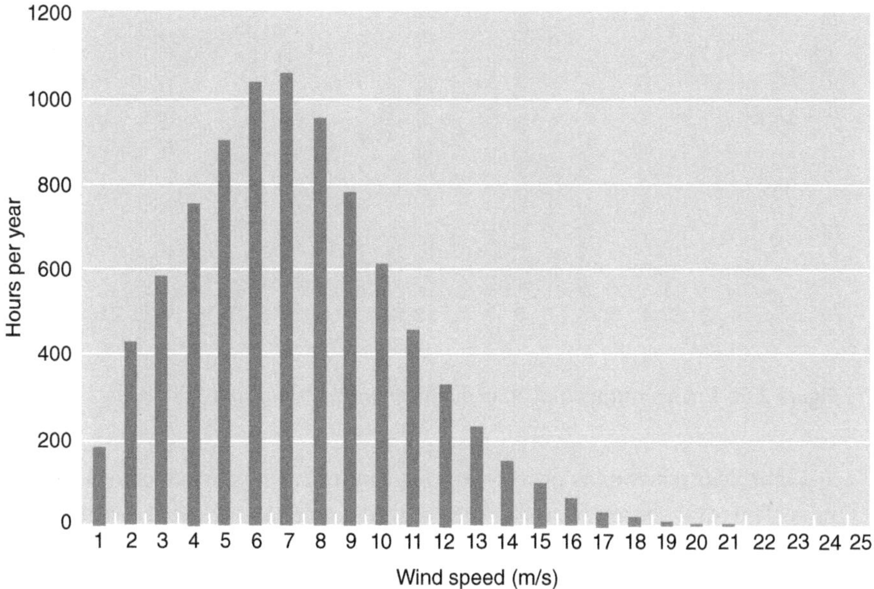

Figure 2.6a Measured hourly mean wind speeds of a UK site in 1 m/s bins.

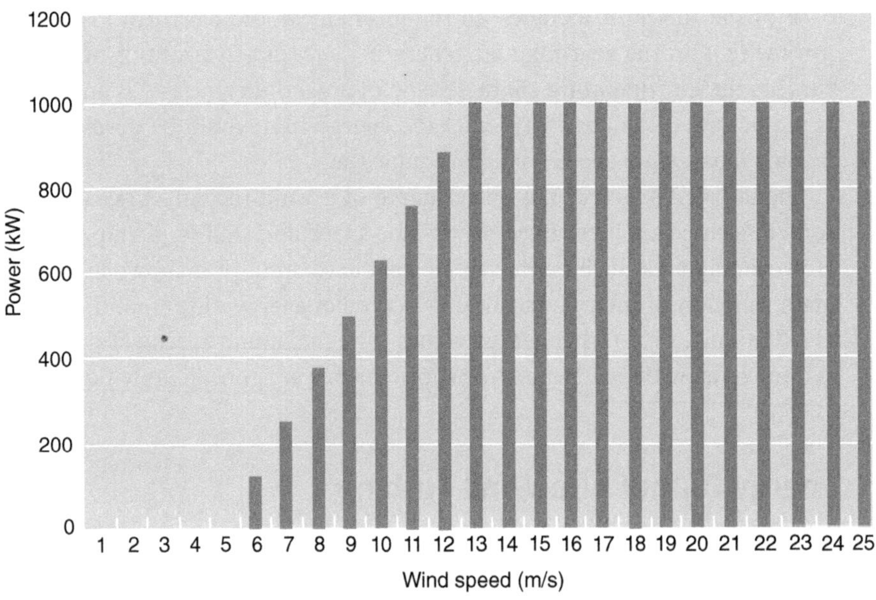

Figure 2.6b Power curve of a 50 m diameter wind turbine in 1 m/s bins.

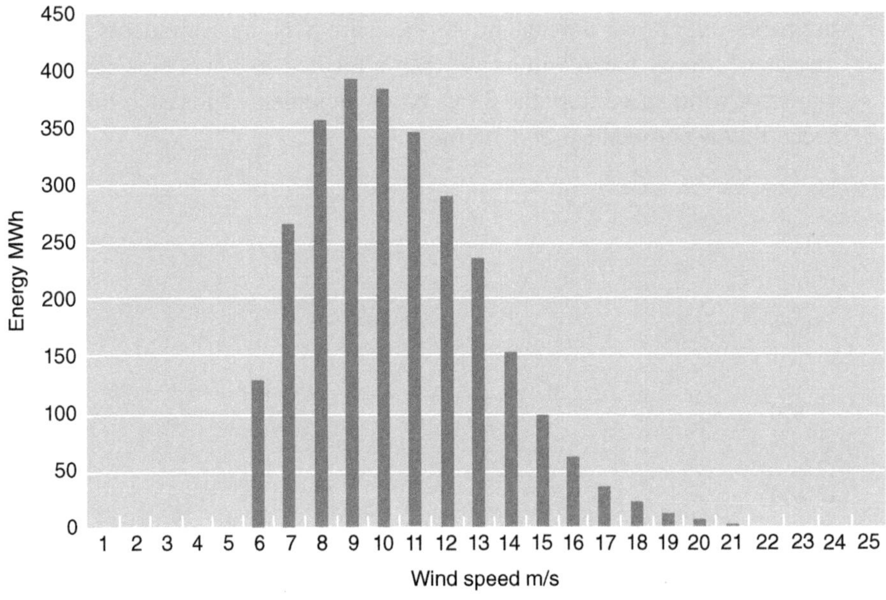

Figure 2.6c Energy output of a 50 m diameter wind turbine at a UK site in 1 m/s bins.

Figure 2.6b shows the power curve of Figure 2.5 expressed using 1 m/s wind speed bins. The total energy produced per year at a given wind speed is obtained by multiplying the hours per year and power output of each wind speed bin; see Figure 2.6c.

The total annual energy output is calculated by summing the energy calculated in each bin of Figure 2.6c. Thus the annual energy is:

$$\text{Annual energy} = \sum_{i=1}^{i=25} H(i) \times P(i) \quad [\text{Wh}] \tag{2.2}$$

where

$H(i)$ is hours per year in wind speed bin i,
$P(i)$ is the power at wind speed bin i.

Although the power curve includes all the losses within each turbine, if the energy produced by a wind farm is to be estimated it is also necessary to take account of:

- different wind speeds at each turbine rotor caused by local variations in site topography and the wakes of upstream turbines,
- losses in the wind farm electrical power collection system.

2.4 Linear Momentum Theory of a Wind Turbine

Students not familiar with the basic ideas of fluid mechanics may find Tutorial III helpful.

All wind turbines operate by extracting energy from the wind that passes through their rotor disk, and a simple way to describe their operation is to consider the rotor as an ideal element that extracts energy from a one-dimensional air flow. This is the Linear Momentum or Actuator Disk theory of wind turbine operation.

Assuming a constant mass flow, as the wind slows from far upstream through the disk to the wake, the area of flow within the streamlines increases. This is shown in Figure 2.7.

Abrupt changes in the velocity of the wind are not possible because of the large accelerations and forces that would result. However, it is possible to extract energy from a step change in pressure and it is by this principle that wind turbines operate.

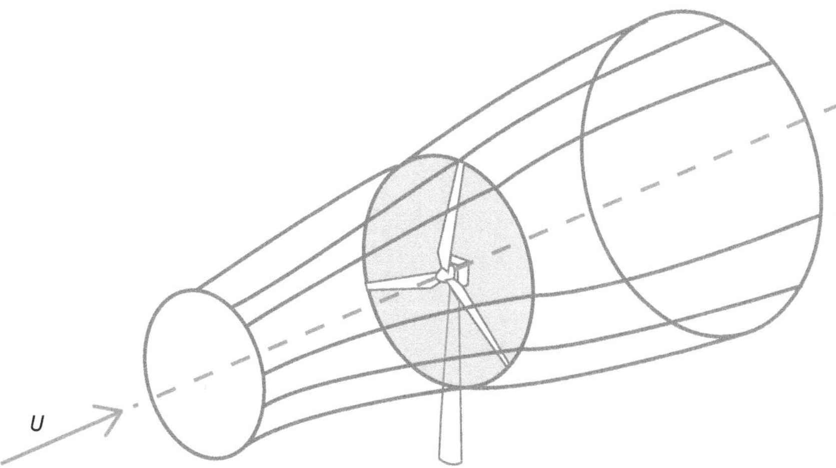

Figure 2.7 Stream-tube of air flowing through the actuator disk.

Thus, immediately in front of the turbine rotor the pressure rises, while behind the rotor there is a sharp pressure drop. As the air moves downstream, the pressure returns to its atmospheric value. The pressure distribution along an axis through the centre of a rotor disk is shown in Figure 2.8a.

The velocity reduces as the wind approaches the rotor disk and continues to fall in the wake; see Figure 2.8b.

The mass flow rate of the air in the stream-tube is constant. Because the velocity of the wind is higher upstream, the stream-tube has a smaller cross-section at this point, while downstream in the slower wake the stream-tube area is larger; see Figure 2.8c. For a stream-tube with constant mass flow rate:

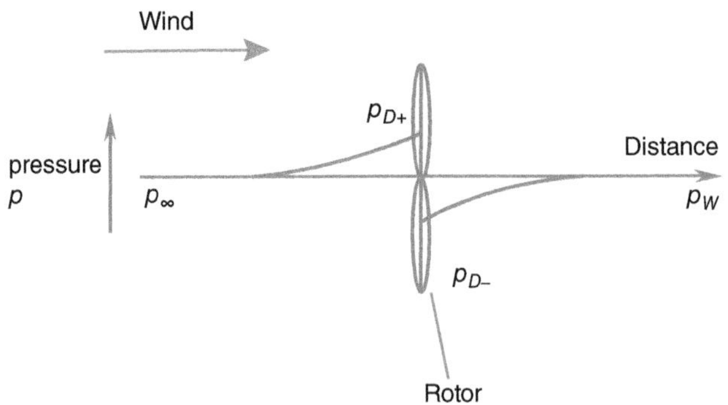

Figure 2.8a Pressure along the rotor axis.

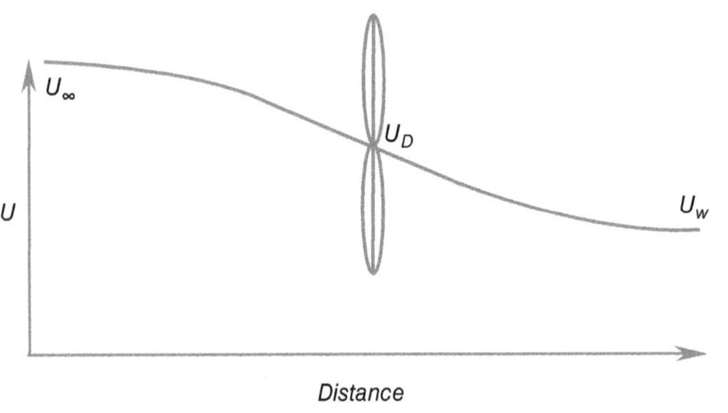

Figure 2.8b Velocity of the wind along the rotor axis.

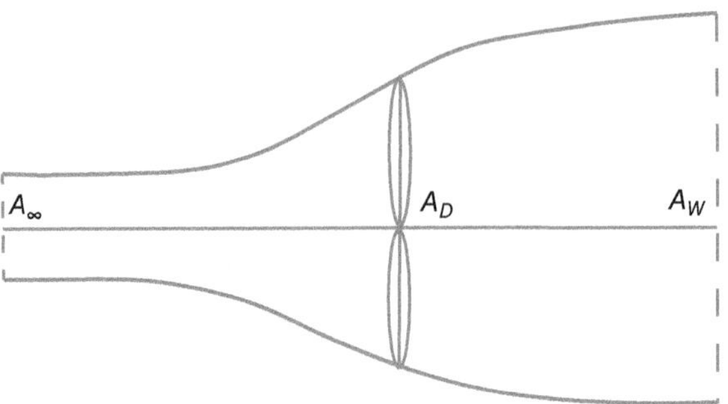

Figure 2.8c Area of the stream-tube of air flowing through the rotor.

$$\rho A_\infty U_\infty = \rho A_D U_D = \rho A_W U_W$$

where

A is the area of the stream-tube,
U is the wind velocity,
ρ is density of air,
subscript ∞ represents 'far upstream (free-stream)',
subscript D represents 'at the disk',
subscript W represents 'in the far wake'.

It is usual to define an axial flow induction factor a where aU is the reduction in axial velocity at the disk. This gives

$$U_D = (1-a)U_\infty$$

The rate of change of momentum of the air passing through the disk is the overall change of velocity $(U_\infty - U_W)$ multiplied by the mass flow rate at the disk $(\rho A_D U_D)$ and so

$$\text{Rate of change of momentum} = (U_\infty - U_W)\rho A_D (1-a)U_\infty$$

The force caused by this rate of change of momentum is equal to the force created by the pressure difference across the rotor disk. Therefore

$$(p_D^+ - p_D^-)A_D = (U_\infty - U_W)\rho A_D (1-a)U_\infty \tag{2.3}$$

where

p_D^+ is the pressure just upstream of the rotor,
p_D^- is the pressure just downstream of the rotor.

The pressure difference is obtained by using Bernoulli's theorem, which states that the total energy in a flow within a control volume is constant (i.e. the sum of the kinetic energy, static pressure and gravitational potential energy is constant). Thus for a unit mass of air

$$\frac{1}{2}\rho U^2 + p + \rho g h = const. \tag{2.4}$$

and assuming horizontal flow (i.e. h is constant)

$$\frac{1}{2}\rho U^2 + p = const.$$

Defining a control volume upstream of the rotor disk we can write

$$\frac{1}{2}\rho U_\infty^2 + p_\infty = \frac{1}{2}\rho U_D^2 + p_D^+ \tag{2.5}$$

and for a control volume downstream of the rotor

$$\frac{1}{2}\rho U_D^2 + p_D^- = \frac{1}{2}\rho U_W^2 + p_W \tag{2.6}$$

Separate control volumes upstream and downstream of the rotor are used in this simple application of Bernoulli's theorem.

At a long distance upstream and in the far wake, the static pressures are the same, $p_\infty = p_W$. Subtracting Equation 2.6 from Equation 2.5 gives

$$p_D^+ - p_D^- = \frac{1}{2}\rho\left(U_\infty^2 - U_W^2\right) \tag{2.7}$$

Substituting Equation 2.7 into Equation 2.3 gives

$$\frac{1}{2}\rho(U_\infty^2 - U_W^2)A_D = (U_\infty - U_W)\rho A_D(1-a)U_\infty$$

$$\frac{1}{2}(U_\infty - U_W)(U_\infty + U_W) = (U_\infty - U_W)(1-a)U_\infty$$

and so

$$U_W = (1-2a)U_\infty \tag{2.8}$$

This important result shows that half the axial speed loss takes place upstream of the rotor and half downstream.

2.4.1 The Betz Limit

The thrust on the rotor is equal to the force exerted by the air, and eliminating U_W from Equation 2.3, we have

$$F = (p_D^+ - p_D^-)A_D = 2\rho A_D U_\infty^2 a(1-a) \tag{2.9}$$

and so we can calculate the power extracted by the rotor disk as FU_D or

$$P_{WT} = FU_D = 2\rho A_D U_\infty^3 a\left(1-a\right)^2 \tag{2.10}$$

Recalling the definition of the Power Coefficient from Equation 2.1, we have

$$C_P = \frac{P_{WT}}{P_{AIR}} = \frac{P_{WT}}{\frac{1}{2}\rho U_\infty^3 A_D}$$

Therefore

$$C_P = 4a\left(1-a\right)^2$$

By differentiating C_P with respect to a, we obtain

$$\frac{dC_P}{dt} = 4\left(1-a\right)\left(1-3a\right)$$

And, for maximum C_P, $a = 0.333$ or 1.

The solution of $a = 1$ is discarded as this implies the wind is stopped at the rotor, so generating no power.

At $a = 0.333$, $C_P = 16/27 = 0.593$. This is usually known as the Betz limit.

The Betz limit is the maximum proportion of the available power in the wind that may be extracted by any wind turbine rotor. In practice, values of C_P between 0.4 and 0.5 are common for the peak performance of modern electricity-generating wind turbine rotors.

The derivation of the Betz limit is based on the Actuator Disk model of a generalized wind turbine that extracts power from the flow of wind across an idealized rotor disk; this simple theory applies to both horizontal- and vertical-axis wind turbines and to rotors with any number of blades.

2.4.2 Thrust Coefficient

A thrust (axial force) coefficient may also be defined as

$$C_T = \frac{F}{T_{MAX}}$$

where the maximum axial thrust that could be applied by the wind on the rotor area is

$$T_{MAX} = \frac{1}{2}\rho U_\infty^2 A_D$$

Thus, from Equation 2.9,

$$C_T = 4a(1-a)$$

The variation of C_P and C_T with the axial induction factor a is shown in Figure 2.9.

2.4.3 Limitations of the Momentum Theory

As $a \rightarrow ½$ the momentum theory breaks down as values of a above ½ would imply the wake had reversed. Hence for high values of a, e.g. $a > 0.3$, an empirical modification is made to the momentum theory.

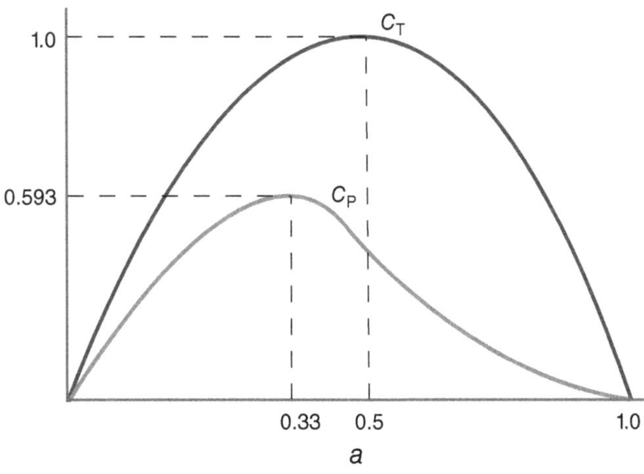

Figure 2.9 Variation of power coefficient (C_P) and thrust coefficient (C_T) with axial induction factor (a).

Example 2.1 – Use of Linear Momentum Theory

A horizontal-axis wind turbine with a 70 m diameter rotor operates in a wind speed of 10 m/s with a thrust coefficient C_T of 0.45.

Use linear momentum theory to estimate:

(a) the thrust on the wind turbine rotor in kN,
(b) the wind speed at the rotor disk,
(c) the power extracted by the rotor in kW.

Take the density of air as 1.225 kg/m^3.

Solution

(a) Thrust on rotor

$$\text{Thrust} = C_T \times \tfrac{1}{2}\rho A_D U_\infty^2$$
$$\text{Thrust} = 0.45 \times \tfrac{1}{2} \times 1.225 \times \pi \times 35^2 \times 10^2 = 106\,073\text{ N} = 106.1\text{ kN}$$

(b) Wind speed at rotor disk
Using linear momentum theory to find the axial induction factor, a:

$$C_T = 4a(1-a)$$

$$4a^2 - 4a + C_T = 0$$

$$a = \frac{4 \pm \sqrt{4^2 - 4 \times 4 \times C_T}}{2 \times 4}$$

$$a = \frac{4 \pm \sqrt{8.8}}{8} = 0.129, 0.871$$

Use $a < 0.5$ and so $a = 0.129$ or 12.9%.
Therefore the wind velocity at the disk is $U_D = U_\infty(1-a) = 8.7$ m/s.

(c) Power extracted by rotor
Calculate the power coefficient, C_P

$$C_P = 4a(1-a)^2 = 4 \times 0.129 \times (1 - 0.129)^2 = 0.391$$

Power extracted by rotor is given by

$$\text{Power} = C_P \tfrac{1}{2}\rho A_D U_\infty^3$$

$$\text{Power} = 0.391 \times \tfrac{1}{2} \times 1.225 \times \pi \times 35^2 \times 10^3 = 921\,676\text{ W} = 922\text{ kW}$$

2.5 Fixed-Speed Wind Turbines

Early designs of electricity-generating wind turbines were relatively simple devices in which the speed of rotation of the drive train, and hence of the aerodynamic rotor, was fixed by the generator locking on to the 50 Hz (in Europe) or 60 Hz (in

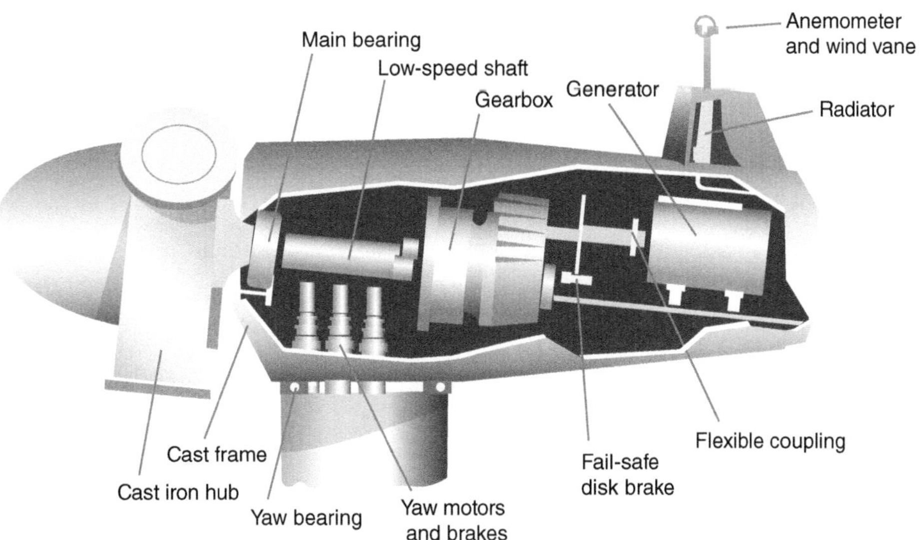

Figure 2.10 Nacelle of a fixed-speed wind turbine.

North America) of the grid. This simple design is still used for some turbines up to 2 MW in rating but is not used for modern very large turbines that now operate at variable speed and are described in Section 2.7. In this book, fixed-speed operation is described first to explain the principles of operation of a wind turbine and particularly the control of power, without considering the complexity of also varying the speed of rotation of the rotor.

Figure 2.10 shows a cross-section through the nacelle of a 700 kW, fixed-speed wind turbine. It consists of three blades mounted upwind of the tower and a low-speed main shaft driving an induction generator through a speed-increasing gearbox. The blades stall when the wind speed and output power exceed their rated values. The blades are either fixed rigidly to the hub (stall regulation) or rotated slowly about their longitudinal axis by an actuator (assisted stall regulation). The nacelle is mounted on a large, slewing-ring yaw bearing at the top of the tower and is orientated into the wind by yaw motors controlled from a wind vane located on top of the nacelle. The wind vane gives a signal showing if the nacelle is not facing directly into the wind. Power is taken from the generator to the bottom of the tower across the yaw bearing by flexible pendant cables that can accommodate two to three rotations of the nacelle. The anemometer that is adjacent to the wind vane is used only to control the starting and stopping of the turbine.

The control system is very simple; it releases a shaft brake to allow the turbine to start when cut-in wind speed is reached and then stops the turbine when the wind speed is above the shutdown value. The control system monitors the operation of the turbine and orientates the nacelle into the wind (and unwinds the pendant cable occasionally) but otherwise there is no active control. The aerodynamic performance of the blades is arranged so that the rotor stalls at wind speeds higher than rated and the lift forces and rotor torque then decrease. The wind speed measurement from

Figure 2.11 Lifting the rotor of a wind turbine to attach it to the low-speed shaft.
Photo: RES.
(A black and white version of this figure will appear in some formats. For the colour version, refer to the plate section.)

the anemometer is not a good representation of the wind speed across the rotor disk as it is a single-point measurement within the turbulent wake of the rotor. The wind speed measurement is used only to control the shaft brake and initiate the start sequence of the turbine.

Figure 2.11 shows a rotor that has been assembled on the ground being lifted into position. Note the large diameter coupling to take the high torque of the low-speed shaft.

2.5.1 The Generator of a Fixed-Speed Wind Turbine

Readers not familiar with electrical generators may find Tutorial I helpful.

The rotational speed of the wind-turbine drive train and rotor is fixed by the generator that is locked on to the 50 (or 60) Hz of the electricity grid. The wind turbine can operate only when it is connected to a strong 50 (or 60) Hz power system. The power network provides a voltage that gives electrical stability to the generator and fixes its rotational speed. Power for the wind turbine auxiliaries is drawn from the grid.

One way to consider a fixed-speed wind turbine is as a large fan connected to a motor but with torque applied to the main drive-train shaft from the wind acting on the blades. The rotational speed of the drive train is fixed by the 50 (or 60) Hz frequency of the electricity network, and the torque (not the rotational speed) varies as the wind speed changes.

Large power stations using fossil fuel, water power or nuclear reactors as their energy source always have synchronous generators. In a synchronous generator the rotor is locked precisely to the 50 (or 60) Hz voltage of the power system network. Synchronous generators have low losses and allow independent control of real and reactive power. However, in fixed-speed wind turbines, an induction generator, rather than a synchronous generator, is always used. The speed of an induction generator varies slightly as the torque from the aerodynamic rotor changes. A speed change of 1–2% from zero to full generating power would be typical. This change in rotational speed with torque provides damping to the drive train but because the speed variations are so small the turbine is known as a 'fixed-speed turbine'. Variations in wind speed cause a change in drive-train torque while the speed is substantially fixed by the electrical generator that is locked on to the constant frequency of the power system.

As each wind turbine blade passes the tower, the rotor experiences a reduction in torque. If a directly connected synchronous generator were to be used in a fixed-speed wind turbine, large and damaging torque oscillations would occur in the drive train. Hence induction generators are always used in fixed-speed wind turbines, even though they present some difficulties to the operation of the electric power system. Large wind farms of fixed-speed turbines using induction generators need additional power electronic equipment to ensure compliance with the grid connection requirements of the electric power companies.

Example 2.2 – Wind Turbine Operation

Consider a 700 kW, 45 m diameter, horizontal-axis, fixed-speed wind turbine operating in a wind speed of 12 m/s. Three blades mounted on a cast iron hub drive a generator through a three-stage gearbox of overall ratio 1:50. The induction generator is connected directly to the 50 Hz network and keeps the high-speed shaft at close to 1500 rpm and hence the low-speed shaft rotates at close to 30 rpm. Calculate (a) C_P, (b) the rotor tip speed and (c) the torque on the shafts.

Solution

At a wind speed of 12 m/s the power in the wind flow passing through the rotor disk is

$$P_{AIR} = \frac{1}{2}\rho A_D U_\infty^3 = \frac{1}{2} \times 1.225 \times 2\pi \times 22.5^2 \times 12^3 = 1683\,312 \text{ W} = 1683 \text{ kW}$$

(a) If the turbine develops its rated power at this wind speed,

$$C_P = \frac{P_{WT}}{P_{AIR}} = \frac{700}{1683} = 0.416$$

(b) The speed of the blade tip is approximately

$$\omega R = 30 \times \left(\frac{2\pi}{60}\right) \times 22.5 = 71 \text{ m/s}$$

(c) If losses in the gearbox and generator are ignored, the torque on the low-speed shaft is

$$Q_{LS} = \frac{700 \times 10^3}{\frac{30}{60} \times 2\pi} = 223\ \text{kNm}$$

and on the high-speed shaft it is

$$Q_{HS} = \frac{700 \times 10^3}{\frac{1500}{60} \times 2\pi} = 4.46\ \text{kNm}$$

2.6 Operation of a Wind Turbine Rotor

Wind turbine blades use aerofoil sections to generate lift and hence shaft torque, operating in a similar manner to aircraft wings. A typical section through an aerofoil in a horizontal air flow is shown in Figure 2.12. Note that V is the incident wind speed making an angle α with the chord line.

As the wind passes over the aerofoil, the pressure on the top surface is reduced and that on the lower surface is increased. This can be understood simply by considering Bernoulli's theorem (Equation 2.4) and that the velocity of the flow is greater over the upper surface of the aerofoil and so the pressure is reduced. This pressure difference and the resulting force is shown in Figure 2.13.

The force F resulting from the pressure difference across the aerofoil can be split into two orthogonal components F_l and F_d, the lift and drag forces. The lift force is perpendicular to the incident wind velocity V and the drag force is parallel to it.

These lift and drag forces are described using lift and drag coefficients C_l and C_d.

$$C_l = \frac{F_l}{\frac{1}{2}\rho V^2 A_a}$$

$$C_d = \frac{F_d}{\frac{1}{2}\rho V^2 A_a}$$

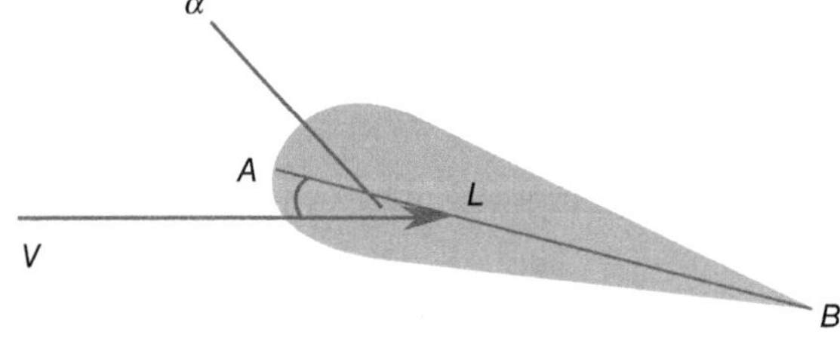

Figure 2.12 Aerofoil section in horizontal flow: A is the leading edge, B is the trailing edge, L is the chord line (a straight line between the leading and trailing edges), α is the angle of incidence (attack) between the incident wind stream and the chord line and V is the wind speed incident on the aerofoil.

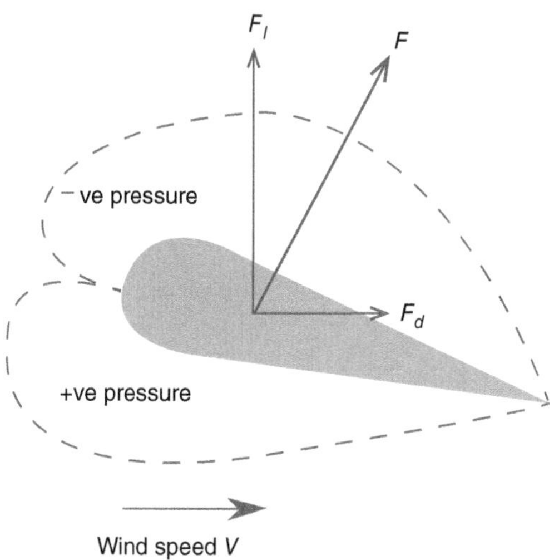

Figure 2.13 Forces on an aerofoil section in horizontal wind flow.

where A_a is the plan area of the aerofoil, the product of the length of the chord line (L in Figure 2.12) and the length of the blade element being considered, dr (dr is shown in Figure 2.15).

The lift and drag coefficients depend on the shape of the aerofoil and the angle of incidence, α; see Figure 2.14. As the angle of incidence increases, the lift coefficient C_l increases almost linearly up to its maximum and the aerofoil then stalls. After stall the lift coefficient reduces while the drag coefficient increases. Stalling is caused by the boundary layer on the upper surface of the aerofoil separating and a circular wake forming. It is a complex phenomenon and hard to predict.

So far, we have been viewing the aerofoil section as if it were an aircraft wing with the incident wind speed V horizontal. In the case of a wind turbine rotor, the wind speed experienced by the blades is created by the combination of the wind speed and the speed of the wind created by the rotational movement.

Figure 2.15 shows the general arrangement of a wind turbine rotor and a radial element of a blade.

The wind speed and forces on a blade element viewed from the tip of the blade are shown in Figure 2.16. Figure 2.16a shows the axial wind speed at the rotor, U_D, and the speed of rotation of the blade element, ωr; α is the angle of incidence, the angle between the incident wind V and the chord line, and ϕ is the angle between the incident wind V and the plane of rotation of the rotor.

The lift force F_l is perpendicular to the incident wind V, while the drag force F_d is parallel to V.

The useful force that provides torque to rotate the blades, F_r, is given by

$$F_r = F_l \sin\phi - F_d \cos\phi$$

while the thrust on the rotor, F_t, is

$$F_t = F_l \cos\phi + F_d \sin\phi$$

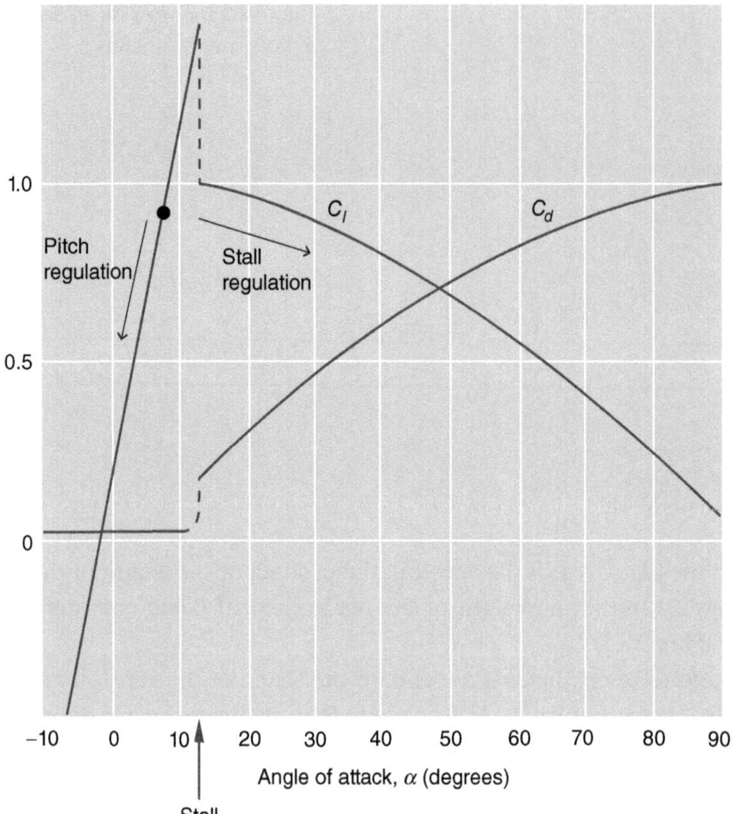

Figure 2.14 Typical aerofoil characteristics. (Pitch and stall regulation are discussed in Section 2.6.1.)

Figure 2.15 Wind turbine rotor showing blade element and the direction of the view of the element from the tip of the blade.

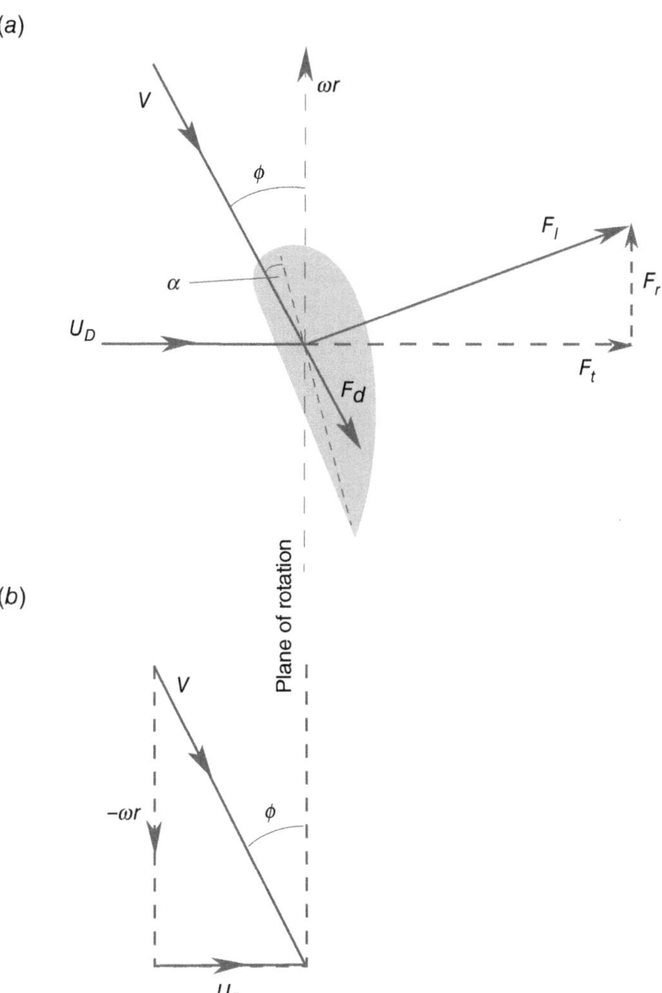

Figure 2.16 (a) Forces on a wind turbine blade element; (b) triangle of wind velocities.

The incident wind, V, is the vector sum of the wind speed, U_D, and the speed of the wind passing over the rotating blade element, $-\omega r$, as shown in Figure 2.16b.

2.6.1 Control of Power Above Rated Wind Speed, Pitch and Stall Regulation

Figure 2.5 showed the power curve of a wind turbine with its output increasing from cut-in to rated wind speed. The maximum power available in the wind is generated at these wind speeds. Between rated and shut-down wind speeds some wind power is rejected and the power captured by the rotor is kept approximately constant, limited through either pitch or stall regulation. Fixed-speed turbines are usually regulated through stall regulation, while variable-speed turbines are usually regulated through pitch regulation (although both techniques of power regulation have been applied to fixed- and variable-speed turbines).

The power output of a wind turbine depends on the rotational force, F_r, created by the lift from the blades. Consider the aerofoil characteristics with the blade operating at an initial angle of attack of, say, 8° (see Figure 2.14). C_d is very small and so can be ignored, but the C_l characteristic is steeply rising. With pitch regulation the angle of incidence α is reduced by mechanically turning the blade about its longitudinal axis. This reduces C_l, F_r and hence the power generated by the wind turbine.

Pitch regulation requires a control mechanism to alter the pitch angle of the blades, as shown in Figure 2.17. A power transducer measures the voltage and current of the generator to calculate the output power of the turbine. This measurement is compared with a set point of rated power and the error signal passed through a control system to operate an actuator that changes the angle of the blades which are mounted on the rotor hub with bearings. The blade pitch actuator may be hydraulic or electrical and operate on individual blades or collectively on all the blades of the rotor.

In a fixed-speed wind turbine, the angular velocity ω of the rotor is held constant by the electrical generator, which is locked on to the 50 or 60 Hz frequency of the power network. Consider the triangle of velocities V, U_D and $-\omega r$ shown in Figure 2.16b. The velocity of the wind caused by the rotation of the blade $-\omega r$ is constant and so an increase of the wind speed U_D will increase ϕ. This increases α as V swings round towards the axial direction. Consider Figure 2.14 with an initial operating angle of attack of 8°. Once the wind speed has increased sufficiently for the angle of attack to exceed around 13° the blade stalls, and the lift coefficient, and hence the torque, decrease. This is stall regulation that does not require any physical change in the pitch angle of the blades.

It is obvious that stall regulation is attractive as no moving parts are required. Unfortunately, it can be difficult to predict the wind speed at which blades stall as in practice aerodynamic stall is a complex three-dimensional phenomenon exhibiting a degree of hysteresis.

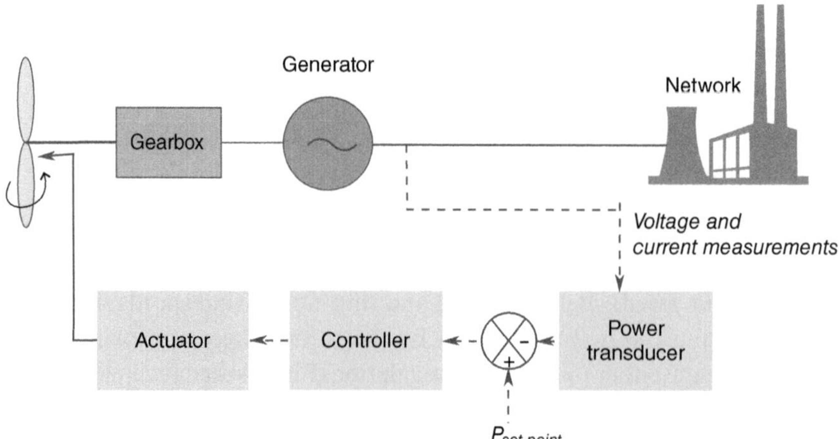

Figure 2.17 Pitch control of a wind turbine.

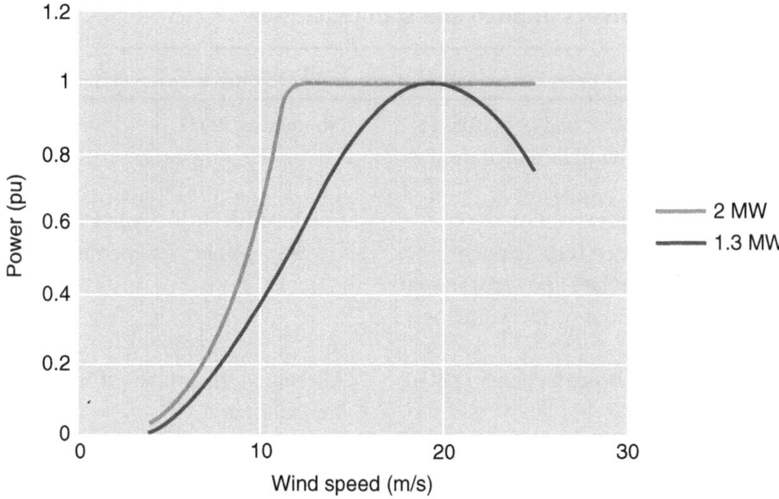

Figure 2.18 Power curves of a 2 MW pitch and 1.3 MW stall-regulated wind turbine. Power is expressed in per unit (pu) on the rating of each turbine.

Because stall occurs at various wind speeds at differing radial locations along the blade, the power curve of a stall-regulated wind turbine is less steep than that of a pitch-regulated turbine with a consequent loss in energy capture. This is shown in Figure 2.18. Pitch-regulated rotors experience lower peak thrust forces as the blades are rotated about their axis in high wind speeds. The power curve describes only the steady-state performance of the rotor and fixed-speed pitch-regulated turbines can experience significant overpower transients (up to an additional 100% of rated power). Both stall- and pitch-regulated fixed-speed wind turbines up to 80 m diameter and 2 MW electrical rating have been offered commercially.

Assisted stall control (sometimes known as active stall) is a development of stall control. In this technique a slow blade pitch actuator is used to position the blades, but the main control mechanism remains passive aerodynamic stall (an increase of α with wind speed). One reason for using assisted stall is that large wind turbines require blade pitch actuators for emergency braking and these can then also be used for power regulation.

The characteristics of pitch and stall regulation are compared in Table 2.1.

2.6.2 Rotor Performance and the C_P/λ Curve

The performance of either a fixed- or variable-speed wind turbine rotor can be calculated by combining principles of the momentum theory (Section 2.4) and blade theory (Section 2.6). This is known as Blade Element Momentum Theory (BEMT) and is described in many textbooks on wind turbines. This theory is used to create the power curve (Figures 2.5 and 2.17) and the C_P/λ curve (Figure 2.19).

The power curve relates the ambient free-stream wind speed U_∞(m/s) to the electrical output power of the entire wind turbine P(kW). It includes all the losses within

Table 2.1 Characteristics of pitch and stall regulation

Pitch regulation	Stall regulation
Requires a control system and actuators to rotate the blades which are mounted on the hub with bearings	No moving parts
Gives maximum power extraction	Can be difficult to predict stall along blade
If used with a fixed-speed turbine pitch regulation can result in large power transients due to dynamic behaviour of the wind flow over the blades	Gives smoother power control
Rotor can be braked through changing blade pitch	Alternative techniques of braking the rotor are required
Pitching of the blades reduces thrust forces at high wind speeds	At high wind speeds large thrust forces are applied to the rotor

the wind-turbine drive train and generator and is used for predicting the energy production of a wind turbine on a site.

The C_P/λ curve is dimensionless and relates the power coefficient C_P to the tip speed ratio λ of the rotor. C_P is the ratio of the mechanical power captured by the rotor and transferred to the turbine low-speed shaft to the power in the wind flow (Equation 2.2). λ is the ratio of the speed of the blade tip to the free-stream wind speed:

$$\lambda = \frac{\omega R}{U_\infty}$$

where

ω is the rotational speed in rad/s,
R is the rotor radius in m,
U_∞ is the free-stream wind speed in m/s.

Figure 2.19 shows the C_P/λ curve of a three-bladed, fixed-pitch rotor. In this example, the maximum C_P of 0.48 is reached at a tip speed ratio of 7. The maximum

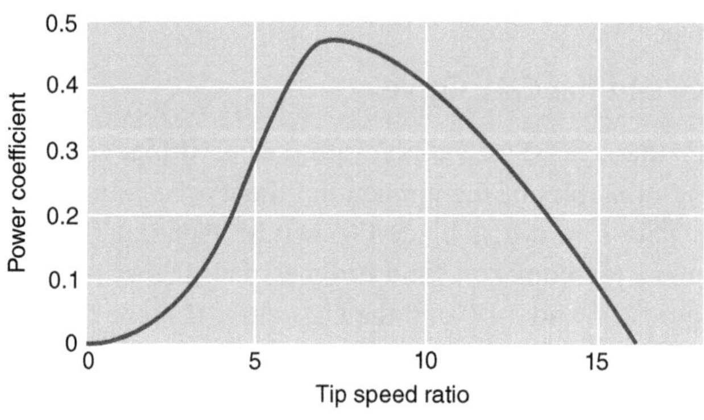

Figure 2.19 C_P/λ curve of a three-bladed rotor.

C_P occurs only at this one ratio of blade tip speed to free-stream wind speed. Varying the pitch angle of the blades results in a family of these curves at various pitch angles.

2.7 Variable-Speed Wind Turbines

The C_P/λ curve (Figure 2.19) shows that for any wind speed only one speed of rotation of the rotor extracts the maximum power from the wind. Thus it is desirable to operate a wind turbine at variable rotational speed, provided the cost and losses of the variable speed equipment are not excessive. An additional important feature of variable speed operation is that the rotor and drive-train rotational speed are allowed to vary slightly when the rotor is hit by a gust of wind, thereby reducing mechanical loads on the turbine. This reduction in loads is very important for larger wind turbines of ratings greater than 2 MW where variable-speed operation is essential.

The frequency of the electrical power system (constant at 50 or 60 Hz) determines the rotational speed of any generator that is connected directly to it. Generators manufactured with two magnetic poles always operate at around 3000 rpm on a 50 Hz system, whereas if they are constructed with four magnetic poles the speed of rotation will be close to 1500 rpm. For variable-speed operation of the drive train, the generator must be decoupled from the grid and connected to the network through power electronic converters to alter this frequency.

There have been attempts to connect the aerodynamic rotor to the generator through a mechanical variable-speed gearbox by controlling the speed of the annulus of an epicyclic gear stage or by using hydraulic systems. Mechanical variable-speed operation of wind turbines is being pursued by several manufacturers but is much less widely used than power electronic systems to vary the speed of rotation.

2.7.1 Full-Power-Converter Variable-Speed Generators

When the aerodynamic rotor of a turbine and the generator rotate at variable speed the electrical power from the generator will be at variable frequency and cannot be directly connected to the grid. Conceptually, the simplest way to obtain variable speed operation of a wind turbine is to convert all the variable frequency output of the generator to direct current (dc) and then use an inverter to convert this power to the 50 Hz (ac) of the electrical network. This rectification to direct current and inversion to 50 Hz allows the generator to rotate at any speed and produce electrical energy at any frequency. It de-couples the generator speed from the network frequency (Figure 2.20).

Two power converters are used to connect the generator operating at varying speed and frequency to the 50 Hz of the network. The generator side converter takes the power at the variable frequency from the generator and converts it to direct current, dc. The grid-side converter inverts the dc to the 50 Hz of

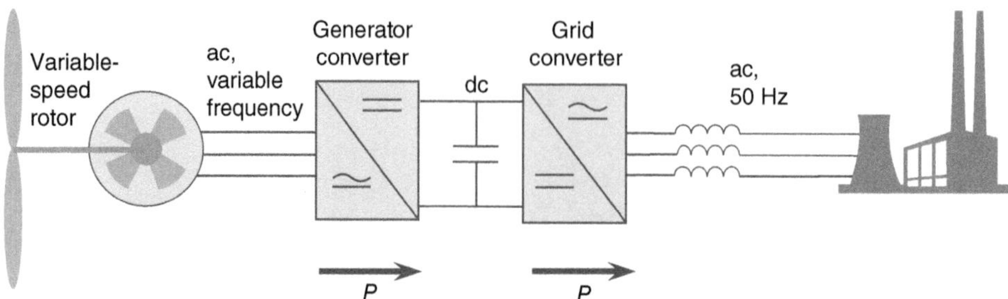

Figure 2.20 Variable-speed operation through full power conversion.

the network. Both converters use large transistors, Insulated Gate Bipolar Transistors (IGBTs), in a bridge connection. These IGBTs can be considered as large switches that can be turned on and off very rapidly. The generator converter switches in such a way as always to produce dc while the network converter synthesizes a 50 Hz sine wave current that is injected into the grid through the coupling reactors.

All the power from the generator is converted to dc and then inverted to 50 Hz. This requires two converters, each rated at the same power as the generator and each with electrical losses of around 1–2% of their power rating. These losses appear as heat and so considerable cooling of the converters is required.

2.7.2 Variable-Speed Wind Turbine Control

The conventional control system of a variable-speed wind turbine uses the rotational speed of the drive train as its input signal. Rotational speed can be measured conveniently using an optical encoder. Wind speed is not used as a measured signal as a measurement representative of the wind speed across the rotor disk cannot be made easily with an anemometer.

Figure 2.21 shows the locus of operation of a variable-speed wind turbine. Below a wind speed of about 4 m/s the turbine remains shut down. From cut-in to rated wind speed (4–12 m/s) the power drawn from the generator is controlled to be proportional to the cube of the rotational speed, $P \propto \omega^3$.

The blade pitch angle remains fixed until rated wind speed is reached (12 m/s in this example) when the blades are pitched to reduce the angle of incidence α and so limit the power extracted from the wind by the blades.

Figure 2.22 shows the control system of a variable-speed wind turbine using fully rated power converters. The speed of the generator is measured and passed through a look-up table containing the characteristic of Figure 2.21 that has been calculated during the turbine design. The power that is to be extracted from the generator (the restraining torque that is to be applied to it) is found from the look-up table and applied to the generator through the machine-side converter. The control system is arranged so that when a gust hits the rotor it speeds up slightly, limiting the mechanical loads. Only a very small amount of wind energy is lost by not following

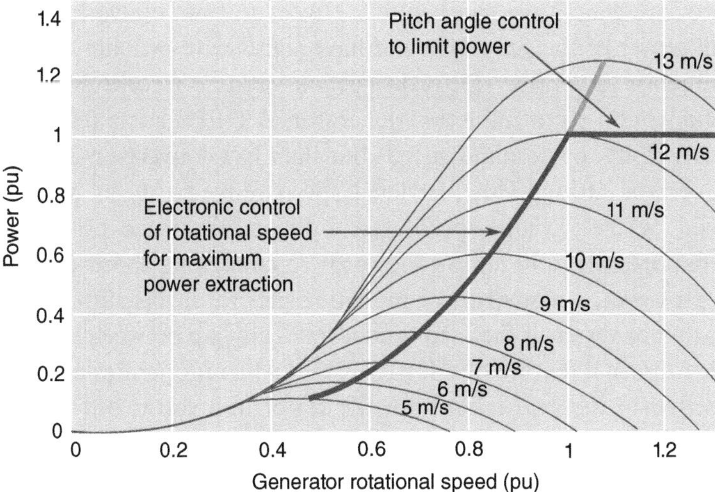

Figure 2.21 Locus of variable-speed operation.

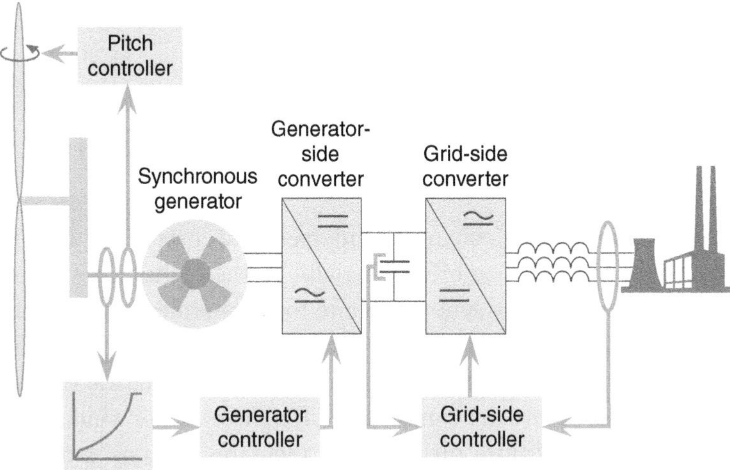

Figure 2.22 Control of a variable-speed wind turbine using full-power converters.

the optimal trajectory exactly and a significant reduction in mechanical loads can be obtained. Although industrial practice varies, a simple arrangement is that all the control of the generator is done by the generator-side converter. The grid-side converter merely takes power from the capacitor and inverts it into the grid.

2.7.3 Doubly Fed Induction Generators

Control of wind turbines using two fully rated power converters is conceptually simple and they can be used with any type of generator (synchronous or induction). High-speed generators driven through a gearbox or low-speed, direct-drive synchronous generators can be used. However, both converters must be rated at the full power of the generator and have significant switching losses. The doubly fed induction generator (DFIG) injects a variable frequency into the rotor circuit of a wound-rotor induction generator. A DFIG has a restricted speed range (typically ±30% of nominal speed) but then has converters sized at only 30% of the generator rating. This restricted speed range captures much of the energy obtained by the variable-speed operation of fully rated power converters but, more importantly, still allows the rotor to change its speed in response to wind gusts. However, a wound-rotor induction generator has slip rings to connect the generator converter to the rotor and a small air-gap between the rotor and stator. A rather unusual feature of a DFIG is that when it operates at high speed, power is exported to the grid from both the stator and rotor, but slower operation is obtained by injecting a fraction of the power generated back into the rotor.

2.8 Offshore Wind Energy

There is increasing development of wind farms offshore. The reasons for moving offshore are:

- reduced visual impact,
- higher mean wind speeds,
- lower turbulence of the wind offshore.

However, there are higher capital and operating costs in the demanding offshore environment which may be offset partially by the higher mean wind speeds (and hence increased electricity generated) and the economies of scale made possible by very large projects. The principles of operation of an offshore turbine are the same as those onshore and the early offshore turbines were marinized versions of land designs. However, there are significant differences in the wind farm environment on- and offshore. Offshore turbines and their foundations experience significant loading from the waves, as well as from the wind. The mean wind speeds offshore are usually higher than onshore with lower levels of wind turbulence. The lower wind turbulence reduces loading on the structure of the turbines but also creates longer-lasting wakes that can create significant loss of power within the offshore wind turbine array if the turbines are spaced closely together and the wind direction is aligned with a row of turbines. It can be difficult to transport very large turbines on rural roads but it is easier to ship them directly from the factory to an offshore site. The submarine cables that are used to connect the turbines within the offshore array and from the offshore wind farm to the shore are more expensive than land cables and their installation is more costly.

Figure 2.23 shows the 90 MW Barrow offshore wind farm off the north-west coast of England, which was commissioned in 2008. The thirty 3 MW, 90 m diameter rotor, variable-speed turbines are in water depth of 15–20 m. Each turbine generates at 33 kV and the voltage is increased to 132 kV by the transformer located on

Figure 2.23 Barrow offshore wind farm showing turbines and offshore substation.
Photo: Rob Arnold/Alamy Stock photo.
(A black and white version of this figure will appear in some formats. For the colour version, refer to the plate section.)

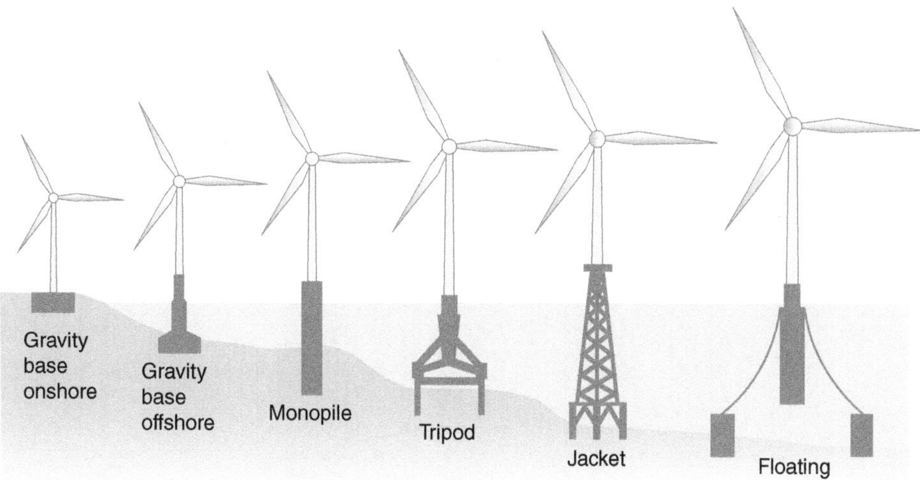

Figure 2.24 Foundations of offshore wind turbines.

the offshore platform for export to the land. More recent offshore wind farms have used larger-diameter wind turbines in deeper water.

A number of approaches have been used for the foundations of offshore wind turbines, depending on the depth of the water (see Figure 2.24). In shallow waters either a gravity base foundation, similar to that used on land, or a monopile can be used.

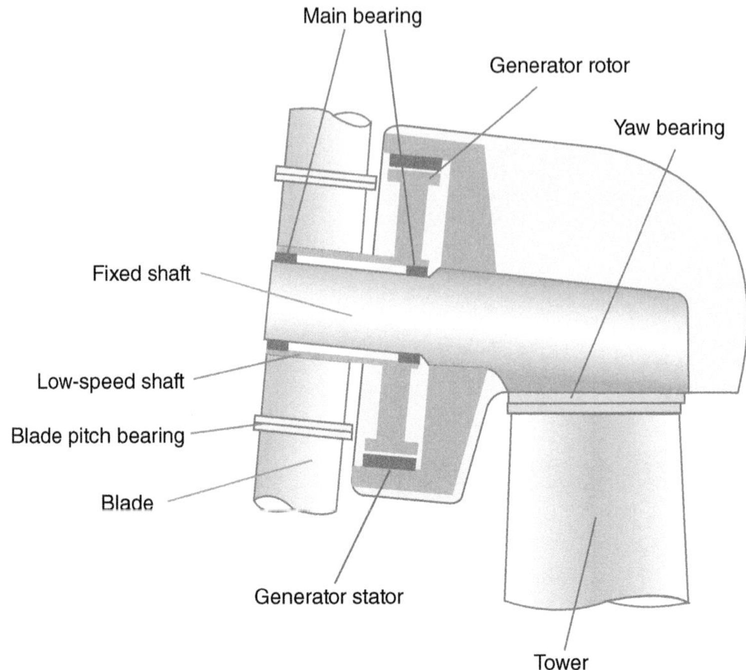

Figure 2.25 Direct-drive wind turbine.

Figure 2.26 Construction of an offshore wind turbine using a jack-up barge to lift the final blade. Photo: RES.
(A black and white version of this figure will appear in some formats. For the colour version, refer to the plate section.)

Around 80% of the offshore turbines installed so far have used monopile foundations. These are driven into the seabed and the turbine tower mounted on a transition piece. Turbines can be installed in deeper water using either tripod or jacket foundations. In much deeper water, floating turbines tethered to the seabed are used.

Some of the very large turbines used offshore do not use a gearbox but instead use a direct-drive electrical generator. This avoids the weight and maintenance of the gearbox, although the direct-drive generator is heavier than a high-speed or medium-speed generator. Direct-drive generators are used on some land wind turbines, but their advantages become greater with the very large turbines used offshore. Figure 2.25 shows a cross-section of a permanent-magnet, radial-flux, variable-speed, direct-drive wind turbine. This would be connected to the grid using the converters shown in Figure 2.20.

Construction and maintenance costs increase offshore, and specialized vessels are used for personnel access and for construction. Figure 2.26 shows a jack-up barge with feet extended to the seabed lifting the final blade of an offshore wind turbine during construction.

2.9 Wind Structure and Statistics

The power in the wind is a function of wind speed cubed while the forces developed by the wind are proportional to the square of the wind speed. Hence small variations in wind speeds make a large difference to generated power and energy yield, as well as to the forces on the turbine. Variations in wind speed are randomly distributed and it is necessary to use statistical descriptions of the wind, although wind turbine components may be designed to withstand extreme loads using simple deterministic representations of extreme gusts (e.g. an assumed shape of wind gusts).

The wind speeds experienced by a wind turbine can be considered to be made up of mean wind speeds which vary gradually from hour to hour, and high-frequency fluctuations in wind speed about the mean, i.e. the turbulence of the wind. The mean wind speeds determine the energy generated and hence the profitability of any project, while the turbulence drives fatigue loading. The wind turbine must also be designed for extreme loading due to wind gusts, which result from the combination of extreme wind speed fluctuations and the highest mean wind speeds encountered at the site.

Records of wind speed taken at various heights demonstrate the following characteristics of wind flow near the ground:

(1) the wind speed varies constantly,
(2) the variations are spread over a range of frequencies,
(3) the wind speed increases with height.

It is usual to consider the variations in wind speed as falling into two time-scales. At shorter time-scales (higher frequencies) the variations in speed are caused by the wind passing over undulating or rough ground and local obstructions. This mechanically induced turbulence causes fatigue loading of wind turbine components and also creates higher-frequency variations of the power generated, particularly in fixed-speed wind turbines. Longer time-scale variations (lower frequencies) are caused by the passage of weather fronts, although in some countries there may also be significant daily variations caused by differential heating and cooling of the land and sea.

For estimating the likely energy yield of a site, mean wind speeds are usually measured with an averaging period of 1 hour. With a 1 hour mean, the higher-frequency variations caused by turbulence are averaged out while the longer-time-scale effects caused by weather patterns are represented in the measurements. This is illustrated by the well-known Van der Hoven spectrum that was measured in New York State in the 1950s (Figure 2.27). From a time series of wind speed measurements, the power spectral density $S(f)$ was calculated and $fS(f)$ plotted against frequency where the x-axis is a log scale. The power spectral density shows the kinetic energy in the wind at each frequency. Three peaks may be seen: the synoptic variations that are caused by weather fronts, the diurnal or daily variations and the higher-frequency turbulence. There is little energy in variations of the wind speed with return periods between 2 hours and 10 minutes, the so-called spectral gap. The spectral gap allows the lower-frequency variations to be considered independently from the higher frequency turbulence. Although some researchers dispute the general applicability of the Van der Hoven spectrum, it does show how variations in wind speed can be considered in various time frames.

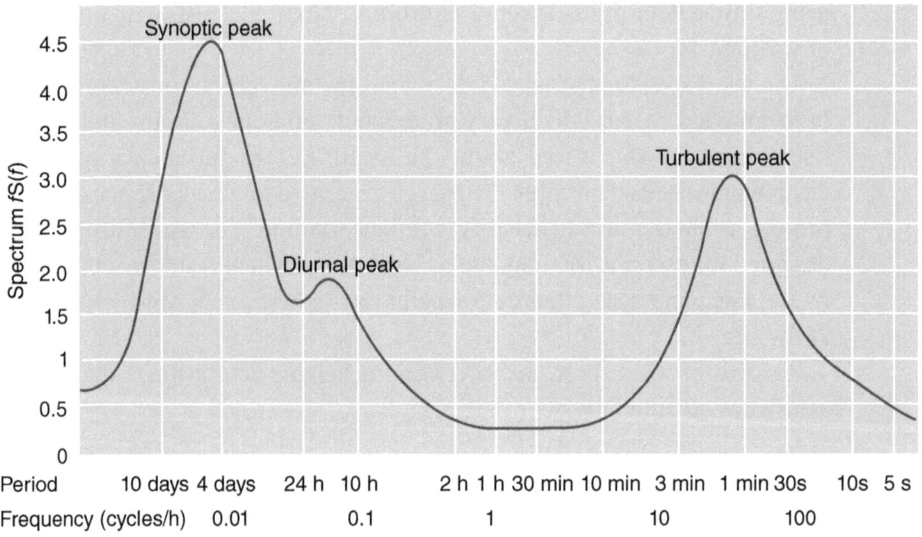

Figure 2.27 Van der Hoven spectrum.
Source: A. Burton, N. Jenkins, D. Sharpe and E. Bossanyi. *Wind Energy Handbook*, 3rd ed., 2021, © John Wiley and Sons.

The wind turbine power curve (e.g. Figures 2.5 and 2.18) is measured using 10-minute averaged values of wind speed and power output. Using a 10-minute averaging period allows the largest amount of data to be captured during the shortest possible test period while staying within the spectral gap where the high-frequency turbulence is averaged out.

Consider a set of measured average wind speeds, U_i. The averaging period chosen is typically 10 minutes or 1 hour.

The mean of this set of wind speed measurements is given by

$$\bar{U} = \frac{1}{n}\sum_{i=1}^{n} U_i$$

where

n is the total number of measurements.

The variance of the wind speed data is given by

$$\sigma^2 = \frac{1}{n-1}\sum_{i=1}^{n}(U_i - \bar{U})^2$$

which may be rewritten as

$$\sigma^2 = \frac{1}{n-1}\left[\sum_{i=1}^{n} U_i^2 - n\bar{U}^2\right]$$

The standard deviation of the wind speed is simply the square root of the variance.

The probability of a discrete wind speed being observed is

$$p(U_i) = \frac{f_i}{n}$$

where f_i is the number of observations of wind speed i.

And the discrete cumulative distribution function $F(U_l)$ is the probability that a measured wind speed will be less than or equal to U_l:

$$F(U_l) = \sum_{i=1}^{l} p(U_i)$$

It is often more convenient to model the wind speed as a continuous function rather than a table of discrete values and then $p(U_i)$ becomes a density function $f(U)$ and represents the probability that the wind speed is in a 1 m/s interval centred on U.

The continuous cumulative distribution function $F(U)$ is given by

$$F(U) = \int f(U)\, dU$$

and

$$f(U) = \frac{dF(U)}{dU}$$

The general shapes of $f(U)$ and $F(U)$ are shown in Figure 2.28.

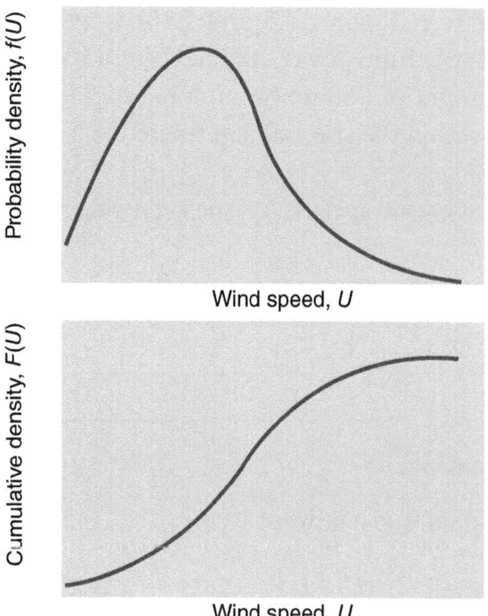

Figure 2.28 Shape of probability density function $f(U)$ and cumulative distribution function $F(U)$.

This section may be omitted at first reading.

The Method of Bins

Wind speeds are measured many times and rather than record and process the analogue values of each average wind speed it is often more convenient to use the 'Method of Bins'. The wind speed range of interest (typically 0–25 m/s) is separated into bins (typically 0–1, 1–2, 2–3, ... m/s). Each measurement of a 10 minute or 1 hour mean wind speed increments a counter of the number of measurements that fall into a particular 1 m/s bin. This provides a convenient way to summarize the wind speed data with reduced computation burden. A similar approach is used for wind speed and power measurements to measure the power curve of a wind turbine.

Using the Method of Bins, the mean and variance of the wind speed are given by

$$\bar{U} = \frac{1}{n}\sum_{j=1}^{B} m_j f_j$$

$$\sigma^2 = \frac{1}{n-1}\left[\sum_{j=1}^{B} m_j^2 f_j - n\bar{U}^2\right] = \frac{1}{n-1}\left[\sum_{j=1}^{B} m_j^2 f_j - n\left(\frac{1}{n}\sum_{j=1}^{B} m_j f_j\right)^2\right]$$

where

m_j is the central wind speed of each bin,
f_j is the number of observations in each bin,
B is the number of wind speed bins

$$n = \sum_{j=1}^{B} f_j$$

2.9.1 Weibull and Raleigh Statistics

It has been found from experience that either the Weibull or the simpler Raleigh distribution can be used to describe the shape of the probabilities of long-term records of mean wind speeds.

The probability density function of a Weibull distribution is given by

$$f(U) = \frac{k}{c}\left(\frac{U}{c}\right)^{k-1} \exp\left[-\left(\frac{U}{c}\right)^{k}\right]$$

where

c is the scale parameter in m/s,
k is the shape parameter,

and the cumulative distribution function (that the wind speed will be less than or equal to U) is given by

$$F(U) = 1 - \exp\left[-\left(\frac{U}{c}\right)^{k}\right]$$

The cumulative distribution function that the wind speed is greater than U is then given by

$$Q(U) = 1 - F(U) = \exp\left[-\left(\frac{U}{c}\right)^{k}\right] \tag{2.13}$$

The Raleigh distribution is a special case of the Weibull distribution with $k = 2$.

For the range of the values of k usually of interest for wind turbine applications ($1.5 < k < 3$) the properties of the Weibull distribution allow c to be estimated with reasonable accuracy from

$$\mathrm{c} = 1.12\overline{U}$$

Therefore an initial preliminary estimate would be to use a Raleigh distribution $(k = 2)$ with $\mathrm{c} = 1.12\overline{U}$.

A more accurate method for calculating Weibull parameters from a dataset is either to use a least-squares fitting program or to consider Equation 2.13 and take logarithms twice to give:

$$\ln\,[-\ln(1 - F(U))] = k \ln U - k \ln c$$

Plotting $\ln U$ versus $\ln\,[-\ln\,(1 - F(U))]$ gives a graph similar to that of Figure 2.29 and so allows estimation of c and k.

2.9.2 Variations of Wind Speed with Height

The wind in the boundary layer is retarded by surface roughness. This is shown diagrammatically in Figure 2.30. In the surface layer the variation of mean wind speed with height may be represented by an expression of the form:

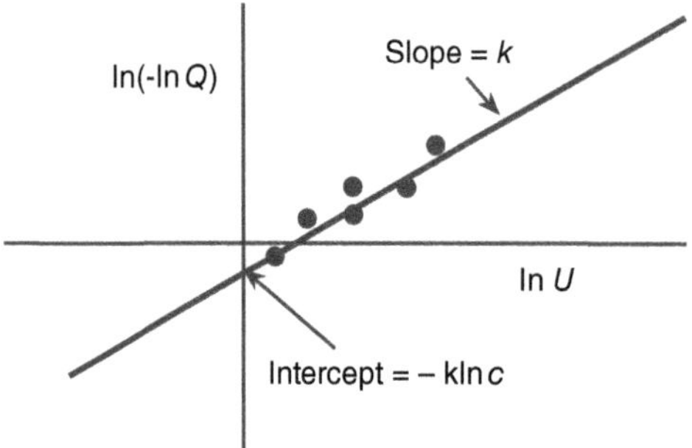

Figure 2.29 Graphical determination of Weibull parameters. Q is defined in Equation 2.13.

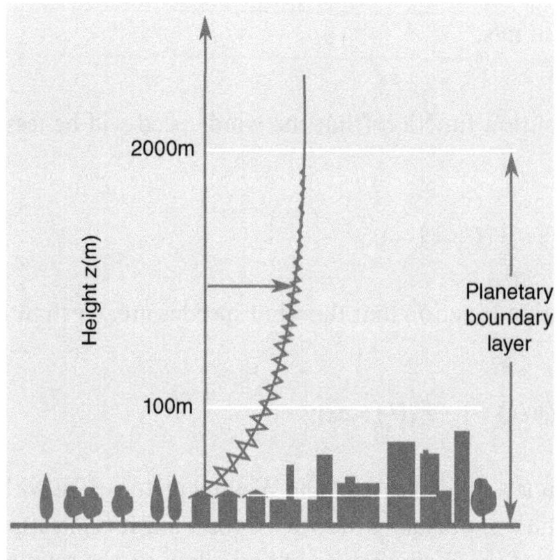

Figure 2.30 Variation of wind speed and turbulence with height.

$$\bar{U}(z) = \frac{U^*}{k} \ln\left(\frac{z}{z_0}\right)$$

where

$\overline{U}(z)$ is mean wind speed at height z above ground,
U^* is friction velocity,
k is the Von Karman Constant (approx. 0.4),
z_0 is the roughness length characteristic of the terrain.

Typical values of z_0 are shown in Table 2.2.

If the wind speed at a reference height (typically 10 m) is known, then the unknown friction velocity may be eliminated and an expression of the form

$$\frac{\bar{U}(z)}{\bar{U}(h)} = \frac{\ln\left(\frac{z}{z_0}\right)}{\ln\left(\frac{h}{z_0}\right)}$$

obtained, where h is the reference height.

Table 2.2 Roughness lengths for various types of terrain

Type of terrain	Roughness length z_0 (m)
Cities, forests	0.7
Suburbs, wooded countryside	0.3
Villages, countryside with trees and hedges	0.1
Open farmland, few trees and buildings	0.03
Flat grassy plains	0.01
Flat desert, rough seas	0.001

Source: A. Burton *et al.*, *Wind Energy Handbook*, 3rd ed., 2021, © John Wiley and Sons.

For approximate calculations it is common to use a simpler empirical power law

$$\frac{\bar{U}(z)}{\bar{U}(h)} = \left(\frac{z}{h}\right)^{\alpha} \tag{2.14}$$

where the exponent α depends on the surface roughness and the range of heights being considered. This is sometimes referred to as the one-seventh power law, where α is chosen to be 1/7. However, it must be emphasized that any of these simple equations give only a very approximate indication of the variation of wind speed with height and for accurate estimates of the energy output of a wind farm it is usual to use a high anemometer mast that allows measurement to be made at the hub height of the wind turbines that are to be used. A LiDAR remote sensing system can be used for measuring wind speeds at the heights of very large wind turbines.

Example 2.3 – Use of Weibull Parameters

The annual distribution of hourly mean wind speeds at a site hub height may be described by a Weibull distribution with

scale parameter, $c = 8.5$,
shape parameter, $k = 2$.

Calculate the gross annual energy yield of a wind turbine installed at this site with a power curve as shown in the table.

Wind speed (m/s)	Output (kW)
0–5	0
6	50
7	120
8	280
9	440
10–25	600

Solution

The calculation is made in 1 m/s bins.

The time that the wind speed is below 5.5 m/s is calculated from

$$F(U) = 1 - \exp\left[-\left(\frac{U}{c}\right)^k\right] = 1 - \exp\left[-\left(\frac{5.5}{8.5}\right)^2\right]$$

$$F(U) = 0.342 \text{ or } 2997 \text{ hours/year}$$

There is no power generated during this time.

The time that the wind speed is above 9.5 m/s and the turbine produces 600 kW is calculated from

$$Q(U) = \exp\left[-\left(\frac{U}{c}\right)^k\right]$$

$$Q(U) = 0.286 \text{ or } 2512 \text{ hours/year}$$

$$\text{Energy generated above rated wind speed} = 0.6 \times 2512 = 1507 \text{ MWh}$$

Using $f(U) = \frac{k}{c}\left(\frac{U}{c}\right)^{k-1} \exp\left[-\left(\frac{U}{c}\right)^k\right]$ the number of hours that the wind was in 1 m/s bins around 6, 7, 8, 9 m/s can be calculated.

Bin (m/s)	P (kW)	U (m/s)	$f(U)$	Hours/year	Energy (MWh)
5.5–6.5	50	6	0.101	884	44.2
6.5–7.5	120	7	0.098	861	103.3
7.5–8.5	240	8	0.091	800	224
8.5–9.5	440	9	0.081	711	312.8

Summing the energy, the total energy generated at wind speeds between 6 and 10 m/s and above 10 m/s is 2.19 GWh.

2.9.3 Turbulence

The wind can be thought of as consisting of high-frequency turbulent fluctuations imposed on a slower varying, quasi-steady-state mean. Therefore, as well as defining the mean wind speed the turbulent fluctuations must also be described. Turbulence $V(t)$ may be defined as the deviations of the instantaneous wind speed $U(t)$ from the mean $\bar{U}$.

$$V(t) = U(t) - \bar{U}$$

The variance of the wind speed may be calculated from

$$\sigma^2 = \frac{1}{n-1}\sum_{i=1}^{n}(U_i - \bar{U})^2$$

as noted above, and a measure of the variation is given by the turbulence intensity

$$I_u = \frac{\sigma}{\bar{U}}$$

To measure I_u it would be typical to take a number of wind speed samples over, say, 10 or 30 minutes measured at 10 Hz. Turbulence intensity usually decreases with height and if measurements are made at a reference height z_0 then the following is reasonable.

$$I_u(z) = I_u(z_0)\frac{1}{\ln\left(\frac{z}{z_0}\right)}$$

The turbulence intensity varies from site to site and with atmospheric conditions, but for the design of wind turbines a value of 20% would be typical. On land the turbulence intensity reduces with increasing wind speed.

2.9.4 Extreme Wind Speeds

The Weibull distribution is a good fit of the hourly mean wind speeds at many sites but is not a useful method of estimating extreme values, which occur very infrequently and for much shorter lengths of time. Although alternative frequency distributions (e.g. the Gumbel distribution) give better representations of extreme wind speeds, the design and application of wind turbines is usually based on defined deterministic standards. Extreme wind speeds together with their duration are defined for a site and the turbine class chosen accordingly. A typical design standard might define a reference 10-minute mean wind speed V_{REF} of five times the long-term mean wind speed, and the once in 50-year extreme wind speed, V_{50}, is then calculated as $1.4V_{REF}$. The more frequent annual extreme wind speed is taken to be $1.12V_{REF}$. A gust duration of either 3 or 5 seconds is assumed.

The wind turbine will be shut down in these high wind speeds. If all the safety systems are operating correctly, for example the turbine is yawed and the blades pitched correctly for this high-wind-speed condition, the V_{50} value is assumed for design calculations. However, if, for example, the grid connection has been lost and so the yaw motors are without power then the design standard requires only that the annual extreme wind speed is considered. This is based on the assumption that the 50-year extreme gust is unlikely to occur at the same time as a fault on the electrical network or wind turbine.

The gust and mean wind speeds, as well as the turbulence intensity, are governed by the class of the wind farm site. Thus the developer of a wind farm project has to define only the particular class applicable to the site and is then not concerned further with the details of the design calculations.

2.10 Wind Farm Development

In an onshore wind farm project, the turbines and their erection will account for around 65–70% of the total cost, with the remainder consisting of balance-of-plant costs (foundations, site roads and lay-down/crane areas, electrical systems) and project development costs. Offshore the increased cost of foundations, submarine cables and access can increase the balance-of-plant fraction to more than half of the (higher) project costs. Commercial developers of wind farms will often prefer larger projects as the fixed costs, particularly grid connection and project set-up, management and financing costs, may then be spread over a bigger investment. However, there are individuals and community groups that develop smaller wind farms and single-turbine projects. The advantage of community involvement in the project is, of course, that planning permission is more readily obtained if it is seen that there is tangible benefit to local people.

Although lattice steel towers are structurally efficient because of their wide base and were used historically, tubular towers are invariably used for modern wind turbines as they are aesthetically more pleasing and provide greater clearance for the blades, which deflect considerably in high winds. They also allow an internal vertical ladder, or lift, to be fitted and an enclosure for the tower base electrical equipment. The foundation of a wind turbine supports the vertical load from the tower and wind turbine nacelle and resists the transverse load from the rotor and the resulting overturning moment. Usually the foundation design is driven by the overturning moment experienced during extreme wind speeds or caused by an emergency stop. If good ground conditions are found close to the surface, a reinforced concrete slab foundation is used, while if there is poor ground near the surface it may be necessary to use a piled foundation. A ring of bolts or an interface can is cast into the reinforced concrete foundation and the tubular tower is bolted to it (see Figure 9.2). The loads on the foundation may come from any direction, depending on the wind, and so the most efficient foundation is circular or octagonal but square foundations are sometimes used for ease of construction.

Site roads are installed for the wind turbine components and cranes. These are typically crushed rock tracks with floating roads using geo-textiles sometimes placed over weak ground. An area of hard standing is created close to the turbine to allow the blades and nacelle to be laid down and for the crane.

Most medium-sized wind turbines generate three-phase power at 690 V (line–line). This is a surprisingly low voltage for generators as large as 2 MW or 3 MW. The choice of such a low voltage is driven by the safety regulations, which become much more stringent for voltages greater than 1000 V, and the wide availability of electrical components at this standard voltage. However, to reduce electrical losses it is necessary to locate a transformer in the nacelle, within the tower or adjacent to it. The transformer will typically increase the 690 V to 10 kV or 30 kV, and power from the wind farm will be collected by underground cables at this voltage. Large wind farms will have another site transformer to increase the voltage further (up to 50 kV or 150 kV) for onward transmission to the main electricity grid. The power generated from a wind farm is exported to the network of the local electricity utility. As the wind farm will be located in an area of high wind speeds, away from centres of population, it is common for there not to be a strong existing electrical power network. A detailed electrical study will be required before a wind farm is

Table 2.3 Connection of wind farms (UK standard voltages)

Location of connection	Maximum capacity of wind farm (MW)
On 11 kV network	1–2
On 33 kV network	12–15
On 132 kV network	30–60

built, but to assess if a connection is likely to be possible, rules-of-thumb as indicated in Table 2.3 may be used.

The development of a wind farm follows a broadly similar process to that of any renewable energy generation project, but with the particular requirement that the wind turbines must be located where there are high winds to maximize energy production. The size of modern wind turbines and the need for a high wind speed makes visual effects particularly important. There are three main elements of the development of a wind farm project: (1) technical and commercial issues, (2) environmental considerations and (3) dialogue and consultation. The technical and commercial issues are often the more straightforward, and project success or failure hinges critically on environmental considerations and the dialogue and consultation process with local residents and planning authorities.

Initially a search is undertaken to locate a suitable site and to confirm it as a potential candidate for the location of a wind farm. A good wind resource is essential, but considerations of environmental impact and public acceptability are of equal importance. The landowner must be an enthusiastic supporter of the project but there must also be suitable access. Blades of the wind turbines sited on land can be more than 70 m long and very large cranes are needed for the construction. These requirements can pose difficulties for transport on minor roads with small bend radii. It is not uncommon for some improvement of roads and removal of street furniture to be necessary when the turbines are transported to site. For a large wind farm, the heaviest piece of equipment is likely to be the main transformer if a substation is located at the site.

It is clearly sensible to review the main environmental considerations as early as possible as experience in many countries is that environmental restrictions are a major cause of projects not proceeding. An early examination of environmental issues will reduce the likelihood of wasted work and costs. The most important environmental constraints include avoidance of National Parks or other areas designated as being of particular amenity or scientific value. It is also necessary to ensure that no turbine is located so close to dwellings that a nuisance will be caused by noise, visual domination or light-shadow flicker. A preliminary assessment of visual effects is also required considering the visual appearance of the wind farm particularly from important public viewpoints. If, within the wind farm perimeter, there are areas of particular ecological value due to flora or fauna, then these need to be avoided as well as any locations of particular archaeological or historical interest. Communication systems such as microwave links, TV, radio and particularly aviation radar may be adversely affected by wind turbines and this also needs to be considered at an early stage. In parallel with the technical and environmental assessments it is normal to open discussion with the local civic planning (permitting) authorities to identify and agree the major potential issues which will need to be addressed in more detail if the project development is to continue.

When a possible site has been identified, more detailed and hence expensive investigations are required to confirm that the project will be profitable.

2.10.1 Wind Farm Power Output

The power output of a wind turbine is proportional to the cube of the wind speed and so the wind farm energy output, and hence the financial viability of the scheme, is very sensitive to the wind speeds seen by the turbines over the life of the project. Wind atlases and similar databases of wind speeds give only a general indication of wind speeds. Wind speeds at a site will vary significantly with local topography but also from one year to the next. Hence it is usual to use a long-term record of wind speeds, such as are held by a local meteorological station, but to take account of the relationship between the wind speed at the site and that at the meteorological station. Therefore a technique known as measure–correlate–predict is used to estimate the long-term wind resource at a wind farm site.

A mast with anemometers and wind vanes at various heights is erected near the centre of the wind farm site. It is usual for the height of the mast to be sufficient to allow one set of measurements to be made at the hub height of the wind turbines that are proposed to be used, thus avoiding the need to account for vertical wind shear. Measurement at hub height may require a very high mast whose installation will attract local attention and may require its own planning application. Measurements of wind speed at the site are made over at least a six-month period including the windiest season. The more data obtained the greater confidence there will be in the result.

The measure–correlate–predict technique takes a series of measurements of wind speed and direction at the mast erected at the wind farm site and correlates them with simultaneous wind speed measurements made at a local meteorological station, which may be at a local airport. The averaging period of the site-measured data is synchronized with that of the meteorological station data. This gives a scatter plot (Figure 2.31) with the mast measurements on one axis and the simultaneous meteorological station measurements on the other. Linear regression is used to establish a relationship between the measured site wind speed and the long-term meteorological wind-speed data of the form:

$$U_{site} = a + bU_{met\,station}$$

The coefficients relating the two sets of measurements, sometimes known as speed-up factors, are calculated for the twelve 30° directional sectors. Then these coefficients are applied to transfer the long-term data record of the meteorological

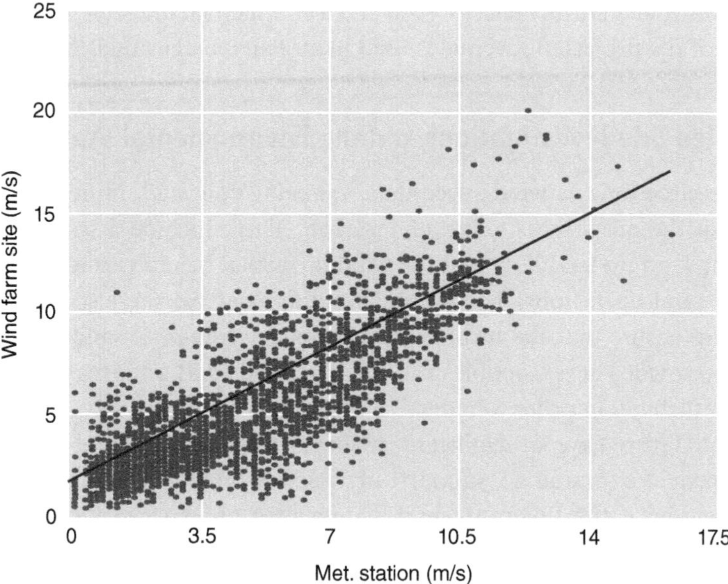

Figure 2.31 Measure–correlate–predict scatter plot.

station to the site. The speed-up factors allow the long-term wind-speed record held by the meteorological station to be used as an estimate of what the wind speed at the wind farm site would have been over that period. It is then assumed that this long-term wind-speed record at the site is representative of the wind speed over the projected life of the wind farm. It is common to use a long-term record of 20 years.

The drawbacks of using the measure–correlate–predict technique include that high site masts or a LiDAR (light detection and ranging) remote sensing system is necessary if large wind turbines are planned. There may not be a suitable meteorological station nearby (within say 50 km) or with a similar exposure and wind climate. The data obtained from the meteorological station may not always be of good quality and may include gaps. Therefore it may be time-consuming to ensure that it is properly correlated with the site data. Critically the technique assumes that the previous long-term record provides a good estimate of the future wind resource over the lifetime of the wind farm.

The estimate of the future long-term wind speed obtained from measure–correlate–predict is only at the location of the site mast. A wind farm design program is then used to transfer these wind speeds to the hub heights of the wind turbines within the wind farm. These programs account for topographic variations of the wind speed together with the effect of the wakes of the other wind turbines to generate the energy yield of any particular turbine layout. The programs may use Computational Fluid Dynamics or analytical solvers to determine the wind flow. The estimates of the wind speeds at the various wind turbine locations are then used with the power curve of the turbines to estimate the energy yield and hence income of the project; see Section 2.3. As may be imagined, there are uncertainties in this estimated future energy yield and a commercial judgement is made as to how much of the projected revenue is used in any financial evaluation of the project.

2.10.2 Detailed Site Investigations and the Environmental Statement

At the same time as wind-speed data are being collected, more detailed investigations of the proposed site are undertaken. These include a careful assessment of existing land use and how best the wind farm may be integrated with, for example, agricultural operations. The ground conditions at the site also need to be investigated to ensure that the turbine foundations, access roads and construction areas can be provided at reasonable cost. Local ground conditions may influence the position of turbines in order to reduce foundation costs.

Wind farms have a significant environmental impact, and an Environmental Statement is required to support any application for planning permission. The preparation of the Environmental Statement is an expensive and time-consuming undertaking and usually requires the input of specialists (e.g. landscape architects and ornithologists).

Any extensions to overhead line networks for the export of the electrical power will also be covered by an Environmental Statement. Recent experience suggests that the visual appearance of new overhead lines, particularly those built with lattice steel towers, are considered by the public to be at least as important as the environmental impact of the wind farm itself.

Visual and landscape assessment is often the largest part of the Environmental Statement and is certainly the most open to subjective judgement. The main techniques that are used include: zones of theoretical visibility (ZTV) to indicate the locations that the wind farm will be visible from; wireframes, which show the location of the turbines from particular viewpoints; and photomontages, which are computer-generated images overlaid on a photograph of the site. Virtual-reality techniques and video footage can also be used.

PROBLEMS

Take the density of air to be 1.225 kg/m^3.

1. Sketch and explain the power curve of a wind turbine.
2. Sketch and explain the C_P / λ curve of a wind turbine rotor. Define C_P and λ. What is the Betz limit?
3. Explain, using appropriate diagrams and sketches, how power may be regulated in a wind turbine by pitch regulation and stall regulation. Why is stall regulation used only for fixed-speed wind turbines?
4. Why are large wind turbines operated at variable speed? Explain the principle of operation of a full-power-converter variable-speed wind turbine.
5. Describe briefly what is meant by the measure–correlate–predict technique for establishing the long-term wind resource at a wind farm site.

6. Discuss how Weibull parameters may be used to describe hourly mean wind-speed data at a site.
7. A wind turbine has a rotor diameter of 33 m. At a wind speed of 13 m/s,
   (a) what is the power in the wind,
   (b) what is the power in the shaft of the turbine if the rotor has a power coefficient C_P of 0.35,
   (c) if the speed of rotation of the shaft is 35 rpm, calculate the tip speed and torque on the shaft.

   [**Answer:** (a) 1.15 MW; (b) 402.8 kW; (c) 60.5 m/s, 109.8 kNm.]
8. A wind turbine has a rotor of 55 m diameter, mounted at a hub height of 60 m. The gearbox ratio is 1:28. If it operates at a tip speed ratio (λ) of 8 in a wind speed of 10 m/s, with a power coefficient of 0.35 and a thrust coefficient of 0.55, calculate:
   (a) the torque on the generator shaft,
   (b) the overturning moment at the base of the tower.

   [**Answer:** (a) 6.25 kNm; (b) 4.8 MNm.]
9. A large horizontal-axis wind turbine has a rotor of diameter 80 m at a hub height of 90 m. It operates at a rotational speed of 15 rpm with a combined efficiency of the gearbox and generator of 95% to produce 2 MW of electrical output power.
   (a) If the hub-height wind speed is 12 m/s, calculate C_P (the power coefficient) of the rotor.
   (b) Write down the tip speed ratio.
   (c) Estimate the wind speed at a 10 m-high met mast adjacent to the wind turbine.

   [**Answer:** (a) 0.395; (b) 5.23; (c) 8.8 m/s.]
10. A horizontal-axis wind turbine has a rotor diameter of 20 m and reaches its full load output at the generator terminals of 150 kW at a wind speed of 13 m/s. The overall efficiency of the gearbox and generator is 94%.
    (a) Calculate the power in the air passing through the rotor disk and hence the power coefficient (C_P) of the aerodynamic rotor.
    (b) The turbine operates at a tip speed ratio (λ) of 4. Calculate the rotational speed of the aerodynamic rotor and hence the torque on the rotor shaft.
    (c) The inertia of the complete rotor, gearbox and generator is 50×10^3 kgm^2. Calculate the speed of the rotor 0.5 seconds after the connection to the electrical power network is broken and write down the energy that is absorbed by the brake when it brings the rotor rapidly to rest.

    [**Answer:** (a) 422.75 kW, 0.377; (b) 5.2 rad/s, 30.7 kNm, (c) 5.5 rad/s, 756 kJ.]
11. A wind turbine is rated at 400 kW at 12.5 m/s. The annual distribution of hourly mean wind speeds at the site may be represented by a Weibull distribution with scale parameter of 8.5, shape parameter of 2.5.

Calculate the annual energy production in kWh per year that the wind turbine will produce at rated power.

[**Answer:** 254 MWh.]

12. A wind farm consists of 10 wind turbines each with:

Hub height	50 m
Rated power	2 MW
Cut-in wind speed	5 m/s
Rated wind speed	13 m/s
Cut-out wind speed	20 m/s

Long-term monitoring of wind speeds has been undertaken using an anemometer at 20 m height and the site wind speeds can be represented by a Weibull distribution with:

scale parameter 8,

shape parameter 2 (Raleigh distribution),

Assume that the scale parameter increases with height according to the one-seventh power law.

Calculate:

(a) the number of hours each year the turbines operate,

(b) the annual energy production (in MWh per year) that the wind turbines will produce when operating above rated power.

Note: the following are likely to be useful.

$$\frac{U(z)}{U(h)} = \left(\frac{z}{h}\right)^{a}$$

$$F(U) = 1 - \exp\left[-\left(\frac{U}{c}\right)^{k}\right]$$

[**Answer:** (a) 6414 hours; (b) 21.5 GWh.]

13. Estimate the Weibull parameters of the following set of hourly mean wind speeds measured over one year.

Wind speed (m/s)	Hours
0–4	2050
4–8	4000
8–12	2200
12–16	460
16–20	50

Note: graph paper and the following may be useful.

$$F(U) = 1 - \exp\left[-\left(\frac{U}{c}\right)^{k}\right]$$

[**Answer:** $c = 7.2$, $k = 2$.]

14. The annual distribution of hub-height hourly mean wind speeds at a site may be described by a Weibull distribution with:

scale parameter, c, of 8.5,

shape parameter, k, of 2.

Calculate the gross annual energy yield of a wind turbine with a power curve shown in the table, installed at this site.

Wind speed (m/s)	Output (kW)
0–5	0
6	50
7	120
8	280
9	440
10–25	600

Note: the following are likely to be useful.

$$f(U) = \frac{k}{c}\left(\frac{U}{c}\right)^{k-1} \exp\left[\left(\frac{U}{c}\right)^{k}\right]$$

$$F(U) = 1 - \exp\left[-\left(\frac{U}{c}\right)^{k}\right]$$

[**Answer:** 2191 MWh.]

FURTHER READING

Anderson C., *Wind Turbines: Theory and Practice*, 2020, Cambridge University Press.
A very accessible, clear textbook dealing with many aspects of wind energy.

Burton A. *et al.*, *Wind Energy Handbook*, 3rd ed., 2021, Wiley.
An advanced book for practitioners and researchers.

Hansen M.O.L., *Aerodynamics of Wind Turbines*, 2nd ed., 2008, Earthscan.
An authoritative treatment of wind turbine aerodynamics.

Jamieson P., *Innovation in Wind Turbine Design*, 2nd ed., 2018, Wiley.
Provides interesting insights into wind turbine design.

Manwell J.F, McGowan J.G. and Rogers A.L., *Wind Energy Explained*, 2nd ed., 2009, Wiley.
Comprehensive, accessible explanation of many aspects of wind energy.

ONLINE RESOURCES

Exercise 2.1 Wind resource assessment: wind speed data are used to demonstrate probability density functions, cumulative density functions and a wind rose.

Exercise 2.2 Wind shear and Weibull parameters: wind speed data are used to demonstrate wind shear and the estimation of Weibull parameters.

Exercise 2.3 Power production: wind speed data and a wind turbine power curve are used to calculate the power produced by a wind turbine.

Exercise 2.4 Wind turbine performance: BLADED software is used to simulate the performance of a wind turbine.

3 Hydro Power

INTRODUCTION

Hydro power produces around 16% of the world's electrical energy (~4300 TWh annually) from more than 1300 GW of generating plant. The power that can be generated from a hydro power scheme depends on the flow of water and the height through which the water falls. In Europe most of the suitable sites for large hydro power schemes have been exploited and any future developments are likely to be small. In less populated parts of the world (e.g. parts of Asia, South America and Africa) there remain sites suitable for large hydro power schemes that have yet to be developed. The technical resource worldwide is estimated to be more than 16 400 TWh annually (4 times present output) although much of this resource will never be developed because of environmental and cost constraints.

The flow of water into a hydro scheme varies season by season so the output of hydro schemes is not constant but varies throughout the year. Although water can be stored in the reservoirs of large schemes, its extraction has to be managed to maintain suitable environmental conditions. Flows in rivers and the height of water in reservoirs must be maintained in order to support fish, plants and wildlife. Worldwide the annual capacity factor[1] of hydro schemes is ~44% and they often operate at only part of their capacity or are shut down. The hydro schemes in the UK have an annual capacity factor of ~39%, depending on the rainfall that year. This rather low utilization of the generating plant is partly due to the availability of water but is also because the output power of a hydro generator can be altered very quickly, and so hydro power stations are often used to balance the varying electrical load demand of the power system.

Several very large hydro power schemes have been developed, such as the 22 GW Three Gorges scheme in China and the 14 GW Itaipu scheme on the border of Brazil and Paraguay. Around 10% of the worldwide capacity of hydro power is from a large number of small hydro schemes (less than 5–10 MW). Large hydro power schemes can have a significant environmental and social impact and may displace rural people from their land as well as destroy important habitats of wildlife and plants. Smaller hydro schemes, particularly those without large reservoirs, can

[1] Capacity factor is defined in Section 1.4.1.

be arranged to have fewer environmental and social consequences and are being actively encouraged in many countries.

Large hydro generating sets invariably use synchronous generators, while small units (<10 MW) can use induction generators (see Tutorial I). Variable-speed generators are occasionally used in order to maintain high efficiency of the turbine over the wide range of hydraulic conditions that are experienced in some small low-head schemes.

The hydro resource, turbines and hydro power plants are discussed in this chapter. Section 3.1 reviews the history, advantages and disadvantages of hydro power. Sections 3.2 and 3.3 present the principle of operation of a hydro power station and its power output. Section 3.4 gives a brief introduction to the types of hydro power scheme. Sections 3.5–3.7 discuss the different types of turbine, specific speed and operation at reduced flows and variable speed. Sections 3.8 and 3.9 explain the calculation of net head and give a short discussion of transient conditions. Finally in Section 3.10 the development of a small hydro power scheme is discussed.

3.1 Hydro Power

3.1.1 History

Vertical-axis waterwheels have been used for thousands of years to provide mechanical power to mill grain and are still used in remote mountainous areas of the Himalayas. Sometimes known as Norse wheels, these simple wooden vertical-axis paddle wheels are struck by the flow of water on one side only and take kinetic energy from a running stream to drive a vertical shaft that turns a millstone directly.

Horizontal-axis waterwheels were widely used in Europe and the USA until the middle of the nineteenth century for grinding grain and then to provide mechanical power for the factories of the industrial revolution. They were the subject of considerable technical development, and a number of variations were built.

Figure 3.1 shows the main types of waterwheel. The undershot wheel (Figure 3.1a) is placed in the flow of a channel diverted from a river; it uses straight wooden blades to take kinetic energy from the flow. Simple undershot wheels generate limited amounts of power as they mainly use the kinetic energy of the stream and there is no significant change in the height of the flow across the wheel. Practical undershot wheels are only up to 20% efficient.

Overshot wheels (Figure 3.1b) use buckets to contain the water, and their operation relies mainly on the weight of water filling the buckets. The effect of the kinetic energy of the flow filling the buckets is negligible. Overshot wheels require a supply of water with a head at least equal to the diameter of the wheel and a strong support structure for the wheel. Large metal overshot wheels with wheel diameters of 10 m using heads of up to 12 m were constructed historically and practical efficiencies of up to 40% were achieved.

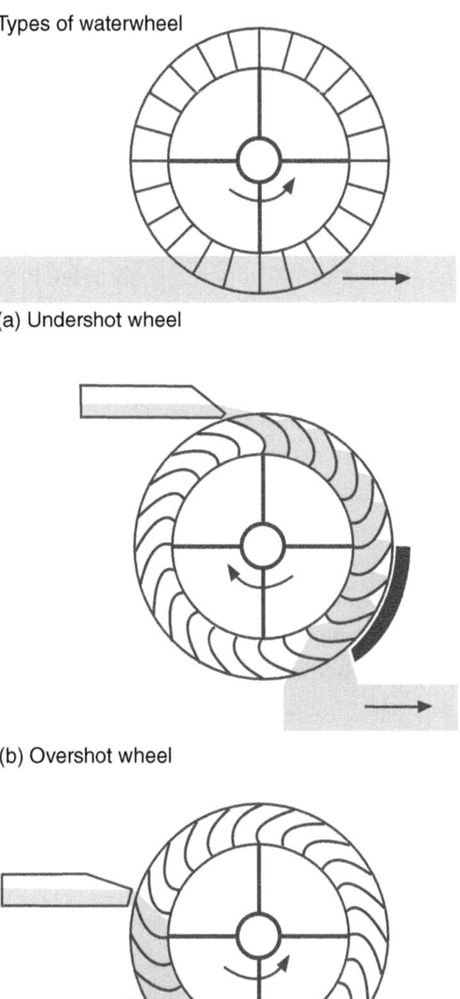

Figure 3.1 Types of waterwheel.

A compromise is the breast wheel (Figure 3.1c), which uses both the effect of gravity and the kinetic energy of the flow. Variable flow rates are accommodated by closing or opening a series of intake gates. The design of breast wheels and their fabrication in metal was highly developed during the industrial revolution and efficiencies of up to 70% were achieved but only at low rotational speeds of 3–7 rpm.

Waterwheels, with their low heads, require large flows to develop significant power, and from the middle of the nineteenth century they began to be replaced by water turbines. Turbines can use higher heads with smaller flows to generate power.

Importantly, turbines operate at much higher rotational speeds than waterwheels. Electrical generators work most effectively with a high speed across the airgap, i.e. a high speed between the electrical stator and rotor. A high-speed turbine allows the generator to be connected directly to a shaft and avoids the use of a gearbox.

Fourneyron's turbine of 1832 is usually considered to be the first hydro turbine. This was a vertical-axis device with fixed central guide vanes that directed the flow of water radially outwards to an external runner. The turbine was positioned below the level of the tail water and operated completely submerged. This early turbine was an important step on the path of development that led to the inward-flow reaction[2] turbine of Francis in 1849. In parallel, Pelton was developing the impulse turbine, first produced commercially around 1890. Kaplan patented the fixed-blade propeller turbine in 1913 and shortly afterwards developed the propeller turbine with variable-blade pitch. Propeller turbines are particularly effective for the utilization of low heads and while Kaplan made important contributions to the developments of both fixed- and variable-blade types only variable-blade pitch propeller turbines are today known as Kaplan turbines. Francis, fixed-blade propeller and Kaplan reaction turbines as well as Pelton impulse turbines are all in widespread use today for hydro generation.

There has been considerable recent interest in the use of novel devices for small-scale hydro generation including the use of Archimedes screws to exploit low-head hydro resources. Archimedes screws are described in Section 3.10.4.

3.1.2 Advantages and Disadvantages of Hydro Power

Hydro power has been used to generate electricity since the late nineteenth century and was the source of energy for many of the early electrification schemes. Because its output can be controlled rapidly, hydro power is very useful when operating an electrical power system and is often used to balance the load demand as it varies throughout the day.

The advantages of hydro power for generating electricity are:

- Operating costs are very low. Many hydro power stations are operated remotely with few staff permanently on site.
- The power output of the turbine generators can be adjusted rapidly.
- If a reservoir is used then water can be stored and used when required.
- The civil works have a long lifetime (up to 100 years).

The main disadvantages are:

- The capital costs of a hydro power scheme, particularly the civil works, are often high.
- Schemes with large reservoirs can result in a significant loss of land with major impacts on the environment and local populations.

[2] Reaction and impulse turbines are described in Section 3.5.

- The power output of a run-of-river scheme varies and depends on the precipitation (rain) and the vegetation and soil of the catchment area. Run-of-river hydro schemes without energy storage can have capacity factors as low as 30%.
- When large reservoirs are first flooded, decomposing submerged vegetation can lead to significant emissions of methane (a powerful greenhouse gas).
- The failure of a large dam can lead to catastrophic loss of life.

3.2 Operation of a Hydro Scheme

Figure 3.2 is a simple schematic of a hydro scheme. The main components are:

- intake (including trash rack to remove debris)
- penstock
- surge tank to limit transient pressures in the penstock
- spillway to divert excess water and avoid overtopping of the dam
- turbine (including governor system to control power output)
- generator
- draft tube (on reaction turbines).

A pipe able to withstand a high pressure takes water from a high-level intake to a turbine at a lower level. The flow of water then operates the turbine to which a generator is coupled either directly or, for smaller generators, through a gearbox. After passing through the turbine, the water exits into the tail waters and re-joins the river system.

The high-pressure pipe is known as the penstock. In earlier times the term penstock was used to describe the gate that controlled the flow of water into the intake,

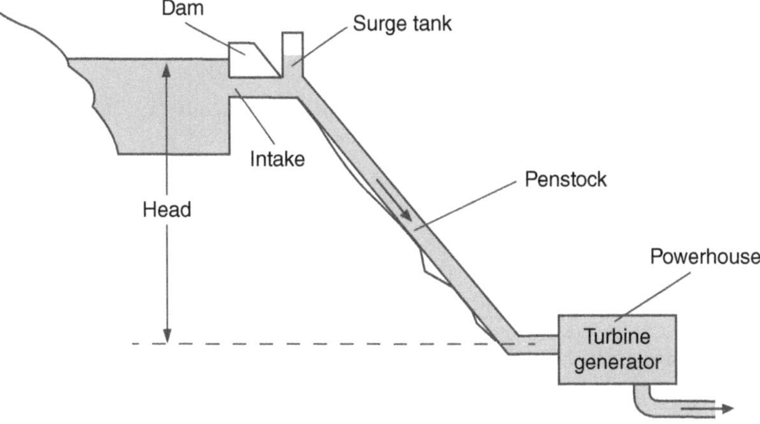

Figure 3.2 Schematic diagram of a high-head hydro scheme.

but the term now refers to the high-pressure pipe. In many situations the penstock is buried to avoid visual intrusion.

The static pressure of the water in the pipe, p, is simply

$$p = \rho g h \quad \left[\text{Pa, N/m}^2\right]$$

where

h is the head in m,
ρ is the water density, 1000 kg/m^3,
g is the acceleration due to gravity, 9.81 m/s^2.

In some hydraulic calculations the specific weight of water (γ) is used:

$$\gamma = \rho g \approx 9.81\,\text{kN/m}^3$$

resulting simply in

$$p = \gamma h$$

The static pressure is proportional to the head h. The gross head is the vertical height between the surface of the water at the high-level intake and, for an impulse turbine, the height of the turbine jet. For a reaction turbine the gross head is measured between the surfaces of the water at the high-level intake and at the tail waters.

Atmospheric pressure is approximately 10^5 Pa and so every additional 10 m of head increases the pressure in the penstock by about 1 atmosphere. Thus, for high head schemes, the penstock has to withstand very high pressures.

The power in a flow of water is the product of the static pressure and flow rate

$$P_{WATER} = p \times Q = \rho g h \times Q \quad [\text{W}] \tag{3.1}$$

where

Q is the flow rate in m^3/s.

There are, of course, losses caused by friction in the flow in the penstock as well as losses in the turbine and generator. However, unlike in a wind turbine, large hydro turbines can achieve efficiencies of 90–95%. A typical overall efficiency of an entire hydro scheme operating at rated power output is 60–70%.

If the high-level intake is fed from a reservoir then, assuming the depth of stored water remains constant, the energy stored E is

$$E = \rho V g h \quad [\text{J}]$$

where

V is the volume of useable water in the upper reservoir in m^3.

Not all the water in a river supplying a hydro scheme or stored in a reservoir can be used for generating electricity. It is necessary to leave a certain residual flow in rivers and height of water in reservoirs for environmental reasons. This is known as compensation flow. Many hydro power stations are important elements of irrigation and public water supply schemes, and the operation of the scheme will then be determined primarily by the use of water for these other purposes.

Example 3.1 – Operation of a Hydro Power Scheme

Consider a small hydro scheme with a head of 30 m and a flow rate of 3 m^3/s. The efficiencies of the main components at rated output are:

penstock efficiency 0.9,
turbine efficiency 0.85,
generator efficiency 0.93.

Calculate the electrical output.

The reservoir supplying the scheme has a useable storage capacity of 10^5 m^3. How much electrical energy can be generated without the reservoir being replenished? Ignore the effect of changes in the depth of the reservoir.

Solution
The power in the water flow is

$$P_{WATER} = \rho g h Q = 1000 \times 9.81 \times 30 \times 3 = 883 \text{ kW}$$

The overall efficiency of the hydro power scheme is

$$\eta_{OVERALL} = 0.9 \times 0.85 \times 0.93 = 0.71$$

Therefore the electrical output is

$$P_{ELECTRICAL} = 883 \times 0.71 = 628 \text{ kW}$$

The potential energy stored in the reservoir is

$$PE = \rho V g h = 1000 \times 10^5 \times 9.81 \times 30 = 2.943 \times 10^{10} \text{ J}$$

At an efficiency of conversion of 0.71 this would allow electrical energy to be generated of

$$E_{ELECTRICAL} = 2.943 \times 10^{10} \times \frac{0.71}{3600} = 5.8 \text{ MWh}$$

or full load output to be generated for

$$\frac{5.8 \times 10^6}{628 \times 10^3} = 9.2 \text{ hours}$$

Figure 3.3 shows a small hydro scheme in Africa. Note the three penstocks taking water to the powerhouse. The electrical switchyard beyond contains transformers to increase the 6 kV voltage of the generators to 33 kV and masts at each corner to protect it against lightning.

Figure 3.3 Small hydro scheme in Africa. The net head is 176 m with a flow rate of each penstock of 3000 litres/s. Each turbine generator unit is rated at 4.2 MW.
Photo: Gilkes.
(A black and white version of this figure will appear in some formats. For the colour version, refer to the plate section.)

3.3 Power Output of a Hydro Scheme

The energy of a hydro power scheme comes from precipitation driven by the hydrological cycle, illustrated in Figure 3.4. Solar radiation causes evaporation from both land and sea, clouds form and rainfall results.

Some of the rainfall is effectively lost due to evaporation, transpiration by vegetation and surface absorption. The remaining runoff flows into streams and then rivers. The catchment area is the surface area that contributes to the discharge of a particular stream or river. Watersheds divide catchment areas from each other. The catchment areas act to concentrate the precipitation into a useful flow of water and, depending on their characteristics, will store water and result in more or less uniform flows in the rivers.

The runoff of a catchment area is given by:

$$\text{Runoff} = (\text{rainfall} - \text{evaporation and transpiration} - \text{surface absorption}) \times \text{catchment area}$$

Across the UK the average rainfall is around 1000 mm/year but with considerable variations by region. Around 40% of the rainfall evaporates on average but this reaches 80% in parts of southern England. There are significant variations in both rainfall and evaporation throughout the year.

The flows in a river are described either by a time series of average daily flows (known as a hydrograph) or by a flow duration curve showing the percentage of time a particular flow is exceeded. Flow duration curves are often plotted with a

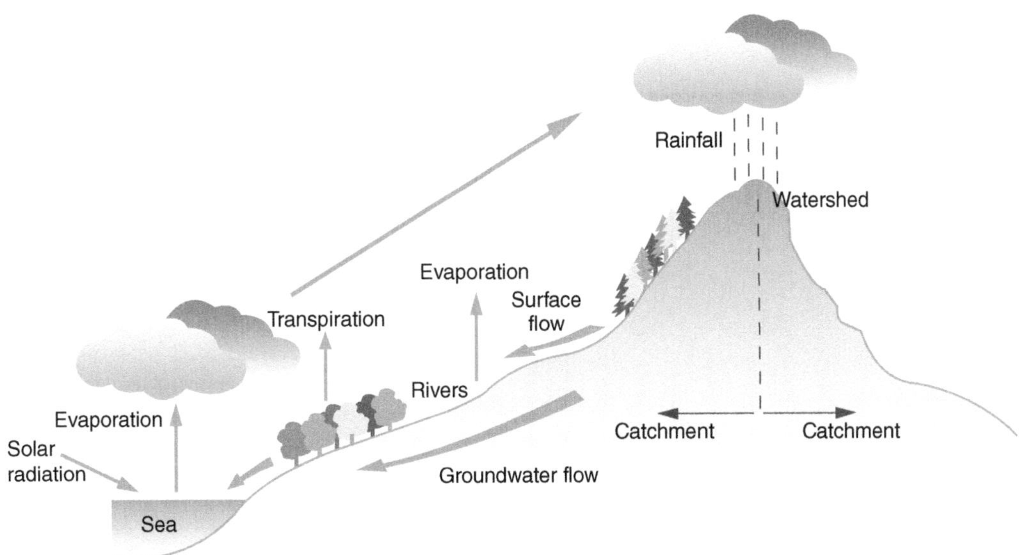

Figure 3.4 The hydrological cycle.
Source: E.M. Wilson, *Engineering Hydrology*, 1990, Red Globe Press, an imprint of Bloomsbury Publishing Plc.

normal probability scale on the x-axis and a logarithmic scale on the y-axis. If the log of the flows is normally distributed, as is often approximately the case, then the resulting curve is a straight line. A more uniform flow gives a flat line while a highly variable flow gives a steep flow duration curve.

Figure 3.5 shows the hydrograph of daily flows and the flow duration curve of a river in the south of England. The river has a long, gently sloping bed and a catchment area with deep soil and considerable vegetation. The flow is approximately constant throughout the year.

Figure 3.6 shows a similar hydrograph and flow duration curve for a river in Wales. The soil of the steeper catchment area is rocky with little vegetation. It may be seen that the daily flows are much more variable resulting in a steeper flow duration curve. Figure 3.6 illustrates clearly that, unless a large reservoir is included in the scheme, the output power of a hydro generator will vary with rainfall. This is particularly the case with small hydro schemes that have smaller catchments or limited storage of the water provided by the soil or vegetation in the catchment area.

The power output of a hydro scheme is determined by the head and the flow of water (see Equation 3.1). The head of a potential hydro power scheme can be found fairly easily either from maps or by surveying the site and, once determined, other than for some low-head schemes, is effectively constant. However, the flow can vary day to day and throughout the year. Unless the river that will feed the scheme has been instrumented for many years, the flows over the anticipated life of a potential scheme are difficult to estimate and it is necessary to correlate the flows of the river under investigation with long-term records of either other nearby rivers or rainfall.

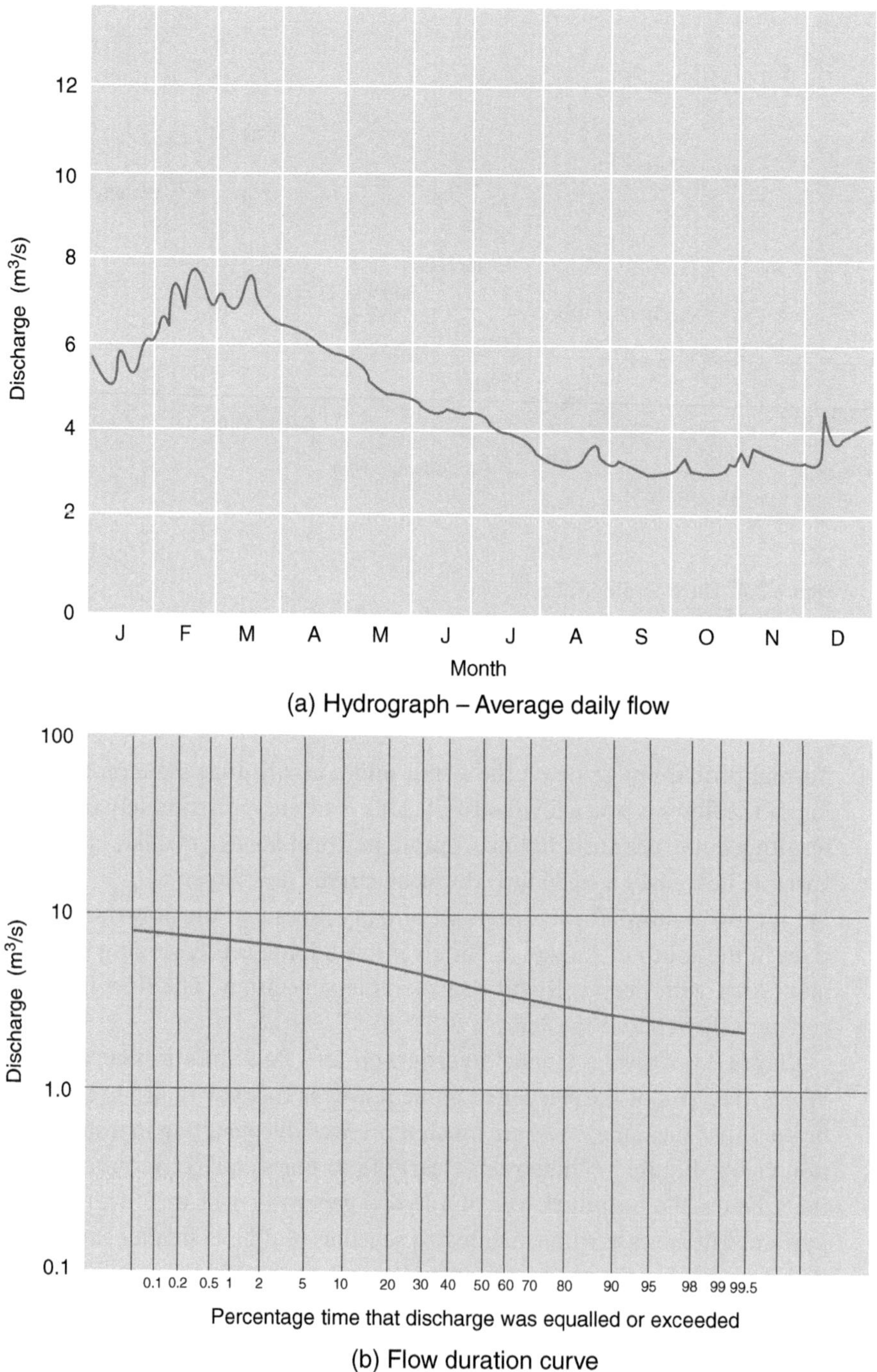

Figure 3.5 (a) Average daily flow and (b) flow duration curve of a river with an even discharge over the year.

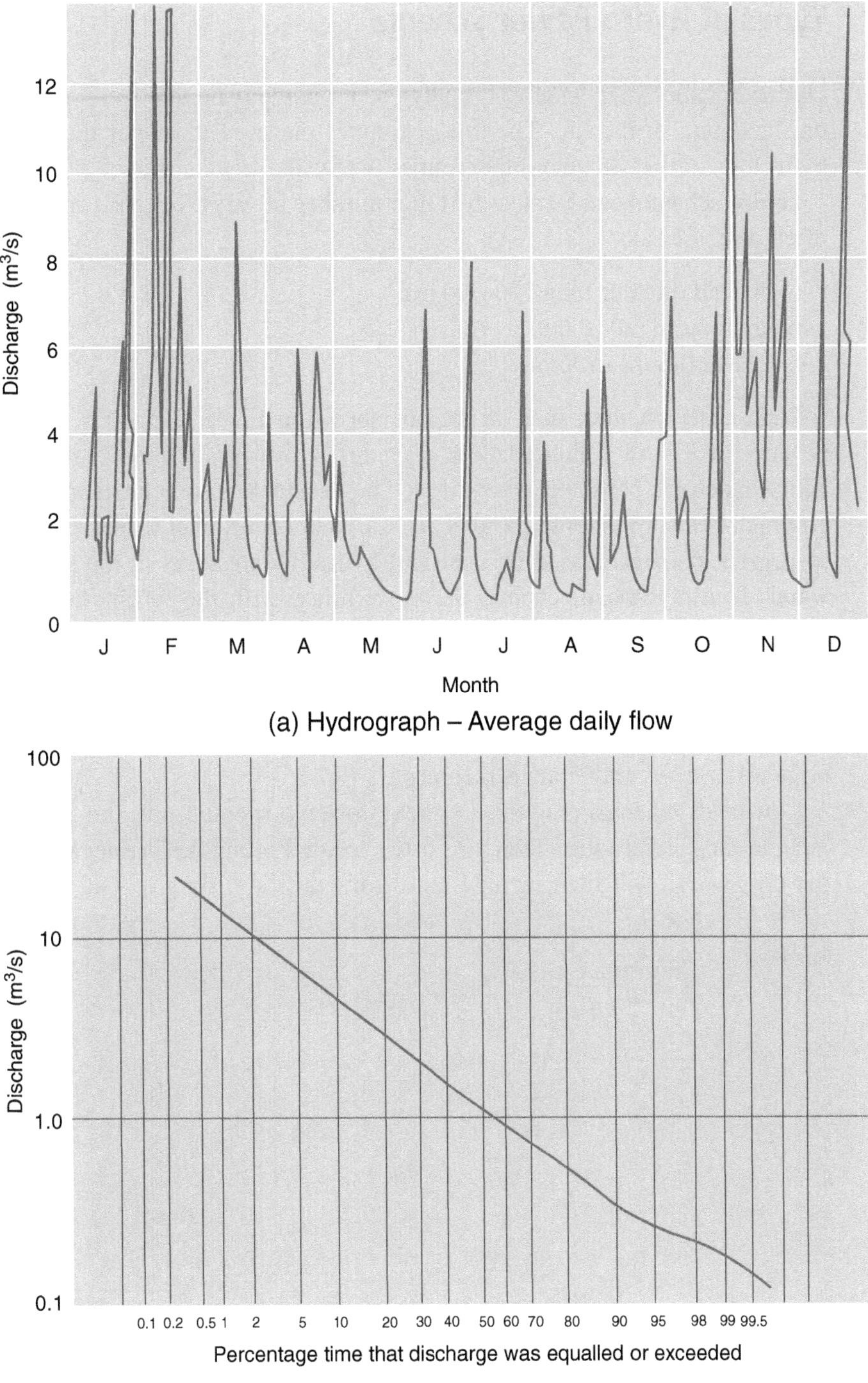

Figure 3.6 (a) Average daily flow and (b) flow duration curve of a river with variable discharge over the year.

3.4 Types of Hydro Power Scheme

Hydro schemes vary widely. Usually each scheme is unique as its design depends on the details of the site. The topography of the site determines the head, and the hydrology of the catchment determines the available water flow.

Hydro schemes can be classified in a number of ways. In terms of head they are often described as:

high head (greater than 200–300 m),
medium head (30–200 m),
low head (less than 30 m).

High-head schemes, such as the one shown in Figure 3.2, allow high power to be generated with a limited flow of water. However, a long, high-pressure and hence expensive penstock is required. The penstock is sometimes made up of several high-pressure pipes in parallel and surge tanks or relief valves may be required to prevent damage caused by transient hydraulic pressures in the long column of water. In a high-head scheme, the powerhouse with the turbine is often located some distance from the intake.

Medium-head schemes often have the powerhouse located at the base of the dam (Figure 3.7) with a limited length of penstock and use a high flow rate to generate high power. Some of the largest hydro power schemes use this arrangement with large water flows and Francis turbines.

Low-head schemes (Figure 3.8) have lower power output and usually do not include large reservoirs. They are often located along navigable waterways, and the civil works will then include locks for boats and barges. The draft tube is a

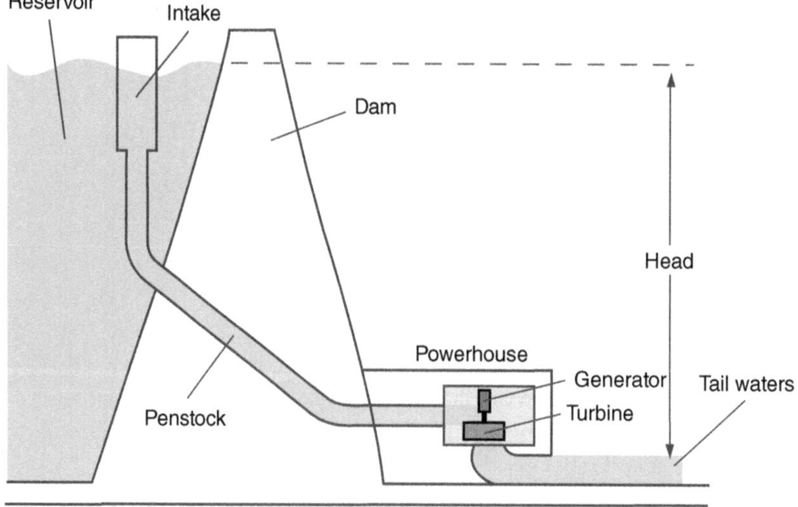

Figure 3.7 Schematic diagram of a medium-head hydro scheme using a reaction turbine. Source: Godfrey Boyle, *Renewable Energy: Power for a Sustainable Future*, 1996. By permission of Oxford University Press.

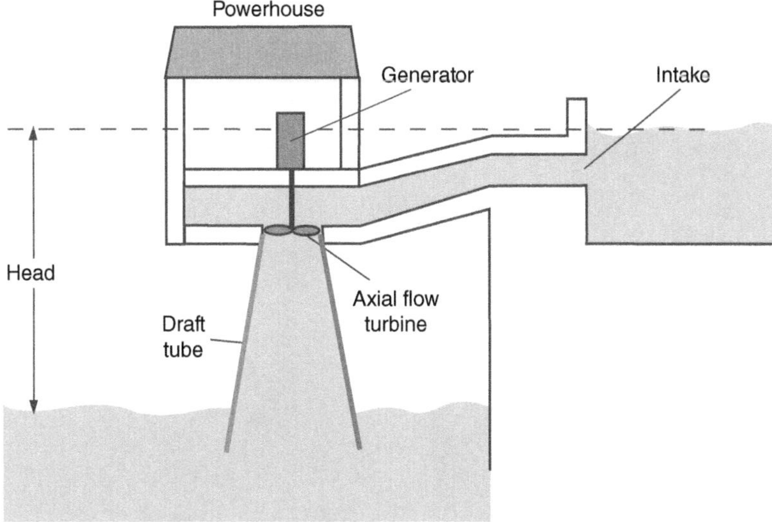

Figure 3.8 Schematic diagram of a low-head hydro scheme using a reaction turbine.

particularly important element of low-head schemes as it can significantly increase the head available.

An important distinction is whether a hydro power scheme uses a reservoir to provide storage of the water or relies on the flow within a river. The output of run-of-river schemes depends immediately on the flow available within the river. This changes with the season, so the output of a run-of-river scheme can vary considerably.

In terms of power output rating, hydro power schemes are often described as:

large hydro (more than 50 MW),
small hydro (5–50 MW),
mini hydro (25–250 kW),
micro hydro (less than 25 kW).

These definitions of a hydro scheme either by head or by power output are not precise and the meaning of the terms differs from country to country.

3.5 Hydro Power Turbines

Hydro power turbines may be divided into two main types, depending on the pressure drop across the turbine runner.

Impulse turbines operate by extracting kinetic energy from one or more high-speed jets of water that strike a runner at atmospheric pressure. There is no pressure drop across the runner, and energy is extracted by changing the direction of the water jet. The pressure energy in the penstock is converted into the kinetic energy of the jet at the nozzle. Impulse turbines are used for high heads. Some small hydro units use designs of impulse turbines that can be made with limited manufacturing facilities.

Figure 3.9 Small hydro turbines in the factory.
Photo: Gilkes. (Annotations by the author.)
(A black and white version of this figure will appear in some formats. For the colour version, refer to the plate section.)

Types of impulse turbines include Pelton, Turgo and crossflow.

Reaction turbines have a runner completely immersed in the flow of water and operate by the hydrodynamic lift forces generated on the runner. The water arriving at the runner is still under pressure and a significant fraction of the power is generated by the pressure drop across the turbine. The runner is located in a pressurized casing with a carefully designed clearance between the runner and casing. Hence precision is required in the manufacture of both the runner and casing. Reaction turbines are used for medium and low heads.

Types of reaction turbines include Francis and propeller type. The Kaplan turbine is a development of the propeller type with variable-pitch blades.

Figure 3.9 shows examples of complete small hydro turbines with their runners, spiral casing and spear valves.

3.5.1 Impulse Turbines

Figure 3.10 shows a single-jet, horizontal-axis Pelton turbine. Water is brought from a high-level intake by the pressurized penstock to form a high-velocity jet. After it exits the nozzle, the jet of water is at atmospheric pressure. The jet strikes the buckets of the turbine, is divided by the splitters on the buckets and turned back

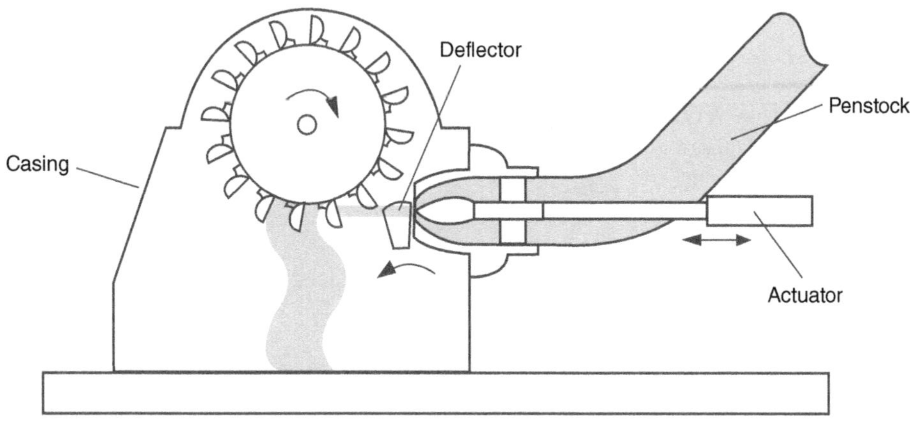

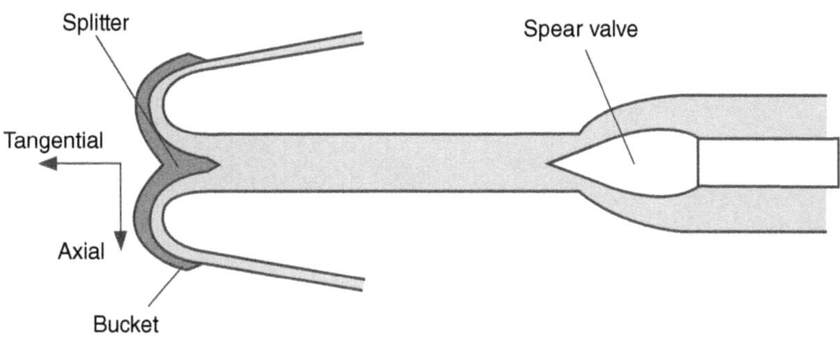

Figure 3.10 Schematic of a Pelton turbine.

on itself. By this change of direction (and hence change of momentum) the jet creates a force on the turbine runner to which a generator is coupled. After striking the runner, the water, from which the energy has been extracted, falls vertically to the tail waters under gravity. The function of the casing is to control splashes and there is atmospheric pressure within the casing.

Figure 3.11 shows two Pelton runners in the factory. The splitters of the buckets are shown clearly. The flow of water and hence the output power of the generator is controlled by adjusting the spear valve. As the spear valve is inserted into the nozzle, the flow of the jet is reduced. The spear valve is operated slowly to avoid high transient pressures (water hammer) in the long, high-inertia column of water in the penstock.

If the generator suddenly loses its electrical load, it will overspeed and be damaged unless the torque from the turbine is reduced rapidly. Rapid control is achieved through rotation of the deflector, which diverts the jet away from the buckets but does not change the flow in the penstock. Any attempt to change the flow rapidly would damage the penstock. The directions of movement of the spear valve and rotation of the deflector are shown in Figure 3.10. Figure 3.12 shows the control devices of a Pelton turbine.

Figure 3.11 Two Pelton runners.
Photo: Gilkes.
(A black and white version of this figure will appear in some formats. For the colour version, refer to the plate section.)

Figure 3.12 Horizontal-axis Pelton wheel in housing showing controls.
Photo: New Mills Engineering. (Annotations by the author.)
(A black and white version of this figure will appear in some formats. For the colour version, refer to the plate section.)

3.5.2 Analysis of a Pelton Turbine

The operation of a Pelton wheel can be understood by considering the velocities of the water before and after the jet strikes the turbine runner. The jet with an absolute velocity V strikes the runner which is moving at a velocity U with a velocity relative to the bucket R. In vector notation:

$$\boldsymbol{V} = \boldsymbol{R} + \boldsymbol{U}$$

Any radial flow is ignored, and these velocity vectors are split into components in two directions:

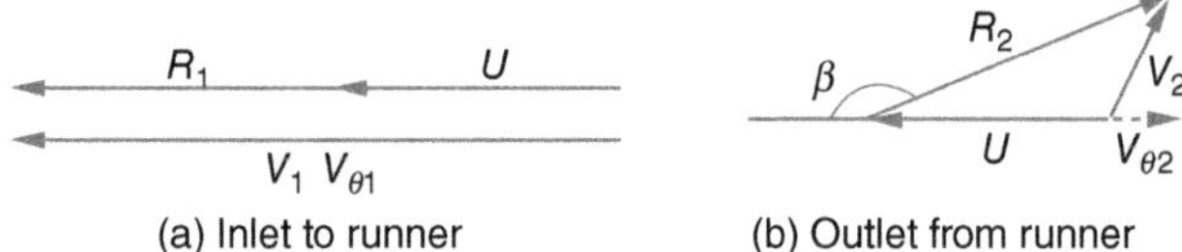

Figure 3.13 Velocity diagram of the jet before and after it strikes the runner of a Pelton turbine: V_1 is the absolute velocity of the jet before it strikes the runner; V_2 is the absolute velocity of the jet after it leaves the runner; U is the velocity of the runner, $U = \omega r$, where r is the effective radius of the runner and ω is the angular velocity in rad/s; R_1 is the relative velocity of the jet to the bucket before it strikes the runner and R_2 is the relative velocity of the jet to the bucket after it leaves the runner.

(1) flow tangential to the rotation of the runner (V_θ) – the tangential direction,
(2) flow in-line with the axis of rotation of the runner – the axial direction.

A bucket of the runner can move only in the tangential direction and only velocities in the tangential direction contribute to torque and hence power. The jet is divided symmetrically by the splitter (Figure 3.10) and so only the velocities of half of the flow are shown in Figure 3.13.

Figure 3.13a shows the velocities as the jet strikes the runner. As the jet strikes a bucket, all these velocities are in the tangential (circumferential) direction and so using scalar notation in the tangential direction

$$V_1 = R_1 + U$$

After it leaves the runner (Figure 3.13b), the jet has been turned through an angle β (around 165°) and its speed across the bucket (relative velocity) is reduced slightly by the bucket friction velocity coefficient C_B (typical value 0.9).

$$R_2 = C_B R_1$$

Its absolute velocity has been reduced from V_1 to V_2 and, as shown here, its direction reversed. V_2 contains a small residual component in the tangential direction and a small axial component to move the water away so that the jet does not strike the following buckets. For high efficiency V_2 is kept small.

The components of the absolute velocities of the jet that are tangential to the rotating runner (known as the whirl velocities), before and after it strikes the runner, are

$$V_{\theta 1} = V_1 = U + R_1$$
$$V_{\theta 2} = U + R_2 \cos\beta \left(\text{Note } \beta \text{ is} > 90^\circ \text{ and so } \cos\beta \text{ is } -\text{ve}\right)$$

Thus the change in the tangential velocity of the jet is

$$\begin{aligned}\Delta V_\theta &= V_{\theta 1} - V_{\theta 2} = U + R_1 - \left(U + C_B R_1 \cos\beta\right) \\ &= R_1\left(1 - C_B \cos\beta\right) \\ &= \left(V_1 - U\right)\left(1 - C_B \cos\beta\right)\end{aligned} \tag{3.2}$$

The mass flow rate in the jet $\dot{m}$ is

$$\dot{m} = \rho Q$$

and so the rate of change of momentum in the tangential direction (or force developed) is

$$\Delta V_\theta \dot{m} = \Delta V_\theta \rho Q$$

The effective radius of the runner is r and so the torque is

$$\Delta V_\theta \dot{m} r = \Delta V_\theta \rho Q r$$

With an angular velocity ω the power developed by the runner is

$$P = \Delta V_\theta \dot{m} r \omega = \Delta V_\theta \rho Q r \omega = \Delta V_\theta \rho Q U$$

Using Equation 3.2, we have

$$P = \rho Q U (V_1 - U)(1 - C_B \cos\beta) \tag{3.3}$$

The absolute velocity of the jet V_1 is determined mainly by the head with some reduction due to friction within the penstock and nozzle. However, the speed of the runner U is a design choice. Differentiating Equation 3.3 with respect to U gives

$$\frac{dP}{dU} = \rho Q (1 - C_B \cos\beta)(V_1 - 2U)$$

It may be seen that P is a maximum when $U = \frac{1}{2}V_1$, i.e. when the runner speed is half the speed of the jet. In practice, because of losses, a slightly slower runner speed is chosen and usually

$$U \sim 0.45 V_1$$

The efficiency of the runner is the power developed (Equation 3.3) divided by the power of the jet.

For any uniform fluid flow of area A

$$P = \frac{1}{2}\rho A V^3 = \frac{1}{2}\rho Q V^2$$

Hence

$$\eta_R = \frac{\rho Q U (V_1 - U)(1 - C_B \cos\beta)}{\frac{1}{2}\rho Q V_1^2} = \frac{2U(V_1 - U)(1 - C_B \cos\beta)}{V_1^2}$$

The efficiency of the Pelton wheel runner with respect to speed is plotted in Figure 3.14. Maximum power efficiency of a Pelton turbine occurs when the linear velocity of a bucket approaches half the jet velocity. The practical effect of the jet not being turned completely through 180° and frictional losses as the flow passes over the buckets are shown in the difference between practical and theoretical runner efficiency in Figure 3.14.

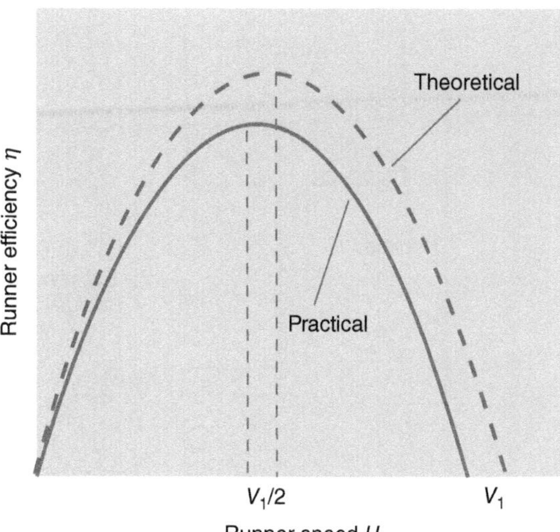

Figure 3.14 Variation of efficiency of a Pelton wheel with runner speed.

The efficiency of the overall turbine is further reduced by the nozzle efficiency

$$\eta_N = C_V^2$$

where C_V is the nozzle velocity coefficient. Overall efficiency of the complete hydro turbine power unit is also reduced by mechanical losses due to friction and windage, η_M, and electrical losses within the generator, η_E.

The maximum speed of the runner occurs when the runner is not loaded (runaway conditions) and then the runner velocity can approach that of the jet. Limiting the runaway speed is an important consideration in hydro turbines as synchronous electrical generators have a limited overspeed capacity before their rotor windings are damaged.

Example 3.2 – Operation of a Pelton Turbine

A hydro scheme with a single-jet impulse turbine is supplied by a penstock with a net head of 200 m. The nozzle velocity coefficient (C_V) is 0.97. Calculate the velocity of the jet as it leaves the nozzle.

The impulse wheel has an effective diameter of 1.1 m and is directly connected to a generator operating at 500 rpm. The diameter of the jet is 100 mm, the bucket friction velocity coefficient (C_B) is 0.9 and the bucket angle (β) is 165°. Calculate the efficiency of the turbine.

Solution

Figure 3.15 shows a simplified representation of the scheme.

Applying Bernoulli's theorem at the surface of the reservoir and at the jet

$$p_R + \frac{1}{2}\rho V_R^2 + \rho g h_R = p_J + \frac{1}{2}\rho V_J^2 + \rho g h_J$$

and we can make the following simplifications.

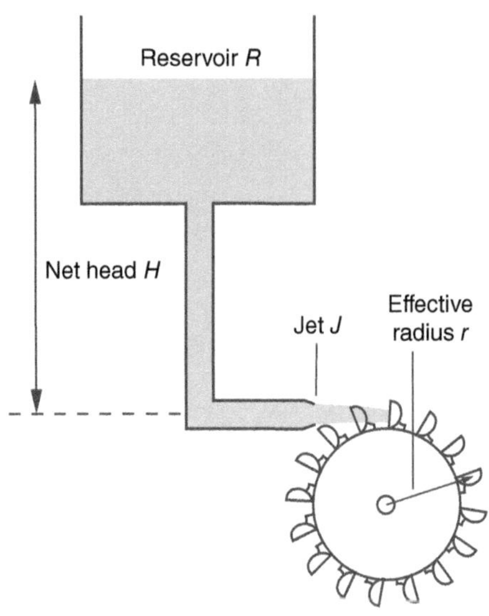

Figure 3.15 Schematic representation of an impulse hydro power scheme.

Both the reservoir and the jet are at atmospheric pressure and the difference of atmospheric pressure with height is ignored. Therefore, using gauge pressures, we have $p_R = p_J = 0$.

h_J is defined as the datum and so $h_J = 0$.

H is the net head and includes penstock friction and other penstock losses, so

$$h_R = H$$

The reservoir is large, so the surface of the water in the reservoir is considered to be stationary:

$$V_R = 0$$

Hence Bernouilli's theorem reduces to

$$\rho g H = \frac{1}{2}\rho V_J^2$$

or

$$V_J = \sqrt{2gH}$$

Including the effect of the nozzle velocity coefficient, the velocity of the jet is

$$V_J = C_V\sqrt{2gH} = 0.97\times\sqrt{2\times 9.81\times 200} = 60.8 \text{ m/s}$$

Using Equation 3.3 with $V_1 = V_J$

$$P = \rho Q U \ \ (V_J - U)(1 - C_B\cos\beta)$$

and

$$Q = V_J \times \pi\frac{D_J^2}{4} = 60.8\times\pi\times 2.5\times 10^{-3} = 0.477 \text{ m}^3\text{/s}$$

$$\omega = \frac{500\times 2\pi}{60} = 52.36 \text{ rad/s}$$

$$U = r\omega = 52.36 \times \frac{1.1}{2} = 28.8 \text{ m/s}$$

Therefore

$$\begin{aligned} P &= \rho Q U (V_J - U)(1 - C_B \cos\beta) \\ &= 1000 \times 0.477 \times 28.8 \times (60.8 - 28.8) \times (1 - 0.9\cos 165) \\ &= 822 \text{ kW} \end{aligned}$$

Check by considering power in the water flow:

$$P = \rho Q H g = 1000 \times 0.477 \times 200 \times 9.81 = 936 \text{ kW}$$

Efficiency = 88%.

Practical considerations limit the diameter of the jet of a Pelton turbine to a quarter of the bucket diameter and to a tenth of the runner diameter. Thus, in order to increase the output power, it is common for Pelton turbines to have more than one jet. Doubling the number of nozzles, and hence the output power, is common in horizontal-axis turbines, while up to six nozzles can be used in vertical-axis turbines where the exhaust water can fall away from the runner without obstructing other jets.

The Turgo turbine (Figures 3.16 and 3.17) is a development of the Pelton impulse turbine. A similar penstock and nozzle are used. Rather than splitting the jet, the Turgo runner diverts the flow to one side and redirects the flow by about 160°. The jet strikes the runner from one side and exits from the other. This arrangement allows a greater flow rate without the exhaust water interfering with the jet. Hence a smaller diameter runner with higher rotational speed can be used and so a gearbox to couple the turbine to the generator can be avoided more often.

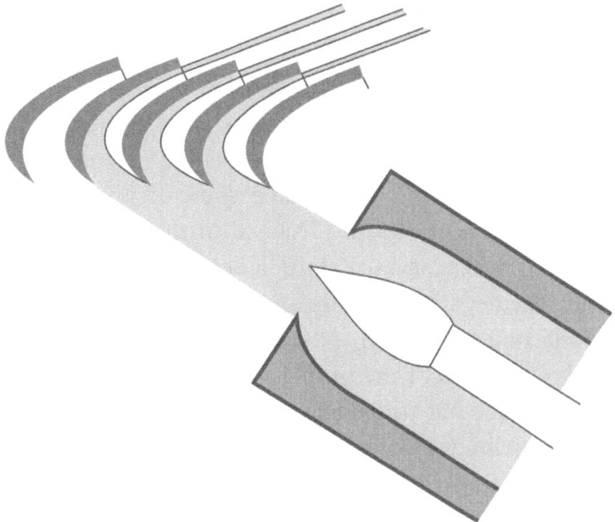

Figure 3.16 Turgo turbine and spear valve.

Figure 3.17 Turgo runner.
Photo: Gilkes.
(A black and white version of this figure will appear in some formats. For the colour version, refer to the plate section.)

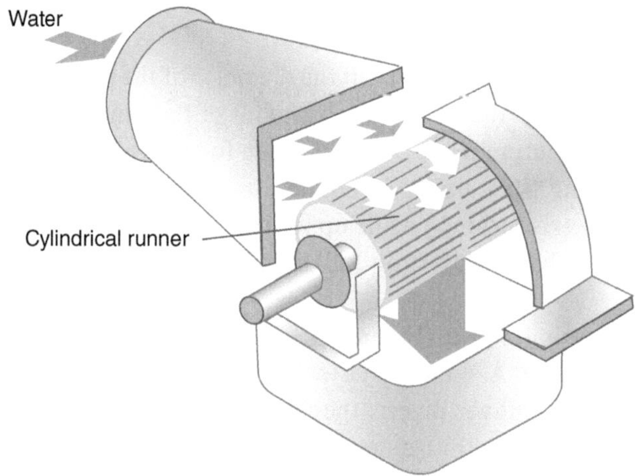

Figure 3.18 General arrangement of a crossflow turbine.

Both the Pelton and Turgo turbines use runners that are manufactured by a metal casting process. Cast runners can be expensive and difficult to source in remote areas of developing countries. An alternative, lower cost impulse turbine used with smaller heads is the crossflow (Michell–Banki) turbine. A crossflow turbine is shown in Figure 3.18, and the path of water through the cylindrical runner is shown in Figure 3.19. The turbine has a cylindrical fabricated runner which is struck twice by an extended sheet jet of water, once as the jet enters and again as it leaves the runner. Unlike a pure impulse turbine, some designs of the crossflow turbine use a sealed casing and a draft tube. The exit of the draft tube is placed under the surface of the tail waters and so creates a negative pressure which increases the power developed. The negative pressure and the height of water in the draft tube are limited by admitting air through a bleed valve that may be automatically or manually controlled. Operation of the turbine is then known as mixed flow and is mainly as an impulse turbine but with some aspects of a reaction turbine. Crossflow turbines are restricted to limited power ratings (<2000 kW) as the maximum practical length of the runner cylinder between bearings limits the rating of the turbine. They can operate with heads of between 3 m and 40 m.

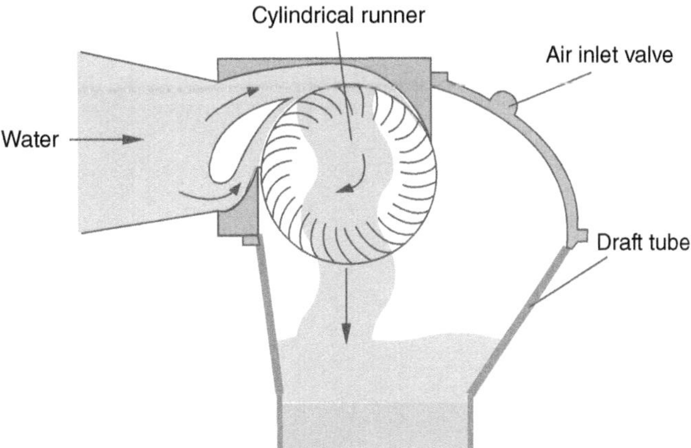

Figure 3.19 Water flow through a crossflow turbine.

3.5.3 Reaction Turbines

An impulse turbine operates by one or more jets of water, at atmospheric pressure, striking a part of the runner. The function of the casing of an impulse turbine is mainly to contain splashes, and the casing does not contribute directly to the operation of the turbine.

In contrast, in a reaction turbine the flow of water around the entire circumference of the runner contributes to the torque. The water enters the turbine from 360° through the spiral casing or volute. The runner of a reaction turbine is always immersed in the flow and there is a significant pressure drop across it. The casing and stator guide vanes of a reaction turbine are an integral part of the turbine. The guide vanes of the stator direct the flow onto the runner. For a given head a reaction turbine operates at a higher rotational speed, and the runner can be smaller than for an impulse turbine with the same power output. The runner and the casing of a reaction turbine, including the guide vanes, are manufactured to a high precision to maintain the clearances between them.

The two common types of reaction turbine are the Francis (Figures 3.20 and 3.21), and the propeller (Figure 3.22) or Kaplan (Figure 3.23) types. They can range from small units of tens of kW up to the highest ratings, ~500 MW. They can use a similar spiral casing to direct the flow from the penstock around the circumference of the stator. The flow in a Francis turbine is predominantly radial while in a propeller and Kaplan turbine it is axial. Small Francis turbines are sometimes mounted with the turbine axis horizontal (as shown in Figure 3.9). Large turbines are usually positioned with the shaft vertical. The generator is then mounted above the turbine and connected directly to it by a single very large shaft.

The Kaplan turbine is a development of the propeller turbine with variable-pitch angle blades on the runner.

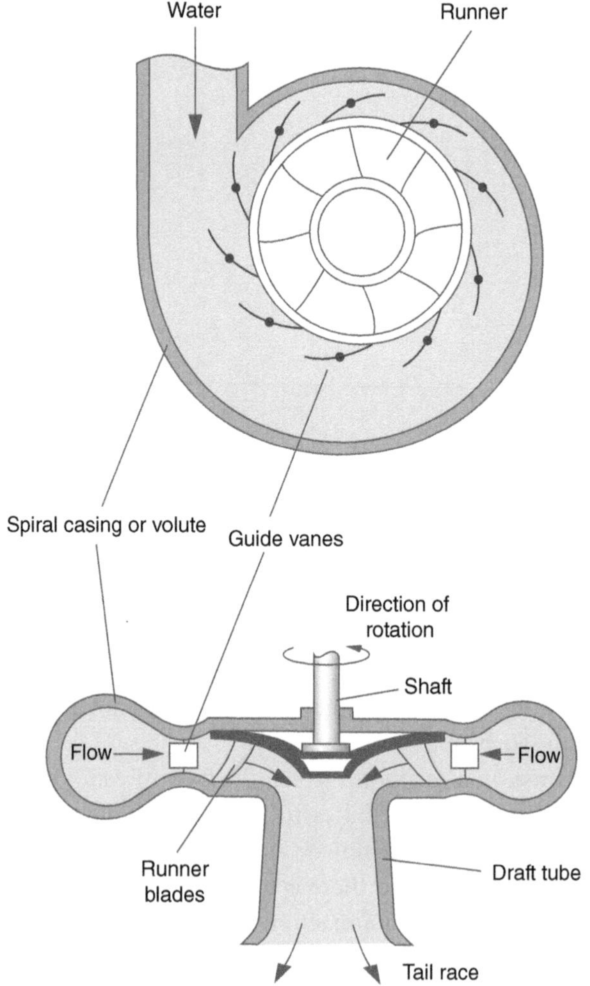

Figure 3.20 Francis turbine.
Source: B.S. Massey and J. Ward-Smith, *Mechanics of Fluids*, 9th ed., © 2012, Spon Press. Reproduced by permission of Taylor & Francis Group.

Figure 3.21 Runner of a Francis turbine in the factory.
Photo: Gilkes.
(A black and white version of this figure will appear in some formats. For the colour version, refer to the plate section.)

Francis, propeller and Kaplan turbines all use adjustable guide vanes in the stator to vary the flow and hence the output power of the turbine. The angle of the guide vanes is varied to control the flow of water and adjust the angle at which it enters the runner. The

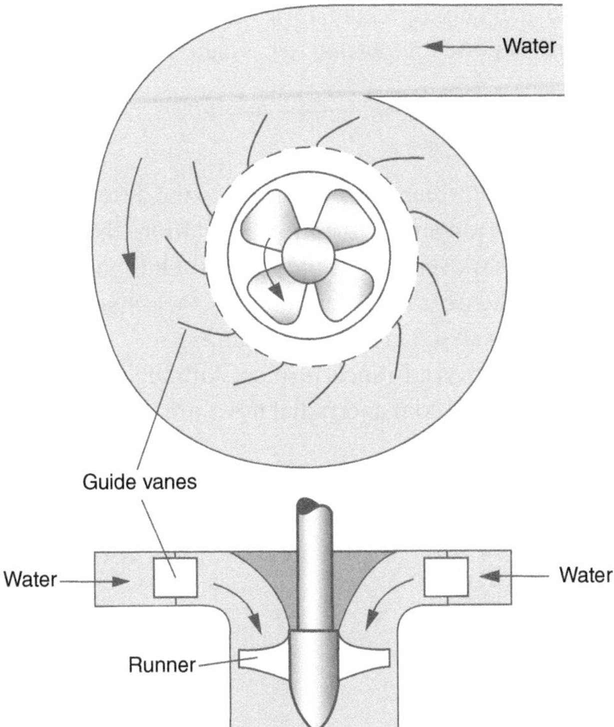

Figure 3.22 Propeller turbine.
Source: B.S. Massey and J. Ward-Smith, *Mechanics of Fluids*, 9th ed., © 2012, Spon Press. Reproduced by permission of Taylor & Francis Group.

Figure 3.23 Runner of a large Kaplan turbine. Note the adjustable angle blades.
Photo: sulaco229/Shutterstock.com.
(A black and white version of this figure will appear in some formats. For the colour version, refer to the plate section.)

assembly of stator guide vanes is sometimes referred to as the distributor or wicket gate. In addition, the Kaplan turbine uses adjustable pitch blades on the runner to provide additional control. By using two control elements (one on the runner and one on the stator) the Kaplan turbine maintains optimum angles of the flow entering and leaving the runner over a wide range of power output and so has high efficiency at reduced flows.

3.5.4 Analysis of a Francis Turbine

The power output of a reaction turbine is proportional to the difference in the absolute velocities of the flow in the tangential direction of rotation (the whirl velocities) before and after the flow passes over the turbine runner. Only the whirl velocities contribute to the torque, and hence power, developed by the runner. Both a Francis and propeller turbine can be analysed in a similar manner.

Figure 3.24 shows the flow across a Francis turbine. Although the water exits axially from the centre of the runner, axial and radial flows are ignored for this simple derivation of the power output.

The water from the penstock is distributed around the periphery of the stator by the volute or spiral casing (see Figure 3.9). The flow is then directed on to the runner by the guide vanes of the stator with an absolute velocity V_1 at radius r_1 and exits with an absolute velocity V_2 at r_2.

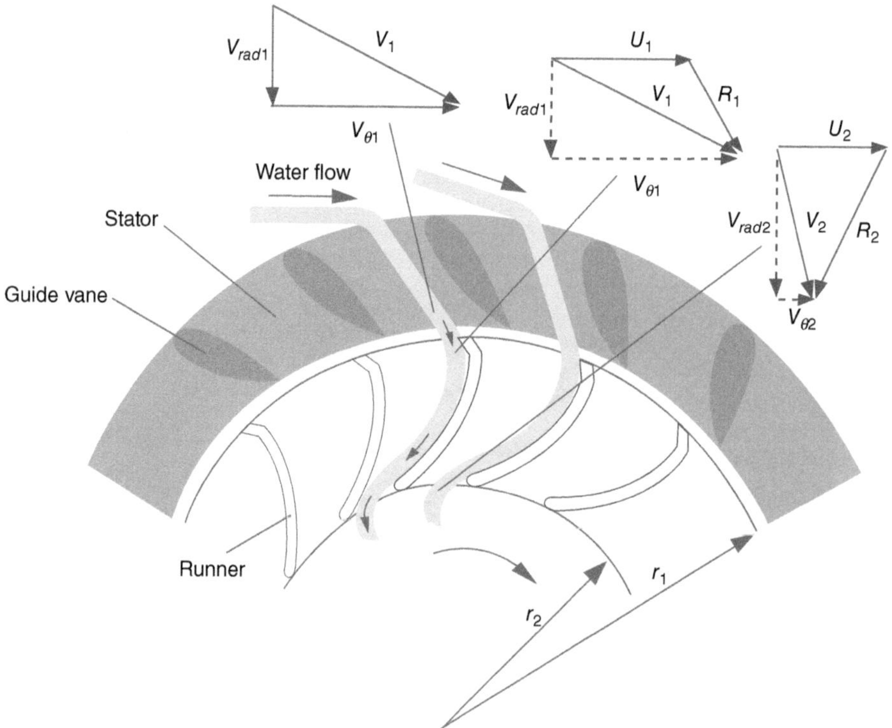

Figure 3.24 Flows and velocities of a Francis turbine: V_1 is the absolute velocity of the flow as it leaves the stator and enters the runner; V_2 is the absolute velocity of the flow as it leaves the runner; U_1 is the velocity of the runner at radius r_1 and U_2 is the velocity of the runner at radius r_2, where r_1 and r_2 are the outer and inner radii of the runner ($U = \omega r$, where ω is the angular velocity in rad/s and r is the radius); R_1 is the relative velocity of the flow as it enters the runner and R_2 is the relative velocity of the flow as it leaves the runner.

In vector notation, the velocity triangles of the flow at the input and output of the runner are

$$\boldsymbol{V}_1 = \boldsymbol{R}_1 + \boldsymbol{U}_1$$

and

$$\boldsymbol{V}_2 = \boldsymbol{R}_2 + \boldsymbol{U}_2$$

The only possible motion of the runner is in the tangential (whirl) direction and so

$$U_{1,2} = U_{\theta 1,\theta 2} = \omega r_{1,2}$$

The components of the absolute velocities in the tangential (whirl) direction are $V_{\theta 1}$ and $V_{\theta 2}$. The mass flow rate $\dot{m}$ is

$$\dot{m} = \rho Q$$

and so the rate of change of the angular momentum (sometimes known as the moment of momentum) in the whirl direction at the inlet is

$$\frac{dM_1}{dt} = V_{\theta 1}\dot{m}r_1 = V_{\theta 1}\rho Q r_1$$

and at the outlet it is

$$\frac{dM_2}{dt} = V_{\theta 2}\rho Q r_2$$

The torque developed by the runner is the difference between the rates of change of the angular momentum in the whirl direction of the water entering and leaving the runner

$$T = \frac{dM_1}{dt} - \frac{dM_2}{dt} = V_{\theta 1}\rho Q r_1 - V_{\theta 1}\rho Q r_2$$

And as $P = T\omega$,

$$P = \dot{m}\left(V_{\theta 1}U_1 - V_{\theta 2}U_2\right)$$

For a well-designed turbine the whirl velocity of the exiting flow is zero, $V_{\theta 2} = 0$, and so

$$T = \rho Q V_{\theta 1} r_1 \text{ and } P = \dot{m} V_{\theta 1} U_1$$

Figure 3.24 shows the flow directed by the guide vanes on to the runner. If the turbine is well designed, the flow passes smoothly over the guide vanes and blades and the angle of the flow is very close to that of the angle of the blades of the runner.

Example 3.3 – Operation of a Francis Turbine

A hydro power site has a net head of 150 m and a flow rate of 12 m^3/s. A 24-pole synchronous generator is to be used with a speed of 250 rpm. The external radius of the Francis runner is 1.2 m. Neglecting losses and assuming the tangential velocity of the flow at the exit to be zero, calculate:

the tangential velocity of the flow at entry to the runner,
the shaft torque,
the shaft power.

Solution

Equating the power in the water flow to that developed by the runner

$$P = \rho gHQ = \dot{m}gH = \dot{m}V_{\theta 1}U_1 = \dot{m}V_{\theta 1}\omega r_1$$

Thus the tangential velocity of the flow entering the runner is

$$V_{\theta 1} = \frac{gH}{\omega r_1} = \frac{9.81 \times 150}{\frac{250 \times 2\pi}{60} \times 1.2} = 46.8 \text{ m/s}$$

The torque is

$$T = \rho Q V_{\theta 1} r_1 = 1000 \times 12 \times 46.8 \times 1.2 = 673.9 \text{ kNm}$$

And the power is

$$P = T\omega = 673.9 \times 10^3 \times \frac{250 \times 2\pi}{60} = 17.6 \text{ MW}$$

Check using $P = \rho gHQ = 1000 \times 9.81 \times 150 \times 12 = 17.6$ MW

3.5.5 The Draft Tube and Cavitation

Reaction turbines operate with their runner completely flooded, and a draft tube is used to take the water exiting from the turbine to the tail water. The draft tube maintains a column of water from the tail water and recovers energy from the velocity of the water exiting the turbine (see Figure 3.8). With the use of a draft tube, the head of a reaction turbine extends from the surface of the upper reservoir to the surface of the tail water.

The draft tube has two purposes:

- to utilize the difference in height between the exit from the turbine and the surface of the tail water,
- to regain part of the kinetic energy from the discharge leaving the turbine by expanding the flow and so reducing its velocity.

The draft tube has a vertical section that can be extended to a bend to direct the exit flow. The sides of the draft tube diverge at an angle of around 10° to ensure smooth flow as it expands, and that the velocity of the water is reduced. The exit of the draft tube is always kept submerged to maintain the column of water.

A negative pressure is created at the exit of the turbine by the draft tube and so there is the possibility of cavitation. Cavitation occurs when the pressure in a liquid falls to its vapour pressure at that temperature. The liquid then boils and forms a large number of small bubbles. These bubbles are carried by the flow and when they reach a region of higher pressure they collapse suddenly and cause repetitive

Figure 3.25 Draft tube of 500 kW low-head axial turbine leaving the factory.
Photo: New Mills Engineering.
(A black and white version of this figure will appear in some formats. For the colour version, refer to the plate section.)

very high transient pressures that can lead to loss of efficiency, vibration and noise as well as severe damage to the material of the turbine or draft tube. In a reaction turbine the point of lowest pressure is usually at the outlet of the runner. Cavitation can be very damaging, so a hydro turbine is designed to avoid it.

The implication of cavitation is to limit the height of the draft tube that can be used with turbines with a high specific speed (e.g. propeller types). If these turbines are used with large heads they must be positioned close to the surface of the tail water with a short vertical section of draft tube.

The draft tube of a reaction turbine is such an important component it is usually considered to be an integral part of the reaction turbine and supplied by the manufacturer. Figure 3.25 shows the draft tube of an axial flow turbine leaving the factory. The flared draft tube is on the left with an additional section to change the direction of the flow on the right.

3.5.6 Open-Flume, Inclined-Shaft and Bulb Turbines

A wide range of positions of the turbines and generators have been used historically and there continues to be considerable diversity in the physical arrangement of small hydro units. The drive train of the generator, coupling and gearbox if used can be positioned horizontally or vertically. In the past, small low-head Francis and Kaplan turbines were placed directly in a flume as an alternative to the use of a spiral casing. This arrangement is now unusual as it is less efficient than using a spiral casing and can involve considerable civil works. Figure 3.26 shows an open-flume Francis turbine with the generator mounted vertically above the turbine out of the water on a long shaft.

An alternative arrangement used for small low-head units has been an inclined shaft to transfer power to a generator mounted out of the water. Another arrangement is the bulb turbine, where the electrical generator is fitted in a watertight

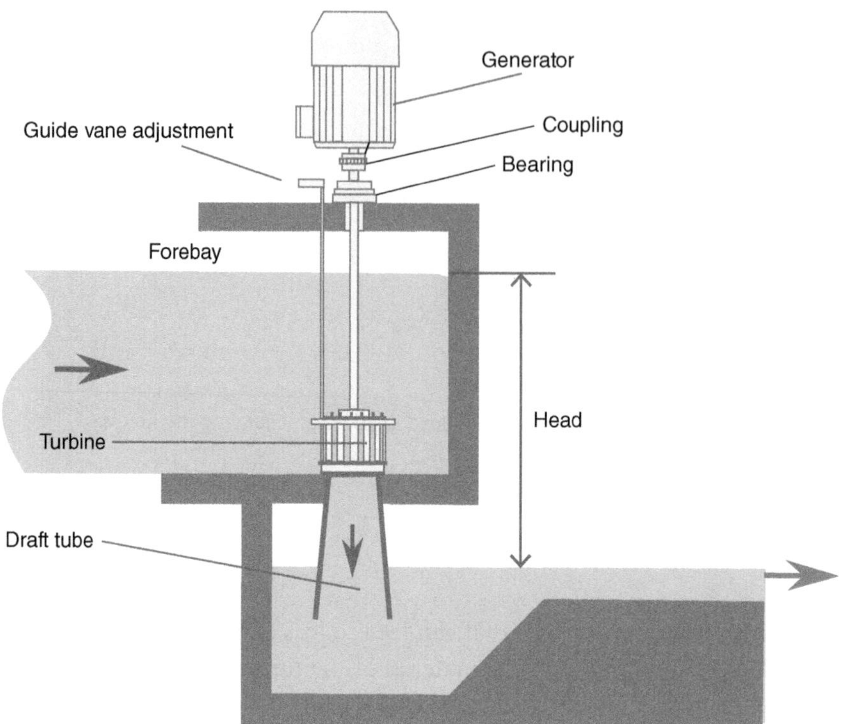

Figure 3.26 Open-flume Francis turbine with vertical drive train.

enclosure which is positioned in the flow. Bulb turbines are used in some run-of-river schemes and in tidal range schemes (Chapter 7).

3.6 Specific Speed of a Hydro Turbine

The various types of hydro turbines work efficiently over different ranges of head and flow rate. Figure 3.27 shows the regions of application of the various turbine types. Pelton wheels can be used with very high heads but limited flow rates. Crossflow turbines are restricted in their power rating as they use cylindrical fabricated runners of limited strength. Propeller types, including Kaplan, are most effective at low heads although they can use large flow rates. Francis turbines are used for a wide range of design head and flow conditions.

The effective operating regime of a hydro turbine is described by its specific speed. This specific speed is defined by the manufacturer of the turbine and shows the conditions of power, head and rotational speed under which the turbine operates at peak efficiency. For engineering studies using metric units it is common to define specific speed as

$$N_S = \frac{n_T P^{1/2}}{H^{5/4}} \tag{3.4}$$

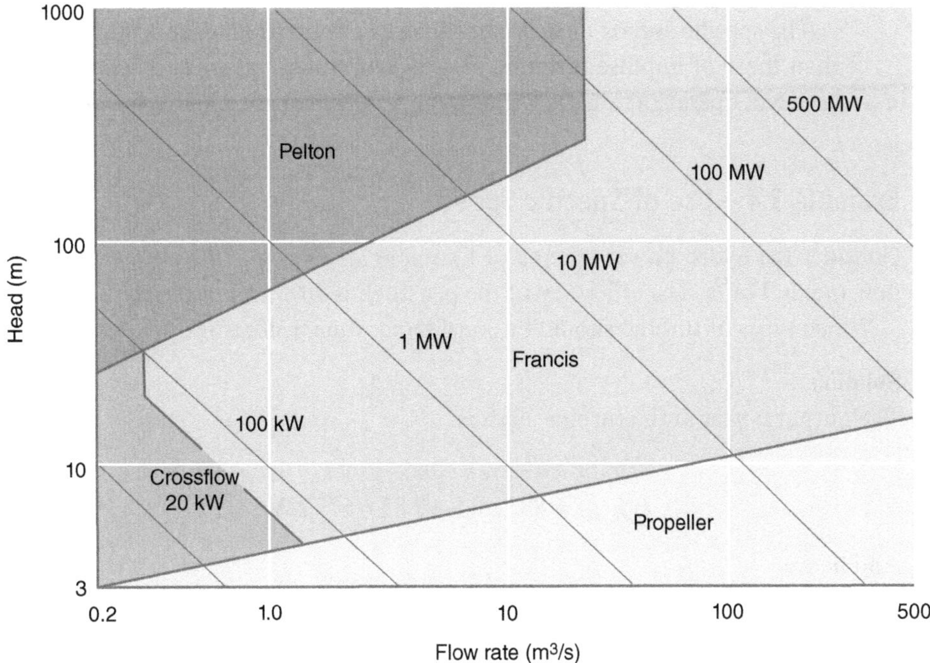

Figure 3.27 Application of hydro turbines.
Source: Godfrey Boyle, *Renewable Energy: Power for a Sustainable Future*, Godfrey Boyle, *1996*. By permission of Oxford University Press.

Table 3.1 Typical values of specific speed of turbines

Turbine type	Specific speed range
Impulse turbines	
Single-jet Pelton	10–35
Two-jet Pelton	10–45
Turgo	20–80
Crossflow	20–90
Reaction turbines	
Francis	70–500
Kaplan	350–1000
Propeller	600–900

where

N_s is the specific speed,
n_T is the turbine speed in rpm,
P is the power in kW,
H is the net head in m.

Typical values of specific speed (N_S) quoted by turbine manufacturers are in the ranges shown in Table 3.1.

The specific speeds of reaction turbines are up to an order of magnitude greater than those of impulse turbines. For a given power rating and head, a reaction turbine will operate at a higher rotational speed than an impulse turbine.

Example 3.4 – Use of Specific Speed

Consider the hydro power scheme of Example 3.1. Assume the net head is 30 m and flow rate is 3 m^3/s. The efficiency of the penstock is 90% and of the turbine is 85%.

Which types of turbine should be considered if the generator operates at 500 rpm?

Solution

The output power at the turbine shaft is

$$P_{SHAFT} = P_{WATER} \times \eta_{PENSTOCK} \times \eta_{TURBINE}$$
$$P_{SHAFT} = 883 \times 0.9 \times 0.85 = 675 \text{ kW}$$

And hence

$$N_S = \frac{n_T \sqrt{P}}{H^{5/4}} = \frac{500 \times 675^{1/2}}{30^{5/4}} = 185$$

These conditions give a specific speed within the range of a Francis turbine. The suitability of a Francis turbine is confirmed by Figure 3.27.

Alternatively at this modest power rating a gearbox can be used. If a 3:1 gearbox is used, then the turbine speed is reduced to 167 rpm while maintaining the generator speed of 500 rpm. The specific speed then reduces to 62. At this specific speed, it is possible to use a four- or six-jet Pelton or a Turgo turbine.

The specific speed (N_s) defined in Equation 3.4 is an engineering simplification of the formally correct dimensionless equation of power specific speed (Equation 3.5). It has become common engineering practice to neglect the constants of water density (ρ) and acceleration due to gravity (g) as well as to express the runner speed in rpm and the power in kW. Equation 3.5 then reduces to Equation 3.4 but the resulting Equation 3.4 is not dimensionless.

$$\Omega_S = \frac{\omega \left(\frac{P}{\rho} \right)^{1/2}}{g H^{5/4}} \tag{3.5}$$

3.7 Operation of a Hydro Turbine at Reduced Flows and Variable Speed

For a given flow of water and power of the turbine, the specific speed (Equation 3.4) defines the operating conditions to give maximum efficiency. Large hydro turbines can be made to have very high peak efficiencies (greater than 90–95%) and even

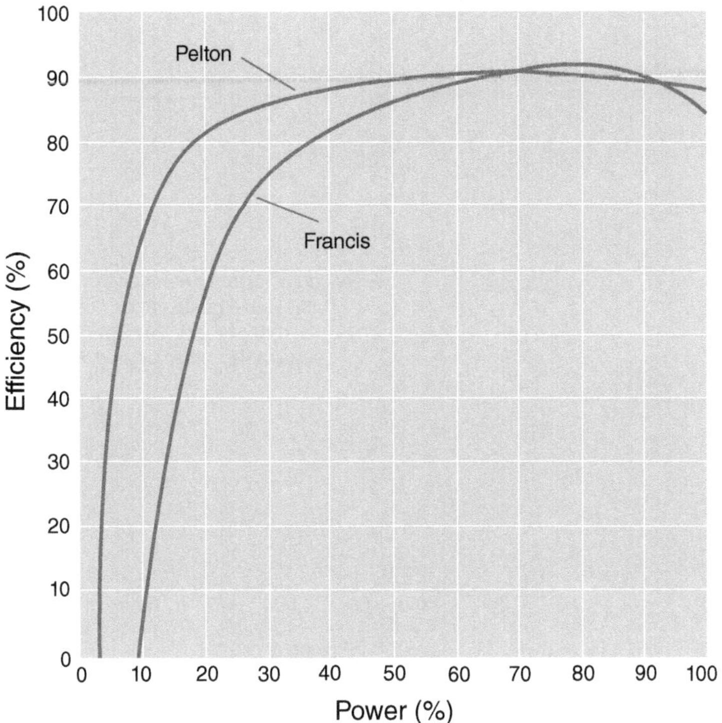

Figure 3.28 Efficiency of Pelton and Francis turbines (operating at constant head and rotational speed).
Source: E. Mosanyi, *Water Power Development*, Vols I and II, 1957, Hungarian Academy of Sciences.

smaller turbines can have peak efficiencies above 80–85%. However, these high efficiencies are obtained only when the conditions of head, power and rotational speed of the scheme match the specific speed of the turbine.

If the supply of water from the catchment is reduced and the flow of water to the turbine drops, then the power and efficiency of the turbine are reduced. Figure 3.28 compares the efficiency of a Pelton and Francis turbine designed for the same conditions of head and flow and operating at a constant head and rotational speed. Both turbines are more than 80% efficient when the output power and flow are greater than 40% of rated. The Francis turbine has slightly higher peak efficiency than the Pelton turbine, but its efficiency drops more quickly at low power output.

Figure 3.29 shows the efficiency of Propeller and Kaplan turbines operating at constant head and rotational speed. The fixed-pitch Propeller turbine has a high maximum efficiency but only at a single power level (near 100% of power output). As the power drops (due to reduced flow) the efficiency drops almost linearly. The Kaplan turbine, with adjustable turbine blades as well as adjustable stator guide vanes, maintains its efficiency above 80% over a much wider range of power and flow.

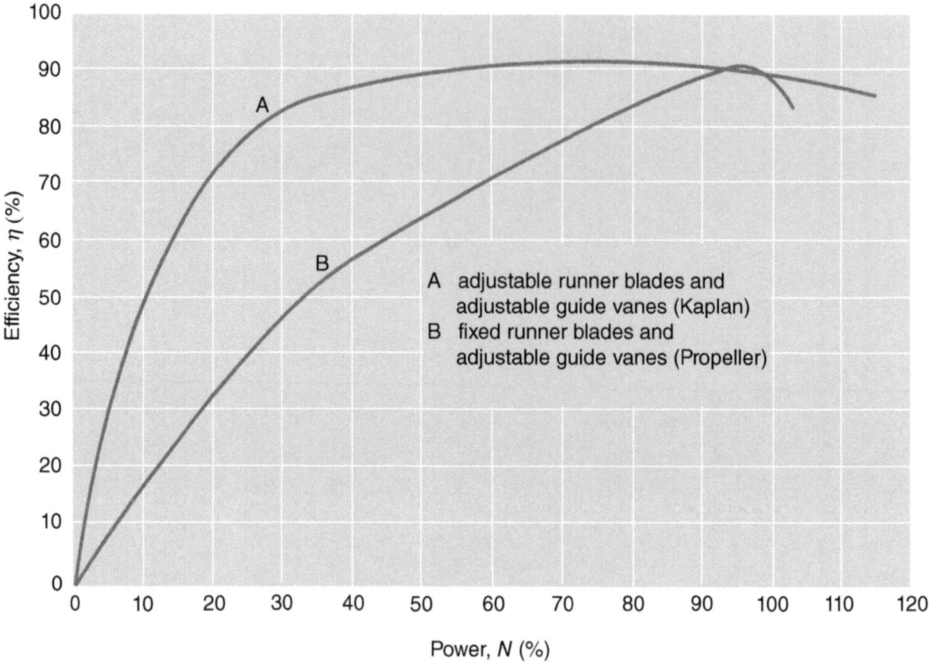

Figure 3.29 Efficiency of propeller and Kaplan turbines (operating at constant head and rotational speed).
Source: E. Mosanyi, *Water Power Development*, Vols I and II, 1957, Hungarian Academy of Sciences.

The efficiency of a synchronous generator will also reduce at lower power outputs as the power required for the rotor excitation field remains approximately constant.

The implication of the reduced efficiency of the turbine generator with low flows is that it is not sufficient to calculate the annual electrical energy output based on an annual flow or average daily flow and a single efficiency. Rather, it is necessary to use the flow duration curves (e.g. Figures 3.5 and 3.6) to calculate how the flow and power output varies during the year and apply the appropriate efficiency at that power level.

The efficiency of a hydro turbine operating at constant head but variable speed and flow rate is described by its Hill chart. Figure 3.30 shows a typical Hill chart of a scaled laboratory model of a hydro turbine. The model is geometrically similar to a large turbine, and scaling laws are applied to allow results of the model tests to be translated to a large unit. The model and large turbine have similar flow conditions and are said to be homologous.

The normalized flow Q_{11} and normalized speed N_{11} of Figure 3.30 are given by:

$$Q_{11} = \frac{Q}{D^2\sqrt{H}}$$

$$N_{11} = \frac{ND}{\sqrt{H}}$$

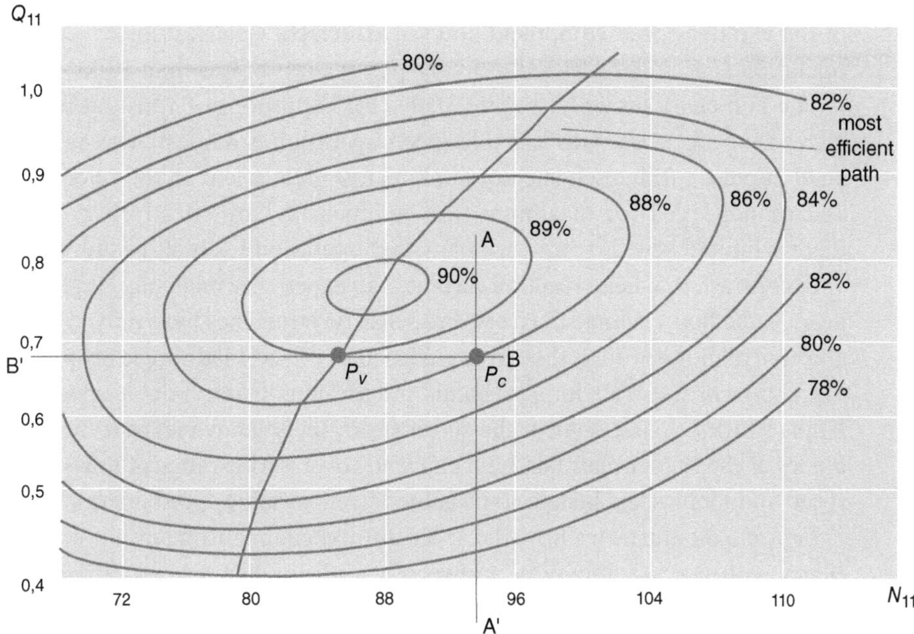

Figure 3.30 Typical normalized Hill chart of a hydro turbine. (N_{11} and Q_{11} refer to the speed and flow of a 1 m diameter model operating under a 1 m head.)
Source: Figure 3 of JWG 11/14.09 'Adjustable speed operation of hydroelectric turbine generators', *Electra* No 167, August 1996 © CIGRE.

where

Q is the flow,
N is the rotational speed,
D is the runner diameter,
H is the net head.

The vertical line A′–A shows operation at constant speed and indicates an efficiency of 88% at point P_c. Maintaining the flow while reducing the speed along B–B′ to point P_v increases the efficiency to 89%. Similar improvements in efficiency can be obtained by increasing the speed at higher flows. The most efficient speed for any flow is found where a horizontal line of constant flow just touches the efficiency contour. In addition to improving efficiency, varying the speed of operation of a hydro turbine with flow gives lower turbulence and reduces the likelihood of cavitation.

A fixed-speed hydro generator operates at a speed dictated by the frequency of the power system to which it is connected (50 or 60 Hz) and the number of poles with which the generator is manufactured. A synchronous generator operates at exactly this speed, while an induction generator will have a small variation of slip speed (~1–2%) but is also essentially fixed speed. Variable-speed operation of a small hydro generator uses similar power conversion equipment to that of a variable-speed wind turbine (Section 2.7). Either all the power is converted to dc and then inverted into the mains or a doubly fed induction generator is used. Because of the costs of the variable-speed equipment and the associated electrical losses, variable-speed operation is usually restricted to hydro power units of less than 10 MW.

The Hill chart shown in Figure 3.30 is for conditions of constant head, and the increase of efficiency with variable speed operation is modest. The head of a high-head impulse turbine scheme does not change greatly and so the velocity of the jet, and runner speed for maximum turbine efficiency, are effectively constant. Thus there is limited benefit from variable-speed operation for high-head hydro units.

In contrast, low-head reaction turbines can experience relatively large variations in head as the flows change in rivers. Propeller type turbines have a fixed angle runner and control flow through the stator guide vanes. Fixed-blade angle propeller turbines can show considerable improvements in efficiency from variable-speed operation. Kaplan turbines that control the stator head, the guide vanes and the angle of the blades of the runner maintain high efficiency over a wide range of flows even at fixed speed and there is less benefit to be gained from variable-speed operation.

Figure 3.30 illustrates how the speed should change to maintain maximum efficiency with varying flow rates and speeds, and a constant head. If the head is also changing, this becomes a multi-dimensional control problem requiring optimization or other heuristic (trial and error) techniques to solve it.

3.8 Net or Effective Head

The gross head (z) of a hydro scheme that uses a reaction turbine with a draft tube is the difference in elevation between the surface of the intake water and the surface of the tail water (see Figures 3.7 and 3.8). An impulse turbine operates at atmospheric pressure and the gross head is taken from the height of the intake water to the height of the nozzle of the turbine (see Figure 3.2).

In practice, the gross head is modified by the energy change due to the velocities of the intake and exhaust flows as well as the penstock friction and other losses to give the net or effective head (H). The gross head is

$$z = z_I - z_E$$

where

z is the gross head in m,
z_I is the elevation of the inlet in m,
z_E is the elevation of the exhaust in m.

The net head is

$$H = z + \frac{v_I^2}{2g} - \frac{v_E^2}{2g} - \sum h$$

$$= z_I - z_E + \frac{v_I^2}{2g} - \frac{v_E^2}{2g} - \sum h$$

where

H is the net or effective head in m,
v_I is the velocity of the inlet flow in m/s,
v_E is the velocity of the exhaust flow in m/s,
$\sum h$ is the penstock friction and other losses in m.

For a high-head scheme using an impulse turbine the velocity of the inlet and exhaust flows can usually be ignored. Impulse turbines use modest flows and hence the velocity of the surface of the inlet water is small. When designed well, the exhaust velocity at the exit of an impulse turbine is also very small. However, the friction losses in a long penstock are likely to be significant.

For a low-head scheme that uses a reaction turbine, the velocities of the flow at the inlet and of the exhaust at the exit of the draft tube can be significant. If a reaction turbine is mounted directly in a flume without a penstock there will still be some hydraulic losses associated with the trash rack and changes in direction of the water flow.

For a high-head hydro scheme, the penstock can be a major expense costing as much as the turbine. Increasing the diameter of the penstock reduces the frictional losses but increases its cost. Hence the penstock diameter is optimized during the design of the scheme by comparing the annual cost of the penstock with the value of the electrical energy lost due to frictional losses.

The head loss created by the frictional losses in the penstock can be calculated using the formula:

$$h_{friction} = \frac{v^2 L f}{2gD}$$

where

v is the mean flow velocity $\frac{4Q}{\pi D^2}$ in m/s,
L is the length of the penstock in m,
f is the friction factor,
D is the diameter of the penstock in m.

The friction factor (f) depends on the roughness of the pipe and on the Reynolds number of the flow; f can be obtained either from the manufacturer of the pipe or by using Moody's chart, which is shown in Tutorial III.

The other losses, sometimes known as minor losses, are caused by the intake, bends and valves, and are calculated by

$$h_{other} = \sum K_l \frac{v^2}{2g}$$

where

v is the mean flow velocity in m/s,
K_l are other (minor) loss coefficients.

Pipes for penstocks are manufactured in discrete sizes and it is convenient to use a spreadsheet for this optimization starting with an assumed frictional loss head (say 5–10% of the gross head). As the penstock diameter increases, the value of the losses will reduce but at additional cost. The optimum diameter is found when increasing the diameter further results in no additional benefit, i.e. the marginal cost of increasing the diameter and marginal benefit of reducing losses are equal.

Note that the alternative definition of the friction factor as $f' = \frac{f}{4}$ results in

$$h_{friction} = \frac{2v^2 L f'}{gD}$$

Example 3.5 – Determination of Penstock Diameter

A hydro power scheme with a gross head of 300 m has a penstock of length 500 m and a flow rate of 2.5 m^3/s. Assume a pipe friction factor f of 0.03 m and coefficients of the other (minor) losses as follows.

Component	Loss coefficient
Intake	0.2
Bends	0.15
Valves	0.2
Total	0.55

Choose the diameter of the penstock.

Solution

Available pipe diameters (m)	Velocity of flow (m/s)	Head loss (%)
0.7	6.5	15.8
0.8	5.0	8.1
0.9	3.9	4.5
1.0	3.2	2.7

As an example, consider a pipe diameter of 0.8 m.

The velocity of the flow is calculated from

$$v = \frac{4Q}{\pi D^2} \text{ m/s}$$

which gives

$$v = \frac{4 \times 2.5}{\pi \times 0.8^2} = 4.98 \text{ m/s}$$

At this velocity the friction head loss is

$$h_{friction} = \frac{v^2 L f}{2gD} = \frac{4.98^2 \times 500 \times 0.03}{2 \times 9.81 \times 0.8} = 23.7 \text{ m}$$

The other losses are given by

$$h_{other} = \sum K_L \frac{v^2}{2g} = 0.55 \times \frac{4.97^2}{2 \times 9.81} = 0.69 \text{ m}$$

Hence the total losses with penstock diameter of 0.8 m are 24.4 m or 8.1% of the gross head. This is a reasonable level of losses, and the water velocity is within the desired range of 3–5 m/s to limit abrasion of the pipe caused by suspended particles.

The losses with other pipe diameters are calculated in a similar manner. If the variation in cost with penstock diameter and the value of electrical losses are known, an economic optimization can be performed.

3.9 Transient Conditions

The lower sections of the penstock can be subject to very high static pressures caused by the height of the column of water and so a strong and expensive pipe is needed. The penstock must also be designed for dynamic pressure changes caused when the flow into the turbine changes quickly. A critical case is the control of overspeed of the turbine generator following a sudden loss of electrical load. Overspeed of a Pelton turbine can be controlled by rotating the deflector and diverting the flow of water away from the buckets while maintaining the flow in the penstock. In a reaction turbine it is necessary to restrict the flow of water through the penstock either by adjusting the guide vanes of the stator (known as the wicket gate) or by an auxiliary valve. Rapid adjustment of the flow in the penstock can lead to the phenomenon of water hammer when the inertia of the long column of water in the penstock and the sudden reduction of flow into the turbine can cause a succession of pressure waves. The pressure in the penstock then increases dynamically.

The effects of water hammer can be controlled either by limiting the speed of operation of the guide vanes or by the use of pressure relief valves and surge tanks to control the pressure waves. A related problem can arise when the guide vanes (or spear valve) are opened too rapidly leading to transient under-pressures in the penstock.

If the electrical load on the generator is lost, the turbine generator will overspeed as the turbine continues to receive energy but the generator cannot export it to the electrical system. The rate of change of the angular speed $\frac{d\omega}{dt}$ is determined by the torque from the turbine and the inertia of the turbine generator.

$$\frac{d\omega}{dt} = \frac{T_M - T_E}{I} \quad \left[\text{rad/s}^2\right]$$

where

$T_M - T_E$ is the difference between the mechanical turbine and electrical generator torques in Nm,

I is the inertia of turbine generator in kgm^2.

The generator can lose its electrical load either by the generator circuit breaker being opened by electrical protection or by a fault depressing the voltage on the electrical network. Such faults are inevitable on any electrical network and must be considered in the design of the hydro scheme.

The rate of change of angular speed depends on the difference in the torque supplied by the turbine to the retarding torque offered by the generator and the inertia of the turbine generator. A high-inertia turbine generator allows slower operation of the guide vanes before an overspeed limit is reached, and some hydro sets (including the Turgo turbines in the scheme shown in Figure 3.3) were fitted with an additional flywheel in their drive shafts to increase inertia.

Example 3.6 – Load Rejection of a Turbine Generator

An 800 kW hydro turbine generator operates at 750 rpm and suddenly loses its electrical load. The inertia of the combined turbine and generator is 1500 kgm^2.

If the maximum permissible overspeed before the generator is damaged is 20%, how rapidly must the flow of water to the turbine be stopped?

Solution

The kinetic energy of the turbine generator during normal operation is

$$KE = \frac{1}{2}I\omega^2 = \frac{1}{2}\times 1500 \times \left(\frac{750\times 2\pi}{60}\right)^2 = 4.63\ \text{MJ}$$

The maximum permissible speed is 900 rpm or 94.25 rad/s and so the maximum permissible kinetic energy is

$$KE' = \frac{1}{2}I\omega^2 = \frac{1}{2}\times 1500 \times \left(\frac{900\times 2\pi}{60}\right)^2 = 6.66\ \text{MJ}$$

The increase in energy ΔE when the electrical load is zero and the turbine continues to develop 800 kW is

$$\Delta E = P \times t = 0.8 \times t\ \text{MJ}$$

and so

$$t = \frac{6.66 - 4.63}{0.8} = 2.54\ \text{s}$$

Control of the intake of water into the hydro turbine within 2½ s is required while at the same time avoiding water hammer in the penstock.

3.10 Development of Small Hydro Schemes

In many countries the sites that are suitable for large hydro schemes have already been developed or have been investigated and discarded for some reason. Even if suitable sites for large hydro schemes can be identified, very careful assessment of the impact of the scheme on the environment and local population is required. Also, a very large investment is needed for major hydro power schemes and, even if the overall economics can be shown to be favourable, such a large investment of capital may be difficult to obtain. Small hydro schemes are potentially easier to develop, although their development requires many of the same steps as for large schemes but with limited funds available for a smaller project.

The development of small hydro schemes follows a similar path to any other renewable energy project. A successful scheme will rely critically on the hydro resource (head and flow) and so an initial layout is proposed to identify the gross head. Figure 3.31 shows a number of layouts of small hydro schemes.

The choice of layout will be influenced by topography, land ownership, access, availability and cost of construction materials as well as the requirements of other users of the water. The anticipated output depends on the flow, which is determined by the hydrology of the catchment. For an initial assessment of the scheme the annual mean flow can be used, but for any detailed appraisal daily flows are required. The annual mean flow can be estimated from:

(1) direct measurements if the river is gauged (instrumented),
(2) measurements from a gauging station on a nearby river provided the catchment areas of the two rivers are similar and their respective areas are within 10–20% of each other,
(3) annual rainfall records held on national databases using

$$Q = \frac{(SAAR - AE) \times A \times 10^3}{8760 \times 60^2}$$

where

$SAAR$ is the Standard Average Annual Rainfall in mm,
AE is the Actual Evapotranspiration in mm,
A is the area of the catchment in km^2.

These annual estimates of the output of the scheme are only approximate but do indicate whether further expenditure on investigating the proposed scheme is justified.

Typical examples of more detailed records of daily flows are shown in Figures 3.5 and 3.6. Flows can vary significantly from day to day as well as year to year and it is necessary to use long-term records to predict the performance of a scheme. If long-term records of flows are not available, one approach is to correlate the short-term flows at the site (measurements made over 1–2 years) with long-term records (at least 10 years) of either rainfall or flows in an adjacent river with a similar catchment. An analogue flow duration curve is created from the short-term measurements and scaled by the ratio of the annual mean flows of the river being investigated and the reference river or rainfall. The short-term measurements of flow can be made either by installing a temporary weir or by a succession of dilution gauging tests on the river under investigation.

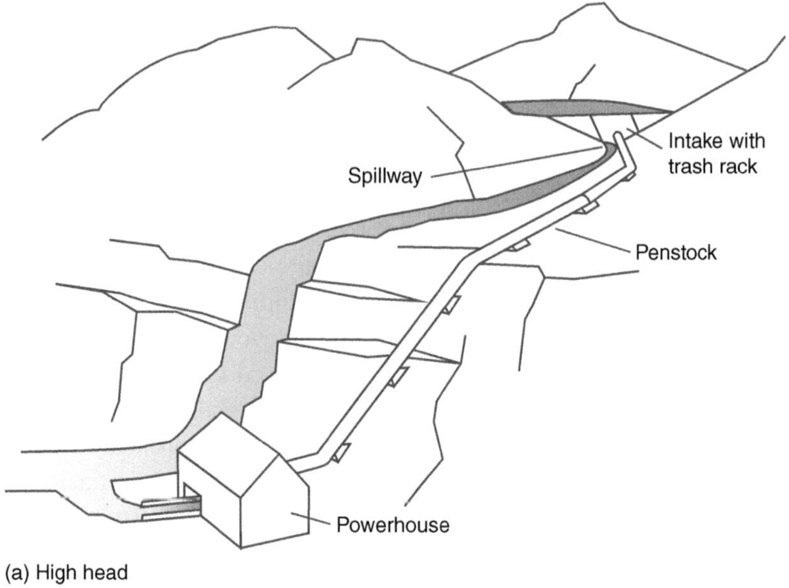

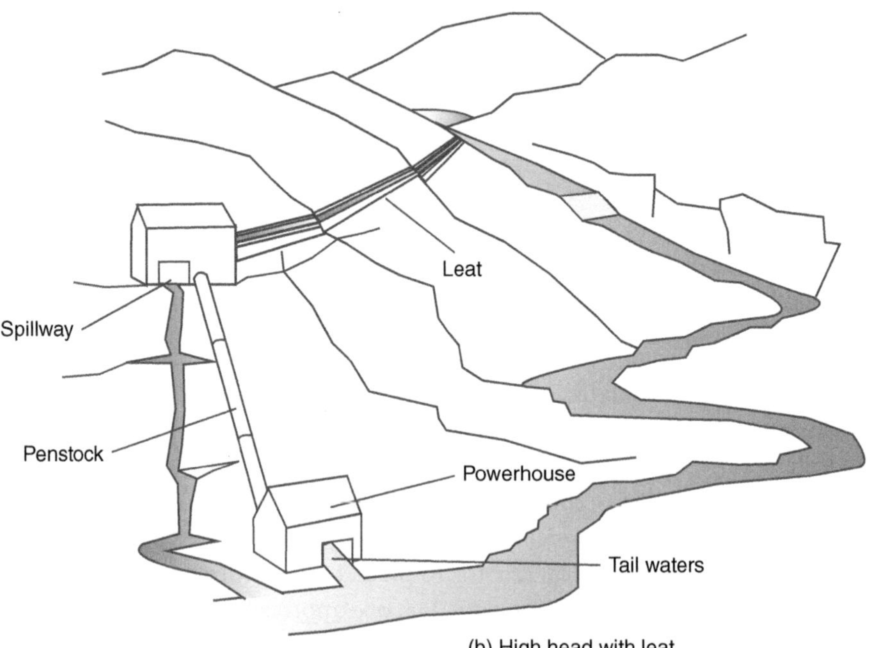

Figure 3.31 (a), (b) High-head small hydro schemes.

Once a flow duration curve of the river is obtained, it is modified to take account of the flow that must remain in the river, known as the compensation flow. This modified flow duration curve then gives the flows available to be used in the hydro scheme.

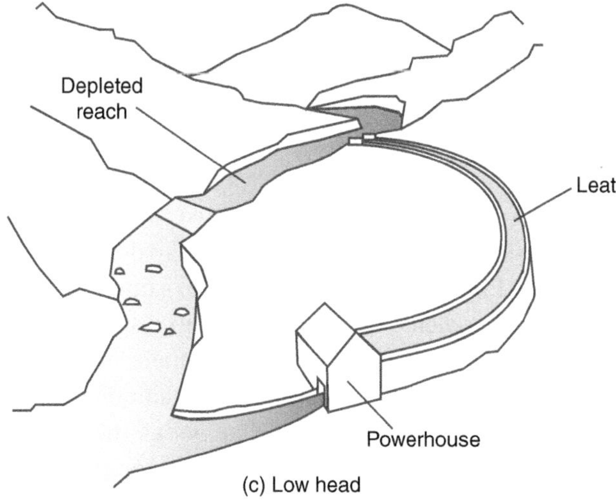

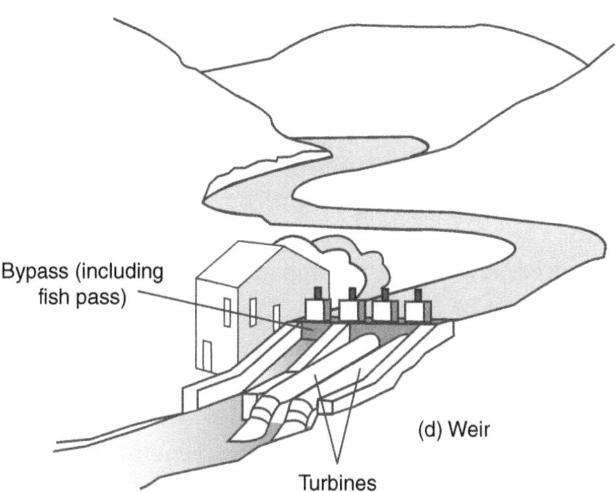

Figure 3.31 (c), (d) Low-head small hydro schemes.

Some rivers (such as that of Figure 3.6) show considerable variation of flow, and the efficiency of the turbine generator will drop at low flows. Because of this variation of efficiency with flow, the annual energy is calculated by dividing the flow duration curve into bands, determining the efficiency at that flow, and summing the energy generated at each flow band. Depending on the environmental conditions and type of turbine generator to be used, a minimum flow is defined below which the turbine is stopped.

3.10.1 Environmental Impact Assessment

The development of any hydro power scheme will require permission from both the local planning authority and those responsible for the water resources. In England, water resources are managed by the Environment Agency. Licences are required to abstract or impound water, and approval must be gained for any fish passes. Any proposed hydro scheme will also be evaluated for its impact on the risk of flooding.

In order to be cost-effective, any new small hydro scheme in a heavily populated country such as England is likely to use an existing weir or civil works from a historic mill or hydro scheme that has fallen into disrepair. If it is proposed that flow is diverted into a leat, or secondary channel, there will be concern over the reduced flow in the main river and a requirement to maintain a level of residual flow in the depleted section of the main channel (known as the depleted reach). Regulations impose limits on the flow that must remain in the original main channel. In England, Environment Agency guidance indicates that, for turbines located on a weir, the design flow through the hydro power scheme should be the mean of the flow duration curve of the river (Q_{mean}), while if the residual flow drops below the minimum flow of the river (assumed to be Q_{95}, i.e. the flow that is exceeded 95% of the time) the hydro turbine is required to stop. For a longer depleted reach with greater environmental impact, the design flow of the scheme is limited to Q_{40} (the river flow exceeded 40% of the time) and the turbine is stopped if the residual flow drops below Q_{85} (the flow exceeded 85% of the time). Similar regulations apply in most European countries.

In addition to the regulations covering the abstraction and impoundment of water, any hydro power scheme will need permission from the local planning (permitting) authority. For a scheme of any size, an Environmental Impact Assessment is likely to be required. Environmental impacts during both construction and operation of the scheme must be considered.

3.10.2 Generators for Small Hydro Schemes

Large hydro schemes invariably use synchronous generators as they have high efficiencies and provide independent control over real and reactive power. Historically many small hydro schemes used induction generators. Induction machines have simple squirrel cage rotors that are able to withstand higher overspeeds than synchronous generators. Although their speed varies with slip, this variation with load is only around 2% and induction generators effectively operate at a fixed speed. Induction generators take their reactive power from the grid, and care is necessary to avoid voltage collapse if they are used in rural areas with weak grids.

Variable-speed operation of small hydro schemes (see Section 3.7) requires a power electronic converter to connect the generator to the national power system.

3.10.3 Governors for Standalone Schemes

Some very small hydro schemes (<50 kW) are not connected to the national power system but supply an isolated load. In this case the expense of the mechanical governor can be avoided using an electronically controlled load. The rotational speed of a generator, and hence the frequency of the electrical power produced, is controlled by maintaining the balance between its mechanical and electrical torque. With these standalone schemes, rather than adjusting the mechanical torque of the turbine, the electrical torque of the generator is controlled by varying the load. The rotational speed of the generator is measured, and more load connected when the rotational speed, and frequency, rises. The use of a dynamic load avoids the cost of a mechanical governor but some energy is lost in the dynamic dump load.

Although interesting, and technically challenging, these standalone schemes are not commercially significant, and the vast majority of small hydro generators take their voltage and frequency from a connection to the main power system.

3.10.4 Archimedes Screw Generators

The last few years have seen an increasing use of Archimedes screws for low-head mini hydro generation, and it is believed that more than 500 schemes are in operation worldwide. Often they make use of an existing water control structure such as a weir or the site of a disused hydro scheme to create a low head. The generation unit consists of an inclined, coarse multi-thread screw typically up to 3.5 m in diameter rotating at between 20 and 50 rpm in a concrete or steel trough.

Single screws can have a power output of up to 500 kW although more usually they are rated at 10–250 kW with heads of 1–6 m and flow rates of 1.5–6 m^3/s. The speed of rotation of the screw is increased to ~1000 rpm through a multi-stage gearbox to drive either a fixed-speed induction generator connected directly to the electricity network or a variable-speed, electronic full-power converter. Both fixed- and variable-speed operation are used. Variable-speed generators have a higher capital cost and a slightly reduced efficiency. However, there is then less need for a sluice gate to control the flow of water, and audible noise can be reduced.

An Archimedes screw is operated by the static pressure of the weight of water in 'buckets' created between the flights of the screw. Thus in some ways its principle of operation is similar to that of an overshot wheel. An infinitely long, frictionless Archimedes screw without leaks can, in theory, exploit all the potential energy of the water. In practice, the water to wire efficiency of a complete Archimedes Screw Generator unit is typically around 70% and is approximately constant above 20% of rated flow. A spillway is arranged to divert water away from the screw at times of excess flow while a fish pass allows the passage of fish and mammals both upstream and downstream.

The key advantages of an Archimedes screw over other forms of mini hydro generator are as follows:

- The environmental impact is limited because of its slow speed of operation and large clearances.
- A high efficiency is maintained over varying flow rates.

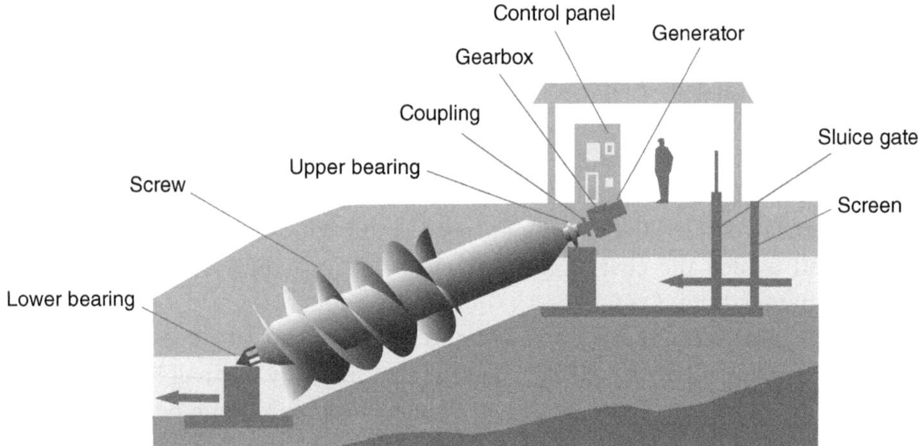

Figure 3.32 Diagram of an Archimedes screw generator.
Source: Diagram developed from one kindly provided by Spaans Babcock Ltd.

Figure 3.33 Archimedes screw generator being delivered to site. Note the coupling, gearbox and generator towards the front of the load and the upper and lower bearings of the screw.
Photo: Stephen Fleming/Alamy. Rights Managed.
(A black and white version of this figure will appear in some formats. For the colour version, refer to the plate section.)

- The robust construction of the screw means it is insensitive to floating debris and requires only a simple coarse screen.

Figure 3.32 shows the main elements of an Archimedes screw generator. Archimedes screws are normally installed at an angle of 22° to the horizontal. The screw is supported by bearings at the upper and lower ends, while a flexible coupling connects the low-speed shaft of the gearbox. There is a coarse screen with typical spacing of its bars of 100–150 mm to catch large items of floating debris and deter large fish. A sluice gate controls the flow of water and is particularly necessary with operation at fixed rotational speed.

Figures 3.33 and 3.34 show photos of an Archimedes screw generator.

Figure 3.34 200 kW Archimedes screw unit being installed on a site in Wales. Photo: Renewables First Ltd, www.renewablesfirst.co.uk.

PROBLEMS

1. Even though hydro power plants with large reservoirs do not produce CO_2, they have negative impact on the environment. Discuss.
2. Draw a schematic diagram of a high-head hydro scheme and discuss the purpose of each component.
3. Distinguish clearly between the principle of operation of impulse hydro turbines and that of reaction hydro turbines. Give one example of each.
4. A hydro scheme is proposed for a site with a gross head of 100 m and a flow rate of 5 m^3/s. What is the gross power in this hydro resource and what turbines might be considered?

   [**Answer:** 4.9 MW, either Francis or multi-jet Pelton turbines.]

5. A hydro scheme with an overall efficiency of 84% and an effective head of 200 m is required to generate 150 MW. What is the volumetric flow rate required?

   [**Answer:** 91 m^3/s.]

6. The reservoir of the scheme described in Problem 5 contains 10^8 m^3 of useable water. How much energy is stored and how long can it generate at 150 MW?

   [**Answer:** 196 TJ, 305 h.]

7. A hydro power plant operates in the following manner.

   During the months January to March and October to December, the reservoir has adequate water to operate the hydro power plant at its full capacity.

   During the months April to September, in order to maintain adequate water level in the reservoir, the hydro power plant operates at half of its capacity.

   Calculate the annual capacity factor (ACF).

   If the generator is shut down for 15 days in the month of January, what will be the new ACF?

   [**Answer:** 75%, 71%.]

8. A 5 MW turbine drives a 1200 rpm generator through a 4:1 gearbox. If the head of the scheme is 180 m, what is the specific speed?

[**Answer:** 32.]

9. A penstock of 0.8 m diameter, 100 m long, has a friction factor of 0.03. The discharge is 3 m^3/s. The viscosity of water is 1.0 mPas.
Calculate
(a) the head loss of the penstock,
(b) the relative roughness of the penstock.

[**Answer:** (a) 6.8 m; (b) 0.05.]

10. A Pelton turbine has the following data:
jet diameter 48 mm,
net head 720 m,
effective radius of runner 0.7 m.
Assuming ideal operation, calculate:
(a) the jet velocity,
(b) the force on the runner
(c) the torque on the runner,
(d) the runner rotational speed (in rpm),
(e) the power developed.

[**Answer:** (a) 119 m/s; (b) 25.6 kN; (c)17.9 kNm; (d) 812 rpm; (e) 1.5 MW.]

11. The nozzles of a two-jet Pelton wheel are located 150 m below the surface of a reservoir. The diameter of each nozzle is 65 mm. The wheel rotates at 750 rpm. Ignore all losses.
Calculate:
(a) the diameter of the bucket,
(b) the shaft power,
(c) the torque if the operating speed is changed to use a 500 rpm generator,
(d) the runaway speed.

[**Answer:** (a) 0.69 m; (b) 530 kW; (c) 8990 kNm; (d) 1501 rpm.]

12. A Francis turbine has the following data:

Volumetric flow rate	20 m^3/s
Rotor height	300 mm
Inlet radius	800 mm
Runner speed	52 rad/s
Stator blade angle	70°

Calculate the output power.

[**Answer:** 30.3 MW.]

13. A Francis turbine of 92% efficiency produces 50 MW at a speed of 250 rpm. The rotor height is 1.5 m and the inlet radius is 2.5 m. The flow enters the runner at an angle of 45° to the radial and exits the turbine in a radial direction. Determine the volumetric flow rate.

[**Answer:** 140 m^3/s.]

14. A proposed site for a high-head hydro scheme has a gross head of 200 m and a flow rate of 0.5 m^3/s. The length of the penstock is 450 m and the diameter of the available pipes is 400 mm. Assume a friction coefficient of 0.03 m and a minor loss coefficient of 0.5. Calculate:
    (a) the net head,
    (b) the power available at the turbine.

    [**Answer:** (a) 172 m; (b) 845.4 kW.]
15. A 1 MW hydro turbine generator operates at 90% efficiency and 1000 rpm. The inertia of the combined turbine and generator is 750 kgm^2.
    What will be the overspeed 1 second after it loses the electrical load?

    [**Answer:** 1126 rpm.]
16. Derive the expression for specific speed given in Equation 3.5.

FURTHER READING

Elliott C., *Planning and Installing Micro-Hydro Systems*, 2014, Routledge.
Practical handbook of micro hydro systems.

Fraenkel P., Parish O. and Edward R., *Micro-Hydro Power: A Guide for Development Workers*, 1991, Intermediate Technology Publications.
Practical handbook of micro hydro systems.

Massey B.S. and Ward-Smith J., *Mechanics of Fluids*, 9th ed., 2012, Spon Press.
Classic undergraduate textbook.

Mosanyi E., *Water Power Development Volumes I and II*, 1957, Hungarian Academy of Sciences.
Historic two-volume book covering all aspects of water power development.

Peak S. (ed.), *Renewable Energy: Power for a Sustainable Future*, 3rd ed., 2012, Oxford University Press.
Comprehensive textbook covering many aspects of renewable energy.

Twidell J. and Weir A., *Renewable Energy Resources*, 3rd ed., 2015, Routledge.
Comprehensive textbook covering many aspects of renewable energy.

Wilson E.N., *Engineering Hydrology*, 4th ed., 1990, Palgrave McMillan.
Accessible introductory textbook on hydrology.

ONLINE RESOURCES

Exercise 3.1. Performance of a hydro scheme: ordered flow data is used to develop a simple Flow Duration Curve and estimate the output of a hydro scheme.

Exercise 3.2. Performance of a hydro scheme: monthly flow data are used to develop a hydrograph and Flow Duration Curve, to estimate the output of a hydro scheme including varying turbine efficiency.

Exercise 3.3. Design of a small hydro scheme: this exercise uses a large set of flow data to predict the performance of a small hydro scheme.

4 The Solar Energy Resource

INTRODUCTION

The sun is essential for life on earth; plant and animal life in their current form would not exist without its power. Radiation from the sun ultimately drives all the forms of renewable energy discussed in this book (the only exception being tidal energy). The sun's radiation has a maximum power ~1000 W/m^2 at the earth's surface, and the natural processes that create the earth's wind and waves, rainfall and biomass act to concentrate the energy from the sun, making its subsequent conversion into mechanical or electrical power easier.

An understanding of the solar energy resource is necessary for the study of both photovoltaic electricity generation and solar thermal processes. This short chapter gives a simple engineering description of the solar resource and is a prerequisite for the study of Chapters 5 and 6. In Sections 4.1 and 4.2 the solar resource is described with examples. Section 4.3 presents a simple view of sun–earth geometry. Section 4.4 discusses the orientation of solar panels. Section 4.5 describes the solar spectrum and air mass. Finally, in Section 4.6, a brief introduction is given to the wave–particle description of light.

4.1 The Solar Resource

The sun is a spherical collection of hot gases with a diameter of 1.4×10^6 km and an internal temperature of up to 20×10^6 K. The earth orbits the sun in a slightly elliptical orbit at a distance of approximately 150×10^6 km. The centre of the sun can be thought of simply as a nuclear fusion reactor converting hydrogen into helium and so releasing energy. The energy generated by the fusion reaction at the centre of the sun is transmitted through its hot gases to the surface with increasing wavelength. Finally, radiation is emitted from the surface of the sun with a spectrum similar to that of a blackbody at a temperature of 6000 K. This radiation then travels through space to arrive at the outer surface of the earth's atmosphere, where the intensity of the radiation is around 1367 W/m^2.

Once the radiation enters the earth's atmosphere it is reflected and scattered by a number of processes (Figure 4.1). The scattering and reflection reduce the overall

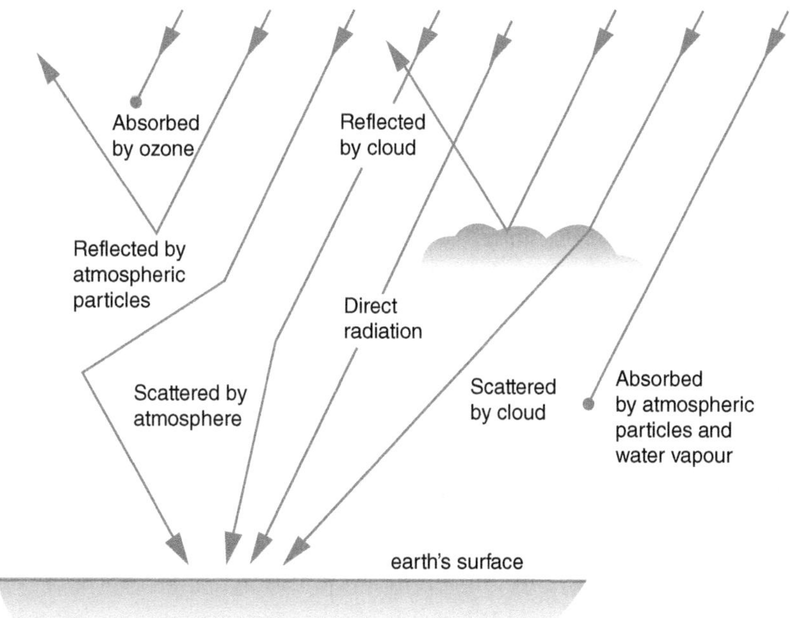

Figure 4.1 Solar irradiance entering the atmosphere.

intensity of the irradiance and deflect some of the radiation from its direct path from the sun. At the earth's surface, the intensity of irradiance drops at sea level to less than ~1000 W/m^2 with its spectrum modified as gases and particles in the atmosphere selectively absorb certain wavelengths.

For engineering purposes, the solar resource is considered in two parts.

- *Direct or beam radiation* is the solar radiation received directly from the sun without it having been scattered or deflected by the atmosphere.
- *Diffuse radiation* is the solar radiation received from the sun after its direction has been changed through scattering in the earth's atmosphere caused by clouds and particles.

In some circumstances there may also be some reflection of radiation from the ground. This is known as *albedo* and depends on local ground conditions. Although albedo can be important when evaluating the solar heating of buildings, its effect on electricity-generating solar energy systems is usually small and is ignored. Ignoring albedo is a conservative (or safe) assumption resulting in a slight underestimate of the electricity generated from the solar resource.

Flat-plate solar modules (either photovoltaic or solar thermal) make use of both direct and diffuse solar radiation as a source of energy. However, it is also possible to concentrate the solar radiation using mirrors or Fresnel lenses. Concentrator systems allow high temperatures to be created and power to be generated from a small area at the focus of the concentrator. Solar systems that concentrate the solar irradiance by more than 5–10 times can make use only of direct irradiance. They must use either single- or two-axis tracking systems to orientate the lenses or concentrating mirrors towards the sun. Mechanically changing the orientation of the concentrating equipment or solar modules to track the sun increases output but the very attractive simplicity offered by fixed orientation devices is then lost. If high concentration ratios are used, forced cooling of the focus of the collector to maintain its performance may also be required. Concentrating solar energy systems are less common than flat-plate modules and not effective in temperate latitudes where much of the radiation is diffuse.

4.2 Examples of the Solar Resource

The units used to describe the solar resource are as follows.

- *Irradiance* (W/m^2) is the rate at which solar radiant energy (i.e. solar power) is incident on a surface. The symbol G is normally used.
- *Insolation*[1] (J/m^2 or kWh/m^2) is the incident energy per unit area on a surface. This is found by integrating the irradiance over a specified time, typically a day although periods of a month or a year are also used. The symbol H is normally used for insolation over a day.

Figure 4.2 shows the total solar irradiance measured on a horizontal surface in California over five days of January. The daily insolation for each of the five days is shown below the x-axis. The daily variation of irradiance from sunrise to sunset due to the position of the sun in the sky as well as the effect of cloud cover can be seen

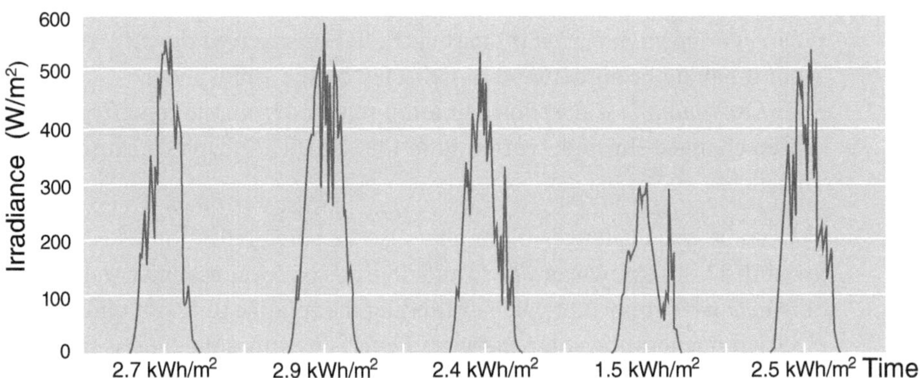

Figure 4.2 Solar irradiance (G) on a horizontal surface in California over five days of January. The daily insolation (H) is given for each day below the daily irradiance curve.

[1] Some international standards use the term *irradiation* to describe incident solar energy. In this book we will use *insolation* to describe incident solar energy (J/m^2 or kWh/m^2) to distinguish it clearly from *irradiance* or solar power (W/m^2).

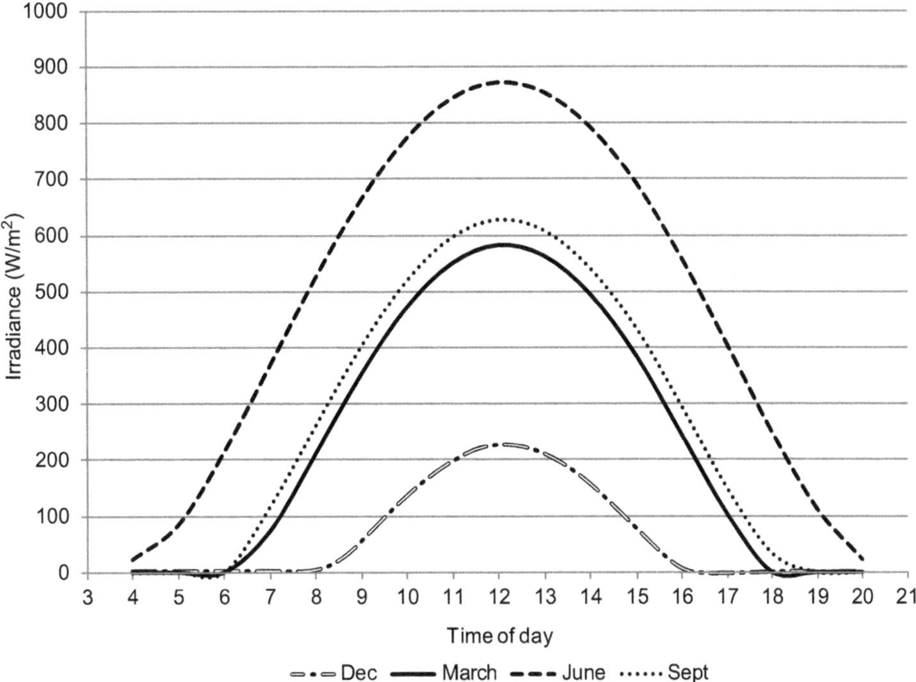

Figure 4.3 Solar irradiance (G) on a horizontal surface at Cardiff, assuming a clear sky. Source of data: EU Joint Research Centre.

clearly. It can be seen that the weather on the fourth day was particularly cloudy. In this week, the irradiance peaked at just less than 600 W/m^2 at midday, while the daily insolation was modest, ranging from 1.5 to 2.9 kWh/m^2 over these days of a winter month.

It can be useful to represent the solar resource by assuming a clear sky, eliminating the effects of clouds. Figure 4.3 shows the variation of total, or global, irradiance (G), assuming a clear sky, on a horizontal surface throughout typical days in four different months of the year for Cardiff, which is at latitude 51½° N. The values are from a combination of ground measurements and calculations based on satellite data.

It can be seen that in Cardiff there is a considerable difference between the solar resource in winter days and in the summer. In winter days the irradiance peaks at only around 200 W/m^2 and there is sun that would be useable by a solar energy system for around 6 hours (from 09:00 to 15:00), while in summer the irradiance reaches 850 W/m^2 with useable sun for almost 15 hours. In Figure 4.3 the effect of clouds is ignored but attenuation of the irradiance by atmospheric absorption, including by water vapour, is included in these data. The irradiance in March is slightly less than that in September when the atmosphere is clearer.

Figure 4.4 shows the average daily direct and diffuse insolation (H) on an inclined surface in Cardiff for each month of the year. A tilt angle of 38° was chosen for maximum annual energy production. Because of the low angle of the sun in the

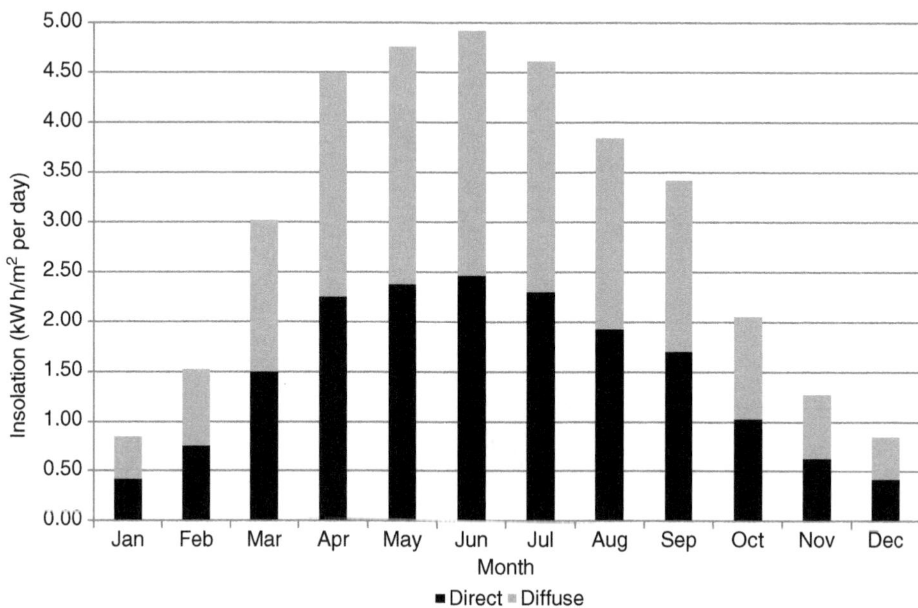

Figure 4.4 Average direct and diffuse daily insolation (*H*) on an inclined surface in Cardiff. Source of data: EU Joint Research Centre.

sky in this northerly latitude and significant cloud, the average diffuse insolation is approximately equal to the direct insolation. The graph shows the average daily insolation over each month but there will be significant variations of the daily insolation within each month. Typical maximum values of the daily insolation in Cardiff range from 1–1.5 kWh/m^2 in January to 7.5–8 kWh/m^2 in June.

4.3 Sun–Earth Geometry

Figure 4.5 illustrates the motion of the earth around the sun. The earth moves around the sun over a year in a slightly elliptical orbit. The earth itself rotates around its polar axis over a period of 24 hours. The angle between the polar axis and a line normal to the plane of rotation of the earth around the sun (the ecliptic plane) is maintained at an angle of 23.5°. If effects of the atmosphere are ignored, a location in the northern hemisphere will receive its maximum solar energy on the longest day when the earth's axis is tilted towards the sun. This is the day of the summer solstice. Similarly, it will receive the minimum solar energy on the shortest day, the winter solstice when the northern hemisphere is tilted away from the sun. On the days of the spring and autumn equinoxes, the earth's axis is perpendicular to a line between the earth and sun, and day and night are of equal length.

The tilted axis of rotation of the earth and its annual orbit results in the apparent motion of the sun in the sky as seen from a point on the earth. This is shown in Figure 4.6 for an observer in the northern hemisphere facing due south. At the

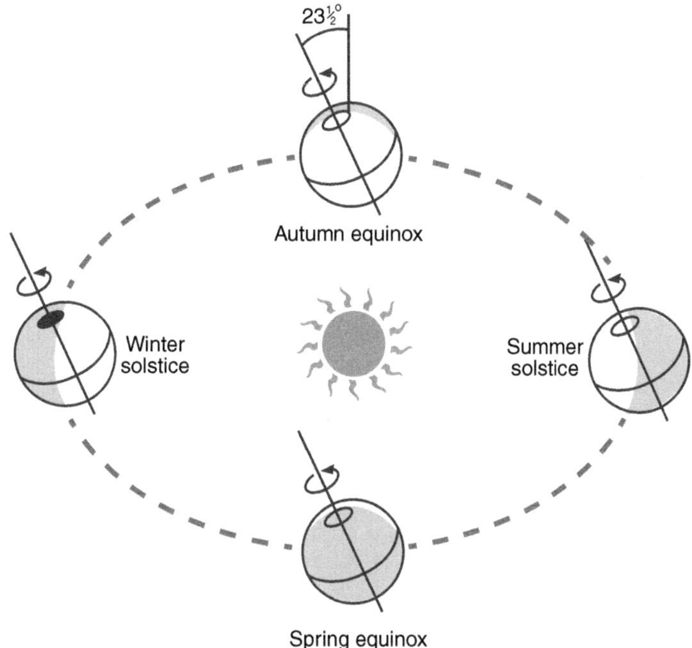

Figure 4.5 Motion of the earth around the sun (not to scale).

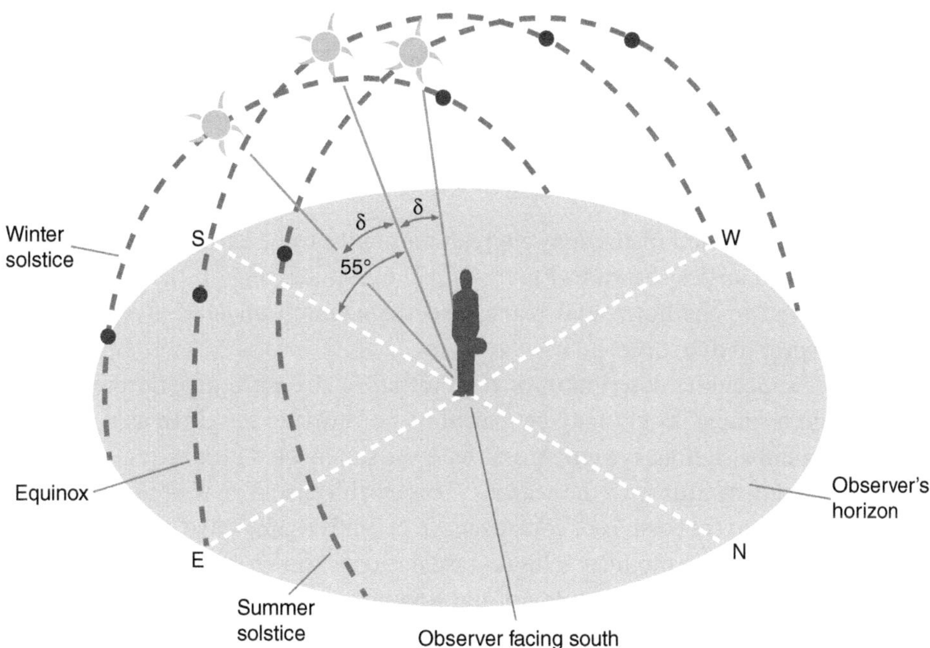

Figure 4.6 Apparent motion of the sun at a latitude of 35°. The position of the sun is shown at solar noon and the dots show the position 3 hours before and after solar noon. The angle of declination (δ) has a maximum of 23½°.
Source: M.A. Green, *Solar Cells: Operating Principles, Technology and System Applications*, 1998, University of New South Wales.

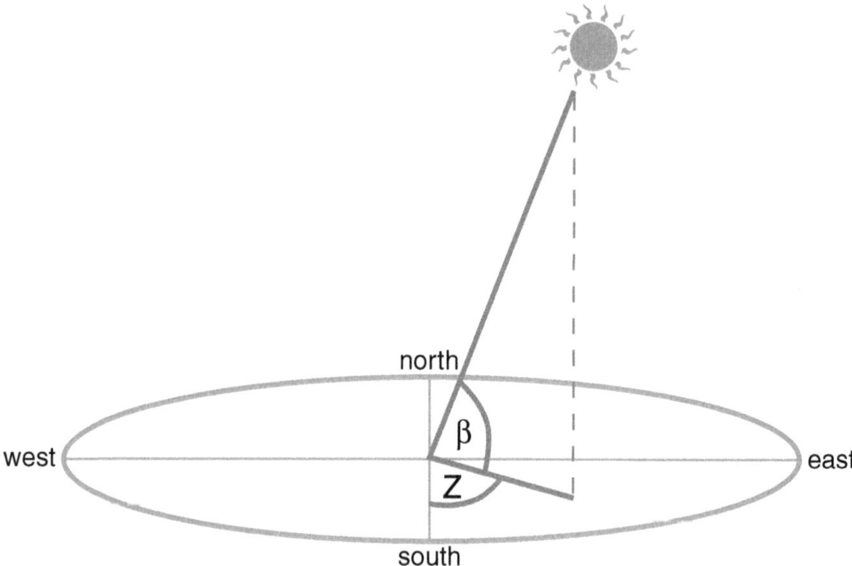

Figure 4.7 Position of the sun viewed from a point on the earth. β is the solar altitude angle relative to the horizontal and Z is the azimuth relative to the south (Although conventions differ, in this book solar azimuth angles to the east are positive.)

spring and autumn equinoxes (21 March and 23 September), the sun appears to rise due east and set due west. At solar noon on the days of the equinoxes the altitude of the sun is equal to (90° – latitude) and the solar declination (δ) is zero. At the winter solstice (22 December) the solar declination is −23½°, while at the summer solstice (21 June) it is +23½°.

The position of the sun relative to a point on the earth's surface is described by the two angles shown in Figure 4.7. The altitude angle β defines the altitude with respect to the horizontal plane, and the azimuth angle Z gives its position with respect to due south.

An accurate description of the motion of the earth around the sun is complex and a simple insight may be gained by considering the earth as if it were spinning around a stationary vertical axis with the sun located in space and moving upwards and downwards with the seasons. This earth-centred view of the motion of the sun relative to the earth is of course incorrect but is useful when considering how best to orientate solar modules. This representation is shown in Figure 4.8. The sun reaches its highest point on the day of the summer solstice, its lowest point on the day of the winter solstice and its rays are parallel to the lines of latitude on the days of the equinoxes.

Using this earth-centred representation, a line drawn between the sun and the earth gives the angle of declination, δ, with respect to the horizontal. At solar noon of any day a point on the earth faces the sun at the angle of declination. The angle of declination, δ, varies between +23.5° at the summer solstice to −23.5° at the winter solstice. At the equinoxes the angle of declination is zero.

Table 4.1 Day numbers of first day of the month

Month	n for first day of the month
January	1
February	32
March	60
April	91
May	121
June	152
July	182
August	213
September	244
October	274
November	305
December	335

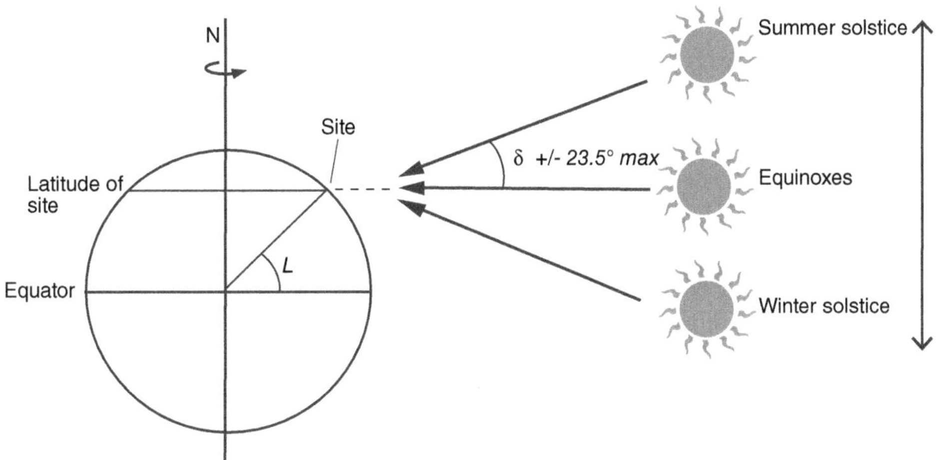

Figure 4.8 Variation of angle of declination, δ, through the year (earth-centred view). Source: G.M. Masters, *Renewable and Efficient Electric Power Systems*, 2004, Wiley.

The angle of declination, δ, varies throughout the year and can be estimated from

$$\delta = 23.45 \times \sin\left[\frac{360}{365}(n+284)\right] \quad [\text{degrees}] \tag{4.1}$$

where

n is the day of the year (counting from 1 January as day 1).

For most engineering purposes the angle of declination δ is assumed to be constant throughout a day. Table 4.1 gives the day numbers n for the first day of each month for use in Equation 4.1.

Continuing with the earth-centred view of the motion of the sun, consider a point on the earth's surface at latitude L facing the sun. Then from Figure 4.9 it may be seen that the altitude angle at solar noon, β_{noon}, is simply

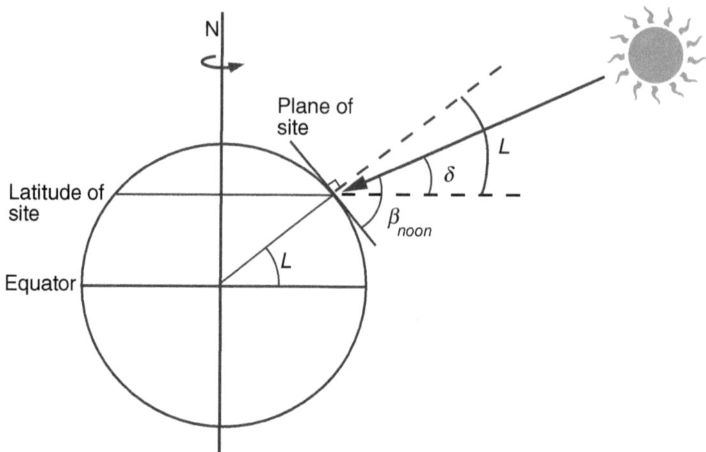

Figure 4.9 Altitude angle β_{noon} at solar noon.
Source: G.M. Masters, *Renewable and Efficient Electric Power Systems*, 2004, Wiley.

$$\beta_{noon} = 90° - L + \delta \tag{4.2}$$

where

β_{noon} is the altitude at solar noon,
L is the latitude,
δ is the angle of declination.

Example 4.1 – Altitude of the Sun at Solar Noon

Calculate the altitude of the sun at solar noon at latitude 35° on 26 March.

Solution
26 March is day 85 (after 1 January).
From Equation 4.1 the angle of declination δ is

$$\delta = 23.45° \sin\left(\frac{360}{365}(85 + 284)\right) = 1.6°$$

From Equation 4.2

$$\beta_{noon} = 90° - L + \delta$$
$$\beta_{noon} = 90° - 35 + 1.6 = 56.6°$$

Because of the earth's rotation, the sun appears to revolve around the earth over 24 hours and so an hour angle, H, is also defined. The hour angle is the number of degrees of longitude the earth must rotate before solar noon and is positive towards the east.

$$H = \frac{t}{24} \times 360° \quad [\text{degrees}]$$

where

t is the hours before solar noon.

The position of the sun as seen from earth using the angles of Figure 4.7 (stated here but not proved) is given by

$$\sin\beta = \cos\delta\cos H\cos L + \sin\delta\sin L \tag{4.3}$$

and

$$\sin Z = \frac{\cos\delta\sin H}{\cos\beta} \tag{4.4}$$

where

β is the altitude angle,
Z is the azimuth,
δ is the declination,
H is the hour angle,
L is the latitude.

Care is necessary when, in early morning or late evening, Equation 4.4 leads to an azimuth angle greater than 90°. An additional test is then required to determine the inverse sine.

Example 4.2 – Location of the Sun

(a) Calculate the position of the sun two hours before solar noon at Cardiff (latitude 51.5°) on 15 May.
(b) Calculate how many hours of daylight there are on 15 May.

Solution

(a) 15 May is day 135 (from 1 January).

Angle of declination, δ, is given by

$$\delta = 23.45^\circ \sin\left(\frac{360}{365}(135+284)\right) = 18.8^\circ$$

Hour angle, H, is given by

$$H = \frac{2}{24}\times 360^\circ = 30^\circ$$

Therefore

$$\begin{aligned}\sin\beta &= \cos\delta\cos H\cos L + \sin\delta\sin L\\ &= \cos 18.8\times\cos 30\times\cos 51.5 + \sin 18.8\times\sin 51.5\\ &= 0.5104 + 0.2522\\ &= 0.7625\\ \beta &= 49.7^\circ\end{aligned}$$

$$\sin Z = \frac{\cos\delta \sin H}{\cos\beta} = \frac{\cos 18.8 \times \sin 30}{\cos 49.7} = 0.730$$

$$Z = 47°$$

(b) At dawn and dusk, $\beta = 0$.

$$\sin\beta = \cos\delta \cos H \cos L + \sin\delta \sin L = 0$$

$$\cos H = \frac{-\sin\delta \sin L}{\cos\delta \cos L} = -\tan\delta \times \tan L = -\tan 18.8 \times \tan 51.5 = -0.4279$$

$$H = 115.34°$$

$$H = \frac{t}{24} \times 360°$$

$$t = \frac{24 \times H}{360} = \frac{24 \times 115.34}{360} = 7.7\text{h}$$

Therefore there will be 15.4 hours of daylight on 15 May.

4.4 Orientation of Solar Panels

The tilt angle is the angle a solar panel makes with the horizontal. Solar panels are rarely installed on a horizontal plane (a tilt angle of zero) as this would give maximum annual energy output only at the equator; a minimum tilt angle of 10–15° is always desirable to help keep the surface clean and ensure dust and dirt are washed off. Figure 4.9 and Equation 4.2 showed that at solar noon on the days of the equinoxes, the irradiance strikes a horizontal surface at an angle of (90° – latitude). In summer the sun rises higher in the sky at noon, while in the winter it is lower.

Figure 4.10 shows a solar panel located at latitude L. The panel is positioned to face due south, and at solar noon the earth has revolved so that it faces the sun. If the tilt angle of the panel is set equal to the latitude, then on the days of the equinoxes at solar noon the sun's rays will strike it perpendicularly giving maximum energy. At noon on the summer solstice the sun will be higher in the sky (by +23.5°) and conversely on the winter solstice it will be lower (by −23.5°). Thus a good first approximation, if maximum annual energy capture is required, is to choose the tilt angle of a solar panel to be equal to the latitude of its location. If additional energy is required in the winter, as is often the case with solar thermal systems, then the tilt angle should be increased (made steeper).

Local meteorological conditions cause considerable changes in the solar resource in addition to the variation that is due to the relative position of the sun in the sky. The effect of cloud cover is shown clearly in Figure 4.1. In the tropics the effect of the monsoon can significantly reduce the insolation in some months when otherwise a good solar resource might be expected.

In the higher latitudes, maximum annual energy output may often be obtained with a tilt angle some 10–15° lower (flatter) than the latitude. This is because of

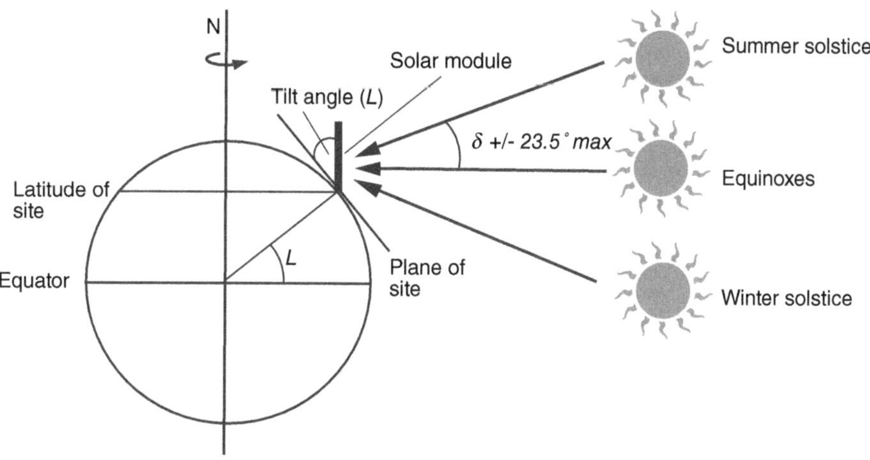

Figure 4.10 Tilt angle of a solar module.
Source: G.M. Masters, *Renewable and Efficient Electric Power Systems*, 2004, Wiley.

increased absorption of the solar irradiance in the atmosphere in the winter, because of the large air mass (see Figure 4.12) and the bulk of the annual energy coming from increased output during the summer.

For maximum energy the solar panels should be orientated within ± 5° of south in the northern hemisphere (± 5° of north in the southern hemisphere). Orientation towards the west will increase output in the evenings, and orientation towards the east will increase output during the mornings.

In practice, both the prediction of the performance of solar systems and the selection of the tilt angle is undertaken with computer programs that use measured and/or satellite data of the solar resource that includes local meteorological effects. This solar resource data is then combined with a mathematical model of the solar energy system to predict its performance. In many circumstances the practical tilt angle will be determined by the angle of roofs or other existing support structures, and modest variations from the optimum tilt angle have a limited effect on the annual system energy output of flat-plate modules.

4.5 Solar Spectrum and Air Mass

The transfer of energy from the sun across the vacuum of space is through radiation. Figure 4.11 shows the spectrum of sunlight, i.e. the variation of solar energy with wavelength. Outside the earth's atmosphere, the light from the sun has a spectral distribution close to that emitted by a blackbody at a temperature of 6000 K, the effective temperature of the external surface of the sun. The AM0 (air mass zero) curve shows the solar spectrum outside the earth's atmosphere on a plane perpendicular to the sun at the mean earth–sun distance. The power density of the sun's radiation outside the earth's atmosphere is usually taken to be 1367 W/m^2 and is known as the solar constant.

Table 4.2 Distribution of energy in sunlight outside the earth's atmosphere (AM0)

Region	Ultraviolet	Visible	Infrared
Wavelength (μm)	0–0.38	0.38–0.78	0.78–3
Energy (%)	9	45	46

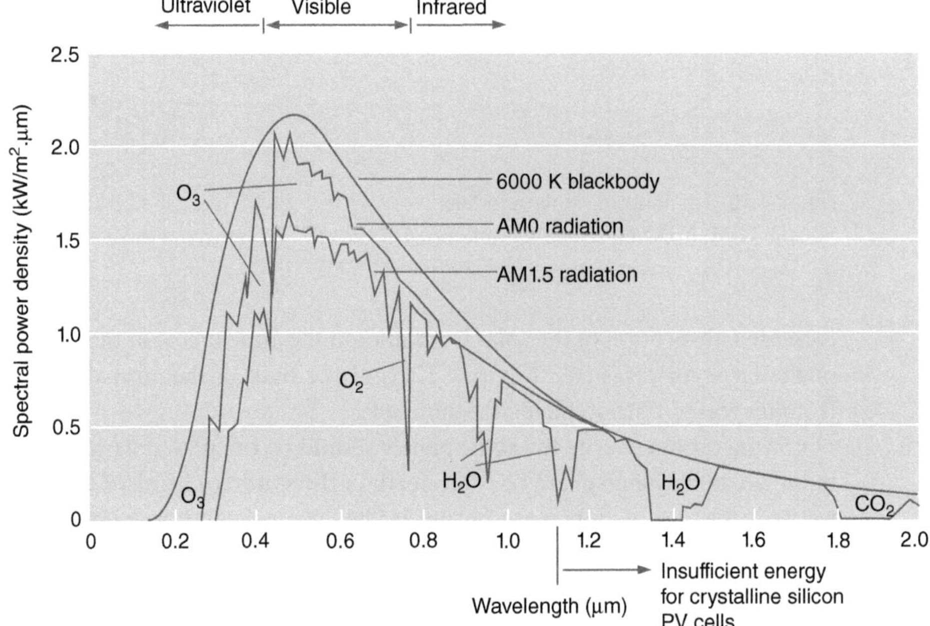

Figure 4.11 The spectrum of sunlight.

Table 4.2 shows that most of the energy of the light striking the earth is in the visible and infrared regions of the spectrum. There is little in the ultraviolet region or at wavelengths greater than 3 μm.

Solar radiation reaching the earth is attenuated as it passes through the atmosphere. Figure 4.11 shows that there is considerable attenuation of the solar radiation by the various gases and aerosols within the atmosphere (e.g. O_3, O_2 and H_2O). Ozone (O_3) absorbs wavelengths of the spectrum in the ultraviolet region and water vapour and carbon dioxide reduce the energy of certain frequencies in the infrared region.

With the sun vertically overhead, i.e. at an altitude angle of 90° or a zenith angle of zero, the resulting spectrum is known as the AM1 spectrum (Figure 4.12). However, the times of the year when the sun is vertically overhead and the radiation passes only through the vertical height of the atmosphere are limited, so the standard test conditions (STC) under which photovoltaic panels are tested assume a zenith angle of 48.2° and hence a spectrum from an air mass of 1.5. It is conventional to scale the AM1.5 spectrum to give a total power density of 1 kW/m^2 in order to establish the standard test conditions for photovoltaic modules.

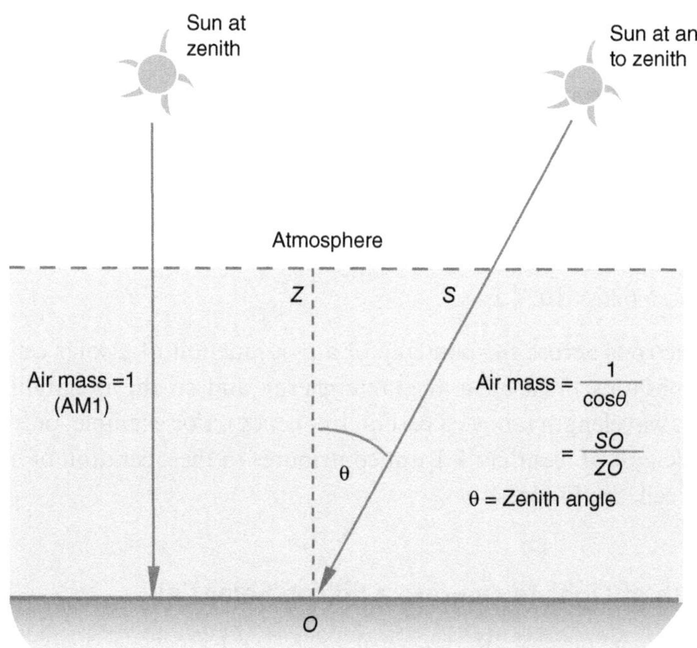

Figure 4.12 Air mass concept.

Example 4.3 – Air Mass at Solar Noon

What is the zenith angle and hence the air mass at solar noon at latitude 35° on 26 March?

Solution

From Example 4.1 the altitude angle is

$$\beta_{noon} = 56.6°$$

At solar noon the zenith angle θ (see Figure 4.12) is $90 - \beta_{noon} = 33.4°$

$$\text{Air mass at solar noon } = \frac{1}{\cos 33.4°} = 1.2$$

4.6 Wave–Particle Duality of Light

The behaviour of light is understood by considering it partly as a wave phenomenon and partly as consisting of particles. Using this idea of wave–particle duality, sunlight is thought of as consisting of electromagnetic waves composed of photons of different energies which travel at the same speed. The wavelength (λ) of solar radiation is inversely proportional to the energy of each photon (E) with two constants: the velocity of light and Planck's constant. Thus

$$\lambda = \frac{hc}{E} \quad [\mathrm{m}] \tag{4.5}$$

The frequency of any electromagnetic radiation (υ) is related to wavelength by

$$\upsilon = \frac{c}{\lambda} \quad [\mathrm{Hz}]$$

where

c is the velocity of light, 2.998×10^8 m/s,
h is Planck's constant, 6.626×10^{-34} Js.

In order to excite electrons across the bandgap of a p–n junction of a solar cell (see Section 5.4), the photons must have adequate energy and so the radiation must be below a certain wavelength (above a certain frequency). For example, only irradiance with a wavelength of less than 1.1 μm contributes to the operation of a crystalline silicon solar cell (see Figure 4.11).

Example 4.4 – Wavelength of Light to Operate a Silicon Solar Cell

Mono-crystalline silicon solar cells require photons with energy of 1.11 eV to excite electrons to cross the bandgap of the p–n junction. What is the maximum wavelength and minimum frequency of the radiation for electricity to flow?

Solution

By definition, $1\ \mathrm{eV} = 1.602 \times 10^{-19}$ J.

Therefore the energy of each photon required to excite an electron to cross the bandgap is

$$E = 1.11 \times 1.602 \times 10^{-19} = 1.78 \times 10^{-19}\,\mathrm{J}$$

Using Equation 4.5, this requires a wavelength of less than

$$\lambda \le \frac{hc}{E} = \frac{6.626 \times 10^{-34} \times 2.998 \times 10^{8}}{1.78 \times 10^{-19}} = 1.116 \times 10^{-6}\ \mathrm{m}$$

or a frequency greater than

$$\upsilon \ge \frac{c}{\lambda} = \frac{2.998 \times 10^{8}}{1.116 \times 10^{-6}} = 2.69 \times 10^{14}\ \mathrm{Hz} = 269\ \mathrm{THz}$$

Only solar radiation with a wavelength of less than 1.12 μm can contribute to the operation of this type of solar cell.

PROBLEMS

1. Explain the terms describing the components of the solar resource: direct (beam) radiation, diffuse radiation and albedo. Which components are likely to dominate at a site in a tropical desert region, and which are likely to dominate on a snow-covered mountain top?

2. Distinguish between the terms irradiance, insolation and irradiation.
3. Write down a first approximation of the tilt angles and orientations of solar panels to give maximum annual insolation in the following places:

   Cairo

   Lusaka.

   [**Answer:** Cairo: 30°, ± 5° of south; Lusaka: 15°, ± 5° of north.]
4. A solar water heating system is to be located in southern England at a latitude of 50°. The main demand for hot water is at 18:00. Estimate the angle of orientation for the panels.

   [**Answer:** azimuth angle −45°; tilt angle 60°.]
5. What should be the approximate orientation of solar panels in Cardiff to obtain maximum annual insolation?

   [**Answer:** around 35–40°.]
6. Use a solar resource website to estimate the resource at your location. Copy and complete the following table. Choose an optimum tilt angle for maximum annual insolation. Comment on the results.

Parameter	Units value
Annual global insolation (irradiation) on a horizontal surface	kWh/m^2 per year
Tilt angle for maximum annual insolation	°
Maximum annual global insolation at an optimum tilt angle	kWh/m^2 per year
Daily insolation in January at an optimum tilt angle	kWh/m^2 per day
Daily insolation in June at an optimum tilt angle	kWh/m^2 per day
Ratio of direct/diffuse daily insolation	%
Peak irradiance in January at an optimum tilt angle	W/m^2
Peak irradiance in June at an optimum tilt angle	W/m^2

7. Calculate the altitude of the sun at solar noon in Cardiff on 17 July. What is the zenith angle and hence the air mass?

   [**Answer:** 30°, 1.15.]
8. How many hours of daylight are there on 25 December in Cardiff? What is the minimum air mass on that day?

   [**Answer:** 7.7 h; 3.7.]
9. What is the energy of photons of light with a wavelength of 2 μm?

   [**Answer:** 0.59 eV.]
10. Amorphous silicon has a bandgap of 1.7 eV. Calculate the maximum wavelength of solar radiation that will generate power from an amorphous silicon photovoltaic cell.

   [**Answer:** 0.69 μm.]

FURTHER READING

Duffie J.A., Beckman W.A. and Blair N., *Solar Engineering of Thermal Processes*, 5th ed., 2020, Wiley Interscience.
The classic book on solar engineering, particularly for thermal processes.

Masters G., *Renewable and Efficient Electric Power Systems*, 2nd ed., 2013, Wiley.
An accessible textbook covering the generation of electricity from renewables.

Peak S. (ed.), *Renewable Energy: Power for a Sustainable Future,* 3rd ed., 2017, Oxford University Press.
Comprehensive textbook covering many aspects of renewable energy.

Twidell J., *Renewable Energy Resources*, 4th ed., 2021, Routledge.
Comprehensive textbook covering many aspects of renewable energy.

Wenham S.R., Green M.A., Watt M.E., Corkish R. and Sproul A., *Applied Photovoltaics*, 3rd ed., 2011, Earthscan.
An extremely clear and concise textbook on photovoltaics.

ONLINE RESOURCES

Exercise 4.1 The solar resource: a solar resource online tool is used to estimate the solar resource at a location.

Exercise 4.2 PV system sizing: a PV system design tool is used to size a system.

Exercise 4.3 Outline design and assessment of a PV generation scheme: a PV design tool and discounted cash flow analysis is used to assess the financial performance of a solar farm.

5 Photovoltaic Systems

INTRODUCTION

Photovoltaic (PV) systems generate electricity directly from the light of the sun and are making an increasingly important contribution to electricity supply in many countries. The worldwide recent rate of growth of the installed capacity of PV systems is around 38% per year. This rapid growth has been stimulated partly by financial support measures that have been offered by some governments as they de-carbonize their electricity supply systems, but also by a dramatic reduction in the price of photovoltaic modules as a result of technology development and increased manufacturing volumes.

Most solar panels that are installed today use mono- or poly-crystalline silicon solar cells and are connected to the electricity network (the grid[1]). The panels are mounted either on the roofs of buildings or are supported on ground-mounted structures in solar farms. Some innovative buildings have photovoltaic modules integrated into their facades or roof. Solar farms are usually located in areas of high solar radiation where land is cheap but are also being constructed in many other parts of the world, including in higher latitudes. Photovoltaic generation also plays an important role in supplying electrical power in remote areas where there is no grid electricity supply.

As well as photovoltaic cells that use wafers of crystalline silicon, cells made from thin film material are offered by several manufacturers. Thin film cells use a small amount of semiconductor material deposited on an inert substrate and have the potential to be cheaper than the thicker bulk silicon cells. There are also exciting developments of new photovoltaic materials, but these have yet to be deployed widely.

The output from a photovoltaic system depends completely on the solar energy resource and usually peaks around noon. At higher latitudes the solar irradiance drops in the winter, resulting in low capacity factors and poor utilization of the photovoltaic equipment. Of course, at night a photovoltaic system produces no output.

This chapter addresses solar cells, modules and systems. Section 5.1 presents the history and advantages and disadvantages of photovoltaic generation. Section 5.2

[1] 'Grid' strictly describes the extra high-voltage electricity transmission grid, but in this chapter the term is used more generally to include lower-voltage electricity distribution networks.

details the standard test conditions (STC) under which the performance of photovoltaic equipment is measured and explains the use of peak sun hours to give an estimate of output. Section 5.3 describes the first- and second-generation photovoltaic technologies. Sections 5.4 to 5.8 discuss the operation of solar cells, their equivalent circuit and their representation as a diode and current source. Sections 5.9 and 5.10 present photovoltaic modules and their performance. Section 5.11 discusses grid-connected systems, and Section 5.12 reviews the present state of photovoltaic technology.

5.1 Photovoltaic Energy Conversion

Photovoltaic devices convert sunlight directly into electricity. There are several ways in which the photovoltaic effect can be exploited, and some relatively new discoveries such as dye-sensitized, organic polymer and perovskite photovoltaic cells using new materials show great promise. However, at present the commercially important photovoltaic technologies all use solid semiconductor material to form a p–n (positive–negative) junction on which light falls. Solar energy falling on the semiconductor material excites a flow of electrons that are pulled across the p–n junction by the electric field that is created when the junction is formed. When the illuminated semiconductor p–n junction is connected to an external circuit, the flow of electrons across the junction creates direct current (dc) electricity. The current can either be stored in batteries or converted into alternating current (ac) electricity by an inverter that is connected to the grid.

5.1.1 History

Although the photovoltaic effect was identified in the nineteenth century it was only in the 1950s when transistors were being developed that effective photovoltaic devices were made. Initially photovoltaic cells were very expensive, and their main application was in spacecraft. Then in the 1970s they began to be used commonly for the supply of small amounts of electrical power in remote parts of the world where there was no grid supply. One interesting early application was to power small refrigerators to keep vaccines at the correct temperatures in rural health clinics of developing countries. More recently there have been increasing numbers of installations of grid-connected photovoltaic systems.

Where there is no grid electricity connection, the supply of small amounts of electrical power to (for example) telecommunication systems, navigational aids or vaccine fridges is so valuable that photovoltaic systems are economic with the present costs of photovoltaic equipment. The value of providing services for rural people is very high and justifies the costs of the photovoltaic modules and the batteries. Off-grid standalone photovoltaic systems are extremely useful for the supply of electrical power in remote locations, but their cost usually limits their size and the amount of power they can supply. They are becoming increasingly cost-effective in rural areas

Figure 5.1 Array of poly-crystalline silicon solar modules on a roof in a city centre.
Photo: RES.
(A black and white version of this figure will appear in some formats. For the colour version, refer to the plate section.)

for powering loads that provide a valuable service but need only a small amount of energy such as lamps that use light-emitting diodes (LED) and mobile phones.

5.1.2 Advantages and Disadvantages of Photovoltaic Energy Conversion

The advantages of photovoltaic generation are as follows.

- Photovoltaic cells generate electricity with no moving parts and can be packaged into robust modules that require very little maintenance and have a life of more than 20 years.
- Established solar cell technologies are stable and their performance degrades only slowly over time.
- Photovoltaic systems can be distributed on the roofs of buildings or constructed as large ground-mounted solar farms. Figures 5.1 and 5.2 show grid-connected PV arrays in an urban and a rural setting.
- Photovoltaic systems have limited environmental impact, and schemes often enjoy public acceptance.

The main disadvantages are as follows.

- The operation of a photovoltaic system relies entirely on the solar resource. Hence there is limited output in overcast weather and none at night.
- In many countries the peak electricity demand occurs at times when the solar resource is low. For example, in the UK, peak electricity demand occurs in the early evening in January when the solar resource is very low or non-existent.

Figure 5.2 A ground-mounted array in a rural solar farm in England.
Photo: RES.
(A black and white version of this figure will appear in some formats. For the colour version, refer to the plate section.)

- The capital cost of photovoltaic systems remains high.
- When used off-grid, batteries are needed and these are often the weak point of a standalone photovoltaic system.
- Although silicon is available throughout the world, there is increasing concern over the availability of some of the materials used in PV systems and batteries.

5.2 Performance of Photovoltaic Systems

Figure 5.3 is a schematic diagram of a small grid-connected PV system that might be found on the roof of a domestic house. It has two strings of eight modules connected in series and a single dc/ac inverter. The arrangement shown with two strings allows either independent control of each string, or for the strings to be connected in parallel at the inverter. With 250 W modules, this system would generate a power output of 4 kW in bright sunlight.

5.2.1 Standard Test Conditions and Use of Peak Sun Hours

The strength of the solar resource, described in Chapter 4, varies with time and location. Hence it is necessary to define a standard set of environmental conditions under which the performance of photovoltaic equipment is measured.

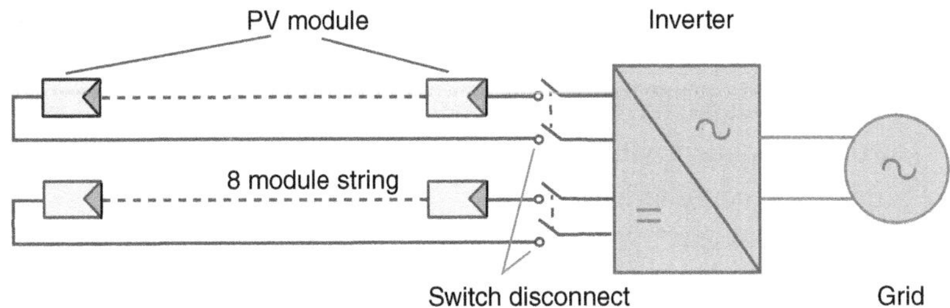

Figure 5.3 Schematic diagram of a small photovoltaic system.

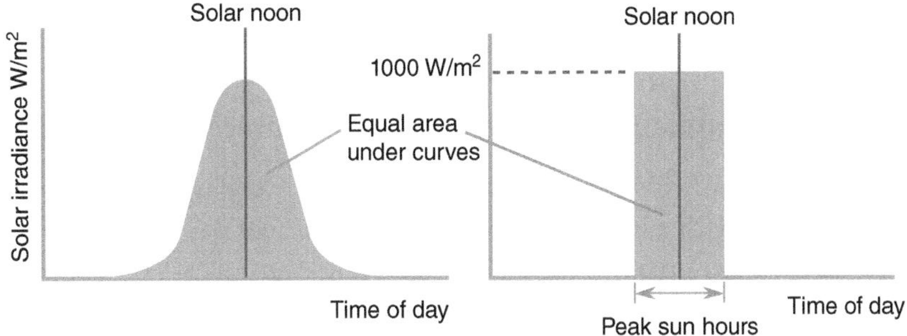

Figure 5.4 Daily peak sun hours.

The peak power output of a photovoltaic module is specified (and tested) under standard test conditions (STC) of:

- irradiance 1000 W/m^2,
- cell temperature 25 °C,
- solar spectrum air mass 1.5 (see Figure 4.12).

The solar energy insolation over a period (year, month or day) with units of kWh/m^2 can be converted into an equivalent number of hours at an irradiance of 1000 W/m^2, known as peak sun hours. This is shown for a day in Figure 5.4. The calculation can be repeated for any number of days.

The output power of a solar cell is proportional to the power of the sun (irradiance) and by defining standard test conditions at a solar irradiance of 1000 W/m^2 an initial estimate of the electricity output energy of a PV system in kWh can be made by multiplying the number of peak sun hours by the rated output power of the PV equipment (defined by the manufacturer at STC). This simple calculation is very convenient but only approximate as it ignores the effect of increased cell temperature reducing the output of photovoltaic modules. Standard test conditions are rarely experienced in practice as an irradiance of 1000 W/m^2 (very bright sunlight) usually increases the cell temperature above 25 °C and a high cell temperature reduces the output power of a photovoltaic module. This simple initial estimate of the energy output of a PV array also neglects electrical losses in the inverter and system wiring. The true output of a photovoltaic system may be 60–80% of this initial simple estimate. More precise calculations can be undertaken using solar PV design software.

The number of peak sun hours of a site shows immediately the approximate capacity factor that can be expected from any photovoltaic equipment installed at the site. The capacity factor can be interpreted as the equivalent fraction of time that the equipment will generate at its rated output and shows how effectively the equipment is being utilized.

The usefulness of peak sun hours to estimate the output of a solar system and check the results of more accurate computer simulations is illustrated in Example 5.1. Example 5.2 demonstrates how peak sun hours can be used to estimate system capacity factors and also illustrates the challenge of using photovoltaic equipment in the winter in higher latitudes. The utilization of PV equipment in winter at high latitudes, as indicated by the capacity factor, can be very low.

Example 5.1 – Estimate the Energy Output of a Photovoltaic System

A photovoltaic system with a rated output of 4 kW at standard test conditions is to be installed at a site with a daily insolation of 5 kWh/m^2 per day. Estimate the electrical energy that the photovoltaic system will generate over a day assuming an overall system efficiency of 75%.

Solution

Insolation (H) at the site over a day is given by

$$H_{SITE} = 5000 \text{ Wh/m}^2 \text{ per day}$$

Irradiance (G) at STC is defined as

$$G_{STC} = 1000 \text{ W/m}^2$$

The peak sun hours are then simply

$$\text{Peak sun hours per day} = \frac{H_{SITE}}{G_{STC}} = 5\,\text{h/day}$$

Electricity generated by PV equipment of 4 kW rating and 75% overall efficiency is

$$\begin{aligned}\text{electricity generated } &= \text{peak sun hours per day} \times \text{generator rating} \times \text{efficiency}\\ &= 5 \times 4 \times 0.75 = 15 \text{ kWh/day}\end{aligned}$$

Example 5.2 – Estimate the Capacity Factor of a Photovoltaic System

The daily insolation at a site in the UK is 5 kWh/m^2 per day in summer and 1 kWh/m^2 per day in winter. The annual insolation is 1200 kWh/m^2 per year. Assume a system efficiency of 75%. Estimate the daily and annual capacity factors of a PV system for this site.

Solution

The capacity factor (CF) of any generator over a period (day, month or year) is defined as

$$CF = \frac{\text{Electrical energy generated over a period } [\text{kWh}]}{\text{Generator rating } [\text{kW}] \times \text{hours in period}} \times 100\%$$

which for a PV system may be rewritten as

$$CF = \frac{\text{Peak sun hours in period} \times \text{PV generator rating} \times \text{efficiency}}{\text{PV generator rating} \times \text{hours in period}} \times 100\%$$

and reduced to

$$CF = \frac{\text{Peak sun hours in period} \times \text{efficiency}}{\text{hours in time period}} \times 100\%$$

Period	Insolation (kWh/m^2)	Time period (hours)	Peak sun hours in period	Capacity factor (%)
Winter day	1	24	1	3.1
Summer day	5	24	5	15.6
Year	1200	8760	1200	10.3

5.3 Photovoltaic Technology

Photovoltaic technology is usually considered in three generations.

- The first-generation technology uses wafers of mono- or poly-crystalline silicon. These bulk silicon devices are produced in large quantities and dominate the commercial market for photovoltaic panels.
- The second generation of technologies uses a thin film of active photovoltaic material on an inert substrate. This generation includes amorphous silicon, which is cheaper but less efficient than crystalline silicon and other thin film semiconductor materials such as copper indium gallium diselenide (CIGS) and cadmium telluride (CdTe). All of these technologies are used in the commercial production of flat-plate photovoltaic modules. In addition, gallium arsenide (GaAs) is used for solar panels in spacecraft and for concentrating terrestrial systems.
- The third generation includes the emerging technologies of dye-sensitized and organic/polymer solar cells sometimes using perovskite materials. The principle of operation of these devices differs from first- and second-generation technologies, and third-generation devices have yet to be produced in large quantities. They are the subject of very active research and development particularly to improve the long-term stability of the materials.

Mono- and poly-crystalline silicon technologies are mature, and the cells are encased in robust modules whose output characteristics are well defined and stable. These first-generation photovoltaic modules have been shown to work effectively for more than 20 years in harsh conditions with little maintenance. However, the cells are relatively thick (100–200 μm) and so use significant quantities of very pure silicon that is expensive and takes considerable energy to produce. The time for a bulk crystalline photovoltaic system to generate the same amount of energy as was used in its manufacture (the energy payback time) is up to five years depending on the solar resource of the site where it is installed. Hence there are active programmes to develop and exploit commercially the second- and third-generation photovoltaic technologies, which use much less semiconductor material, take less energy to produce and have lower costs. There is considerable research being undertaken into the manufacturing processes needed to produce large areas of stable, high-efficiency, thin-film photovoltaic cells.

5.4 The Silicon Solar Cell

Crystalline silicon solar cells continue to be used in most commercial photovoltaic systems. Silicon has a well-defined crystal structure, and the operation of crystalline silicon solar cells is relatively easily explained, so the operation of photovoltaic cells is described by using the simple representation of a first-generation photovoltaic cell shown in Figure 5.5. These photovoltaic cells use mono- or poly-crystalline silicon to form a p–n junction that is illuminated by sunlight. Pure silicon is either grown into a single crystal (mono-crystalline) or cast in ingots containing a number of large crystals (poly-crystalline). The silicon is doped with boron to create the p-type semiconductor as the crystalline silicon is manufactured. The p-type silicon is sliced into wafers some 100–200 μm thick using a diamond-tipped saw, although up to half the silicon may be lost in the cutting process. Then, in a separate process, a thinner layer (~0.3 μm) of n-type semiconductor is created by diffusing phosphorus into the front surface of the wafer. The p–n junction is formed where the two semiconductors meet. Electrical contacts are applied to the front and rear surfaces of the

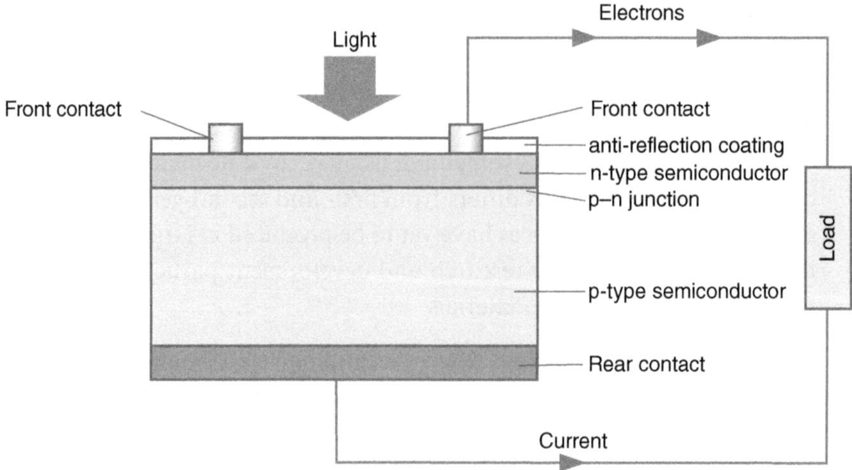

Figure 5.5 Schematic of a crystalline silicon solar cell.

photovoltaic cell, and an anti-reflection coating is placed on the front surface. The cells are connected together and placed into modules to provide mechanical support and protection. These solar cells have been the subject of an intense development effort for more than 30 years and have now reached cell efficiencies of up to 26% for test cells in the laboratory and up to 19% for commercially available complete modules.

5.4.1 The Bond Model of the Silicon Solar Cell

A silicon atom has a nucleus of 14 protons and 14 neutrons. There are 14 electrons orbiting the nucleus in three shells. The outermost shell of electrons contains four electrons and so is half full (the outermost shell can contain up to eight electrons). Each atom forms covalent bonds with four neighbouring atoms. In the dark and at a temperature of absolute zero all the electrons are locked up in the bonds and so there can be no passage of electricity through the semiconductor material. However, if the silicon is exposed to light and so bombarded by photons (or heated) some of the electrons gain enough energy to break free from the bonds. This leads to free electrons and a deficit of electrons in the outer orbits. The deficit of an electron is known as a hole and its movement through the lattice allows current to flow as electrons move to fill it. When a potential difference is applied to pure silicon that has been raised to an excited state by light or heat energy, a current will flow as the electrons and holes move through the lattice. Pure silicon is an intrinsic semiconductor with an equal number of electrons and holes.

An extrinsic semiconductor is made by doping the intrinsic silicon with donor atoms of another material. Phosphorous has five electrons in its outer shell and so produces an excess of free electrons. These free electrons then act as charge carriers in the material.

Similarly, if the intrinsic silicon is doped with boron, which has only three electrons in its outer shell, there is a deficit of electrons, and the semiconductor becomes p-type (positive).

Figures 5.6–5.8 show a simple representation of the atomic structure of doped silicon. This is known as the bond model.

5.4.2 The Band Model of the Silicon Solar Cell

An alternative model, the band model, describes the behaviour of silicon in terms of energy bands. Quantum theory states that the energy of an electron always lies within one of several bands. In pure silicon the electrons in the outer shell that form covalent bonds with neighbouring atoms are described as occupying the valence band. If any of these electrons acquire enough energy, from photons or heat, to enable them to move around within the material and thus to conduct electricity, they are described as being in the conduction band. There is a so-called bandgap between these two energy bands, the magnitude of which varies from material to material and is 1.1 eV (electronvolts) for intrinsic crystalline silicon.

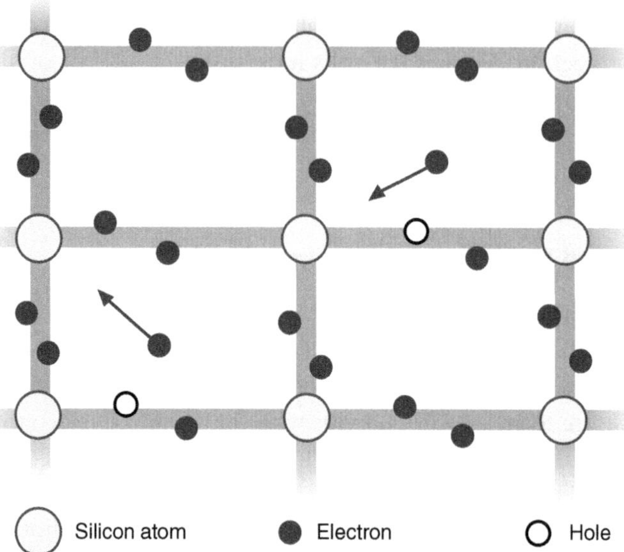

Figure 5.6 Bond model of the structure of pure silicon showing two free electrons leaving two holes. The silicon is either heated or bombarded by photons.

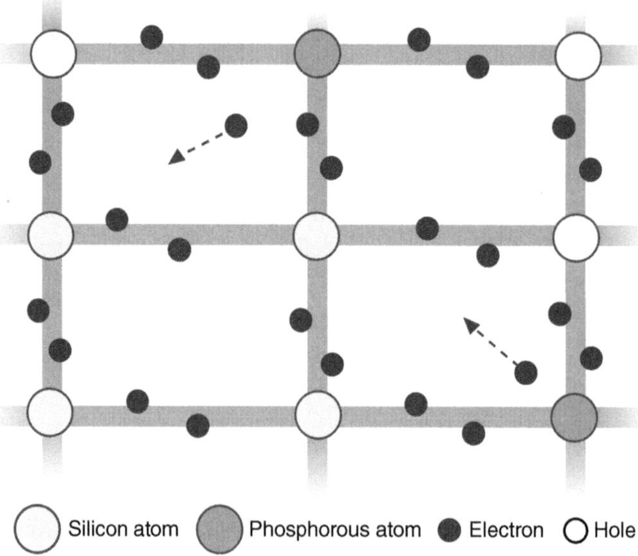

Figure 5.7 Bond model of the structure of n-type silicon doped with phosphorus, showing excess of electrons.

Thus there are two simple models for silicon. Figure 5.6 shows the bond model, and Figure 5.9 shows the band model of intrinsic silicon. Conduction can occur in intrinsic pure silicon only if electrons are excited into the conduction band by heat

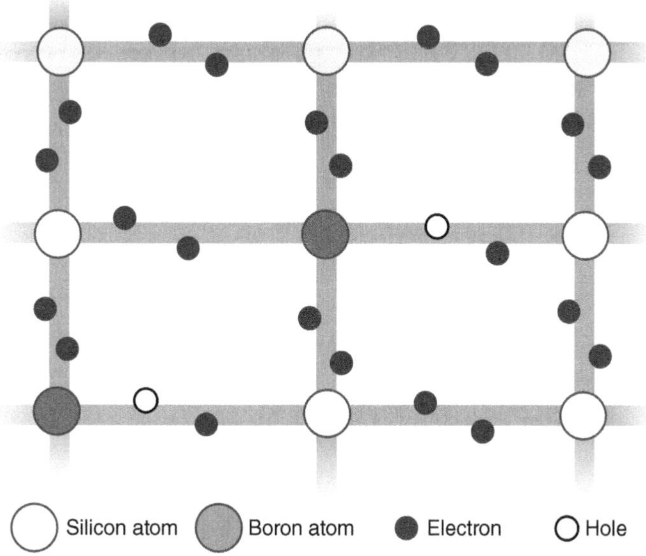

Figure 5.8 Bond model of the structure of p-type silicon doped with boron, showing excess of holes.

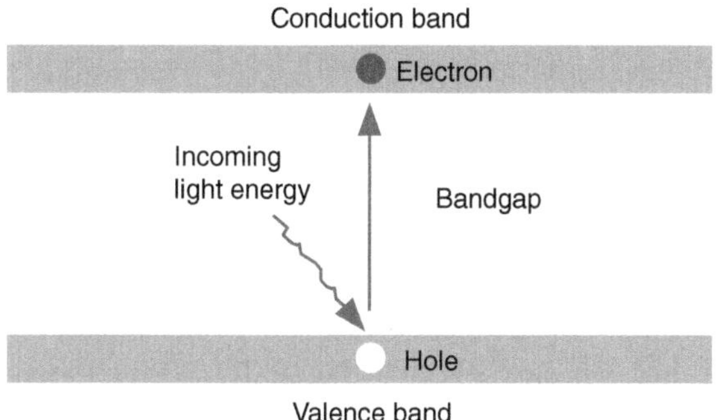

Figure 5.9 Energy bands in an intrinsic semiconductor.

or light energy and are free to move through the lattice. This leaves a hole in the valence band. If a potential difference is applied to ensure a uniform direction of movement of the electrons, and of holes in the opposite direction, then a current will flow.

The two models can also be used to describe extrinsic doped semiconductors. In the bond model, the effect of doping silicon with an n-type material is to create free electrons (Figure 5.7), while doping with a p-type material creates holes (Figure 5.8). In the band model, the effect of doping silicon with an n-type material is to add an additional energy level of the excess electrons just below the conduction band. Similarly doping with a p-type material adds an energy level just above the valence band.

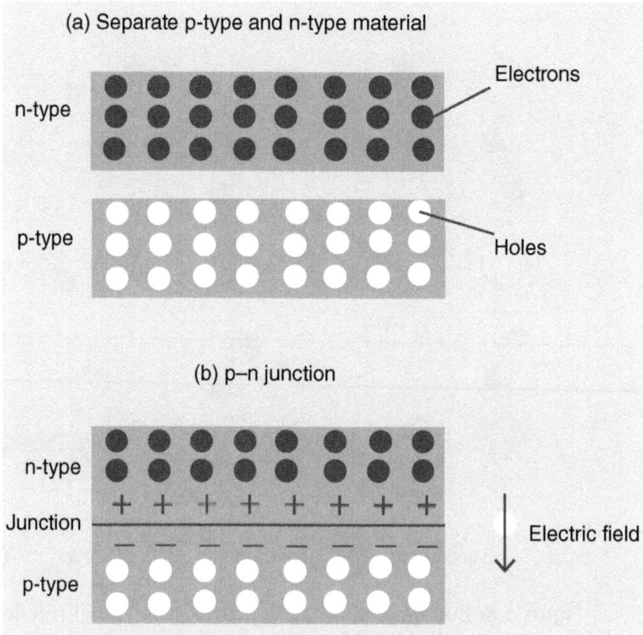

Figure 5.10 Illustration of formation of p–n junction.
(a) Separate p-type and n-type material showing excess holes and electrons, (b) p–n junction and electric field formed by migration of electrons and holes.

5.4.3 The p–n Junction

A p–n junction is created where the p-type and n-type material meet. This is illustrated in Figure 5.10. For simplicity consider that separate pieces of p-type and n-type material are brought together to form the junction, although in practice doping of the silicon is carried out sequentially. At the junction there is a junction region formed as the excess electrons from the n-type material migrate to the excess holes of the p-type material. The charge on the nuclei of the doping atoms does not change, so the junction region has a positive potential in the n-type material and a negative potential in the p-type material. An electric field builds up to limit the movement of excess electrons and holes, and this results in an electric potential across the junction of around 0.6 V for silicon.

When an external voltage is applied that reinforces this electric field, very little current will flow across the junction. When an external voltage greater than 0.6 V of opposite polarity to the electric field is applied, current will flow. This leads to the well-known dark diode characteristic shown in Figure 5.11.

The performance of an ideal dark diode is described by the ideal diode equation:

$$I = I_0\left[e^{\frac{qV}{kT}} - 1\right] = I_0\left[e^{\frac{V}{V_T}} - 1\right] \tag{5.1}$$

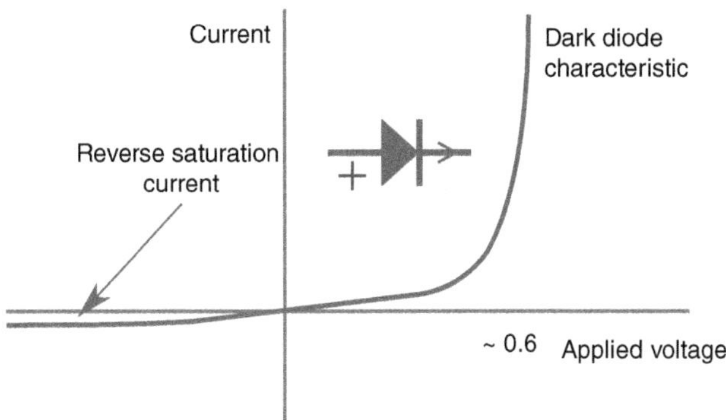

Figure 5.11 Dark diode characteristic.

where

I is the current through the diode in A,
V is the voltage across the diode in V,
I_0 is the reverse saturation current in A,
q is the charge on an electron, 1.602×10^{-19} C,
k is Boltzmann's constant, 1.38×10^{-23} J/K,
T is the absolute temperature,
$V_T = \frac{kT}{q} = 0.026$ V for silicon at room temperature of 300 K (27 °C).

Example 5.3 – Forward Voltage Drop Across a Silicon Diode

Consider an ideal dark diode with a reverse saturation current of 10^{-9} A. What will the forward voltage drop be if a current of 20 A flows at room temperature?

Solution
Equation 5.1 can be rearranged as

$$V = V_T \ln\left(\frac{I}{I_0} + 1\right)$$

$$V = 0.026 \times \ln\left(\frac{20}{10^{-9}} + 1\right) = 0.62 \text{ V}$$

5.5 Operation of a Solar Cell

A solar cell is a p–n junction (a diode) that is exposed to light. When light energy strikes the p–n junction, photons excite electrons from the valence band to the conduction band of the silicon atoms. The electrons migrate across the p–n junction under the influence of the electric field that is created in the junction region when the junction is formed, and current flows in the external circuit. The operation of

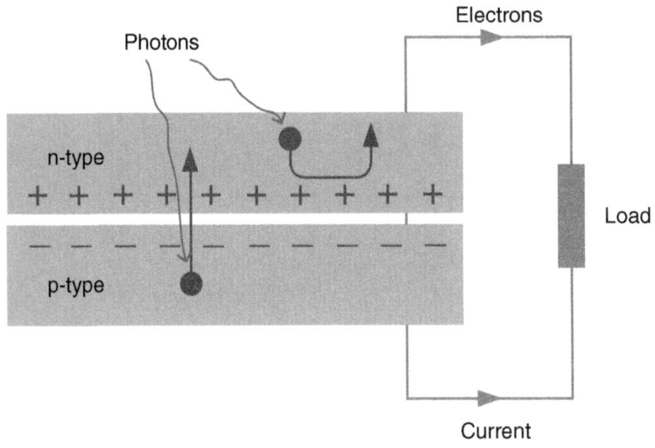

Figure 5.12 Operation of a photovoltaic cell.

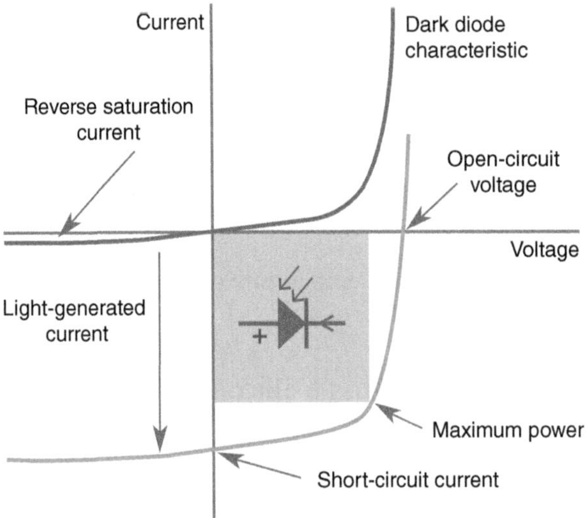

Figure 5.13 *V–I* characteristic of an illuminated diode showing its translation by the light-generated current.

a photovoltaic cell is summarized in Figure 5.12. Light photons excite electrons so that they can move through the crystal lattice. The electrons are influenced by the field of the junction region to move in the same direction. The uniform movement of the electrons creates a current.

A dark p–n junction has the characteristic of a diode (Figure 5.11) but when illuminated the curve is translated downwards as shown in Figure 5.13.

Figure 5.13 shows that electrical power is generated in the lower right-hand quadrant. By convention the sign of the current is reversed to describe the performance of a photovoltaic cell that generates power. This results in the lower right-hand quadrant of Figure 5.13 being reflected in the x-axis and gives the V–I characteristic of a solar cell (Figure 5.14).

The light-generated current is proportional to the number of photons that excite electrons to the conduction band from the valence band. With increased solar

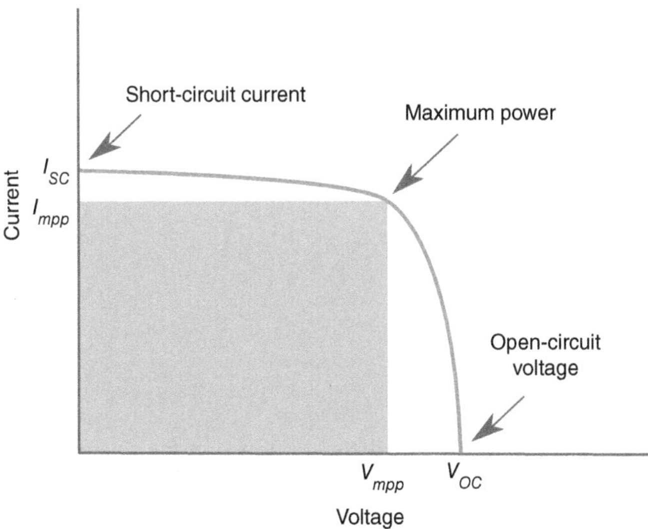

Figure 5.14 *V–I* characteristic of a solar cell: I_{SC} is the short-circuit current, V_{OC} is the open-circuit voltage, I_{mpp} is the current at the maximum power point and V_{mpp} is the voltage at the maximum power point.

irradiance, there are more photons and so more electrons are promoted to the conduction band. These electrons are swept across the junction by the field of the junction region, and as the solar irradiance increases, more current flows. Hence the short-circuit current of a photovoltaic cell is directly proportional to the solar irradiance (G). The open-circuit voltage of a photovoltaic cell is determined by the electric field created in the junction region of the p–n junction and so is largely independent of the solar irradiance. It is equal to the forward voltage drop of a diode and at room temperature is approximately 0.6 V for silicon.

The operation of an ideal solar cell is described by Equation 5.2:

$$I = I_p - I_0 \left[e^{\frac{V}{V_T}} - 1 \right] \tag{5.2}$$

where

I_p is the photocurrent that is directly proportional to solar irradiance (G).

5.6 Equivalent Circuit of a Solar Cell

Equation 5.2 leads to the simple equivalent circuit of a solar cell, which is a current source in parallel with a diode model, shown in Figure 5.15. The current source represents the electrons being promoted from the valence to the conduction band as photons strike the solar cell. The diode model represents the p–n junction of the solar cell and is not an additional physical diode.

If the terminals of the solar cell are connected together, the diode model is short-circuited and only the current source is effective. The output current of the photovoltaic cell is then directly proportional to the solar irradiance G. For modern

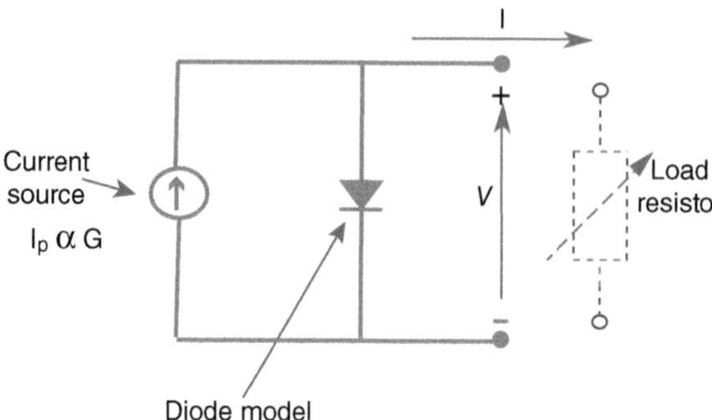

Figure 5.15 Simple equivalent circuit of an ideal solar cell.

mono-crystalline silicon cells at standard test conditions the current generated is approximately 300–400 A/m^2 of cell surface.

If the terminals are open-circuited, the output voltage of the cell will be determined by the forward voltage drop of the diode, which for crystalline silicon cells at standard test conditions is approximately 0.6 V.

Power in a dc circuit is simply the product of voltage and current, $P = V \times I$. Hence when the terminals are short-circuited, no power is produced ($V = 0$), and similarly when the terminals are open-circuited ($I = 0$), the power delivered is zero. Maximum output power occurs at the knee of the V–I curve when the product of $V \times I$ is a maximum.

This may be illustrated by considering a photovoltaic cell connected to a variable-load resistor. The operating point is determined by the intersection of the V–I characteristics of the photovoltaic (PV) cell and the load resistance as shown in Figure 5.16. The load characteristic is a straight line with a slope of $1/R$. As the load resistance increases from zero to infinity, the system operating point moves along the V–I characteristic curve of the cell from I_{SC} (short circuit) to V_{OC} (open circuit). Maximum power is obtained when the operating characteristic is at point A (V_{mpp}, I_{mpp}). At this point, the area of the rectangle under the V–I characteristic curve, which is equivalent to power, is a maximum. Matching the load resistance to the photovoltaic cell V–I characteristic is essential if maximum power is to be generated by the cell.

A fill factor (FF) describes the quality of the cell and is defined as:

$$FF = \frac{V_{mpp}\, I_{mpp}}{V_{OC}\, I_{SC}} = \frac{P_{mpp}}{V_{OC}\, I_{SC}} \tag{5.3}$$

The closer the fill factor is to unity, or the nearer the V–I characteristic of the solar cell is to the shaded rectangle of Figure 5.16, the greater the output power and the better the quality of the cell. Fill factors of up to 80% for crystalline solar cells have been reported.

The equivalent circuit can be extended to include shunt and series resistances to represent losses within the cell, see Figure 5.17. The series resistor R_S represents the resistance of the bulk semiconductor, the metallic contacts and the connection of the contacts to the semiconductor material. The shunt resistor R_{SH} represents

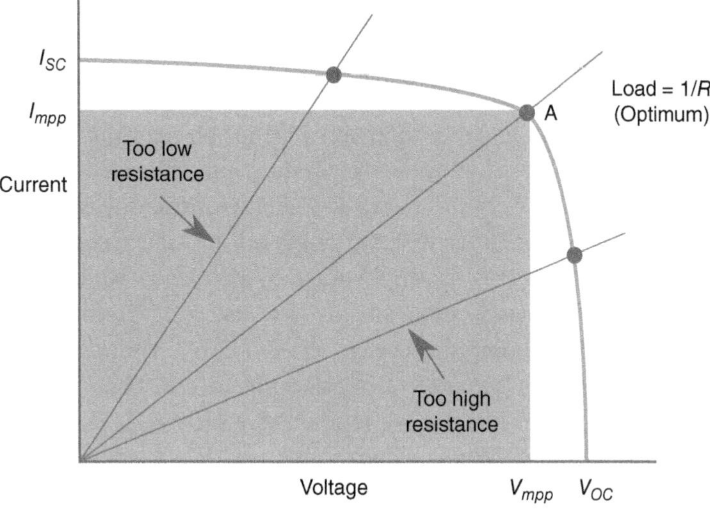

Figure 5.16 Maximum power point operation of a photovoltaic (PV) cell.

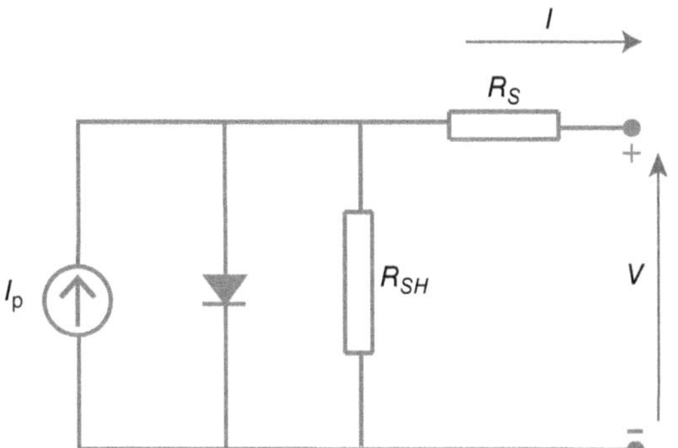

Figure 5.17 Equivalent circuit of a solar cell with series and shunt resistance.

leakage of current across the p–n junction around the edge of the cell and the effect of defects and impurities in the junction region. The effect of these shunt and series losses is to reduce the Fill Factor and the maximum power that can be generated. An ideality factor n is used to represent recombination of electrons and holes at defects in the junction region. The ideality factor varies between 1 and 2. Including the effect of these losses results in Equation 5.4

$$I = I_p - I_0\left[e^{\frac{q(V+R_S I)}{nkT}} - 1\right] - \frac{V + R_S I}{R_{SH}} \tag{5.4}$$

where

n is the ideality factor (between 1 and 2),
R_S is the series resistance in Ω,
R_{SH} is the shunt resistance in Ω.

If R_S and R_{SH} are set to zero and $n = 1$, then Equation 5.4 reduces to Equation 5.2.

5.7 Performance of the Solar Cell with Varying Irradiance and Cell Temperature

The *V–I* characteristic of a solar cell depends on the irradiance and the cell temperature, while the position of the operating point depends on the resistance presented to the cell. Figure 5.18 shows the *V–I* characteristic of a silicon photovoltaic cell at constant cell temperature with varying levels of solar irradiance. It can be seen that the short-circuit current is proportional to irradiance, while the open-circuit voltage remains approximately constant at these levels of irradiance. The voltage of the maximum power point, the knee of the *V–I* curve, varies with irradiance.

Figure 5.19 shows the effect of cell temperature. There is a small increase in the short-circuit current as the cell temperature increases. However, the main effect is on the open-circuit voltage, which reduces with cell temperature.

Standard test conditions are defined at a cell temperature of 25 °C. For crystalline silicon solar cells, the increase in short-circuit current with cell temperature above 25 °C is about 0.06% per °C, while the decrease on open-circuit voltage is about 0.3% per °C. In equation form this is

$$\frac{dI_{SC}}{dT} = I_{SC}^{STC} \times 0.0006 \quad [\text{A/°C}]$$

$$\frac{dV_{OC}}{dT} = V_{OC}^{STC} \times -0.003 \quad [\text{V/°C}]$$

This variation of the performance of a photovoltaic cell with temperature may be understood by considering that the bandgap of the semiconductor junction reduces with temperature. At increased cell temperature, more photons have enough energy

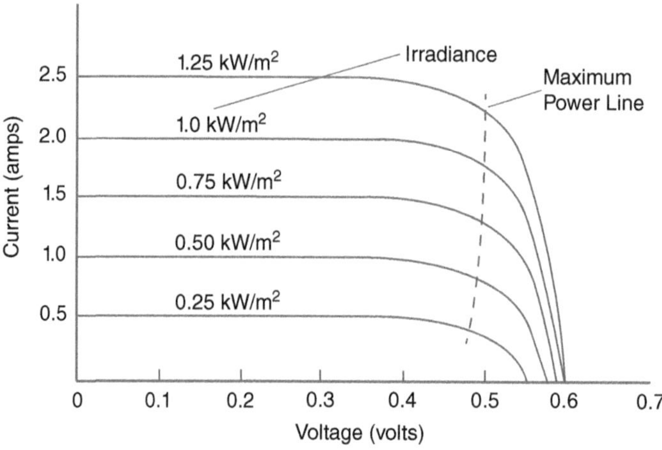

Figure 5.18 *V–I* characteristic of a crystalline solar cell at constant temperature and various levels of irradiance.
Source: Adapted from F. Lasnier, T. Gan Ang and K.S. Lwin, 'Solar photovoltaic handbook', 1988. Unpublished manuscript.

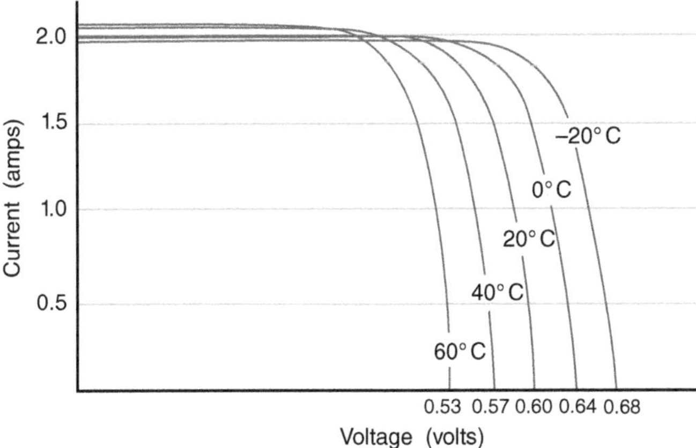

Figure 5.19 *V–I* characteristic of a crystalline solar cell at constant irradiance and various cell temperatures.
Source: Adapted from F. Lasnier, T. Gan Ang and K.S. Lwin, 'Solar photovoltaic handbook', 1988. Unpublished manuscript.

to promote electrons from the valence to the conduction band across the reduced bandgap. Hence more electron hole pairs are formed and increased current flows across the junction. However, the electric field of the p–n junction is lower with a smaller bandgap, so the effect of raised cell temperatures is to reduce the open-circuit voltage.

Example 5.4 – Performance of a Solar Cell at Increased Cell Temperature

A crystalline silicon PV cell has an open-circuit voltage of 0.6 V and a short-circuit current of 2 A at standard test conditions. What is the maximum output power at:

(a) standard test conditions,
(b) a cell temperature of 55 °C?

Assume a fill factor of 0.75.

Solution

(a) Using Equation 5.3, and considering standard test conditions, we have

$$P_{mpp} = FF \times V_{OC} \times I_{SC} = 0.75 \times 2 \times 0.6 = 0.9 \text{ W}$$

(b) At 55 °C and assuming a constant fill factor, we have

$$I_{SC} = I_{SC}^{STC} \times (1 + 0.0006 \times \Delta T) = 2 \times (1 + 0.0006 \times 30) = 2.036 \text{ A}$$
$$V_{OC} = V_{OC}^{STC} \times (1 - 0.003 \times \Delta T) = 0.6 \times (1 - 0.003 \times 30) = 0.546 \text{ V}$$
$$P_{mpp} = FF \times V_{OC} \times I_{SC} = 0.75 \times 2.036 \times 0.546 = 0.834 \text{ W}$$

The fill factor itself has a temperature coefficient of around −0.15% per °C giving a maximum power point at 55 °C of

$$P_{mpp} = FF \times V_{OC} \times I_{SC} = 0.75(1 - 0.0015 \times \Delta T) \times 2.036 \times 0.546 = 0.796 \text{ W}$$

The output power is then reduced by 0.1 W or 11%.

The overall reduction in power is around 0.4% per °C rise in cell temperature above 25 °C but with the maximum power point occurring at a lower voltage.

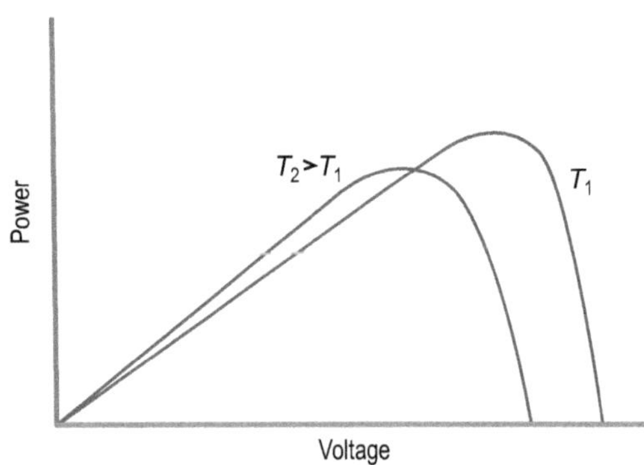

Figure 5.20 *V–P* characteristic of a crystalline solar cell showing the effect of increasing cell temperature.

The effect of increasing cell temperatures on the power developed by a photovoltaic cell is shown in Figure 5.20. Both the maximum power and the voltage at which maximum power is generated are reduced.

5.8 The Solar Cell as a Current Source

Considered simply, most common electrical sources of energy can be represented as a *voltage source* in series with a low source impedance. A lead–acid battery is commonly represented as a dc voltage source in series with a small resistor, while the mains electricity supply is thought of as an ac voltage source in series with a small impedance. A voltage source attempts always to keep its terminal voltage at a constant voltage and will try to pass whatever current is necessary to achieve this. Thus, when the output of a voltage source is short-circuited, and the internal resistance of the source is low, a very large current will flow, and it is common to use a fuse to detect and then cut off this large fault current. For example, a 12 V battery in a car will normally pass a maximum current of around 50 A, supplying load or being charged by the alternator. However, if the terminals of a car battery are short-circuited (a very dangerous occurrence) it will deliver up to 1000–2000 A. Hence a suitable fuse is placed in the circuit to limit this current. Similarly, the electrical installations of buildings have fuses to limit short-circuit currents if a fault occurs in the wiring or connected appliances.

The equivalent circuit of a photovoltaic cell, Figures 5.15 and 5.17, and Equations 5.2 and 5.4, show the very different electrical performance of a photovoltaic cell.

The cell is represented by a current source in parallel with a diode model. A current source attempts always to keep its current at a pre-determined value. The current that is generated by a solar cell is fixed by the solar irradiance and will not exceed I_{SC}, even when short-circuited. Also, when the terminal voltage of a cell approaches 0.6 V, the diode will begin to conduct to limit the voltage to V_{OC}. Thus a photovoltaic cell can be short-circuited or open-circuited without the current and voltage exceeding I_{SC} and V_{OC}. This benign performance is very attractive for small photovoltaic systems where the output voltage and current are limited to safe values by the characteristics of the cells. However, it is difficult to use fuses or other over-current devices to detect short-circuits on the dc wiring of the photovoltaic array as the available short-circuit current will be only slightly greater than the normal operating current.

5.9 Photovoltaic Modules

An individual photovoltaic cell produces only a few watts of electrical power and crystalline silicon wafers are fragile. Therefore the cells are collected into modules to give more useful levels of power and packaged to provide mechanical and environmental protection. A typical rigid module of crystalline cells consists of a sealed sandwich of strong, impact-resistant glass and the cells cushioned on a layer of ethyl vinyl acetate with a layer of tedlar at the rear to provide a barrier against moisture. There is a rigid aluminium frame to hold the module and for fixing it to the supporting structure. Within the module, the cells are connected in series to give an increased output voltage and in parallel to provide a higher module current (see Tutorial I for an explanation of series and parallel connections). A junction box at the rear of the module is used to terminate the connections from the series and parallel circuits of the cells and to connect the modules into larger arrays. An important early application of crystalline silicon photovoltaic modules was to provide power into 12 V batteries. This requires an open-circuit voltage at standard test conditions of around 22 V and so many modules were manufactured with 36 cells in series. This arrangement is shown in Figure 5.21a and is still used. Larger modules for grid-connected applications now often use 60 or 72 cells, as shown in Figure 5.21b.

5.9.1 Module Bypass Diodes

In a practical system, not all the cells in a module can be guaranteed to have the same *V–I* characteristics and work at the same operating point. Differences in performance of a cell can be due to manufacturing defects, cell degradation, shading and dirt as well as variations in cell temperature. This is particularly important when one cell in a series-connected circuit generates less current than the other cells. A reverse voltage then appears across the faulty cell as current is forced though it by the action of the other cells. This condition can reduce the output power of the series circuit of cells significantly and can be extremely damaging to the faulty cell

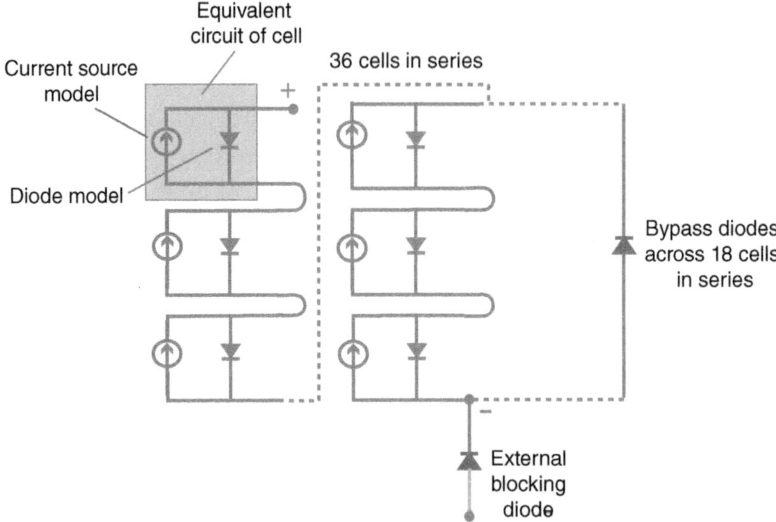

Figure 5.21(a) 36-cell, 100 W photovoltaic module for battery charging. Typically V_{OC} ~ 22 V and I_{SC} ~ 5 A.

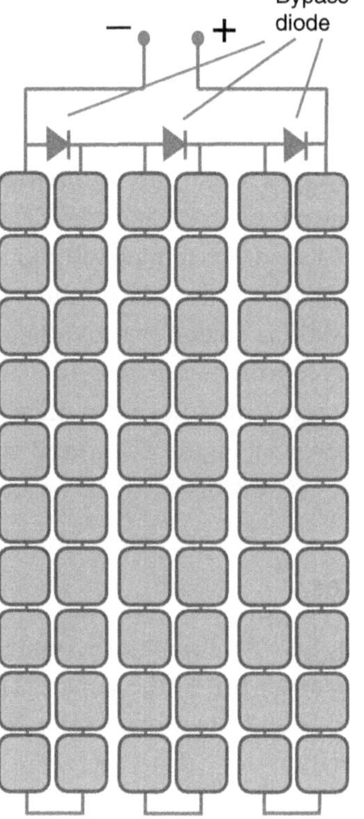

Figure 5.21(b) 60-cell, 360 W photovoltaic module for grid-connected systems. Typically V_{OC} ~ 40 V and I_{SC} ~ 10 A.

Example 5.5 – Performance of a Photovoltaic Module

A photovoltaic module has 72 circular mono-crystalline cells, each of 100 mm diameter, connected in two circuits, each of 36 cells in series. Each cell will produce a short-circuit current of 300 A/m^2 under solar irradiance of 1000 W/m^2. Estimate the open-circuit voltage and short-circuit current of the module at this irradiance.

Solution

Each cell will produce an open-circuit voltage of 0.6 V, so 36 cells in series will give an open-circuit voltage for each series circuit of 21.6 V.

Each cell will produce a short-circuit current of approximately:

$$\left(\frac{100}{2}\right)^2 \times \pi \times 10^{-6} \times 300 = 2.36\ \text{A}$$

There are two parallel circuits, each of 36 cells connected in series. Therefore the module open-circuit voltage will be 21.6 V and the module short-circuit current will be 4.72 A.

if a hot-spot is formed as current is forced through a small area of the p–n junction. The conventional solution is to add bypass diodes to divert current around a cell if it becomes reverse biased. To reduce cost, it is usual to connect a bypass diode across a group of cells, but then if one cell in the group becomes reverse biased the output of the group is reduced (Figure 5.21a). The danger of damage through hot-spots being formed is particularly acute if the output of the module is short-circuited; it can be reduced by limiting the number of cells that are connected in series before a bypass diode is added. A 36-cell module would typically have two bypass diodes, and a 60-cell module often has three bypass diodes.

5.9.2 Blocking Diodes

If no current is being generated by a solar cell, as occurs at night, and if a voltage is left connected to the cell terminals, e.g. from a battery, a reverse current flows through the diodes formed by the p–n junctions. This would lead to a loss of charge in the battery. Thus off-grid battery systems use a blocking diode connected in series with the array output to ensure that current can flow out of the solar array only. Although diodes with a low forward voltage drop are used, usually Schottky diodes, an additional voltage drop and hence power loss is introduced into the dc circuit. Blocking diodes can also be used to ensure that one string of modules in a large array does not inject current into other strings in which one or more cells is not working correctly and in which the output voltage of the faulty string is reduced. Figure 5.21a shows a blocking diode connected to the negative terminal of the module.

Both bypass and blocking diodes are physical diodes, while the diode models of Figure 5.21a represent the p–n junctions of the PV cells.

5.10 Performance of Photovoltaic Modules and Systems

5.10.1 Estimation of Cell Temperature

High cell temperatures significantly reduce the performance of crystalline silicon photovoltaic cells, reducing both the output power and open-circuit voltage. Hence the module and support structure are designed to provide maximum cooling. The module manufacturer will usually declare a nominal operating cell temperature (NOCT), which gives the cell temperature (in °C) if the module is used under the following conditions:

- open circuit,
- irradiance 800 W/m^2,
- air temperature 20 °C,
- wind speed 1 m/s.

Cell temperatures for different levels of irradiance and ambient temperatures can be estimated from Equation 5.5:

$$T_{cell} = T_{air} + \frac{NOCT - 20}{800} \times G \quad [°C] \tag{5.5}$$

where

T_{air} is the air temperature in °C,
$NOCT$ is the nominal operating cell temperature in °C,
G is the irradiance in W/m^2.

Example 5.6 – Reduction of Output with Cell Temperature

Estimate the reduction in power output of a 120 W$_p$ (watt-peak) photovoltaic module with a NOCT of 46 °C that is used under the following conditions:

ambient air temperature 30 °C,
irradiance 600 W/m^2,
variation of cell power output with temperature $\delta P = -0.4\%$ per °C.

Solution

$$T_{cell} = 30 + \frac{46 - 20}{800} \times 600 = 49.5\ °C$$

$$P = 120 \times (1 - 0.004(49.5 - 25) = 108.24\ W$$

5.10.2 Performance Assessment of Photovoltaic Systems

Once a photovoltaic system has been installed, it can be difficult to assess whether it is working satisfactorily as its performance depends on the solar resource and temperature as well as on whether the equipment has been designed, installed and operated correctly. A number of indicators are commonly used to assess the performance of practical photovoltaic systems.

The specific yield is simply the ratio of the electrical energy produced by the photovoltaic system (known as the system yield) to its nominal system power at standard test conditions:

$$\text{Specific yield} = \frac{\text{System yield} \quad [\text{kWh}]}{\text{System power at STC} \quad [\text{kW}]}$$

It is usual to calculate an annual specific yield, although monthly specific yields are sometimes determined. Typical values of annual specific yields for systems installed in the UK would be 750–800 kWh/kW; annual specific yields rise to 1500–2000 kWh/kW for locations near the equator with a very good solar resource. The annual specific yield depends on the insolation, so it will vary as the weather changes year by year even though the equipment is working consistently.

The performance ratio includes the effect of the solar resource through the peak sun hours that are incident on the array (see Section 5.2). It shows the effect of cell temperature and electrical losses within the inverter and wiring.

The dimensionless performance ratio is given by

$$\begin{aligned}\text{Performance ratio} &= \frac{\text{System yield} \quad [\text{kWh}]}{\text{Potential yield} \quad [\text{kWh}]} \\ &= \frac{\text{System yield} \quad [\text{kWh}]}{\text{System power} \quad [\text{kW}] \times \text{Peak sun hours} \quad [\text{h}]} \\ &= \frac{\text{Specific yield} \quad [\text{kWh/kW}]}{\text{Peak sun hours} \quad [\text{h}]}\end{aligned}$$

Again, it is common to calculate the performance ratio of a system over a year. Annual values of performance ratio usually vary between 0.7 and 0.8. If monthly values of the performance ratio are calculated, they are likely to be higher in the winter when ambient temperatures are low. Plotting annual performance ratios over a number of years will show if the performance of the photovoltaic system is degrading with time.

Example 5.7 – Performance of Photovoltaic Systems

A photovoltaic system with an output power of 2.4 kW at STC is installed at a location with a solar resource of 1050 peak sun hours per year. If its electrical output is 1800 kWh what is the specific yield and performance ratio?

Solution

$$\text{Specific yield}(SY) = \frac{\text{System yield}}{\text{System power}} = \frac{1800}{2.4} = 750 \text{ kWh/kW}$$

$$\text{Performance Ratio}(PR) = \frac{\text{System yield kWh/kW}}{\text{Peak sun hours h}} = \frac{750}{1050} = 0.71$$

5.11 Grid-Connected Photovoltaic Systems

Figure 5.22 shows the main functional elements of a grid-connected photovoltaic system: the solar modules in an array, a maximum power point tracker and a dc/ac inverter. The dc output of the photovoltaic panels is fed to a maximum power point tracker. The maximum power point tracker stage is a dc/dc converter which varies the voltage applied to the modules to extract maximum power for the ambient conditions. The dc power is then inverted to ac and injected into the power network (the grid). The maximum power point tracking function may be implemented within the inverter or in a separate device. A large grid-connected system is likely to have multiple maximum power point trackers and either a large central inverter or multiple smaller inverters.

For most grid-connected PV systems, the grid serves as an infinite energy sink and provides a strong frequency and voltage reference for the operation of the inverter. An inverter for a grid-connected photovoltaic system constantly monitors the condition of the electricity network to which it is connected and continues to operate only if the grid network is in its normal operating condition, i.e. the frequency and voltage are normal. Usually a grid-connected photovoltaic inverter cannot operate in a standalone mode and will shut down if the grid loses its supply from the large central generators or if a fault causes a local section of network to become isolated.

5.11.1 Maximum Power Point Tracking

The *V–I* characteristic of a photovoltaic module depends on both irradiance and the temperature of the cells (see Figures 5.18 and 5.19), and to obtain maximum output power it is necessary to operate the module at its maximum power point. This is done by continuously adjusting the resistance that the maximum power point tracker (MPPT) presents to the module so that the product of module voltage and current is a maximum. There is no analytical way to determine the maximum power point and there are a number of maximum power point tracking methods. Some of them give sub-optimum power output but are easy to implement, while others use more complex algorithms.

The simplest method uses the fact that the voltage at the maximum power point of a module is always at between 65% and 80% of the open-circuit voltage. However, the open-circuit voltage itself varies with irradiance and temperature. Thus the

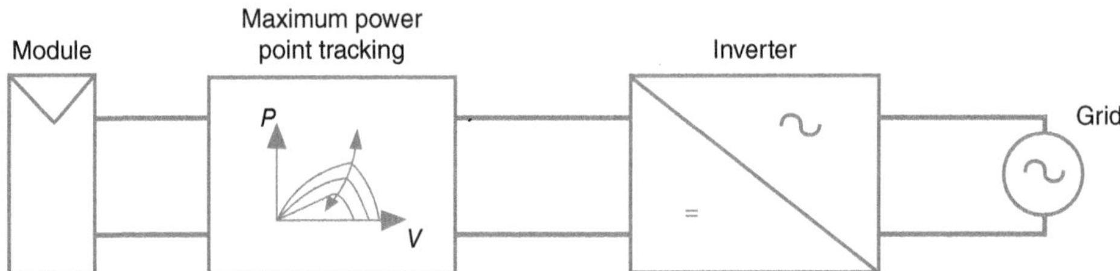

Figure 5.22 Functional blocks of a grid-connected photovoltaic system.

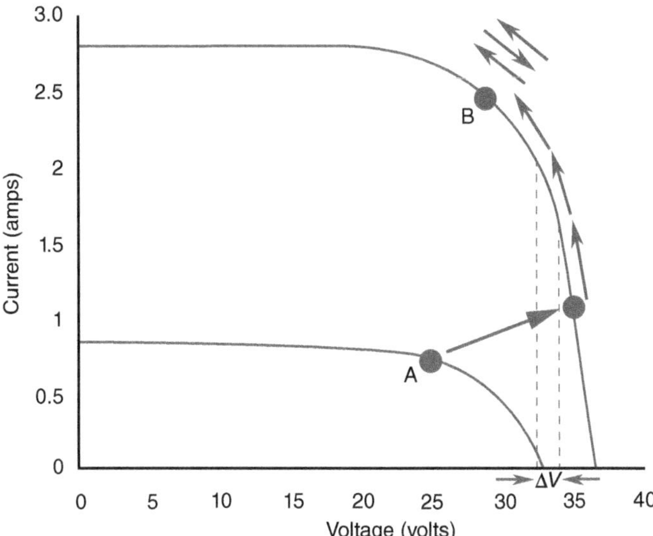

Figure 5.23 Perturb and observe (P&O) or hill climbing method of maximum power point tracking.

open-circuit voltage of the module is measured periodically by switching off the inverter for a short period of time, e.g. for 50 ms every minute. Once the open-circuit voltage is known for the prevailing cell temperature and irradiance, the new operating voltage of the module is set at a fixed percentage of this voltage (say 70%).

An alternative is the hill climbing or perturb and observe (P&O) method. The controller operates by periodically increasing or decreasing by a small amount the voltage applied to the module. If the perturbation of the voltage leads to an increase in module output power, the subsequent perturbation will be made in the same direction. If not, the direction of perturbation is reversed. By this principle, the maximum power point can be determined. Figure 5.23 shows the P&O method. Initially the module experiences low irradiance, perhaps due to cloud, and operates on the lower V–I characteristic at point A. Then the irradiance suddenly increases and the module V–I characteristic changes to the upper trace. The current drawn by the inverter does not change instantaneously and the voltage of the module of 35 V is now too high. By adjusting the voltage presented to the module in steps of ΔV the new maximum power point (point B) is sought. The approach has two possible drawbacks. Firstly, the module operating point oscillates around the maximum power point. This leads to a sub-optimum operation. Reducing the magnitude of the perturbation ΔV can solve this problem. However, a small ΔV requires more perturbation cycles to follow the maximum power point when the atmospheric conditions start to vary, and more power will be lost. The P&O method is readily implemented in a microprocessor, and the frequency and size of the voltage changes are chosen by the designer.

5.11.2 Grid-Connected PV Inverters

A large number of power electronic circuits have been used for the inversion of the output of photovoltaic modules from dc to ac and connection to the grid. These

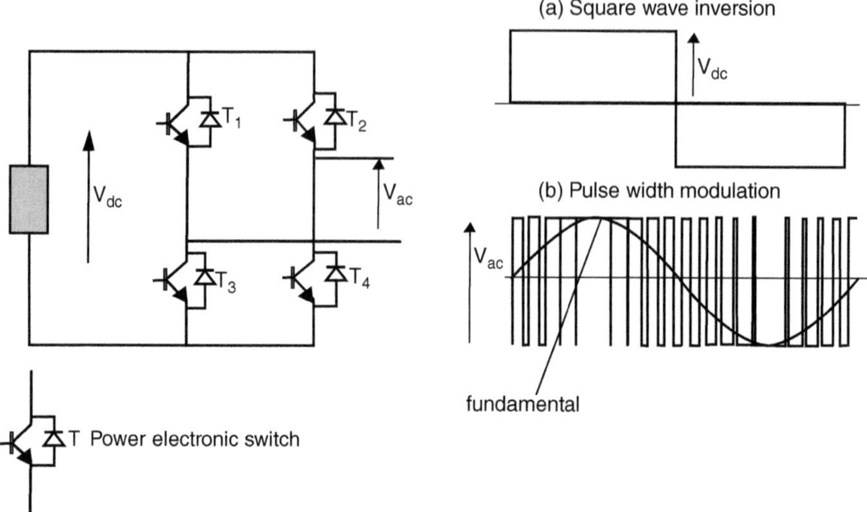

Figure 5.24 Single-phase inverter (reproduced from Figure I.27).

have taken advantage of the operating characteristics of the various semiconductor switching devices that are available and, as improved semiconductor switches are developed, the circuits will continue to evolve.

A simple single-phase inverter is shown conceptually in Figure 5.24. In this arrangement, four controlled power electronic switches are used to synthesize a sine wave. Using present-day technology these switches are likely to be insulated gate bipolar transistors (IGBTs). The switches are turned ON and OFF repeatedly in a complementary manner to obtain a sinusoidal waveform at the output of the inverter. The pulse width modulation (PWM) pattern of switching is arranged so that the output voltage is a close approximation to a sine wave.

The inverter creates ac from the dc of the photovoltaic modules by switching its transistors into either the fully ON or fully OFF state. Figure 5.24 shows how, by varying the switching time of a constant dc voltage, a sine wave can be synthesized. The sine wave is synchronized with the frequency and phase of the grid voltage and so power is injected into the grid.

Switching circuits are always used for power electronic inverters because if a transistor is fully ON, the voltage across it is low and so the losses ($V \times I$) are low. Similarly, if a transistor is fully OFF, the current through it, and hence the losses, will be zero ($V \times 0$). Some losses occur during the transition from ON to OFF (and OFF to ON) when there is both a voltage across the transistor and current flowing through it. These losses are proportional to the frequency of switching. If the transistors are switched rapidly, a closer approximation to a sine wave can be achieved but at the expense of higher switching losses.

Small photovoltaic inverters use sophisticated circuits to minimize losses, particularly at low irradiance, and very high efficiencies of 98–99% at full output are achieved. In practice the MPPT stage is often implemented as an integral part of the inverter which also includes protection circuits to disconnect the inverter if the grid is not in its normal state.

5.12 The Technologies of Photovoltaic Cells

Crystalline silicon solar cells are well understood, can be made into robust modules and have stable performance over their lifetime. Crystalline silicon cells are used for most terrestrial applications and, given the very large investment that has been made in manufacturing facilities, are likely to dominate the market for some years to come. However, they are made from a comparatively thick wafer of very pure silicon and so remain expensive. Considerable energy is used in their manufacture. Alternative materials are being investigated to reduce the cost and the energy used in the manufacturing process.

Despite the application of anti-reflective coating on the top surface of the cell, only around 50% of the available solar energy is absorbed by crystalline silicon, the remainder being lost through reflection or transmission through the cell. One fundamental limitation to increasing the efficiency of photovoltaic cells is the bandgap of the semiconductor material, typically quoted in electronvolts (where $1 \text{ eV} = 1.602 \times 10^{-19}$ J). The bandgap represents the forbidden energy level that exists between the highest energy level of the valence band and the lowest level of the conduction band. Only photons of sufficient energy can overcome the bandgap and generate charge carriers, electrons and holes. These electrons and holes migrate in opposite directions at the junction's electric field and a photo-generated current is produced. A photon with energy higher than the bandgap generates a pair of charge carriers, and any excess energy is converted to waste heat. Low-energy infrared photons (<1.1 eV) are not energetic enough to promote electrons from the valence to the conduction band, while high-energy photons (e.g. blue photons, >3.1 eV) dissipate excess energy as heat. This mismatch between the solar spectrum and the bandgap leads to a maximum theoretical Shockley–Queisser limit (radiative efficiency limit) for a single p–n junction device under AM1.5 solar spectrum of 33.3% for a bandgap of 1.1 eV. Optimum photovoltaic efficiency occurs when the bandgap of the material is well matched to the light spectrum of the solar resource.

Table 5.1 lists the first- and second-generation PV technologies with typical module efficiencies and bandgaps at 302 K.

The first generation of solar cells have p–n junctions formed in crystalline silicon. The cells must be comparatively thick as crystalline silicon is an indirect bandgap material with low absorption coefficient necessitating a significant (100–200 μm) absorber thickness. Indirect bandgap materials absorb photon energy in a process that is accompanied by lattice vibration to enable transition of the electrons from the valence to the conduction band. Owing to the thickness of the cells and the packaging required, crystalline modules are comparatively heavy and this needs to be considered during the structural survey for roof-mounted arrays.

Table 5.1 Photovoltaic materials for flat-plate, terrestrial module (first and second generation)

Generation	Cell type – definition	Typical module power conversion efficiency (%)	Bandgap at 302 K (eV)
First	Monocrystalline silicon – made from a single, continuous cubic silicon crystal.	13–17	1.1
First	Poly-crystalline silicon – contains large crystals (grains) of silicon.	11–15	1.1
First and second	Hetero junction with intrinsic thin layer – contains both monocrystalline and amorphous silicon materials.	14–19	1.1 and 1.7
Second	Amorphous silicon – a network of disordered silicon atoms chemically stabilized by hydrogen atoms.	5–9.5	1.7
Second	Copper indium gallium (di)selenide – a compound semiconductor (hetero-junction) formed with cadmium sulphide.	14	1.0–1.7
Second	Cadmium telluride – a compound semiconductor (hetero-junction) formed with cadmium sulphide.	11	1.45

Mono-crystalline silicon wafers are manufactured by slicing large single crystals that have been formed by the Czochralski process. This process involves dipping a small seed crystal into a hot (~1500 °C) melt of very pure silicon, which has been doped with boron, and slowly withdrawing it to form a large sausage-shaped single crystal. This sausage-shaped crystal is then sliced and polished, losing some 40–50% of the p-type material. Mono-crystalline solar cells are round or cut into approximate squares to increase their packing density and active area within modules.

Poly-crystalline (sometimes known as multi-crystalline) silicon cells are formed by taking the very pure silicon that has been doped p-type and casting it into rectangular blocks, then slicing the blocks into square wafers. The casting process results in a structure of multiple large crystal grains. The recombination of charge carriers at the grain boundaries causes significant reduction in power conversion efficiency. Because of the effect of the grain boundaries, poly- or multi-crystalline solar cells are up to 5% less efficient than mono-crystalline cells but can be produced at lower cost. As poly-crystalline silicon solar cells are square they can cover the entire surface of the solar module and often have an attractive blue 'grainy' appearance.

After the p-type wafers have been formed, the subsequent processing to make mono- and poly-crystalline solar cells is similar. The top surface is treated with an anti-reflection layer to improve absorption of light, and a thin n-type layer is created by diffusing phosphorus (as phosphine gas) into the wafer (see Figure 5.5). Electrical contacts are added to the front and rear of the cell. Other than a difference in efficiency, the performance of modules made from mono- and poly-crystalline solar cells is very similar, and the decision by a project developer to use one or other type of solar cell is usually made on price and availability.

The second generation of thin film photovoltaic cells, deposited on substrates such as glass, has been developed to reduce the quantity of semiconductor material used and reduce manufacturing cost. A number of semiconductor materials are in use in either homo-junctions, where the base material is the same across the junction, or hetero-junctions, where different materials form each side of the p–n junction. Second-generation photovoltaic cells use direct bandgap materials that only need photon energy to promote valence to conduction band transitions. Reasonable levels of efficiency are achieved with much thinner active layers (typically 1–5 μm thick). Thin film materials are typically deposited at low temperatures (200–400 °C) without vacuum techniques. This significantly reduces manufacturing costs and emissions of greenhouse gases compared with the energy-intensive crystalline silicon manufacturing technologies.

Amorphous silicon was used initially only in consumer products, e.g. calculators and watches, but is now used for power generation particularly on the facades of buildings. Amorphous silicon, without a well-defined crystal structure, is deposited on an inert substrate. The cells are usually fabricated with an intermediate intrinsic (i) layer in a p–i–n structure. Low-energy photons cannot be harvested usefully at wavelengths greater than 700 nm, leading to lower efficiency than is possible with crystalline silicon devices, but amorphous silicon photovoltaic modules have been shown to perform well in diffuse sunlight. The efficiency of amorphous silicon solar cells reduces over an initial period of operation as defects are caused by strong sunlight. This is known as the Staebler–Wronski effect and is reduced by the introduction of hydrogen atoms into the silicon network. Amorphous silicon cells have a wide bandgap of 1.7 eV.

Copper indium gallium (di)selenide (CIGS) cells have high photon absorption across the solar spectral range. The bandgap of the material depends on the proportions of indium and gallium in its composition. Thus the bandgap ranges from 1.1 eV for CIS to 1.7 eV for CGS. CIGS cannot, on its own, be doped to form an efficient p–n junction and so it is formed into a hetero-junction cell with cadmium sulphide (CdS). CIGS cells have efficiencies similar to poly-crystalline silicon. However, the cost of CIGS manufacture is currently higher owing to the poor series resistance of the transparent conducting oxide layer necessitating the use of metal pastes to form contact grids. Indium is a rare earth element with 0.05 ppm abundance in the earth's crust. Therefore substitution of indium with zinc, and gallium with tin, to form copper zinc tin sulphide/selenide, is an area of materials research with keen industrial interest.

Cadmium telluride (CdTe)/cadmium sulphide (CdS) is another thin film hetero-junction PV technology where the p-type CdTe compound forms the photon absorber layer. CdS forms a thin, n-type layer. Continuous, extensively automated deposition of large areas of material makes this process commercially attractive. Module recycling is often included as an initial cost of module purchase due to the toxicity and concerns over disposal of the constituent cadmium. Additionally, tellurium is a rare earth material so processes and devices with a minimum thickness CdTe layer are currently being developed.

Gallium arsenide (GaAs) cells are commonly used for power supplies in space as they do not suffer degradation from cosmic radiation. Also, they can be used at relatively high temperatures without appreciable performance degradation, so they are suitable for use in concentrator systems to minimize the device area and therefore system cost. GaAs has a bandgap of around 1.4 eV and can be made into triple junction cells with germanium (Ge) and gallium indium phosphide (GaInP) to extend the frequency over which photons can be absorbed. These complex III:V structures are precisely grown via metal organic vapour phase epitaxy (MOVPE) or molecular beam epitaxy (MBE). Multiple thin layers are grown onto a crystalline substrate (typically Ge) totalling ~8 μm. Interfaces are very precise, and dislocations in the crystal lattice are avoided by carefully balancing tensile and compressively strained layers at the hetero-junctions. Multi-junction cells exhibit the highest power conversion efficiency of all PV technologies (46%). This is due to multiple bandgap materials (1.8–0.67 eV) absorbing in separate regions of the solar spectrum, GaInP in the blue, (In)GaAs mid-spectrum and Ge in the infrared. Owing to the complexity and small volumes of manufacture, the production costs of III:V cells are high, so this technology is not used for flat-plate terrestrial solar modules.

The emerging technologies of dye-sensitized and organic/polymer solar cells are described as the third generation. Dye-sensitized solar cells are well developed and are finding an initial application for use indoors or in other conditions of low diffuse light. Nano-particles of the cheap and widely available semiconductor TiO_2 with particle diameters ~20 nm are coated with a dye and immersed in an electrolyte. Light is absorbed by the dye to form electron–hole pairs. Electrons from the dye enter the conduction band of the TiO_2 and are transported to the cathode. The holes formed in the dye are filled with electrons from negatively charged ions in the electrolyte. Although the electrons enter the conduction band of the TiO_2, equivalent holes are not formed in its valence band. Unlike PV cells that use a p–n junction, the separation of the charge carriers is not created by an internal electric field. It is caused by the reaction kinetics and the speed at which electrons are injected into the conduction band of the TiO_2 compared with the much slower recombination of electrons and holes in the dye. The open-circuit voltage (V_{OC}) of a dye-sensitized solar cell is around 0.7 V with a current density of 200 A/m^2 giving efficiencies ~11.9% with good performance in low intensity/indoor light.

Recently there has been great interest in the use of perovskite materials for solar cells. The name comes from the Russian mineralogist L.A. Perovski who discovered a naturally occurring calcium titanium oxide mineral with the chemical formula $CaTiO_3$. The name perovskite is now used to describe any material with a crystal structure represented as ABX_3, where the A represents a molecular cation, typically methylammonium $CH_3NH_3^+$, B is typically lead or tin and X is a halide, usually an iodide, bromide or chloride. There are vast numbers of possible combinations of atomic species to determine both the optical and electronic properties of the material. One popularly studied perovskite is an organic hybrid halide, methylammonium lead iodide (MAPI). Perovskite cells are being developed in four different architectures: sensitized structures using a TiO_2 scaffold, scaffold structures using Al_2O_3, thin-film planar p–i–n (the perovskite forms the intrinsic layer) and thin-film planar p–n junction (the perovskite forms the p-type semiconductor).

Perovskite materials have considerable promise for use in third-generation PV cells owing to their low cost, tuneable bandgap and potential to form tandem solar cells. The cells can be flexible and semi-transparent and have a high V_{OC} of 1.15 V. These materials have an almost direct bandgap with high and broad optical absorption. Absorber layers of <1 μm can also give low carrier recombination rate and good carrier mobility. Processing and manufacture are comparatively easy. Perovskite solar cells are the subject of very active research to increase lifetime, substitute the toxic lead material and develop scaled-up manufacturing. Potential markets identified for this technology include building integrated photovoltaic (BIPV), smart glass, automotive and portable devices.

New PV materials are currently being designed by computational combinatorial analysis. Ideal PV semiconductor materials should utilize abundant earth elements, making them sustainable. Toxic elements should be avoided wherever possible. A direct optical bandgap of ~1.5 eV for AM1.5 spectrum would be optimum for maximizing efficiency. The material should have a high photon absorption coefficient, low rate of charge carrier recombination, high carrier mobility and efficient charge separation. The cells should be designed to maximize V_{OC}, short-circuit current I_{SC} and fill factor to maximize power conversion efficiency. The final PV devices need to be tolerant to both impurities and defects and be chemically stable at all interfaces. The interconnected cells should exhibit low series and high shunt resistance and minimal contact resistance. To enable high-volume production, the material growth/deposition process needs to be controlled effectively to make consistent and highly reproducible devices.

Finally, there are three essential factors (the 'golden triangle') for PV devices. Of primary importance is efficiency. Lifetime (device reliability) is then considered in order to determine potential applications. Minimal levelized cost of the technology, directly related to both PV cell efficiency and lifetime, is vital for high-volume economically viable production. A satisfactory balance of these three factors has to be achieved to obtain commercial success.

PROBLEMS

1. Describe the three generations of PV cells. Why do first-generation cells continue to dominate the market?
2. Draw a sketch to show the main components of a grid-connected PV system. Which component dominates the cost?

3. Draw a sketch to show the main components of a standalone PV system. Which component dominates the reliability?
4. What is the purpose of a blocking diode and a bypass diode?
5. Sketch the V–I and V–P characteristics of a photovoltaic (PV) module. Show clearly the effect of changes in irradiance and cell temperature.
6. Use sketches of the V–I and V–P characteristics of a photovoltaic (PV) module to explain why maximum power point trackers are used. Using sketches, discuss one common method of maximum power point tracking.
7. Draw and explain the equivalent circuit of a photovoltaic cell.
8. A photovoltaic system with a rated output of 4 kW at standard test conditions is to be installed at a site with a daily insolation of 5 kWh/m^2. Estimate the electrical output of the photovoltaic system and the daily capacity factor.

   [**Answer:** 12–16 kWh, 14.6%.]
9. A solar array is formed by six modules each having 36 cells in series. The modules are arranged in three strings in parallel, each string has two modules in series. At 1000 W/m^2 irradiance, each cell has a short-circuit current of 3.6 A and an open-circuit voltage of 0.6 V. Sketch the V–I characteristic of the array at an irradiance of 500 W/m^2.

   [**Answer:** The curve passes through the (43.2 V, 0 A) and (0 V, 5.4 A) points.]
10. A solar module with a rectangular V–I characteristic is connected to a resistive load of 3 Ω. Under standard conditions, the short-circuit current of the module is 6.75 A and the open-circuit voltage of the module is 21.6 V. Calculate the power generated into the load at an irradiance of 800 and 500 W/m^2.

    [**Answer:** 87.5 W, 34.2 W.]
11. A crystalline silicon PV module has 36 cells connected in series. Each cell is square with sides 150 mm. Estimate the open-circuit voltage of the module, the short-circuit current and the power output at an irradiance of 800 W/m^2. Assume I_{SC} = 300 A/m^2 at standard conditions and the cells have a rectangular V–I characteristic.

    [**Answer:** 21.6 V, 5.4 A, 116 W.]
12. Refer to the solar spectrum shown in Figure 4.11.
    (a) Explain why it is that silicon solar cells can utilize only about 25% of the solar energy available.
    (b) How can a more efficient solar cell that makes use of the full solar spectrum be made? What are such cells called?

    Useful data: the bandgap of Si is 1.1 eV.
13. A 1.5 cm^2 GaAs solar cell has a saturation current of 0.5 nA and a short-circuit current of 100 mA. This cell has a fill factor (FF) of 0.8 and current at its maximum power point of I_m = 90 mA.

    Calculate:

    (a) the open-circuit voltage V_{OC} of the cell,
    (b) the value of the maximum power point voltage, V_{mpp},

(c) the efficiency of the cell if it is illuminated with 100 mW/cm^2.
Assume that the ideality factor is unity, i.e. $n = 1$.
At room temperature:

$$V_T = \frac{kT}{q} = 0.026$$

where k is Boltzmann's constant, T the absolute temperature and q the electronic charge.

[**Answer:** (a) 0.497 V; (b) 0.44 V; (c) 26.4%.]

14. A crystalline silicon PV module has 36 cells connected in series. Each cell is square with sides 100 mm. At an irradiance of 600 W/m^2, estimate
(a) the open-circuit voltage of the module,
(b) the short-circuit current,
(c) the maximum available power.
Two of the modules are connected (in parallel) to a 12 V battery. The battery terminal voltage is described by $V = 12+I$, where I is the current into the battery in amps.
(d) Estimate the power delivered into the battery.
Assume $I_{SC} = 300$ A/m^2 at standard conditions and the cells have a rectangular V–I characteristic (i.e. 100% fill factor).

[**Answer:** (a) 21.6 V; (b) 1.9 A; (c) 39 W; (d) 56 W.]

15. A PV module has 38 poly-crystalline silicon cells connected in series. Each cell is square with sides 120 mm. The module receives irradiance at 600 Wm^2 with a cell temperature of 60 °C.
(a) Calculate the open-circuit voltage and short-circuit current of the module.
(b) Assuming a rectangular V–I characteristic (i.e. 100% fill factor), calculate the resistance the maximum power point tracker should present to the module for maximum power output under these conditions.
(c) What is the percentage power lost if this resistance is applied to the module when it is used on a mountaintop site with irradiance of 1000 W/m^2 and a cell temperature of 25 °C?
Assume I_{SC} is 300 A/m^2 and V_{OC} is 0.6 V/cell at standard conditions of 1000 W/m^2 and 25 °C; δV_{OC} is –2.5 mV/°C per cell.

[**Answer:** (a) 19.5 V, 2.6 A; (b) 7.5 ohm; (c) 30%.]

16. A manufacturer's catalogue of photovoltaic modules advertises a large assembled module supplying a peak output power at 25 V and 12 A at standard test conditions (1000 W/m^2), but has omitted the physical dimensions.
The catalogue mentions that the cells have an efficiency of 10% when mounted within the module and that their active area makes up 80% of the total area of the module.
Each cell has the following characteristics at standard test conditions:
peak power point: 0.5 V, 3 A,
short-circuit current: 3.5 A,
open-circuit voltage: 0.6 V.

(a) How big is the module in square metres?
(b) Calculate the number of cells in series and parallel in each module and hence sketch the *V–I* characteristic of the module.
(c) Using the sketch, estimate the power that could be delivered into a 1 Ω resistive load.

[**Answer:** (a) 3.75 m^2; (b) 50 in series, 3 in parallel, (c) 170 W.]

FURTHER READING

Masters G., *Renewable and Efficient Electric Power Systems*, 2nd ed., 2013, Wiley.
An accessible textbook covering the generation of electricity from renewables.

Peak S. (ed.), *Renewable Energy: Power for a Sustainable Future*, 4th ed., 2017, Oxford University Press.
Comprehensive textbook covering many aspects of renewable energy.

Twidell J., *Renewable Energy Resources*, 4th ed., 2021, Routledge.
Comprehensive textbook covering many aspects of renewable energy.

Wenham S.R., Green M.A., Watt M.E., Corkish R. and Sproul A., *Applied Photovoltaics*, 3rd ed., 2011, Earthscan.
An extremely clear and concise textbook on photovoltaics.

ONLINE RESOURCES

Exercise 5.1 Performance of a PV module: calculation of the performance of a PV module.
Exercise 5.2 Performance of a PV array: demonstration of the performance of a PV array and sizing the dc cable.
Exercise 5.3 Design of a PV project: sizing the dc:ac inverter, calculation of the Specific Yield and Performance Ratio.

6 Solar Thermal Systems

INTRODUCTION

Energy from the sun is radiated across space and enters the earth's atmosphere to provide sunlight and heat. Without solar radiation there would be no life on earth. Photovoltaic cells directly convert sunlight into electrical energy and their operation was discussed in Chapter 5. This chapter addresses the use of the sun's energy to provide useful heat. The particular uses of solar energy considered here are for:

(1) heating of buildings,
(2) heating of water,
(3) supply of high-temperature heat for electricity generation and industrial processes.

Solar thermal energy may be divided into those processes that use a low temperature (less than ~150 °C) and those that use a high temperature (greater than ~150 °C). Although the use of low-temperature solar heat may appear simple, it is of great practical importance. More than 40% of the energy used in the UK is as heat, with 75% of this low-temperature heat; 80% of the heat energy used in the UK is from burning natural gas (methane), and replacing this cost-effectively with heat from sources that do not create greenhouse gases is a key focus of the transition to a net zero carbon energy system.

The design of low-energy buildings depends on natural heat gain from the sun and the retention of heat in winter while avoiding overheating in summer. Windows are essential for the thermal performance of buildings and increasingly are double or triple glazed. Most governments apply building standards that specify the thermal performance of new buildings to save energy and minimize the use of fossil fuels. In addition, many governments have financial support programmes to improve the thermal performance of existing buildings and reduce their heat loss. However, improving the energy performance of old buildings while maintaining their appearance and function is difficult and expensive. Just as with the energy efficiency measures discussed in Chapter 1, improving the energy performance of buildings is not only a technical matter but depends heavily on the behaviour of the occupants.

Solar thermal water heaters are widely used in countries that have a suitable solar resource, e.g. Greece and Japan, where they provide the domestic hot water needs of dwellings for at least part of the year. Solar water heating is a well-established technology with many competing suppliers of equipment.

High-temperature solar thermal systems to generate mechanical power were first demonstrated to operate effectively more than 100 years ago but their use for electricity generation remains expensive and has yet to be adopted widely. Recently the number of prototypes and demonstration plants of large-scale high-temperature solar thermal systems for electricity generation using parabolic troughs or solar towers and heliostats (mirrors that track the sun) has increased. This technology is known as concentrated (or concentrating) solar power (CSP) and is being actively developed for areas of the world with a suitable solar resource but is not as widespread as photovoltaic generation.

This chapter addresses the use of solar energy for the heating of buildings and water as well as for the supply of high-temperature heat for electricity generation and industry. Section 6.1 describes the important role that solar thermal energy plays in our society, along with its advantages and disadvantages. Section 6.2 discusses passive solar heating of buildings. Section 6.3 describes the simple circuit representation of low-temperature heat transfer, while Section 6.4 extends this approach to the heat loss of buildings through ventilation. Section 6.5 defines degree-days and discusses their use in monitoring the thermal performance of buildings. Section 6.6 describes in more detail the important different responses of glass to solar and low-temperature radiation. Section 6.7 presents the application of solar energy for water heating, and Section 6.8 gives a simple analysis of the performance of a flat-plate solar collector. Finally, in Section 6.9, high-temperature concentrating solar power generating systems are discussed.

6.1 Solar Thermal Energy

Figure 6.1 shows how energy is used in the UK (other than for transport). The importance of energy use in buildings is clear, with 45% of energy used for space heating and another 10% for water heating. Low-temperature solar thermal heating can be divided into passive applications where the solar energy is captured and used within a building, and active devices that use a separate collector to heat air or water. British buildings have a notoriously poor thermal performance, with heat being lost by conduction through badly insulated walls and roofs, as well as through convection by drafts and leaks of air. Once the thermal performance of dwellings has been improved, it is anticipated that water heating will become a more significant part of domestic energy use. In hotter climates a greater fraction of energy is used for cooling, and good passive solar design to reduce overheating in hot climates is another important way to reduce the use of fossil fuels.

Many low-temperature solar thermal systems rely on the important property of glass that it transmits the high-frequency (short-wavelength) radiation from the

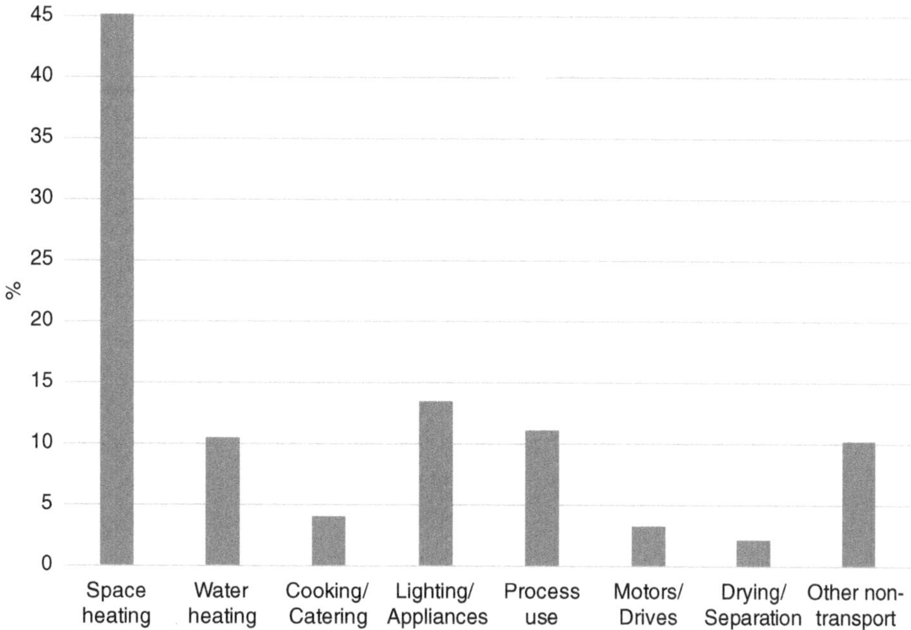

Figure 6.1 Use of energy in the UK.
Source of data: *Energy Consumption in the UK (ECUK)*, 2021, www.gov.uk.

very high-temperature source of the sun (~6000 K) but blocks the retransmission of lower-frequency (longer-wavelength) radiation from the cooler temperatures of warm objects on earth (~300 K). Glass windows, a simple passive solar system component, have been used in buildings since Roman times while glazed flat-plate solar thermal collectors to heat domestic hot water have been manufactured commercially for more than 100 years.

Low-temperature solar thermal systems can make use of both direct and diffuse solar irradiance and so can be fixed in their orientation. They do not need the complex tracking systems of concentrating solar systems that use only direct irradiance and so must be orientated towards the sun. Multiple sheets of glass in a carefully designed flat-plate solar thermal collector can increase water temperatures by up to ~50 °C. Glass and other transparent materials play another important role in solar thermal systems by limiting the movement of air, reducing heat losses by convection.

To achieve temperatures higher than ~150 °C it is necessary to concentrate the solar irradiance using either mirrors or lenses. Over the years a number of plants have been constructed with parabolic mirrors and troughs to focus the sun and obtain high temperatures. A central solar collector system, with the solar irradiance concentrated by moveable mirrors (known as heliostats), has been demonstrated to produce temperatures of up to 3500 °C, over a small area. This exceptionally high temperature was used for materials research that needed a very high temperature without contamination by products of combustion.

Once a high temperature is obtained, the heat can be used for industrial processes or with a turbine to create mechanical power and then generate electricity. Solar thermal turbine units working at up to ~500 °C have used a conventional water/steam turbine cycle for electricity generation, with the heat transfer from the solar concentrator by synthetic oil. Successful solar thermal generating systems using parabolic troughs have been installed in California, and several demonstration plants have been built in Spain. One ambitious proposal that was studied but not pursued was to construct large solar thermal generating systems in North Africa with transmission of the electrical energy to Europe.

The concentrator mirror or lens systems of high-temperature solar thermal systems focus direct irradiance, so they must be regularly orientated towards the sun using one- or two-axis mechanical tracking systems. Solar thermal concentrators need clear skies, so are often located in deserts. One important advantage of solar thermal electricity-generating systems is that heat can be stored more easily than electricity, and heat energy stores using molten salts have been included in some solar thermal plants that generate electricity.

6.1.1 Advantages and Disadvantages of Solar Thermal Energy Systems

Low-temperature solar thermal energy makes a major and cost-effective contribution to the energy used in buildings.

- Effective control of the passive solar gain of a building reduces the need for fuel to heat or cool it. In cold climates, passive solar heating reduces the length of the heating season, the months of the year during which additional heating is required. In hot climates careful design of the building eliminates or reduces the need for air conditioning.
- In those countries that have a good solar resource, active solar heating collectors are used to provide hot water in domestic dwellings for at least part of the year. Solar heating of hot water is particularly useful when, on summer days, it eliminates the need for electric water heating and so helps to reduce the peak of electricity demand that is caused mainly by air conditioning.

However, all solar energy systems rely critically on the solar resource, and in temperate latitudes this is in anti-phase to the need for space heating: heating is most needed at the times of the year when the solar resource is low. This is illustrated in Figure 6.2, which contrasts the need for space heating with the available solar energy in UK. It shows the percentage of the total annual energy needed for space heating and the percentage available from the sun throughout the year. The need for space heating peaks in January and February while the solar resource is a maximum in June. Chapter 4 showed that the maximum irradiance that would fall on a horizontal flat plate in Cardiff in December is just over only 200 W/m^2 and there can be 6 times the daily insolation (kWh/m^2) in the summer as in winter. In most temperate climates, solar systems for either space or water heating can provide only a limited output in the winter months when heating is needed most.

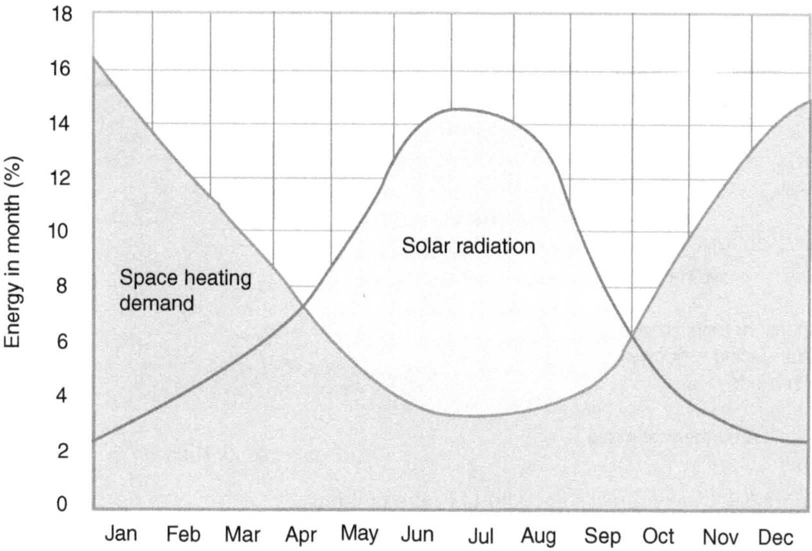

Figure 6.2 Illustration of solar resource and space heating demand in UK. Source of data: *Energy Trends*, ET 71-April 2022, www.gov.uk.

6.2 Passive Solar Thermal Heating of Buildings

The use of the sun to provide space heating is an important aspect of low-energy building design. Passive solar heating, where the sun's radiation directly heats a building, can be used together with high levels of thermal insulation and control of ventilation to reduce to very low levels the additional energy that is required while maintaining a comfortable indoor environment. Good passive solar thermal design is also necessary to avoid overheating of such low-energy buildings during the summer.

An effective passive solar design for a building in a temperate climate consists of five main elements.

(1) A large area of glazing transmits solar radiation into the building. This glazed area is orientated within ± 30° of south (in the northern hemisphere) and is not shaded during that part of the year when heating is required.
(2) Internal surfaces, usually made from hard material of dark colour, absorb the solar radiation and increase in temperature.
(3) A large thermal mass, either as part of the fabric of the building or an additional heat store (e.g. water or phase-change material), is used to maintain an even temperature in the building and prevent overheating.
(4) A distribution system transfers heat from the warmed internal surfaces throughout the building. This heat transfer can be a combination of convection by air movement, conduction through the surfaces dividing the rooms and low-temperature radiation.

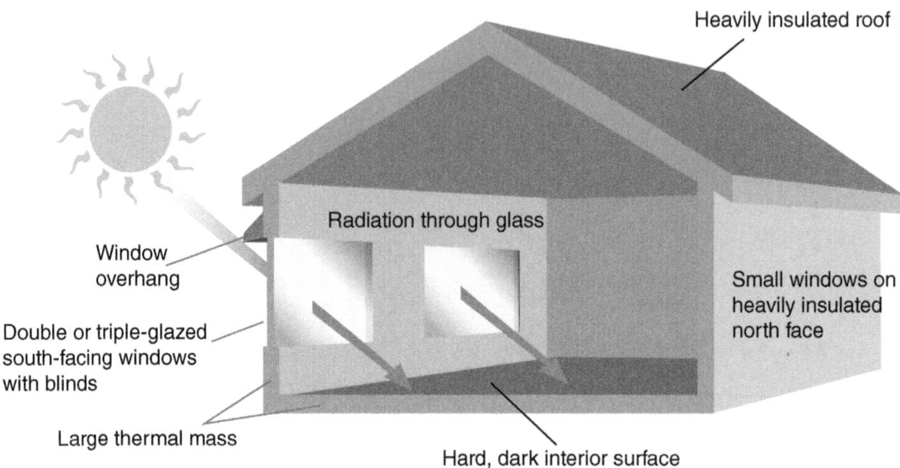

Figure 6.3 Use of direct solar gain to heat a building.

(5) A control strategy ensures that heating is adequate when needed but the building does not overheat in summer. The incoming irradiance can be limited actively by using blinds and external shutters or passively by window overhangs. Air flow through the building is controlled with vents and dampers. Active devices such as blinds and vents may be controlled automatically or manually.

Passive solar heating systems for buildings can be categorized as

- direct gain,
- isolated gain,
- indirect gain.

Figure 6.3 shows the principle of using direct solar gain to provide passive solar thermal heating. The building is orientated east–west with large windows located on the south side (in the northern hemisphere). Any windows on the north side, which receive little irradiance, are kept small to reduce heat loss. Solar irradiance is transmitted into the building through the large, south-facing double- or triple-glazed windows and is absorbed by the dark surfaces of the internal structure of the building. Buildings that use passive solar gain in this way are usually designed with a large thermal mass so that the building does not overheat at times of high irradiance and maintains a reasonably constant temperature.

The use of direct solar gain to heat a building requires careful design. In temperate latitudes the solar resource varies greatly between winter and summer. In the winter the solar resource is limited but heating of the building is then required, leading to a need for large windows. In the summer there is a limited need for heating, but the large windows can cause overheating. A number of approaches are used to allow the use of large windows for adequate solar heating in winter while at the same time controlling overheating in summer. These measures include the use of external shutters or manually operated or automatic internal blinds to reduce the

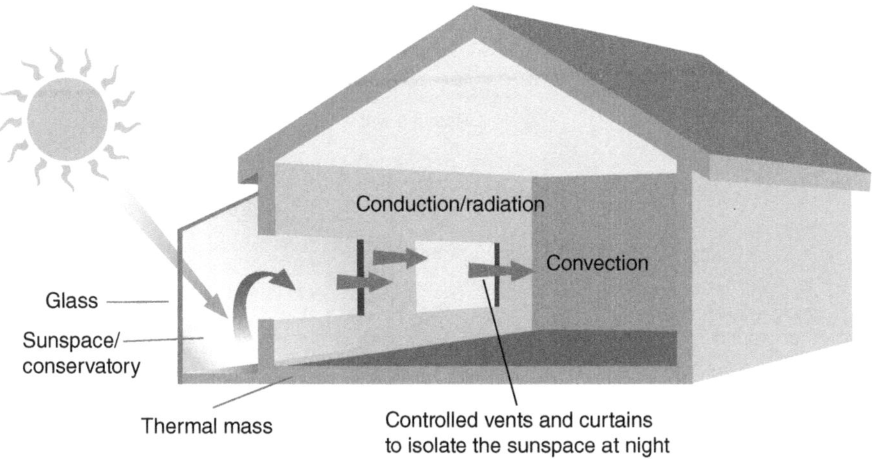

Figure 6.4 Isolated gain solar system using a sunspace (e.g. a conservatory).

irradiance entering the building in summer, and overhangs positioned above the windows, which allow irradiance to enter the building in winter when the sun is low but block the irradiance in summer when the sun is high in the sky.

An alternative approach is to use an isolated sunspace that captures the solar energy (Figure 6.4). A typical example is a conservatory attached to a domestic house. The sunspace captures the solar irradiance, and heat is passed into the building either as warm air (through convection) or through the wall between the sunspace and dwelling (through conduction and low-temperature radiation). Again, a building with large thermal mass is needed to maintain a consistent level of comfort. If convection is the main heat transfer mechanism from the sunspace, then fans may be used to control the flow of air into the building. At night or on cold days the sunspace is isolated from the dwelling to control heat loss.

Conservatories have been added to many houses in the UK. Although they can provide significant passive solar gain, care is required to avoid them overheating in summer. Also, the energy use of a dwelling can be greatly increased if the conservatory is used as an additional room in winter or at night and then requires heating. In cold weather or at night the heat loss through the large glazed surfaces of a conservatory can be very high.

A Trombe wall is a dark wall of large thermal mass located inside the building and separated from a south-facing glazed surface by a small airspace (Figure 6.5). The temperature of the large thermal mass of the wall is raised by the solar irradiance and its heat transferred to the building either by air circulation (convection) through vents in the wall or by conduction and low-temperature radiation. Air circulates in the space between the glazing and the wall by natural convection. At night or on cold days the airspace is isolated by closing the shutters, and there may be active control of convection with fans. A number of innovative low-energy houses have been constructed with Trombe walls, but the large internal wall cannot easily be fitted into a building after it is built, so it is only possible to use this technique

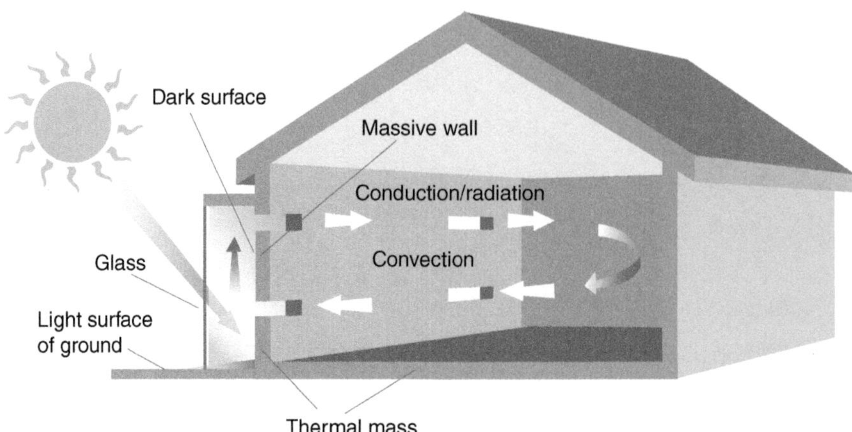

Figure 6.5 Trombe wall (an example of indirect solar gain).

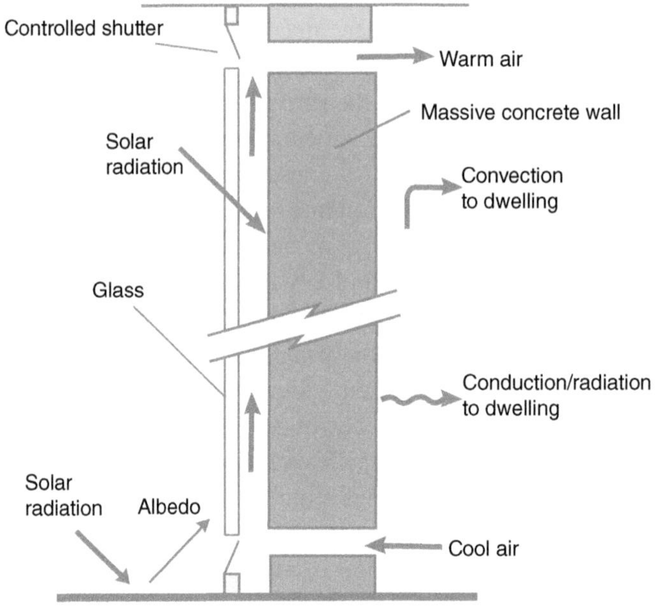

Figure 6.6 Cross-section of a Trombe wall.

when a new house is being built. Figure 6.6 shows a cross-section of a Trombe wall and the main mechanisms of heat transfer.

All the approaches shown in Figures 6.3–6.5 make use of both direct (beam) irradiance from the sun and diffuse radiation scattered by the atmosphere. The irradiance entering the building can be increased by installing a hard, light surface on the ground of the south aspect. This increases the reflected irradiance, which is known as albedo.

Commercial buildings sometimes use an unglazed solar collector to pre-heat air. An unglazed solar collector of perforated metal is added to the roof or to a south-facing vertical surface of the building. Air is taken into the collector through the perforations and then the heated air passed into the building. A large unglazed area is needed; this approach, using what is described as a transpired solar collector, is used mainly on commercial buildings such as warehouses.

6.2.1 Solar Gain from Glazing

The performance of glass plays an essential role in passive solar thermal heating. High-frequency, short-wavelength radiation is emitted by the sun at a surface temperature of ~6000 K and is transmitted through the windows and absorbed on dark surfaces. These surfaces then warm to a much lower temperature, ~300 K; the resulting longer-wavelength radiation is blocked by the glass, and the energy remains within the building.

Solar gain through a window is described by:

$$Q_{solar} = S \times G \times A \quad [\mathrm{W}] \tag{6.1}$$

where

Q_{solar} is the heat gain in W,
S is the solar gain factor (0–1),
G is the irradiance in W/m^2,
A is the area in m^2.

The solar gain factor (between 0 and 1) describes the overall performance of a window. This includes direct transmission of high-frequency radiation through the glass, reflection and retransmission of energy absorbed by the glass, as well as the effect of the frame. It is known as the g-value in Europe and the solar heat gain coefficient in the USA.

Example 6.1 – Heat Gain Through a Window

Estimate the heat gain from solar irradiance of 500 W/m^2 striking a large window of 3 m × 2 m at an angle of 30°. The window has a solar gain coefficient (g-value) of 0.55.

Solution

The heat gain is given by:

$$\begin{aligned} Q_{solar} &= S \times G_{normal} \times A \\ &= 0.55 \times 500 \times \cos 30° \times 6 \\ &= 1429 \text{ W} \end{aligned}$$

In addition to solar gain from radiation through glazed surfaces, the opaque surfaces of walls will also absorb some radiation and transmit heat into the building though conduction. South-facing dark walls may absorb a considerable amount of solar energy in bright sunlight.

6.3 Circuit Representation of Low-Temperature Heat Transfer

Heat transfer processes are complex and are usually non-linear with respect to temperature, i.e. the heat power exchanged per unit of temperature difference is not constant. However, for many low-temperature solar thermal processes (typically in buildings that experience a temperature range of −20 °C to +80 °C) it is usual to assume that heat transfer is directly proportional to temperature difference, and so simple linear equations may be used to represent heat losses and gains. This allows a simple representation of heat flows similar to that of electrical current flows in dc electrical circuits. Practical heat transfer often relies on a combination of conduction, convection and radiation but for simple calculations each process can be considered separately, and the individual effects added.

As discussed in Tutorial II, thermal resistance can be defined for elements of a building under particular conditions of conduction, convection and radiation. Once the thermal resistances for the elements under consideration have been defined, the heat flow model is simply

$$Q = \frac{\Delta T}{R_\phi} = \frac{T_{higher} - T_{lower}}{R_\phi} \quad [\mathrm{W}] \tag{6.2}$$

where

Q is the heat flow in W,
ΔT is the temperature difference in K,
T_{higher} is the higher temperature in K,
T_{lower} is the lower temperature in K,
R_ϕ is the thermal resistance of the element in K/W.

The analogy becomes clear by considering Ohm's law for the current flow through a resistive circuit driven by a voltage difference, as shown in Equation 6.3.

$$I = \frac{\Delta V}{R} = \frac{V_{higher} - V_{lower}}{R} \quad [\mathrm{A}] \tag{6.3}$$

where

I is the current flow in A,
ΔV is the voltage difference in V,
V_{higher} is the higher voltage in V,
V_{lower} is the lower voltage in V,
R is the electrical resistance in Ω.

Comparing Equations 6.2 and 6.3 shows how a simple heat circuit model may be constructed. The thermal resistances can be combined in series and parallel to reflect the resistance of elements to heat flows in the same way that resistors are combined in simple electrical circuits.

This simple model of heat flow uses a set of linear heat transfer coefficients to model the flow of heat by conduction, convection or radiation. These coefficients are constant over only a limited range of temperature and other conditions, such as

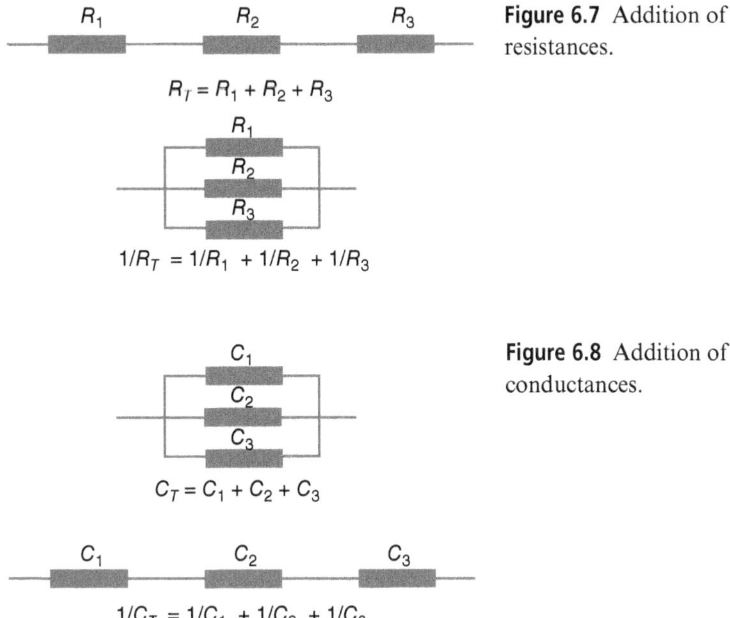

Figure 6.7 Addition of resistances.

Figure 6.8 Addition of conductances.

air flow for convection. Hence they are valid for a particular range of conditions, which may be stated or only implied by the purpose of the calculation.

As the elements of a building have heat flows that are in parallel (e.g. through the floor, roof, windows, etc.) it is often more convenient to base the calculation of the performance of a building on thermal conductance C_ϕ, which is the reciprocal of thermal resistance R_ϕ:

$$C_\phi = \frac{1}{R_\phi}$$

Figures 6.7 and 6.8 show how resistances and conductances are added in series and parallel. Figure 6.7 shows resistances being added. If the combined thermal resistance through a wall is to be calculated, the thermal resistances of the various cross-sections through the wall are added in series. For example, the conductive resistance of the bulk material of the wall is added to the internal and external combined convective/radiative resistances in series to give the overall thermal resistance through the wall. Alternatively, if the thermal resistance of an entire building is to be calculated then the thermal resistances of the various elements are added in parallel. For example, the thermal resistances of the wall, roof and windows are added in parallel using the reciprocals of the resistances of the individual components.

Figure 6.8 shows the use of conductances. Addition in parallel, e.g. of the thermal conductances of a wall, roof and windows, is by simple addition, while the addition of the conductances of elements in series, e.g. the conductances of various layers of materials through a wall, requires the use of the reciprocals of the conductances.

The analogy of these simple thermal circuits with electrical circuits is described in Table 6.1. At a node, the total heat flow is zero, i.e. the heat flow out of a node is balanced by the heat flow into it. The heat flow through a branch is determined

Table 6.1 Electrical and thermal circuits

	Electrical circuit	Thermal circuit
Flows at a node	$\sum I_n = 0$	$\sum Q_n = 0$
Flows in a branch	$I = V/R$	$Q = T/R_\phi$ $Q = T \times C_\phi$
Potential source	Constant voltage	Constant temperature
Flow source	Constant current	Constant heat flux (e.g. solar irradiance)

by the temperature difference across the branch and the thermal resistance (or conductance). Temperatures that drive the network (e.g. internal or external temperatures) are potential sources analogous to voltages. Irradiance falling on a surface is a flow source, analogous to a current source of an electrical circuit.

The composite heat transfer coefficient through a complete component of a building, e.g. a wall or a window, is known as the U-value or thermal transmittance. The thermal transmittance is the thermal conductance of a unit area of the component with units of W/m^{2}K. U-values are widely used in the calculation of the thermal performance of buildings. The lower the U-value the less heat is transferred for a temperature difference.

6.4 Heat Loss of Buildings Due to Ventilation

In addition to the heat loss through the fabric of a building, heat is lost due to air changes either intentionally through ventilation or unintentionally through air leakage. The heat transfer due to air changes is given by

$$Q_V - (T_{\text{int}} - T_{amb})\,\dot{m}c_p \quad [\text{W}] \tag{6.4}$$

where

$$\dot{m} = \rho n V$$

and

Q_V is the heat loss due to air changes in W,
T_{int} is the internal temperature in K,
T_{amb} is the ambient temperature in K,
$\dot{m}$ is the mass flow rate in kg/s,
ρ is the density of air, 1.25 kg/m^3,[1]
c_ρ is the specific heat of air, 1.005×10^3 J/kgK,
n is the air changes per second in s^{-1},
V is the volume in m^3.

[1] This book uses a density of air of 1.25 kg/m^3 for calculations of the heat loss in buildings and 1.225 kg/m^3 for wind turbine calculations. The difference is due to the assumed temperature of the air.

Equation 6.4 may be rewritten approximately as

$$Q_V = (T_{int} - T_{amb})0.33NV \quad [\mathrm{W}]$$

where

N is the air changes per hour in h^{-1}.

Example 6.2 – Estimation of the Heat Loss from a Small Building

Estimate the heat power lost from a small rectangular building of dimensions

length 6 m,
breadth 5 m,
height 3 m,

with a temperature difference of 10 °C between inside and outside. Include the heat loss from ventilation assuming two air changes per hour.

Solution

Building element	Area (m^2)	Assumed U-value of building component (W/m^2K)	Heat loss through fabric (W) Area × U-value × ΔT
Roof	30	0.25	75
Floor	30	0.45	135
Walls	56	0.45	252
Windows (double-glazed)	10	3	300
Windows (single-glazed)	10	6	600

The heat loss through the fabric with a 10°C temperature difference and double-glazed windows is 762 W, which rises to 1062 W if the windows are single-glazed.

The mass flow rate of the exchange of air with the atmosphere is

$$\dot{m} = \rho n V = 1.25 \times \frac{2}{3600} \times 90 = 0.0625 \text{ kg/s}$$

where

ρ is the density of air, 1.25 kg/m^3,
n is the air changes per second in s^{-1},
V is the volume of air in m^3.

The heat loss of this exchange of air is

$$Q_V = \dot{m} c_p \Delta T$$
$$= 0.0625 \times 1.005 \times 10^3 \times 10 = 628 \text{ W}$$

where c_p is the specific heat of air, 1.005×10^3 J/kgK.

Thus the total heat loss if double-glazed windows are fitted is 1390 W.

6.5 Degree-Days

The heat transfer across the fabric of a building and through the exchange of air depends on the outdoor ambient temperature, and this variation in the requirement for heating or cooling energy with outdoor temperature makes it difficult to predict the energy consumption of a building or assess the effect of improvements in insulation as the external temperature changes in different seasons and years. For example, a reduction in the energy used to heat a building could be the result of additional insulation being installed, changes in the behaviour of the occupants or merely because that year the weather was unusually warm.

The concept of degree-days is used to allow the performance of building heating and cooling systems to be evaluated in a simple manner that excludes the effects of outside temperature. It is based on a similar approach that was first used in agriculture to eliminate the effect of year-on-year changes in seasonal temperatures when evaluating variations in crop growth.

Degree-days can be used in two main ways:

- to estimate the future energy consumption for heating or cooling a new or refurbished building,
- to monitor the operation of a building heating or cooling system.

The degree-day is calculated as the difference, averaged over a 24-hour period, between the outside temperature and a reference or base temperature. The base temperature is the outside ambient temperature at which neither heating nor cooling of the building is required (the balance point temperature). In the UK, a base temperature of 15.5 °C is commonly used, although a higher base temperature may be used for hospitals (18.5 °C) or a lower base temperature for well-insulated buildings with considerable solar gain. The base temperature of 15.5 °C was chosen many years ago and assumes a comfortable indoor temperature of 65 °F (18.3 °C) within buildings; it continues to be widely used, although many buildings are now held at a higher indoor temperature, e.g. 20 °C.

There are several ways in which degree-days may be calculated. If data are available, 24 degree-hours can be calculated from hourly mean temperatures for each day and the average taken. Alternative techniques to calculate degree-days, e.g. based on maximum and minimum daily temperatures, are used if hourly temperature data are not available. Degree-days are usually evaluated over a month and expressed as an integer number. Heating degree-days are calculated from deviations of the outside temperature only below the base temperature, while cooling degree-days are calculated from variations only above a different base temperature. For practical purposes, tables of degree-days are made available by national meteorological services and published online.

An example of monthly heating degree-day data for England is shown in Figure 6.9. In this example, the heating requirement peaks in January (as is usual) and the spring for the year of these data appears to have been rather mild. In the summer months (May–October) heating is not required.

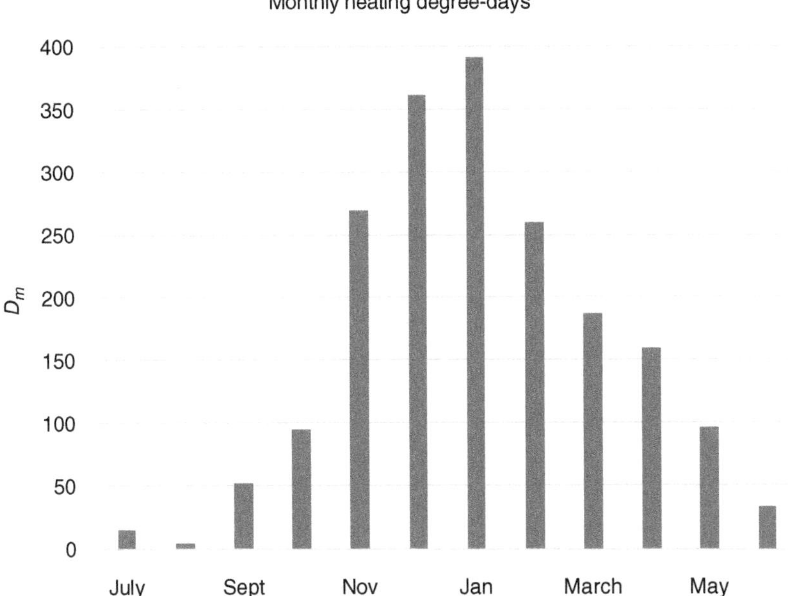

Figure 6.9 Example of monthly heating degree-day (D_m) from a location in England.

6.5.1 Monitoring the Thermal Performance of Buildings using Degree-Days

Figure 6.10 shows a simple representation of the thermal performance of a building, with all the heat losses from the building represented by a thermal resistance. Assume $T_{int} > T_{amb}$ and that heating is required. There are two major components to the heat loss of the building. Heat is lost through the fabric and by air changes.

The rate of heat transfer through the fabric, Q_F, is given by:

$$Q_F = UA(T_{int} - T_{amb}) \quad [\text{W}]$$

where

U is the U-value of the entire fabric in W/m^{2}K,
A is the area of the entire fabric in m^2.
The rate of heat transfer due to air changes, Q_V, is given by

$$Q_V = \dot{m}c_p(T_{int} - T_{amb}) \quad [\text{W}]$$

where

$\dot{m}$ is the mass flow rate of air in kg/s,
c_p is the specific heat of air, 1.005 kJ/kgK.

The mass flow rate is

$$\dot{m} = \rho n V \quad [\text{kg/s}]$$

where

ρ is the density of air, 1.25 kg/m^3,

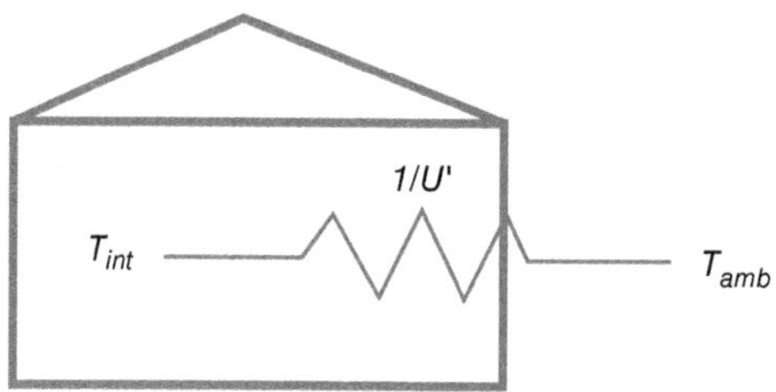

Figure 6.10 Simple representation of the heat loss of a building.

n is the number of air changes per second in s^{-1},
V is volume in m^3.

Defining an overall heat loss coefficient of the building as U':

$$U' = UA + \rho n V c_p \quad [\text{W/K}]$$

The heat transfer across the fabric and by air exchange is then

$$Q_F + Q_V = (UA + \rho n V c_p)(T_{int} - T_{amb}) = U'(T_{int} - T_{amb}) \quad [\text{W}]$$

Within the building, there is a useful internal heat gain from the occupants, solar gain and heat generated by internal equipment, Q_G.

The heat balance of the building, Q_T, is then

$$Q_T = Q_F + Q_V - Q_G = U'(T_{int} - T_{amb}) - Q_G \quad [\text{W}]$$

The energy demand (E) over a period is

$$E = \int Q_T \, dt$$

$$E = \int \{U'(T_{int} - T_{amb}) - Q_G\} dt \quad [\text{J}]$$

or

$$E = U' \int \left\{ (T_{int} - T_{amb}) - \frac{Q_G}{U'} \right\} dt \quad [\text{J}]$$

The term $\frac{Q_G}{U'}$ has the units of temperature and can be considered the internal temperature rise due to internal heat gain.

Thus, defining the base temperature, T_{base}, as

$$T_{base} = T_{int} - \frac{Q_G}{U'}$$

allows us to write

$$E = U' \int (T_{base} - T_{amb}) dt \quad [\text{J}] \tag{6.5}$$

This is a linear relationship between the ambient (outside) temperature and energy demand.

The average over the 24 hours of the difference between the base and outside ambient hourly mean temperatures gives the daily degree-days (D_d) as

$$D_d = \frac{1}{24} \sum_{h=1}^{24} (T_{base} - T_{amb}(h)) \quad [\text{K.day}]$$

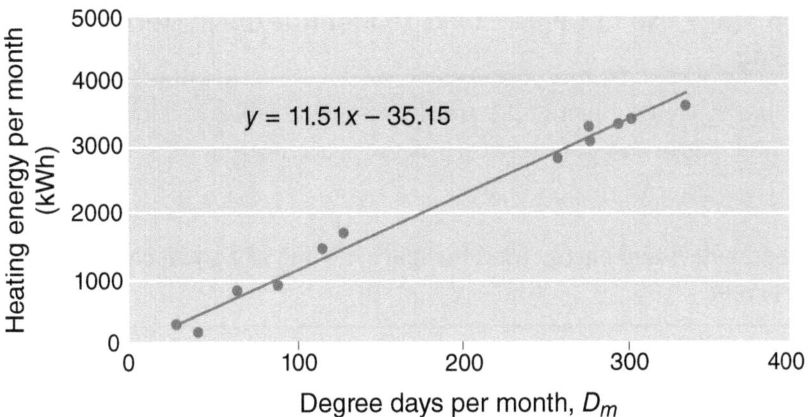

Figure 6.11 Example of a performance line of a small building.

and the estimated fuel consumption required to heat the building over one day (F) is

$$F = \frac{U'D_d}{\eta} \times \frac{24}{1000} \quad [\text{kWh}]$$

where

η is the system heating efficiency.

If monthly degree-days (D_m) are used, then the calculation provides an estimate of the fuel consumption to heat a building over a month.

An important use of degree-days is to check the performance of the heating system of a building. A performance line is created by plotting the number of heating degree-days in a month, D_m, against the energy consumed in that month. The number of monthly heating degree-days for a location is obtained from the national meteorological service and plotted against monthly metered energy use. For accurate results the readings of the energy meter are taken at the start and end of each calendar month as the calculation of the degree-days is based on calendar months. An example of a performance line for a small industrial building is shown in Figure 6.11. The 12 monthly data points are plotted, and a regression line drawn.

As predicted by Equation 6.5, the relationship between the degree-days and energy demand in a month is linear. Malfunctioning heating equipment or changes in the behaviour of the occupants can be detected from any deviations away from the regression line.

It may be useful to sum the deviations from the regression line with a technique known as the CUmulative SUM of the differences (CUSUM). The cumulative sum of the differences between the metered energy and that predicted from the regression line indicates clearly over time if the heating system is operating in a changed manner.

Example 6.3(a) – Use of Degree-Days to Monitor the Performance of a Building

The data used to create Figure 6.11 are shown in the table. It is the heating energy consumed and degree-days of last year for the space heating of a small commercial building.

Monthly degree-days and energy used for space heating of a small commercial building last year

Month	Jan	Feb	Mar	April	May	June	July	Aug	Sept	Oct	Nov	Dec
Degree-days	301	277	257	277	115	88	27	40	64	128	295	336
Energy consumed (kWh)	3420	3150	2860	3310	1450	880	280	124	770	1700	3340	3670

A behaviour change campaign was initiated at the start of the current year to encourage the occupants of the building to reduce their energy use. The changes in behaviour that were encouraged included better use of the controls of the heating system and ensuring doors and windows were not left open. In a cold month of October of 180 degree-days, the measured energy consumption was 1750 kWh. Was the behaviour change campaign successful?

Solution

Using the performance line of Figure 6.11, we have

$$y = 11.51x - 35.15$$

Without the behaviour change campaign, the energy that would be expected to be consumed this October $= 11.51 \times 180 - 35.15 = 2035\,\text{kWh}$.

Therefore, the energy awareness campaign has resulted in a reduction in energy use in this month of

$$\Delta E = \frac{1750 - 2035}{2035} = -14\%$$

A 14% reduction in energy use shows that the behaviour change campaign has been initially very successful, but monitoring over several years would be necessary to be sure the reduction in energy use is sustained.

Example 6.3(b) – Use of Degree-Days to Predict Building Energy Consumption

The building is modelled as having a surface area of 600 m^2, with a composite U-value of 0.45 W/m^2K. It has a volume of 900 m^3 and 0.7 air changes per hour. How useful is this model? What would be the effect of improving the U-value of the fabric to 0.35 W/m^2K and reducing the number of air changes per hour to 0.65 per hour?

Solution

The energy use over the same month of October is predicted by the model from

$$E = \left(UA + \rho NVc_p\right) \times D_m \times \frac{24}{1000} \quad [\text{kWh}]$$

where in this case

U is 0.45 W/m^{2}K,
A is 600 m^2,
ρ is 1.25 kg/m^3,
N is 0.7 h^{-1},
V is 900 m^3,
c_p is 1005 J/kgK,
D_m is the monthly degree-days.

Therefore we have

$$E = \left(0.45 \times 600 + 1.25 \times \frac{0.7}{3600} \times 900 \times 1005\right) \times 180 \times \frac{24}{1000}$$
$$= 2116\ \text{kWh}$$

This shows a difference of the modelled energy consumption from that predicted by the performance line of

$$\Delta E = \frac{2116 - 2035}{2035} = 4\%$$

With this small difference, the model is useful.

The model indicates that the effect of changing the composite U-value to 0.35 W/m^{2}K and the air change rate to 0.65 air changes per hour (a very demanding specification) would be a reduced fuel use of

$$E = \left(0.35 \times 600 + 1.25 \times \frac{0.65}{3600} \times 900 \times 1005\right) \times 180 \times \frac{24}{1000}$$
$$= 1789\ \text{kWh}$$

The modelled effect of these changes in the building fabric and ventilation would be a reduction in fuel use of

$$\Delta E = \frac{1789 - 2116}{2116} = -15.5\%$$

Either improving the building fabric and reducing ventilation, or a successful behaviour change campaign, are effective ways of reducing energy consumption. Physically improving the performance of the building requires an investment of capital and so the behaviour change campaign would be a more cost-effective approach if the reduction in energy use could be sustained. However, sustained energy reduction from behaviour change is likely to require continuous interventions, such as regular publicity campaigns and competitions among staff, with an associated staff cost.

6.6 Radiation and the Behaviour of Glass

When radiation strikes the surface of a transparent material, part will be reflected, part transmitted and part absorbed; see Figure 6.12. For a perfectly transparent material the reflectivity and absorptivity are zero and the transmissivity unity. Note that in the heat transfer literature the terms reflectivity/reflectance, absorptivity/absorptance and transmissivity/transmittance are used interchangeably.

These coefficients sum to unity:

$$\rho + \alpha + \tau = 1$$

where

ρ is the reflectivity or reflectance,
α is the absorptivity or absorptance,
τ is the transmissivity or transmittance.

The values of these coefficients, which are all between 0 and 1, are not constant but depend on the wavelength (frequency) of the radiation. Glass has the important property of allowing transmission of short-wavelength radiation from the high-temperature source of the sun (value of $\tau \sim 0.8$–0.9) while blocking the longer wavelength emissions from lower-temperature sources on earth (value of $\tau \sim 0.02$).

Figure 6.13 shows how the transmissivity of glass varies with wavelength. In the wavelengths of the solar spectrum (0.3–3 μm) the transmissivity is around 0.8–0.9 (80–90%) but this drops to around 0.02 (2%) at wavelengths of 4–10 μm, the wavelength of radiation from objects on the earth at a temperature of 300–400 K.

Table 6.2 Transmissivity of glass

Temperature of source (K)	6000 (surface of the sun)	300–400 (warm object, 25–125 °C)
Wavelength of radiation (μm)	< 3	4–10
Transmissivity of glass	0.8–0.9	0.02

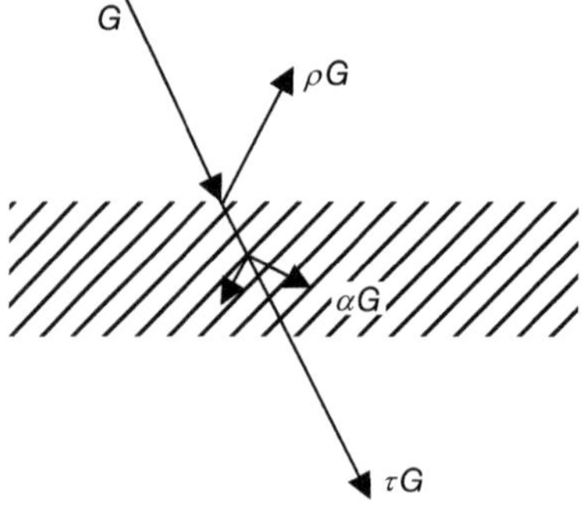

Figure 6.12 Radiation striking a transparent surface.

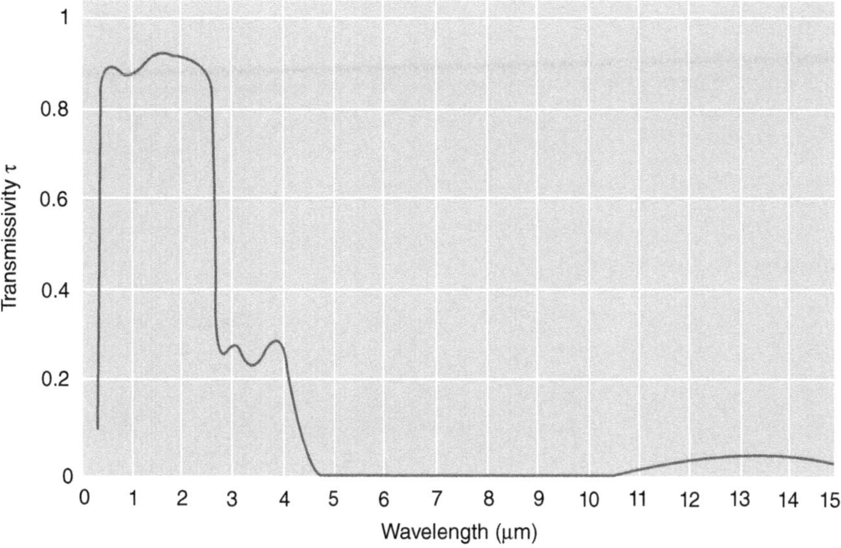

Figure 6.13 Transmissivity of glass.

Table 6.2 gives values of the transmissivity of a sheet of window glass, 3–4 mm thick. Two values are given corresponding to the transmissivity of short-wavelength (high-frequency) radiation from the sun at a temperature of 6000 K and long-wavelength (low-frequency) radiation in the infrared region from objects at 300–400 K.

The change in the transmissivity of glass with wavelength, shown in Figure 6.13 and Table 6.2, is the basis on which windows trap the heat from the sun in a room and how solar thermal heating of buildings works.

6.7 Solar Water Heating

After space heating, the supply of hot water consumes the greatest proportion of heat energy used in buildings in the UK. In warmer countries, where less space heating is required, up to 25% of the energy used in homes may be for water heating. Solar water heating systems using either flat-plate solar collectors or evacuated tubes are offered by a number of suppliers but the limited solar resource in winter, a relatively low price of gas and the high efficiency of condensing gas boilers means the financial case for using solar water heating systems in the UK is only marginal. Hence, although solar water heating is used in some houses in the UK, it is not widespread. This is in contrast to some other countries with a much better solar resource, e.g. Greece, Spain, Japan and Israel, where solar hot water systems are commonly used and required by regulation to be installed on new buildings.

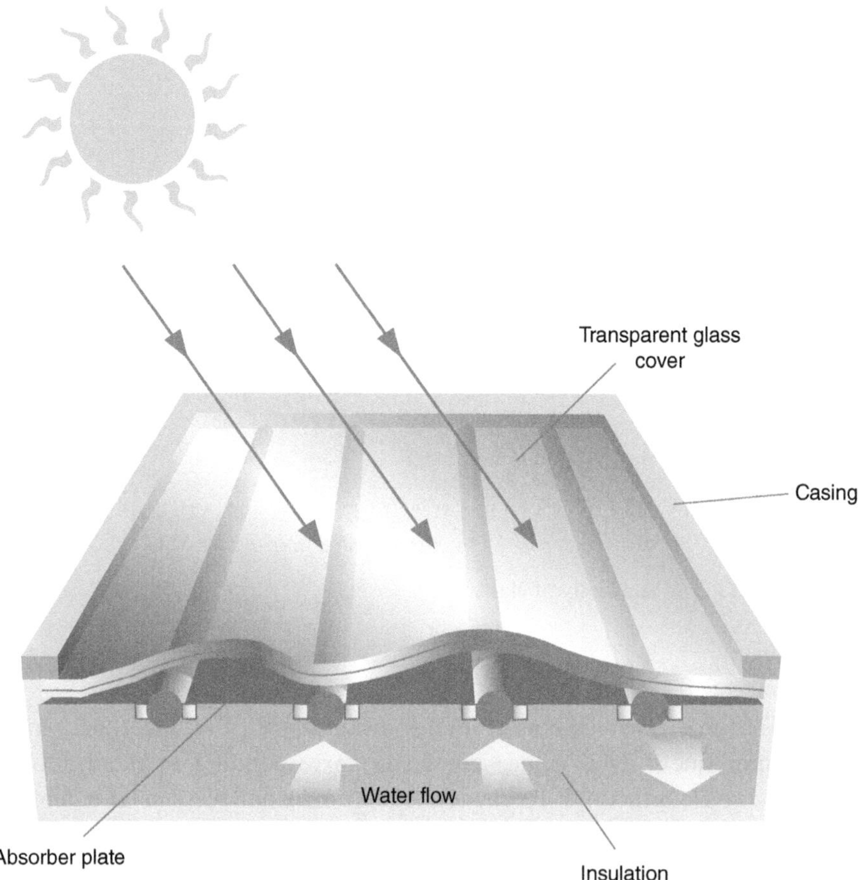

Figure 6.14 Simple flat-plate solar collector.
Source: Godfrey Boyle, *Renewable Energy: Power for a Sustainable Future*, 1996.
By permission of Oxford University Press, USA.

There are a number of designs of solar hot water systems. The collector itself can be a type that is:

- unglazed flat plate,
- glazed flat plate,
- evacuated tube.

An unglazed-flat-plate collector can provide an increase in water temperature of only up to 10 °C above ambient and is generally used to heat outdoor swimming pools. Its performance is very sensitive to wind speeds that cause large convection losses. Glazed flat-plate collectors use one or more sheets of glass to increase the water temperature up to 50 °C above ambient. Temperatures of more than 50 °C above ambient can be obtained with evacuated tube collectors.

Figure 6.14 shows a glazed-flat-plate solar collector which would typically be mounted on the pitched roof of a dwelling facing approximately south (in the northern hemisphere) at the angle of latitude. It has a transparent cover (which

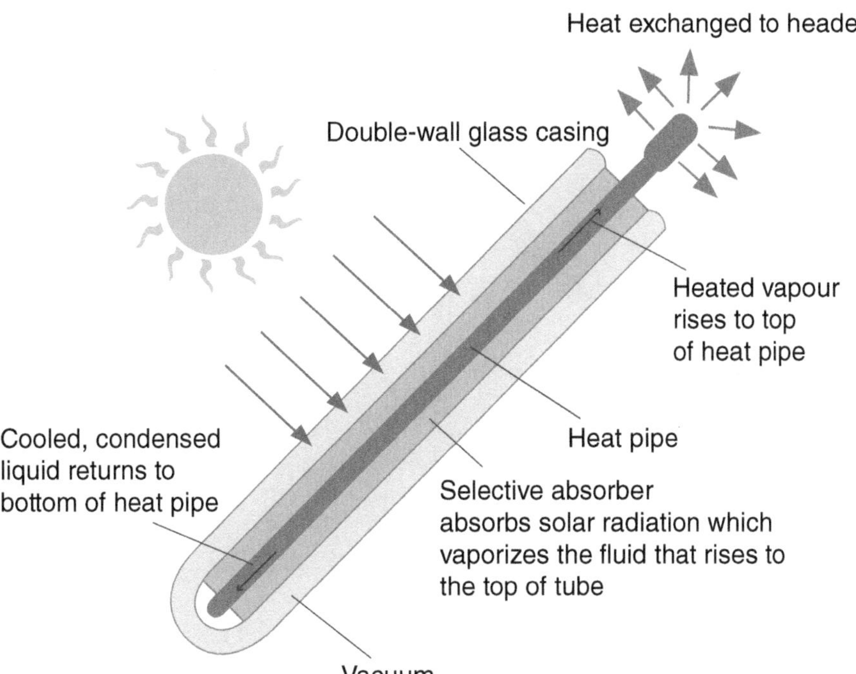

Figure 6.15 Evacuated-tube solar collector, with heat pipe.

may consist of multiple sheets of glass), an absorber plate to which water pipes are bonded, back insulation and a casing. Solar irradiance warms the absorber plate, and heat is extracted by the flow of water in the pipes.

A flat-plate solar collector loses heat from the absorber plate due to convection within the collector and radiation to the sky, although these losses can be reduced by using multiple sheets of glass in the transparent cover. Also, the transparent cover reflects some of the irradiance, particularly that which strikes at an oblique angle. These limitations can be overcome by an evacuated-tube solar collector (see Figure 6.15).

In an evacuated-tube solar collector, a central metallic absorber is heated by the solar irradiance. This heat is transferred from the absorber to a heat pipe by conduction and radiation. The heat pipe contains a fluid with a low boiling point (such as methanol). The vapour within the warmed heat pipe rises upwards, where it is condensed, and heat is transferred to the header. From the header, the heat is transported to the hot water system of the building. The absorber and heat pipe are surrounded by a double-walled glass tube enclosing a vacuum to reduce heat loss from the absorber. Evacuated-tube solar collectors are more efficient than flat-plate collectors and can provide higher-temperature water. They are less sensitive to the angle of incidence of the irradiance. Some alternative designs replace the heat pipe with a fluid circuit through the evacuated tube.

In the UK, solar thermal water heating systems commonly use an indirect pumped water system (see Figure 6.16). A primary circuit transfers heat from the

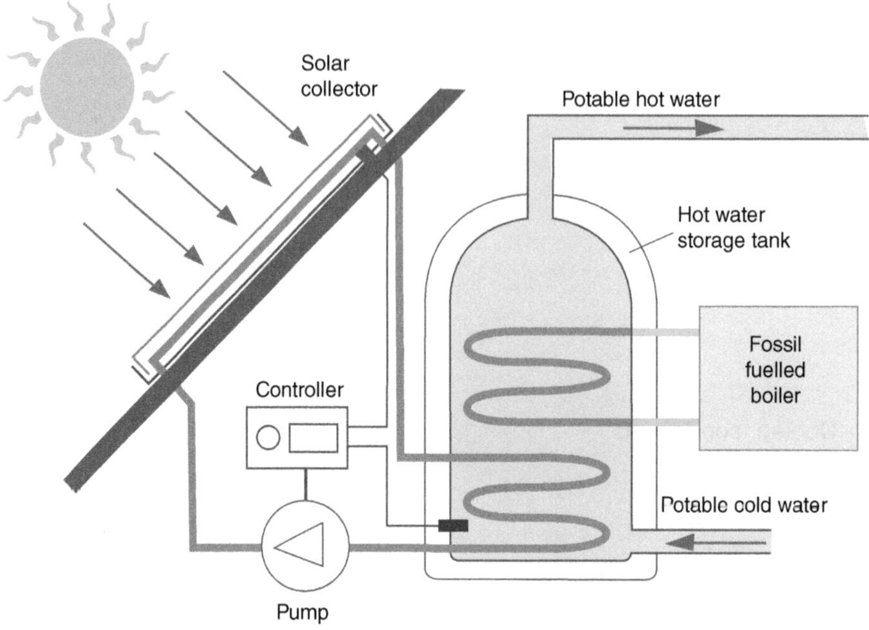

Figure 6.16 Indirect pumped solar water heating system.
Source: Volker Quaschning, *Understanding Renewable Energy Systems*, 2005, Earthscan.

solar collector (either flat-plate or evacuated-tube type) to the hot water storage tank. The fluid in the primary circuit is circulated by a pump that is controlled by measuring the temperature of the outlet of the solar panel and comparing it to the temperature of the storage tank. The pump is operated only when there is an adequate temperature difference so that heat is supplied from the collector to the tank. In some designs the pump is powered from a small photovoltaic panel and the entire installation can be operated without mains electricity. The fluid in the primary circuit can be treated with a non-toxic anti-freeze to ensure that it does not freeze at night or in cold weather. An alternative approach to avoiding damage due to freezing of the primary circuit fluid is the 'drain-back' system that drains the primary circuit when the pump is not operating.

In temperate latitudes, only a part of the annual hot water requirement is supplied by solar thermal energy. Supplementary heating is supplied by either a gas boiler or an electric immersion heater to ensure hot water is available during times of low solar irradiance. With an indirect pumped system, the hot water storage tank can be located anywhere but care is necessary to insulate well any long pipes. Domestic hot water is usually stored at ~65 °C to ensure Salmonella bacteria are destroyed.

In some countries with a very good solar resource and no risk of freezing, a direct system is used where potable water is circulated in the solar collector. This simplifies the hydraulic circuit. A further simplification can be made by using a thermo-syphon to circulate the water and so avoid the use of a pump, but this requires the storage tank to be located higher than the solar collector.

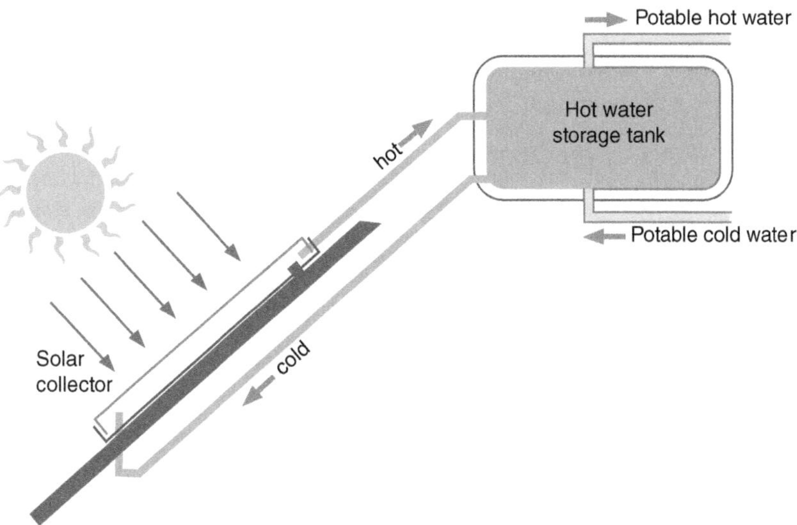

Figure 6.17 Thermo-syphon solar water collector.
Source: Volker Quaschning, *Understanding Renewable Energy Systems*, 2005, Earthscan.

Figure 6.18 Roof-mounted domestic hot water heater using a thermo-syphon flat-plate solar collector.
Photo: Roy Pedersen/ Shutterstock.com.
(A black and white version of this figure will appear in some formats. For the colour version, refer to the plate section.)

Figure 6.17 shows the principle of operation and Figure 6.18 an example of a thermo-syphon flat-plate solar water heating system. Note the tank located above the solar collector and the rather shallow tilt angle of the roof.

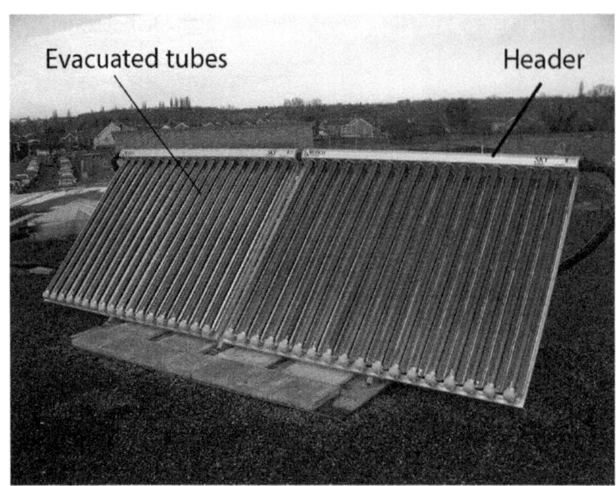

Figure 6.19 Evacuated-tube solar thermal collector on the roof of a school in England. Photo: Rhico. (Annotated by the author.)
(A black and white version of this figure will appear in some formats. For the colour version, refer to the plate section.)

Figure 6.19 shows an example of an evacuated-tube solar thermal collector on the roof of a school in England supplying a pumped hot water system. Note the steeper tilt angle in a latitude of around 50°.

6.8 Performance of a Flat-Plate Solar Collector

The performance of a flat-plate solar collector can be described simply by considering three energy flows:

(1) the solar energy captured by the absorber plate,
(2) the useful heat energy extracted by the water flow,
(3) the losses to the atmosphere.

The irradiance of the sun passes through the transparent cover and strikes the absorber plate. The solar power absorbed by the absorber plate, Q_T, is given by

$$Q_T = AG(\tau\alpha) \quad [\mathrm{W}]$$

where

A is the area of absorber plate in m^2,
G is the irradiance from the sun in W/m^2,
τ is the transmissivity of the cover (dimensionless coefficient, 0–1),
α is the absorptivity of the plate (dimensionless coefficient, 0–1).

It is usual to write the transmissivity and absorptivity as a combined dimensionless coefficient $(\tau\alpha)$. The transmissivity of the glass drops when the irradiance strikes the cover at an angle of greater than ~60° to the normal.

The useful output heat power of the solar collector is the rate of heat energy extracted by the water flow, Q_W:

$$Q_W = \dot{m}c_p\left(T_{out} - T_{in}\right) \quad [\mathrm{W}]$$

where

$\dot{m}$ is the mass flow rate of water in kg/s,
c_p is the specific heat of water in J/kgK,
T_{in} is the inlet water temperature in K,
T_{out} is the outlet water temperature in K.

The heat loss from the collector plate Q_L is from convection and radiation to the glass cover, and from the glass cover to the atmosphere. There are also heat losses from the back of the absorber plate. All these losses are expressed as

$$Q_L = AU\left(T_{mean} - T_{amb}\right) \quad [\mathrm{W}]$$

where

U is the heat loss coefficient in W/m^{2}K; U represents convection, conduction and radiative losses from the absorber plate,
T_{mean} is the mean temperature of the absorber plate; this is a rather difficult parameter to measure although it can be approximated as the mean of the inlet and outlet water temperatures,
T_{amb} is the ambient air temperature.

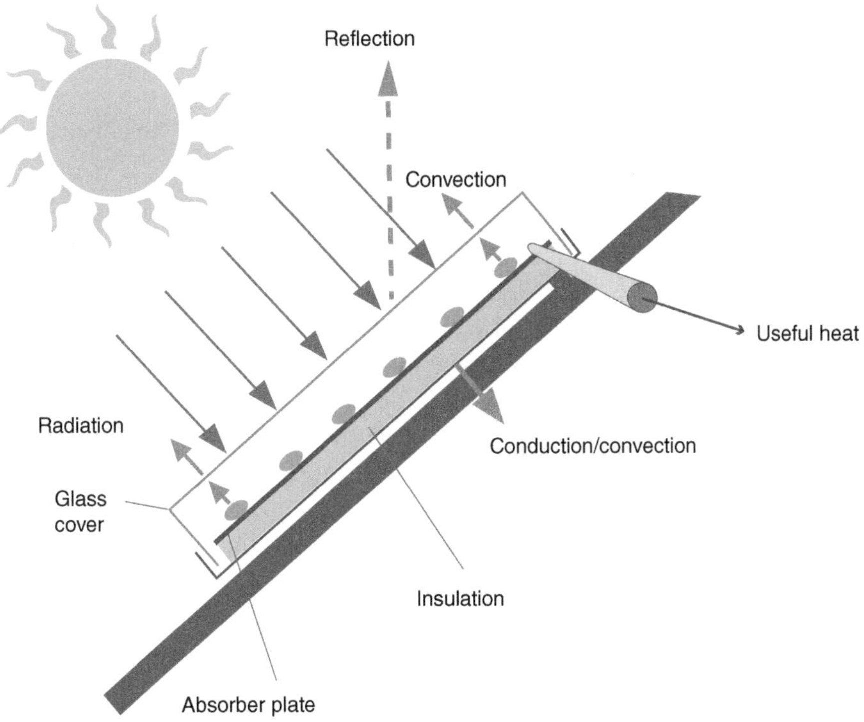

Figure 6.20 Heat flows of a flat-plate solar collector.
Source: Volker Quaschning, *Understanding Renewable Energy Systems*, 2005, Earthscan.

Combining these equations gives the heat extracted by the water, Q_W:

$$Q_W = Q_T - Q_L \quad [W]$$

$$Q_W = \dot{m}c_p\left(T_{out} - T_{in}\right) = A\left(G(\tau\alpha) - U\left(T_{mean} - T_{amb}\right)\right) \quad [\mathrm{W}]$$

The efficiency of the solar collector is the useful heat that is extracted by the water divided by the power of the incident irradiance

$$\eta = \frac{Q_W}{AG} = (\tau\alpha) - \frac{U}{G}\left(T_{mean} - T_{amb}\right) \tag{6.6}$$

Equation 6.6 is of limited practical use as the mean temperature of the absorber plate, T_{mean}, is difficult to determine. Therefore it is common to define a term, F_R, which allows the use of the inlet water temperature. F_R describes the effectiveness of the heat transfer from the collector plate to the water and is the ratio of the useful heat gained by the collector, Q_W, to the possible heat that would be gained if all the collector surface was at the temperature of the inlet water.

$$F_R = \frac{\dot{m}c_p\left(T_{out} - T_{in}\right)}{A\left[G(\tau\alpha) - U\left(T_{in} - T_{amb}\right)\right]}$$

When using the inlet water temperature, this results in:

$$\eta = F_R\left[(\tau\alpha) - \frac{U}{G}\left(T_{in} - T_{amb}\right)\right] \tag{6.7}$$

where

η is the efficiency of the solar collector,
F_R is the collector heat removal factor, accounting for the effectiveness of heat transfer from the collector plate to the water,
T_{in} is the inlet water temperature in K.

Equation 6.7 is often known as the Hottel–Whillier–Bliss equation after the researchers who first derived it.

Equation 6.7 is very useful practically as measuring the inlet water and ambient temperature is relatively straightforward, and for practical purposes the efficiency is independent of reasonable flow rates. Typical efficiency characteristics of an unglazed and a double-glazed flat-plate solar water heater are shown in Figure 6.21. These show that, as the temperature of the water entering the collector (T_{in}) increases, the efficiency drops. The steep slope of the unglazed solar collector shows the significant losses from this type of collector, which will be very sensitive to the speed of the wind passing over it.

The y-axis intercept of the characteristic of Figure 6.21 gives a maximum value of efficiency, known as the optical efficiency ($F_R \times \tau\alpha$), which occurs when $T_{in} = T_{amb}$. The slope of the characteristic shows the heat coefficient ($F_R \times U$). In practice the characteristics of Figure 6.21 may be slightly non-linear as U and F_R change with collector temperature.

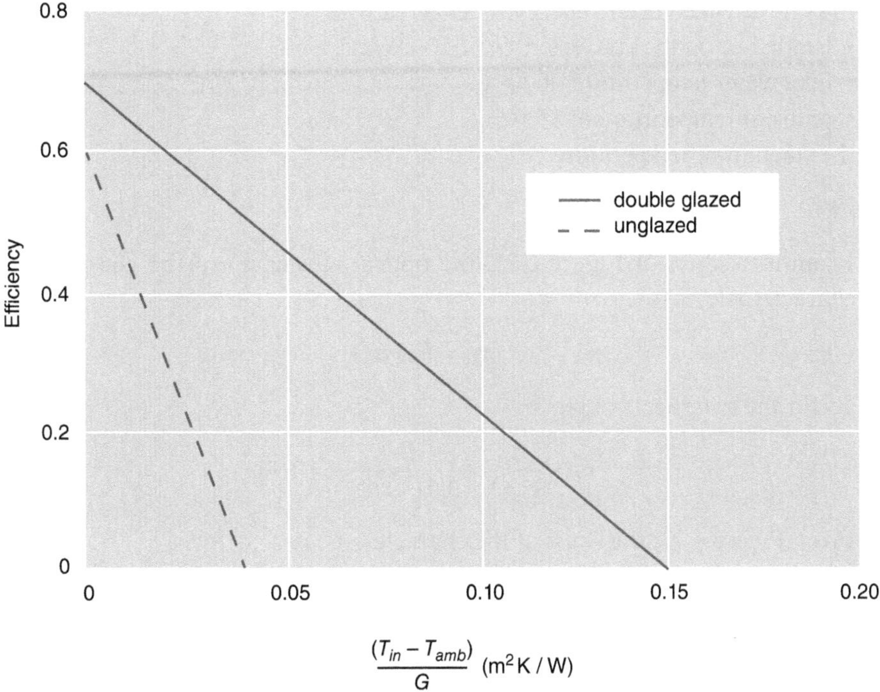

Figure 6.21 Characteristics of flat-plate solar collectors.

If there is no water flow, no useful heat can be extracted and so the efficiency η drops to zero. Hence at zero flow we can find the stagnation temperature from Equation 6.7 with $\eta = 0$.

$$T_{stag} = \frac{G}{U}(\tau\alpha) + T_{amb}$$

where

T_{stag} is the stagnation temperature.

The stagnation temperature is the maximum temperature that the collector will experience under given conditions of irradiance and ambient temperature.

Example 6.4 – Performance of a Flat-Plate Solar Collector

Consider the flat-plate solar collector shown in Figure 6.20 with a characteristic shown in Figure 6.21. Each sheet of glass of the double-glazed solar collector has a transmissivity, τ, of 0.91 and the plate has an absorptivity, α, of 0.93.

Calculate:

(a) the heat removal factor, F_R,
(b) the heat loss coefficient, U,

(c) the rate at which useful energy is delivered with
- irradiance on the converter, G, of 500 W/m^2,
- inlet water temperature of 30 °C,
- ambient temperature of 15 °C,

(d) the stagnation temperature.

Solution

(a) From inspection of Figure 6.21, the optical efficiency η_0 (the y-intercept of the characteristic) is 0.7.

$$\eta_0 = F_R(\tau\alpha)$$

and so for the two sheets of glass

$$F_R = \frac{\eta_0}{(\tau\alpha)} = \frac{0.7}{0.91^2 \times 0.93} = 0.909$$

(b) From Figure 6.21, the slope of the characteristic is

$$\text{Slope} = -\frac{0.7}{0.15} = 4.67 \text{ W/m}^2\text{K}$$

and so from Equation 6.7

$$\text{Slope} = F_R U = 0.909$$

$$U = \frac{4.67}{0.909} = 5.14 \text{ W/m}^2\text{K}$$

(c) Using

$$\eta = F_R\left[(\tau\alpha) - \frac{U}{G}(T_{in} - T_{amb})\right]$$

$$\eta = 0.909\left[(0.91^2 \times 0.93) - \frac{5.14}{500}(30 - 15)\right] = 0.909[0.77 - 0.155] = 0.56$$

Under these conditions, the efficiency of the collector is 56%.

Therefore useful energy is delivered at the rate of

$$q = G \times \eta = 500 \times 0.56 = 280 \text{ W/m}^2$$

(d) At the stagnation temperature $\eta = 0$

$$T_{stag} = \frac{G}{U}(\tau\alpha) + T_{amb}$$

$$T_{stag} = \frac{500}{5.14}(0.91^2 \times 0.93) + T_{amb} = 89.9 \text{ °C}$$

6.8.1 Selective Absorber Surfaces

The principle of operation of a solar thermal collector is that it should absorb the maximum possible radiant energy from the sun while minimizing thermal losses. The convection and conduction losses from a flat-plate collector are minimized by

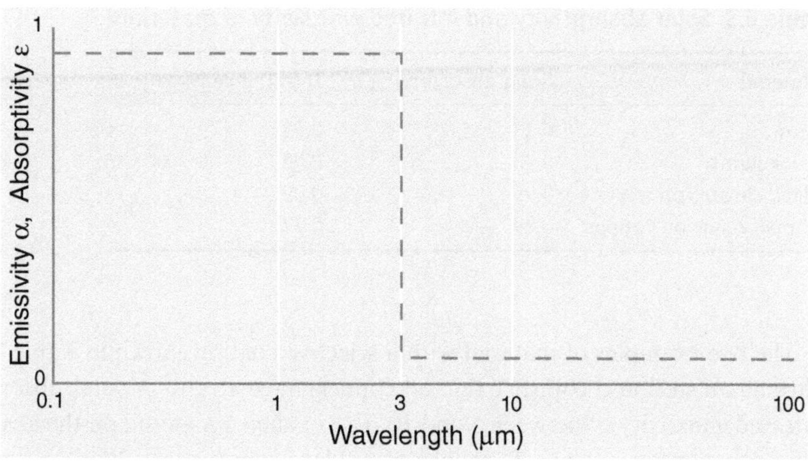

Figure 6.22 Ideal characteristic of a selective absorber surface, $\alpha_\lambda = \varepsilon_\lambda$.

careful choice of the distance between the collector plate and the glass cover(s) as well as by using thick insulation at the back of the collector plate. If evacuated tubes are used, then the vacuum effectively eliminates convection loss from the absorber.

The sun emits 98% of its irradiance at wavelengths of less than 3 μm, while a blackbody at 400 K emits 99% of its radiation at wavelengths greater than 3 μm. This is illustrated in Tutorial II, Figure II.10. Hence the heat gained for a given level of solar irradiance can be increased by careful choice of the material of the absorber so that it has a high absorbtivitiy, α, of the solar irradiance and a low emissivity, ε, at the temperature of the absorber.

An ideal characteristic of the material of the collector is shown in Figure 6.22. This shows the variation of monochromatic emissivity and absorptivity with wavelength of an ideal surface for the heat absorber of a low-temperature solar collector. At wavelengths of up to 3 μm the absorptivity is high, ~0.95, and the maximum solar irradiance is absorbed. At wavelengths above 3 μm the emissivity drops to a low value and little energy is radiated from the absorber.

Table 6.3 shows the absorptivity of radiation from the high-temperature source of the sun (the solar absorptivity) and the emissivity from the low-temperature source of the collector (the infrared emissivity) of a range of materials. Snow has a low absorptivity of the solar irradiance and a high emissivity of infrared radiation. Thus snow, or white paint, is a poor material for a surface that is intended to absorb solar energy. This agrees with our common experience that snow-covered objects take a long time to warm even in bright sunlight, and that houses in Mediterranean climates are often painted white to stay cool. Black paint has an infrared emissivity that is similar to its solar absorptivity and is sometimes used to coat the absorber plate of simple solar thermal water heaters made by hobby enthusiasts. This works well as the absorber plate of a solar thermal collector is at a temperature of less than 400 K and the heat energy radiated is proportional to εT^4. Hence black paint is a reasonable surface for the absorber of a low-performance solar water heater, even though its infrared emissivity is high.

Table 6.3 Solar absorptivity and infrared emissivity of materials

Material	Solar absorptivity, α	Infrared emissivity, ε
Snow	0.2	0.85
Black paint	0.98	0.95
Black chrome on steel	0.95	0.09
Copper oxide on copper	0.89	0.17

The two examples of material with a selective coating shown in Table 6.3 (black chrome on steel and copper oxide on copper) have a ratio of solar absorptivity to infrared emissivity of between 5 and 10. Use of such a material on the absorber significantly improves the performance of a solar water heater, although at increased manufacturing complexity and cost. Set against the improvement in performance, selective absorber coatings tend to be expensive, are difficult to apply, may degrade over time and the low infrared emissivity comes at the cost of some reduction in the solar absorptivity.

Thus the decision to use selective absorber coatings on low-temperature flat-plate solar thermal water heaters is a finely balanced choice made by the manufacturer. However, the use of selective absorber material becomes more obviously justified as the operating temperature of the solar collector increases, and selective absorbtion materials are always used for the absorbers of high-temperature solar thermal devices.

6.9 High-Temperature Concentrating Solar Thermal Systems

Temperatures of up to 150 °C can be developed by flat solar collectors that make use of the solar irradiance at its natural intensity of less than 1000 W/m^2 but the low-temperature heat from these collectors cannot be used effectively to generate electricity. For electricity generation, heat at higher temperatures is needed and it is necessary to concentrate the solar irradiance using either lenses or mirrors. This technology to concentrate solar irradiance, create high-temperature heat and then drive a turbine generator through a thermodynamic cycle is known as concentrating solar power (CSP).

Although the principle of concentrating solar power has been known for a century and plants have been in commercial operation for more than 40 years, at present there is only around 6 GW of CSP-generating capacity in service worldwide. However, CSP has a number of particular attributes that indicate it may, in future, make a major contribution to world electricity supply.

- If parabolic trough collectors are used then large generating plants can be developed, each with electrical power ratings of up to 200 MW.
- The turbine generators used are pieces of conventional electricity-generating equipment and so a CSP unit is easy to integrate into the electric power system.

- In areas with a good solar resource with insolation of 1800–2000 kWh/m^2 per year, the annual capacity factor of CSP can be up to 25%.
- The capacity factor can be improved further by including a heat energy store in the system and/or a supplementary fossil-fuelled boiler. A heat energy store makes the electrical output controllable so that it no longer depends on the instantaneous irradiance. The CSP unit is then able to respond to the demand for electricity and generate at times of the maximum electricity price. Including a fossil-fuelled boiler similarly makes the plant flexible and able to generate when the solar resource is limited.
- The energy payback of a CSP unit located in a good solar resource is around 6–12 months. This is similar to the energy payback of a wind turbine in a good wind resource but shorter than that of crystalline silicon photovoltaic modules.

However, CSP units also have a number of limitations that restrict the areas of the world where they can be deployed effectively.

- CSP units make use only of direct irradiance and are unlikely to be cost-effective in locations with solar insolation of less than 1800 kWh/m^2 per year.
- They use significant areas of land (between 5 and 10 m^2/MWh per year).
- If a steam turbine is used to power the generator, significant amounts of water are needed to cool the condenser. If wet cooling of the steam turbine is used, then this will require around 3–4 m^3/MWh. This requirement for cooling water can be reduced by dry cooling but at the cost of reduced efficiency of the generating plant. There is also a smaller need for water to wash the mirrors.

Hence the existing CSP units have been installed in flat areas with a good solar resource of direct irradiance and where water is available. Suitable locations have been found in the Mojave Desert in southern California and in southern Spain.

The three main technologies for CSP are:

- line focusing, using either a parabolic trough or a linear Fresnel collector,
- point focusing, using heliostats and a solar power tower,
- parabolic dish, using a Stirling engine or small gas turbine at its focus.

These are listed in Table 6.4.

Table 6.4 Concentrating solar power technologies

CSP technology	Concentration ratio of irradiance	Operating temperature of absorber (°C)	Capacity of plant (MW)	Annual solar to electricity conversion efficiency (%)
Line-focusing parabolic trough	70–80	400–500	10–200	15–16
Linear Fresnel collector	25–40	300–400	10–200	8–10
Point-focusing central solar power tower	300–1000	Up to 800	10–150	15–17
Parabolic dish with Stirling engine	1000–3000	Up to 800	0.01–0.4	20–25

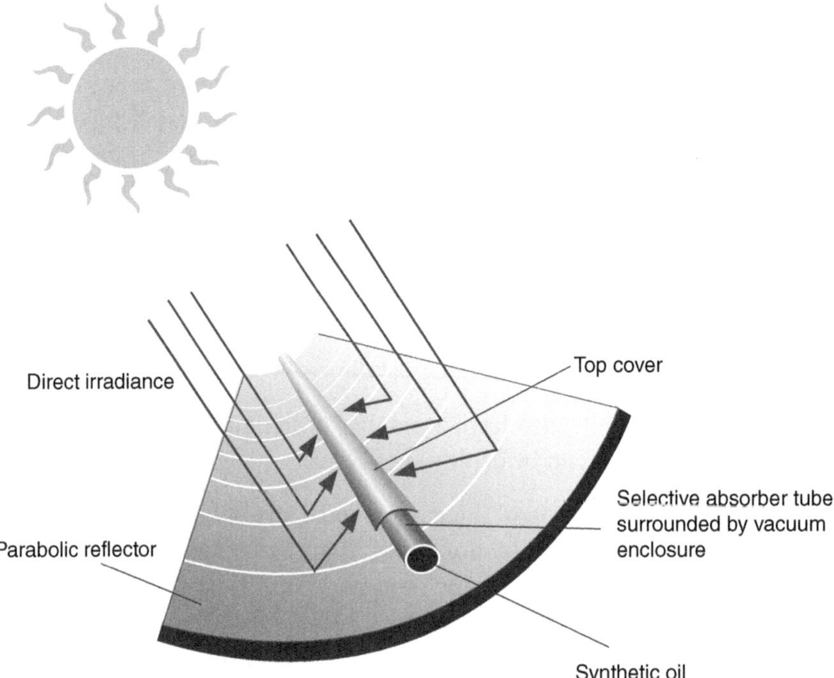

Figure 6.23 Parabolic-trough solar collector.

Figure 6.24 Solar thermal generating plant.
Photo: Tom Grundy/Shutterstock.com.
(A black and white version of this figure will appear in some formats. For the colour version, refer to the plate section.)

Most of the CSP plants that are in commercial operation use line-focusing parabolic troughs to concentrate the irradiance. Parabolic troughs of silver-coated glass mirrors are placed along an axis, usually north–south, and the troughs are rotated through the day by an actuator and control system to track the sun from east to west (Figures 6.23 and 6.24). Each trough assembly of mirrors is up to 150 m long and is typically in 15 m long sections. Each section is made up of curved mirrors with the cross-section of a parabola supported on a space frame to form a rigid rotating structure. The parabolic trough focuses the direct irradiance of the sun onto an absorber pipe, which is located at the focus of the parabola. The focal length of the

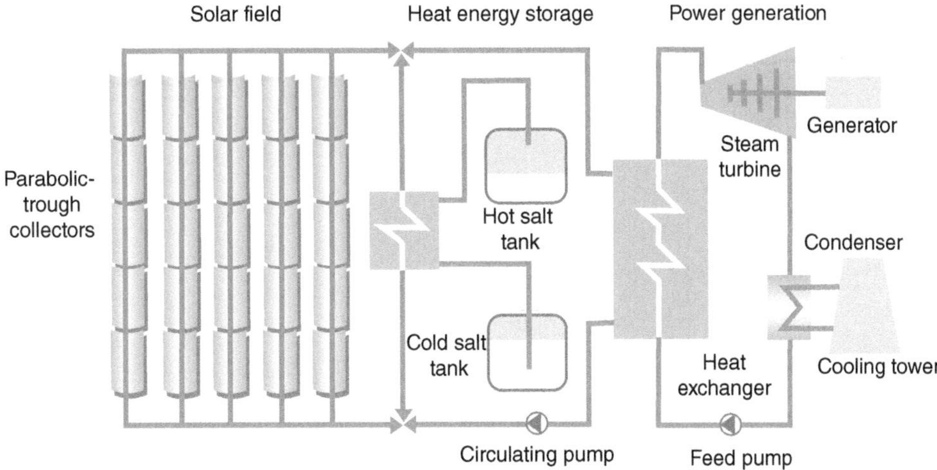

Figure 6.25 Parabolic-trough solar thermal power station including heat storage.
Source: adapted from *The Parabolic Trough Power Plants Andasol 1 to 3*, 2008, Solar Millennium.

parabola is short, less than 3 m, and the absorber pipe is mounted on the mirror assembly. The absorber pipe is surrounded by a glass vacuum enclosure to reduce heat losses and is coated with a selective absorber material (see Section 6.8.1). There may be a cover above the absorber pipe to limit the re-radiation of energy back to the sky. Heat is extracted from the absorber pipes by a flow of synthetic oil, and the hot oil is collected at a central generating plant. Then, through a heat exchanger, the heat is used to power a steam turbine. It is common for CSP plants to consist of a number of trough assemblies to give an electrical output power rating of up to 200 MW using steam at a temperature of up to 400 °C.

In one recent installation, heat is stored in large heavily insulated tanks containing a mixture of potassium, sodium and nitrate salts. The heat store is charged during the day and discharged at night when the price of electricity is high. Figure 6.25 shows a schematic diagram of a CSC plant with energy storage.

Some recent designs of CSP plants have done away with the synthetic oil and associated heat exchanger to pass water/steam through the absorber pipe and power the steam turbine directly at up to 500 °C. Synthetic oil limits the maximum temperature of the collector to around 400 °C. However, direct steam generation is a more difficult process to control as the flow of steam through the absorber then depends on instantaneous irradiance. The use of an intermediate heat transfer material such as synthetic oil makes the CSP unit easier to control and allows a heat energy store to be integrated more easily into the system.

Nine parabolic-trough CSP plants, with a total electrical rating of 354 MW, were built in California between 1980 and 1991 by LUZ International Limited. Each plant had an electrical rating of either 30 or 80 MW. The plants were technically successful, and technical availabilities of up to 99% were claimed. However, the company experienced financial difficulties partly due to the low price of natural gas used for electricity generation and partly because of changes in the power market and support mechanisms for renewable energy projects in California. After

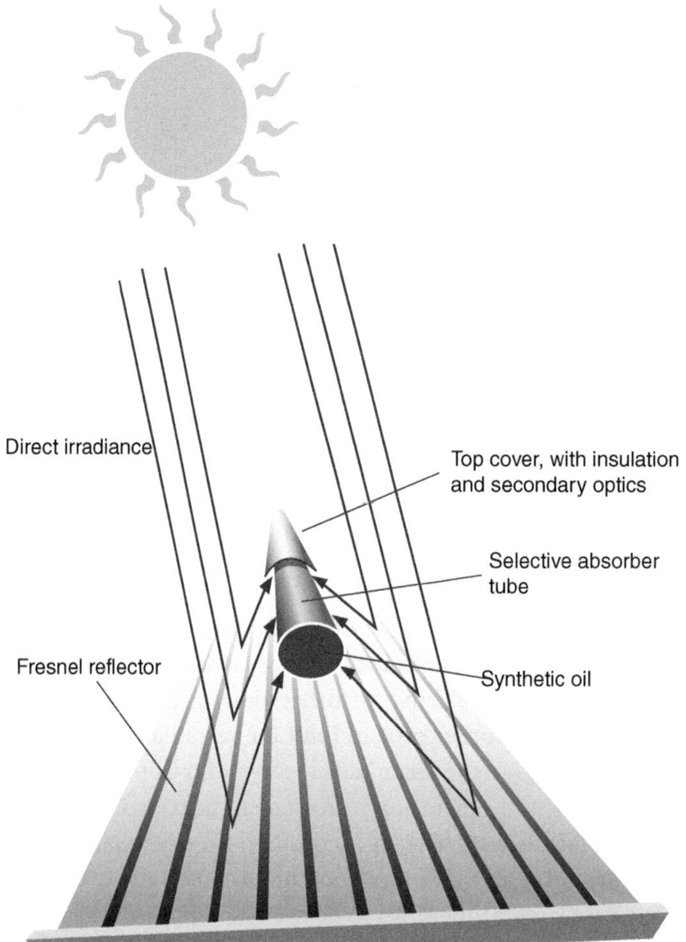

Figure 6.26 Linear Fresnel trough.

financial restructuring and changes in their ownership, the CSP plants continued to be operated successfully. New parabolic-trough CSP units are being commissioned in Spain, California and the Middle East, and parabolic-trough CSP units can now be considered a rapidly maturing technology. There have been ambitious plans by a major industrial consortium, Desertec, to install very large CSP plants in North Africa and transmit the electrical energy to Europe.

The rotating parabolic-trough mirror assemblies are expensive, difficult to keep clean and their large structure leads to them having to be designed to withstand the high forces developed by strong winds. Hence there are initiatives to reduce costs by using linear Fresnel collectors (Figure 6.26), and a number of installations using this technology have been built (Figure 6.27).

A linear Fresnel collector is a line-focusing device that uses a series of controlled flat or slightly concave ground-mounted mirrors with a fixed absorber tube mounted above them. The mirrors are rotated independently about their longitudinal axis by a control system to focus the irradiance on to the absorber tube. The fixed absorber

Figure 6.27 Solar thermal generating plant using Fresnel lenses.
Photo: Eunika Sopotnicka/Shutterstock.com.
(A black and white version of this figure will appear in some formats. For the colour version, refer to the plate section.)

tube is mounted at the focus length of the assembly of the controlled mirrors; this may be at a height of up to 10 m. Heat is extracted from the absorber tube by a flow of synthetic oil or by direct steam generation in the same manner as a parabolic trough. The collector does not need to be orientated along a north–south axis and so there is more flexibility in the layout of the solar collector field. The absorber tube has a secondary concentrator mounted above it and an insulated cover to reduce thermal losses. It has higher optical losses than an equivalent parabolic trough due to the greater distance between the mirrors and absorber tube. The linear Fresnel collector develops lower temperatures than an equivalent parabolic trough (and hence the turbine has a lower thermal efficiency) but it is expected that its fixed absorber tube and ground-mounted mirrors will lead to a lower cost of electricity.

Solar power towers use a field of heliostats (individually controlled large flat mirrors) to focus irradiance on to a collector mounted on a high tower (Figure 6.28). The large field of heliostats provides concentration ratios of 600–1000 and temperatures of the receiver of 800–1000 °C. Heat is extracted from the receiver using either water/steam, molten salts or a flow of air as the heat transfer fluid. After passing the heat transfer fluid through a heat exchanger, a conventional steam turbine is used to generate electricity. The very high temperatures that can be developed by solar power towers also allow a gas turbine cycle to be used and this has been demonstrated at low power (250 kW). The use of molten salts for the heat transfer allows heat energy storage tanks to be included in the system conveniently. A number of demonstration plants at ratings of 10–20 MW have been built and been shown to operate successfully while larger plants are under construction.

Table 6.5 gives details of two solar power tower plants that have been constructed in southern Spain. Note the large number of mirrors and height of the towers. An annual efficiency of around 16% (electrical output/solar insolation striking the heliostats) is claimed.

Table 6.5 Details of two solar power tower plants

Plant name	PS10	PS20
Rated electrical output (MW)	10	20
Size of individual heliostat (m^2)	120	120
Number of heliostats	700	1225
Height of tower (m)	80	163

Source of data: H. Müller-Steinhagan, 'Concentrating solar thermal power', *Proceedings of the Royal Society Philosophical Transactions A*, 371 (2013).

Figure 6.28 Power tower solar generating plant.
Photo: raulbaenacasado/Shutterstock.com.
(A black and white version of this figure will appear in some formats. For the colour version, refer to the plate section.)

Parabolic dishes with a Stirling engine located at the focus have been demonstrated at lower electrical ratings (up to 100 kW). These self-contained units are likely to find their first application to supply power to remote off-grid loads. They can concentrate the solar irradiance by up to 1000–3000 times and achieve high thermal efficiency. However, they are complex mechanical devices, tracking the sun in two dimensions with high-temperature Stirling engines or small gas turbines, and have yet to be demonstrated to be cost-effective and reliable when located in remote areas. Parabolic dishes are likely to face strong competition from concentrating photovoltaic systems using multi-junction cells, while the simplicity of non-concentrating flat photovoltaic modules will be very attractive in areas where maintenance costs are high.

PROBLEMS

1. Distinguish between direct, isolated and indirect solar gain of a building.
2. Explain how glass windows act to heat a building. How can the temperature within a building be controlled using windows?
3. List and discuss the features needed for effective passive solar heating of a building.
4. Why are degree-days used when monitoring the energy performance of the heating and cooling system of a building? Sketch and explain a performance line, labelling the axes.
5. Compare the advantages of electricity generation by solar PV and CSP.
6. The cross-section of a wall is shown in Figure 6.29. The thermal resistance of each material is given in the following table.

Material	Thermal resistance (m^2K/W)
Brick	0.13
Air	0.18
Sheathing	0.07
Mineral wool quilt	3.50
Plaster board	0.13

(a) Calculate the composite U-value for the cross-section shown in Figure 6.29.

(b) In the wall there is a double-glazed window with a U-value of 2.5 W/m^2K. The overall area of the wall is 10 m^2 and within it the area of the window is 3.5 m^2. If the temperature inside the room is 20 °C and that outside is 10 °C calculate the heat loss.

[**Answer:** (a) 0.25 W/m^2K; (b) 103.75 W.]

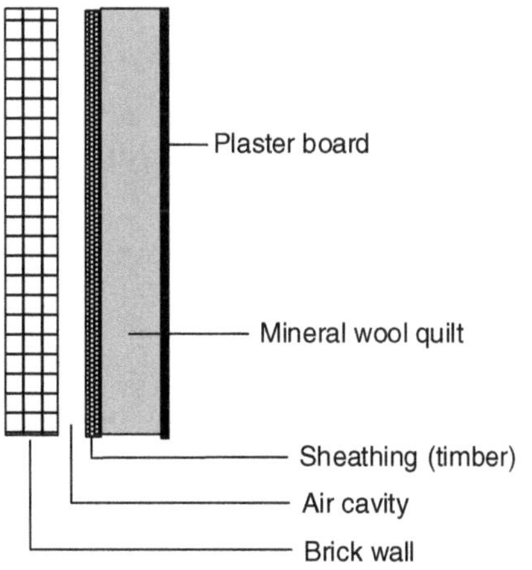

Figure 6.29 Cross-section of a wall.

7. A small rectangular room has a glass window. The following parameters are provided:
   solar gain coefficient of the window, $S = 0.7$,
   heat transfer coefficient of the window, $U = 1.4$ W/m^{2}K,
   temperature difference between the ambient and interior air: $\Delta T = 20$ °C.
   Assume that the room is so well insulated that the heat transfer through the opaque external walls, floor and ceiling is negligible. If the sunlight strikes the window at an angle of 60° to the glass, calculate the irradiance required to maintain a constant temperature within the room when the sun is shining.

   [**Answer:** 46 W/m^2.]

8. The ventilator of the room in Problem 7 is opened to give the following conditions:
   there are two air changes per hour,
   the volume of air within the room is 50 m^3,
   the irradiance is 300 W/m^2 striking the window at an angle of 30°,
   the specific heat of air is 1.005 kJ/kgK,
   the density of air is 1.25 kg/m^3.
   Calculate the minimum area of window required to maintain a constant temperature of 20 °C above ambient within the room.

   [**Answer:** 4.44 m^2.]

9. A building has the following parameters:
   the overall heat loss coefficient of the fabric, U, is 0.5 W/m^{2}K,
   the surface area of the building is 1000 m^2,
   the volume of the building is 800 m^3,
   air changes per hour are 0.75,
   the specific heat of air is 1.005 kJ/kgK.
   The temperature inside the building is maintained at 20 °C and the mean outdoor temperature is 10 °C. Calculate:
   (a) the heat transfer across the fabric,
   (b) the heat transfer by air exchange,
   (c) the overall heat loss coefficient, U'.

   [**Answer:** (a) 5 kW; (b) 2.1 kW; (c) 709.4 W/K.]

10. The building described in Problem 9 is situated in a city where the heating degree-days in each month are given in the table in Example 6.3. The efficiency of the heating system is 75%.
    Calculate the expected energy consumption for a year.

    [**Answer:** 41 MWh.]

11. A glazed flat-plate solar collector to heat water has a transmissivity $\tau = 0.85$ and absorptivity, $\alpha = 0.9$. The following data are given:
    irradiance on the plate, $G = 800$ W/m^2,
    heat loss coefficient, $U = 3$ W/m^{2}K,
    ambient temperature, $T_{amb} = 22$ °C,
    mean temperature of the plate, $T_{mean} = 29$ °C.

Calculate:

(a) the thermal loss of the collector per unit area,

(b) the useable heat derived from the collector per unit area,

(c) the thermal efficiency.

[**Answer:** (a) 21 W/m^2; (b) 591 W/m^2; (c) 74%.]

12. The area of the flat-plate solar collector in Example 6.4 is 1 m^2, the specific heat of water is 4185 J/kgK and the inlet water temperature is T_{in} = 18 °C.

(a) What is the mass flow rate of water required to maintain the outlet water temperature at 25 °C?

(b) What is the outlet water temperature if the mass flow rate of water is doubled?

[**Answer:** (a) 0.02 kg/s; (b) 21.5 °C.]

13. A solar heater with 10 m^2 flat absorber plate is designed such that under the following conditions the outlet water temperature is approximately 50 °C:

the irradiance is 700 W/m^2,
ambient temperature is 22 °C,
inlet water temperature is 18 °C,
water mass flow rate is 0.03 kg/s.

The following data are also given:

heat loss coefficient, U = 3 W/m^{2}K,
heat removal factor, F = 0.975,
specific heat of water, c_p = 4185 J/kgK.

If the transmissivity of the glass cover is τ = 0.8, select a suitable plate for the absorber plate from the following:

Plate 1: absorptivity, a = 0.35,
Plate 2: absorptivity, a = 0.74,
Plate 3: absorptivity, a = 0.95.

[**Answer:** Plate 2.]

FURTHER READING

Peake S. (ed.), *Renewable Energy: Power for a Sustainable Future*, 4th ed., 2017, Oxford University Press.
Comprehensive textbook covering many aspects of renewable energy.

Duffie J.A. and Beckman W.A., *Solar Engineering of Thermal Processes*, 2nd ed., 2013, Wiley Interscience.
The authoritative textbook on solar thermal engineering.

Hagentot C.-E., *Introduction to Building Physics*, 2003, Studentlitteratur.
A most useful textbook of building physics.

Twidell J., *Renewable Energy Resources*, 4th ed., 2021, Routledge.
Comprehensive textbook covering many aspects of renewable energy.

ONLINE RESOURCES

Exercise 6.1 Energy use of a small building: this demonstrates a simple spreadsheet calculation of energy use of a building.

Exercise 6.2 Flat-plate collectors: this investigates the performance of solar thermal collectors.

7 Marine Energy

INTRODUCTION

The oceans cover more than 70% of the earth's surface, and energy from the tides and waves is an important potential source of electrical power. However, the marine environment is extremely demanding, not least because of the wide range of forces that any marine energy conversion device will experience. In normal operating conditions a marine device must extract energy effectively from relatively small and benign movements of water caused by the waves and tides, but it must also survive damaging extreme storms. In addition, installing and gaining access to an offshore marine energy device for maintenance is difficult, while transmitting electrical power to the shore using submarine cables is expensive. Seawater is corrosive, so any materials used must be carefully selected and protected. Hence, although there has been considerable research and development effort over many years and a number of demonstration projects have been installed, marine energy has yet to achieve a commercial breakthrough: only around 1 GW of marine-energy electricity-generating capacity is in service worldwide.

Marine energy describes three distinct technologies that, because of the very different characteristics of the resource and stages of maturity of their technical and commercial development, are considered separately in this book. These are electricity generation by:

- tidal range,
- tidal stream,
- wave power.

The tides are caused by a very-low-frequency water wave that is created by the gravitational attraction of the moon and the sun and the rotation of the earth. This wave, with a period of around 12½ hours, leads to variations in the height of the water and in the creation of marine currents as it approaches the shoreline. In contrast, ocean waves are created by winds, which are ultimately due to the heating of the earth by solar energy and contain a range of higher frequencies. Both tidal range and tidal stream generation have the most desirable attribute that their power output can be predicted years in advance. This is not the case with wave energy.

Table 7.1 Large tidal range power schemes

Location	Country	Power rating (MW)	Year of commissioning
Sihwa Lake	South Korea	254	2011
Annapolis Royal	Canada	20	1984
La Rance	France	240	1966

Tidal range generation exploits the potential energy of the difference in the height of the water level at the edge of the oceans that is caused by the tides. A barrage is constructed across the mouth of an estuary to create a difference in water height inside and outside an impounded area as the height of the tide varies. Low-head reaction hydro turbines are placed in the barrage. The hydro turbines generate electricity from the potential energy of the difference in the height of the water across the barrage. As the tide rises and falls, this difference in height results in flows through the turbines into the basin (flood generation) or out of the basin (ebb generation). The water flowing through these turbines is constrained in a duct, so the turbines are not subject to the Betz limit and can have an efficiency approaching 90% at a particular state of the tide.

The large tidal range power plants that have been constructed, with their power rating and commissioning date, are shown in Table 7.1.

Tidal range technology has been studied extensively and is now well understood. The schemes listed in Table 7.1 have demonstrated that generating electricity from the tidal range can be technically successful and the future exploitation of this source of energy now depends on its cost-effectiveness and environmental impact.

Tidal stream generation is a quite different technology that uses the kinetic energy of a tidal current to drive a submarine turbine. The operation of a tidal stream turbine has some similarities to that of a wind turbine although a submarine rotor for the same power output has a smaller diameter, as water is 837 times denser than air. The speed of water at which a tidal stream device develops peak power is around a quarter of the rated wind speed of a wind turbine and the flow speed varies throughout the tidal cycle. Tidal stream turbines are subject to the Betz limit of a maximum 59% efficiency of conversion of power from the fluid flow to the shaft. As with the wind, a tidal stream is not a simple smooth flow and is likely to contain significant turbulence and eddies as well as variations of flow speed with depth, leading to significant time-varying loads on the tidal stream turbine and its supporting structure. The effect of surface waves adds to the turbulence created by the shape of the seabed to create a complex flow. A number of designs of tidal stream turbines, with both horizontal and vertical axis, have been developed and demonstrated with ratings of each single unit up to ~1 MW. Some designs use ducts or cowls to accelerate the water flow through the turbine. There have also been several prototype devices using hydrofoils to create a reciprocating movement.

The speed of tidal flows varies around the world depending on the behaviour of the tidal wave in the deep oceans. Locally the flows change with the shape of the shoreline and seabed and accelerate around headlands and through narrow channels. There are a number of demonstration projects of small arrays of turbines in fast-flowing tidal streams off the west coast of Europe. Tidal stream generation is an emerging area of technology, and the most effective turbine architecture has yet to be found. Choices being investigated by device developers include horizontal or vertical axis, direct drive or through a gearbox, fixed or variable speed, as well as the means of power regulation. Wind power went through a similar process of refinement of the technology in the 1980s before upwind, three-bladed, variable-speed, horizontal-axis wind turbines became commercially dominant. Tidal stream turbine design has yet to converge to a single preferred architecture.

Wave power is a different approach to marine energy conversion in which the power of the ocean surface waves is captured and converted into electricity. Shoreline, near-shore and offshore floating wave power devices have all been demonstrated but the majority of the wave energy resource is offshore in deep water. Wave power generators are difficult to build cost-effectively, and the extreme forces developed in storms have damaged several prototype devices. Personnel access for maintenance, the mooring system as well as the power take-off and cables all pose particular challenges for floating wave energy converters. Even more than for tidal stream devices, no clear technical favourite architecture has yet emerged and there are a number of wave energy conversion concepts and designs under development. Although prototype wave power devices are being demonstrated, wave power is currently the least mature of the marine energy technologies, and commercial exploitation of the wave power resource is still some years in the future.

This chapter discusses, in turn, tidal range, tidal stream and wave power generation. Section 7.1 presents the tidal range energy resource, the description of the tides using harmonic constituents, tidal range generation, turbine generators for a tidal range generation scheme, the environmental impact of tidal range generation and tidal lagoons. Section 7.2 addresses the tidal stream resource, development of a tidal stream project, tidal stream turbines and compares the loads on a tidal stream turbine with a wind turbine using axial momentum theory described in Chapter 2. Finally, in Section 7.3, wave power generation is discussed. This section covers the basics of water waves, the wave energy resource and the functional requirements of devices for wave power generation.

7.1 Tidal Range Generation

The action of the tides was exploited to power mills to grind corn from the Middle Ages until the steam engine became dominant. Some of these mills used a simple waterwheel placed in the tidal flow of a river. In other places a barrage was constructed across the mouth of a river to trap water at high tide, which was then released when there was sufficient difference in the height of the water inside and outside the barrage to operate a horizontal-axis undershot wheel (see Section 3.1). The mill was operated only as water flowed out of the basin (on the ebb flow), and simple flap sluice gates were used to allow the basin to fill during the incoming flood tide. These tidal mills established the principle of using the tidal range to provide motive power more than 500 years ago.

The modern era of the generation of electricity using tidal range began with the construction of the barrage and 240 MW power generating plant on the La Rance estuary in the west of France in the 1960s by Électricité de France. The La Rance estuary experiences an average tidal range of 8 m with a peak range of 13.5 m. A 750 m long barrage was constructed forming a basin with an impounded area of 22.5 km^2. Twenty-four 5.3 m diameter, 10 MW double-regulated bulb turbine generators were placed across the barrage, which generates around 540 GWh/year, giving an annual capacity factor of ~26%. The turbines were designed for bidirectional operation, generating on both the flood and ebb tides, but over the life of the project mainly ebb generation has been used. Apart from some remedial work being required on the electrical generators early in their life, the plant has operated successfully for more than 50 years and demonstrated convincingly that tidal range power generation schemes can be technically successful. There was a considerable environmental impact at the time of construction as part of the estuary was drained to allow construction of the barrage and locks. This change in the environment of part of the estuary caused considerable damage to the local ecology. The ecosystems of the estuary recovered substantially over the subsequent 10–15 years, but such a construction technique and its environmental impact would be difficult to justify nowadays. Although there have been studies of larger tidal range schemes to exploit the tidal range resource on the western coasts of France, none has yet been built.

The Annapolis Royal Scheme was constructed in Nova Scotia, Canada, in the early 1980s and has now been decommissioned. A caisson containing a single 20 MW straight-flow turbine and rim generator was placed in an existing barrage that had been built for flood defence. The tidal range at the site is around 6 m and the annual energy output of the scheme was 50 GWh giving an annual capacity factor of ~28%. Generation was only on the ebb tide and the level of water within the basin was allowed to vary by around only 1 m. The scheme successfully demonstrated the operation of the straight-flow turbine and the 13 m diameter rim generator, which were being considered for use in larger tidal schemes in the adjacent Bay of Fundy. Although this demonstration unit was technically successful and there is a large tidal range resource in the area, no large-scale schemes were built subsequently. This was partly due to concern over potential changes in the water levels of the area and the environmental impact of any large scheme.

In 2011, a 254 MW tidal range scheme using ten 25.4 MW bulb turbine generators was commissioned at Sihwa Lake in South Korea. The double-regulated bulb turbines had a runner diameter of 7.5 m and a rated head of 5.8 m. The scheme

Table 7.2 Main characteristics of the proposed (but not built) barrage between Wales and England on the Severn estuary

Total length of barrage	15.9 km
Impounded area	480 km^2
Design life	120 years
Construction period to generation	7 years
Operation mode	ebb generation with flood pumping
Sluices	166 with total area 35 000 m^3
Turbines	9 m rotor diameter bulb turbine, single regulated with variable pitch rotor
Generators	216 × 40 MW, giving a total capacity of 8640 MW. Rotational speed 50 rpm
Annual electrical energy production	17 TW h
Capacity factor	22.5%

Source of data: *Energy Paper Number 57*, 1989, The Stationery Office. Contains public sector information licensed under the Open Government Licence v3.0.

generates only on the flood tide as it was designed to increase the exchange of water with the sea from an artificial lake that had been created previously by a flood control barrage and needed an increased flow of water to improve its water quality. The annual energy production is 550 GWh giving an annual capacity factor of ~25%.

In addition to these multi-megawatt tidal generation schemes there have been a number of smaller schemes (each less than 3 MW rating) built in China and the former USSR.

Over the past 100 years, numerous tidal range generation schemes around the world have been studied but not constructed. In the UK, the Severn estuary has been studied in considerable detail and a number of locations for a barrage proposed. In a series of studies that were reported in 1989 a barrage across the Severn estuary between Cardiff (on the Welsh coast) and Weston-super-Mare (on the English coast) was investigated. It was concluded by this detailed study that the project was feasible and would provide around 7% of the then projected electrical energy demand of England and Wales. Table 7.2 gives the main design characteristics of the proposed Cardiff–Weston tidal barrage. A similar scheme was proposed in 2011 but again did not proceed.

The technical success of the initial schemes but the lack of widespread adoption of the technology indicates that the main difficulties of developing large tidal range power schemes are the following.

- *The high initial capital cost and long construction period.* The high discount rates demanded for power-generating projects make it very difficult to obtain finance from commercial lenders for large tidal range schemes. The schemes may take 5–7 years to construct, during which time there is no revenue and, once built, the civil works then last for up to 100 years. Simple discounted cash flow analysis (see Section 9.2.1) shows that, even with a discount rate as low as 5%, any income 30 years or more in the future is of little present value.

- *The anticipated environmental impact.* There is often concern over the effect of the turbines on fish and marine mammals and over the loss of inter-tidal zones at the edge of the estuary that provide the habitat for wading birds. Once a barrage is built, the water level of the impounded area of the basin no longer changes with the tides to the same extent as before and the area of mud flats that are alternately dry and submerged is reduced. In the UK several influential environmental organizations have a policy to oppose any large tidal range schemes.
- *Interference with shipping and other users of the estuary.* Although ship-locks would be included in any large barrage there is likely to be opposition to any scheme that is proposed from shipping interests and other users of the estuary.

7.1.1 The Tidal Energy Resource

The tidal range energy resource is created by the change in water level at the edge of the ocean that is caused by low-frequency tidal waves created by the gravitational attraction of the moon and sun coupled with the rotation of the earth. In some locations the tidal range of the deep oceans (which is only ~ 0.5 m) is amplified many times by local hydrodynamic effects including the shelving of the seabed and the creation of hydraulic resonances by the shape of the shoreline. The Bay of Fundy in Canada experiences tidal ranges with a height of up to ~15 m and the Severn estuary in the UK experiences tidal range variations of up to 12 m.

The water level of the ocean changes regularly twice during each period of a tidal (lunar) day, which has a period of 24 hours, 50 minutes and 28 seconds (1.035 times the solar day or 24.84 hours). The tidal day is slightly longer than the 24 hours of the solar day as the moon orbits the earth in the same direction as the rotation of the earth. The water level increases to its highest point at high water at the conclusion of the flood tide and then drops to low water at the end of the ebb tide. The tidal range is the difference in height between high and low water.

The earth and the moon are held in position by a balance of gravitational and inertial forces. The body of the earth is effectively rigid, but the gravitational force created by the mass of the moon acts to attract the water of the oceans towards the moon with a similar, balancing inertial force acting on the opposite side of the earth. This leads to two bulges in the height of the oceans, towards and away from the moon. Rotation of the earth about its polar axis leads to the bulges appearing to rotate around the earth. If the moon is aligned with the earth's equator, then a point on the earth will see the height of water changing twice, approximately equally, in the period of a tidal day. This gives rise to the twice-daily (semi-diurnal) tides and is illustrated in Figure 7.1.

However, the position of the moon is not generally aligned with the equator but is at an angle determined by the orbit of the moon around the earth. This angle of declination of the moon with respect to the earth's equator results in the two tides each day being unequal, or even in a single diurnal (daily) tide in some locations (Figure 7.2).

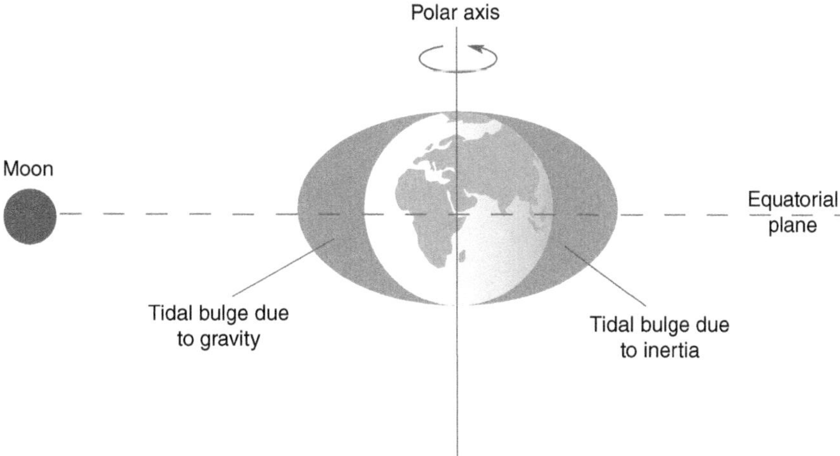

Figure 7.1 Semi-diurnal tides caused by the moon. (The tidal bulges are shown exaggerated for clarity.)

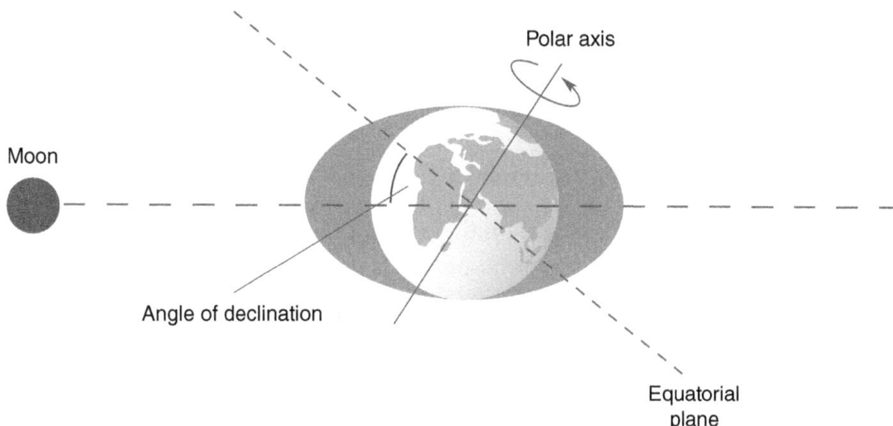

Figure 7.2 Declinational tides caused by the moon. (The tidal bulges are shown exaggerated for clarity.)

The moon's orbit around the earth is not circular but elliptical, with the earth not at the centre of the ellipse. The point at which the moon is nearest to the earth is known as the perigee and the furthest point as the apogee. The period of time from one perigee to the next perigee is the lunar (synodic) month of 27.53 days. This variation of the orbit of the moon is illustrated in an exaggerated manner in Figure 7.3. There is a similar effect created by the elliptical orbit of the earth around the sun, as the sun is not at the centre of this ellipse.

In addition to the effect of the moon, the sun has a significant effect on the tides. The sun has a much larger mass than the moon but is at a much greater distance from the earth. The gravitational forces exerted by the sun and moon on the earth are proportional to their mass and inversely proportional to the square of their distance from the earth:

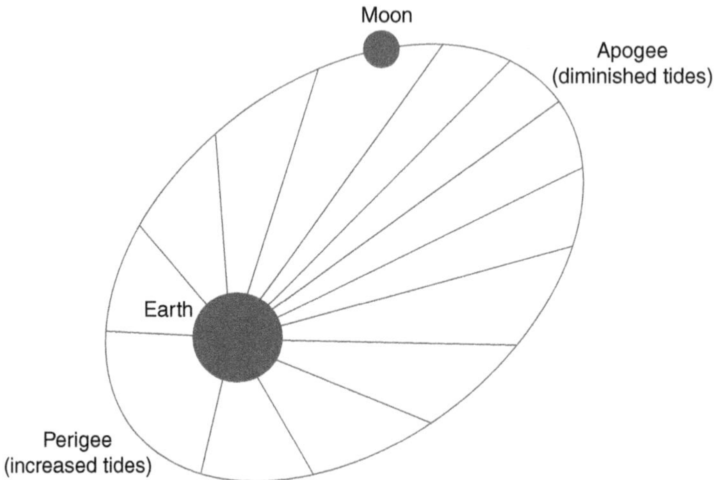

Figure 7.3 Lunar month or anomalistic tides caused by the elliptical orbit of the moon. (The ellipse is exaggerated for clarity.)
Source: R.H. Clarke, *Elements of Tidal-Electric Engineering*, 2007, IEEE Press–Wiley.

$$F \propto \frac{M}{R^2}$$

where for either the sun or moon:

F is the gravitational force on the earth in N,
M is the mass of the sun or moon in kg,
R is the distance of the sun or moon from the earth in m.

The tidal generating force of the moon and sun is proportional to the difference of the gravitational force exerted by these objects across the diameter of the earth. Thus the tidal generating force is proportional to $\partial F/\partial R$. By simple differentiation,

$$\frac{\partial F}{\partial R} \propto \frac{-2M}{R^3}$$

Thus the ratio of the tidal generating force created by the moon and sun is

$$\frac{M_{moon}/R_{moon}^3}{M_{sun}/R_{sun}^3}$$

Substituting the masses of the moon and sun and their distances from earth into this equation, it can be shown that the tidal generating force of the sun is around 46% of that of the moon.

When the earth, sun and moon are aligned, the gravitational effects of the sun and moon act together to cause the high spring tides. When the moon is at 90° to a line from the earth to the sun there is a degree of cancellation of the tidal forces that results in the lower neap tides. This is illustrated in Figure 7.4.

The variation over time in the height of the ocean leads to tidal current flows. The land masses of the continents modify these tidal current flows and the bathymetry (depth) and topography (shape) of the estuaries results in amplification of the deep

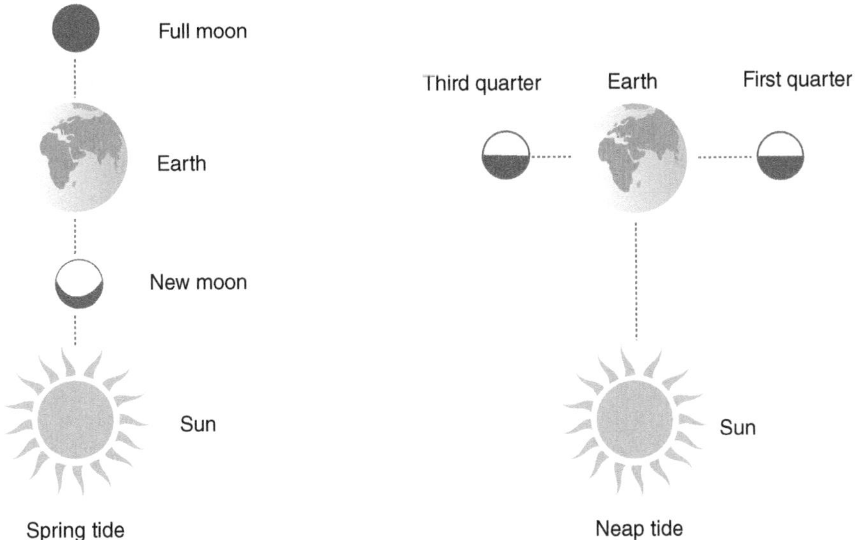

Figure 7.4 Spring and neap (synodic) tides. The relative positions of the sun, earth and moon, and the phases of the moon seen from earth, are shown.

ocean tidal range. The shape of some estuaries leads to resonance effects that play an important role in amplifying both the tidal range and tidal currents.

7.1.2 Description of the Tides Using Harmonic Constituents

The tidal range at a location is the result of a complex combination of multiple astronomical influences as well as effects of the geography of the deep oceans and the bathymetry and topology of the local shoreline. This results in a complex time series of the height of the tide at a location that repeats approximately every 20 years. The tidal height is described by a harmonic series of the form:

$$h(t) = h_0 + \sum_{n=1}^{m} h_n \sin(\omega_n t + \alpha_n) \tag{7.1}$$

where

$h(t)$ is the height of tide at time t,
h_0 is the height of mean water level,
h_n is the amplitude of the nth harmonic constituent,
ω_n is the angular velocity of the nth harmonic constituent,
α_n is the phase displacement of the nth harmonic constituent,
t is time,
m is the number of harmonic constituents.

The harmonic constituents of the tide at a particular location are found from site measurements of the height of the tide that are then decomposed into the harmonic constituents using harmonic analysis. These harmonic constituents are stated in tide tables and can be used to calculate the height of the tide at any time starting from a defined instant.

Around 20 tidal harmonic constituents are needed to create a full representation of the tide at a location but the two semi-diurnal (twice daily) constituents, caused by the moon (M2) and sun (S2), are the most important for describing the tidal range in many locations. Table 7.3 lists five of the most significant tidal constituents.

Table 7.3 Principal harmonic constituents of a tide

Harmonic constituent	Period (h)	Angular velocity (°/h)	Description	Number of figure illustrating main effect
M2	12.42	28.98	Principal lunar semi-diurnal constituent. This constituent represents the rotation of the earth with respect to the moon	7.1
S2	12	30	Principal solar semi-diurnal constituent. This constituent represents the rotation of the earth with respect to the sun. The M2 and S2 constituents act together to create the spring and neap tides	7.4
N2	12.66	28.44	Larger lunar elliptic semi-diurnal constituent to account for the elliptical orbit of the moon	7.3
O1	25.82	13.94	This constituent accounts for the declination of the moon over a day	7.2
K1	23.93	15.04	This constituent combines with O1 to express the effect of the moon's declination	7.2

Table 7.4 Tidal constituent data for Figure 7.5 using Equation 7.1 (*t* is in hours and h_0 is zero)

Harmonic constituent	h_n (m)	α_n (°)	ω_n (°/h)
M2	3.89	188	28.98
S2	1.37	243	30
K1	0.72	120	15.04
O1	0.92	359	13.94

Source of data: G.A. Alcock and D.T. Pugh, 'Observations of tides in the Severn Estuary and Bristol Channel', 1980, Report No. 112 to the UK Department of Energy. Unpublished manuscript.

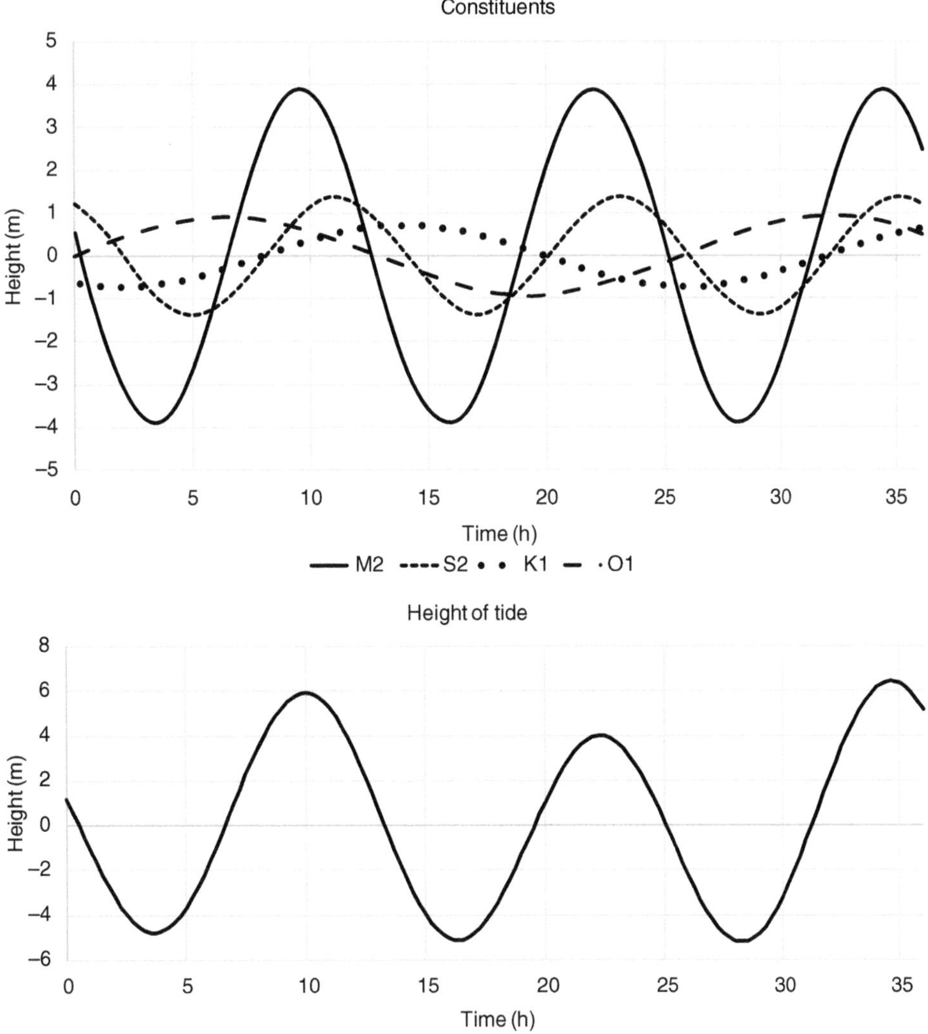

Figure 7.5 Use of harmonic constituents and Equation 7.1 to predict the height of a tide.

Figure 7.5 shows an example of how the constituents are combined to provide a prediction of the tide at a site. Equation 7.1 was used together with four of the tidal constituent data shown in Table 7.4. In this simple example using data from the Severn, the M2 constituent dominates the height of the tide. The result of combining these constituents is a tide with a maximum range of 11 m over the 36 hours shown.

In order to classify the type of tide, a form factor, F, is defined from the amplitudes of the harmonic constituents as

$$F = \frac{h(K1) + h(O1)}{h(M2) + h(S2)}$$

The form factor is used to define the type of tide as shown in Table 7.5.

Table 7.5 Type of tide defined by form factor

Value of form factor	Type of tide
0–0.25	semi-diurnal
0.25–1.5	mixed, mainly semi-diurnal
1.5–3	mixed, mainly diurnal
> 3	diurnal

Example 7.1 – Type of a Tide

Determine the type of tide that has the harmonic constituents given in Table 7.4.

Solution

$$F = \frac{0.72 + 0.92}{3.89 + 1.37} = 0.31$$

This indicates a mixed, mainly semi-diurnal tide.

7.1.3 Tidal Range Generation

Figure 7.6 shows a simple representation of a tidal range generation scheme. A barrage is constructed to separate water in the impounded area (the basin) from that in the sea. The maximum energy available during each tidal cycle is the work done in raising and lowering the water. Hence for each ebb and flow of the tide the maximum available energy (E) is

$$E = \int_0^H \rho A g h \, dh = \rho g \frac{AH^2}{2} \quad [\mathrm{J}] \tag{7.2}$$

where

A is the area of the basin in m^2,
H is the tidal range in m,
ρ is the density of seawater, 1025 kg/m^3,
g is the acceleration due to gravity, 9.81 m/s^2.

The tidal day is 24.84 hours. Hence, assuming energy can be extracted from equal semi-diurnal ebb and flood tides, the total daily energy is

$$E = 4\rho g \frac{AH^2}{2} \times \frac{24}{24.8} \quad [\mathrm{J}]$$

The average power available ($P_{average}$) may be estimated, by dividing the energy produced by the length of the lunar day, as

$$P_{average} = \frac{4 \times \rho g \frac{AH^2}{2}}{24.8 \times 3600} \quad [\mathrm{W}]$$

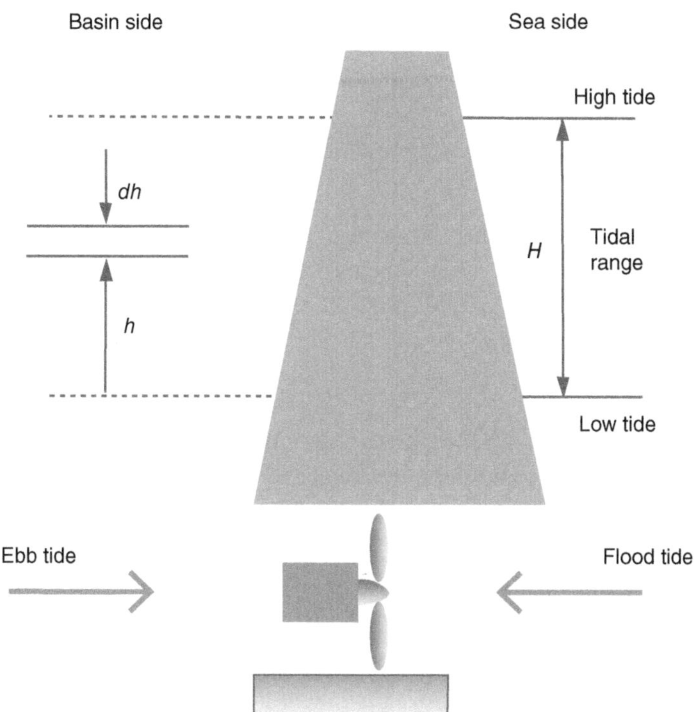

Figure 7.6 Tidal range scheme.

Example 7.2 – Power Available in an Estuary

Estimate the maximum average power that might be generated from an estuary. Assume the average useable tidal range is 6 m and the basin area is 22 km^2 and is constant with depth.

Solution

The potential energy available in each ebb or flow of the tide is

$$E = mgh \quad [\mathrm{J}]$$

The mass of water moved is

$$m = \rho AH \quad [\mathrm{kg}]$$

The mean head is

$$h = \frac{H}{2} \quad [\mathrm{m}]$$

The energy available in each ebb and flood tide is

$$E = mgh = \rho AH \times g \times \frac{H}{2} \quad [\mathrm{J}]$$

The average power over a lunar day is

$$P_{average} = \frac{4 \times 1025 \times 9.81 \times 22 \times 10^6 \times \frac{6^2}{2}}{24.84 \times 3600} \text{ W}$$
$$= 178 \text{ MW}$$

This estimate gives only a very initial indication of the maximum resource but does illustrate how the output of a tidal range scheme is proportional to the area of the basin A and the square of the useable tidal range H.

7.1.4 Ebb Generation

In practice, tidal generation schemes often use only ebb generation. The operation of an ebb generation scheme is shown in Figure 7.7. Water is allowed to enter the basin through large sluice gates during the flood tide until the heights of the water within and outside the basin are equal. The sluices are closed, and the water is then held in the basin as the level of the sea drops. When the minimum head required to operate the generating sets efficiently is reached, the water held in the basin is released through the turbines. The head seen by the turbines varies continuously as the water levels of the sea and in the basin drop at different rates.

The mass of water that enters the basin during each flood tide is

$$m = \rho A \frac{H}{2}$$

During generation, the mean head seen by the turbine is the difference between the water levels inside and outside the basin. It may be calculated from areas 1 and 2 marked in Figure 7.7.

$$\text{Area 1} = \frac{T}{2} \times \frac{H}{2} \times \frac{1}{2}$$
$$\text{Area 2} = \frac{T}{2} \times \frac{H}{2} \times \frac{2}{\pi}$$

where

H is the tidal range,
T is the period of the tide.

Therefore the mean head is

$$\frac{\text{Area 1} + \text{Area 2}}{\text{T/2}} = \frac{H}{2} \times 1.137 = 0.568H \quad [\text{m}]$$

so the work done per cycle is

$$E = m \times g \times 0.568H = \rho A \frac{H}{2} \times g \times 0.568H \quad [\text{J}]$$

and the energy available each tidal day from both of the two ebb tides is

$$E = \rho g A H^2 \times 0.568 \quad [\text{J}]$$

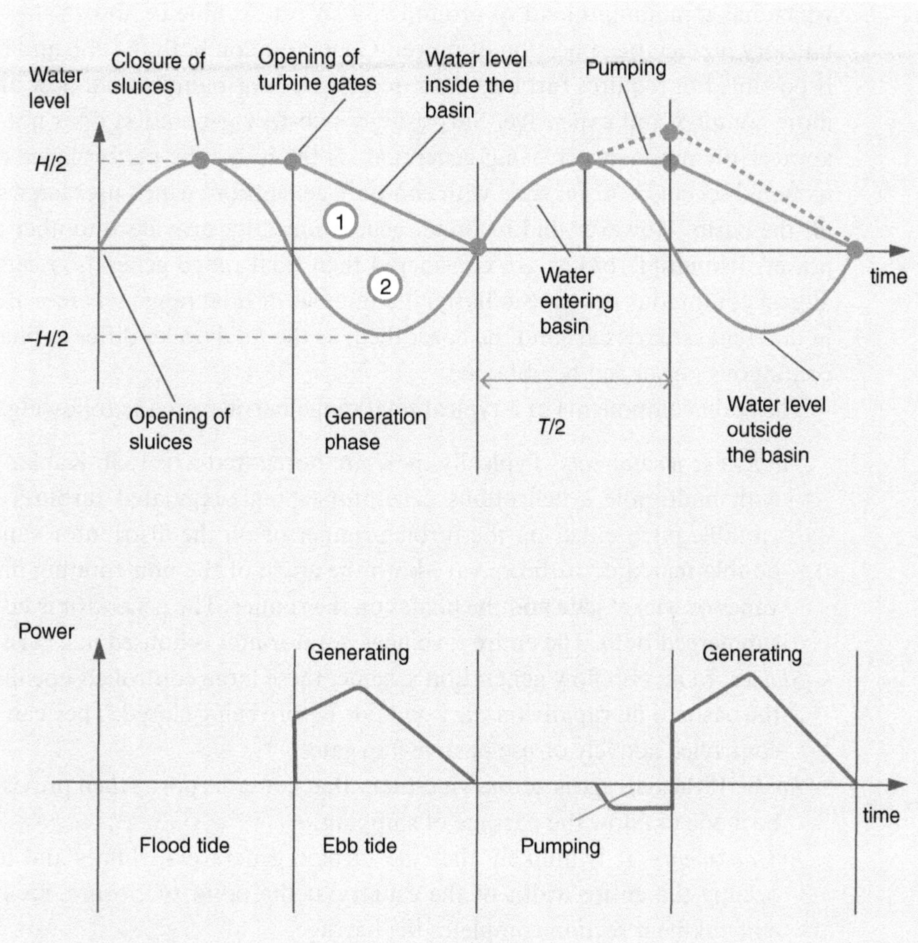

Figure 7.7 Ebb generation.

Ebb generation provides only ~28% of the theoretical resource given by Equation 7.2 with injection of electrical power for a period of only around 4–6 hours twice each day. In any practical scheme it is also necessary to account for hydraulic losses and the minimum head needed to operate the turbines. Any practical scheme that is proposed is evaluated by simulation and its operation optimized.

The turbine generators are rated above the mean power level and give a power output as shown in the lower trace of Figure 7.7.

As shown in the second cycle of Figure 7.7, pumping on the flood tide can be used to increase the volume of water in the basin. However, this needs more complicated and hence expensive turbines and generators. One important advantage of pumping with an ebb generation scheme is that the pumping takes place against only a small head, while generation occurs at a higher head.

A large tidal scheme, such as the proposed Severn Barrage described in Table 7.2 with a peak power output of 8.6 GW, would be connected to the GB power system, which has a minimum load of around 25 GW and is able to absorb this variable, but very predictable, injection of power. Generation on both the ebb and flood tide is possible but requires turbines that are designed for bidirectional flow and so are more complex and expensive. Surprisingly two-way generation does not result in appreciably more energy being generated, as the hydraulic performance of a bidirectional scheme is lower than with ebb-only generation, which uses large sluices to fill the basin. However, bidirectional generation does provide smoother electrical power. If multiple basins are considered then tidal range generation can provide almost continuous power. Similarly, if a number of tidal range schemes are located in different estuaries around the coast then, as the tidal cycles differ in phase, more continuous power can be obtained.

The main components of a typical tidal range barrage are the following.

Turbine generator sets. Typically these are horizontal-axis bulb Kaplan turbines with multi-pole synchronous generators. Single-regulated turbines may use variable pitch either on the turbine runner or on the distributor vanes, while double-regulated turbines vary both the angle of the non-rotating distributor vanes or wicket gate and the blades on the runner. The generator is housed in a submerged bulb. The entire turbine generator unit is housed in a caisson.

Sluices. In an ebb flow generation scheme, these large controlled openings allow the basin to fill rapidly on the flood tide before being closed. They can either be controlled actively or use passive flap gates.

Locks. If the barrage is across an estuary that contains ports, then provision must be made to allow the passage of shipping.

Embankments. It is unlikely that the turbine generators, sluices and locks will occupy the entire width of the estuary at the point of closure, so a cheaper embankment section completes the barrage.

7.1.5 Turbine Generators for a Tidal Range Generation Scheme

The operating head of the turbines of a tidal range generating scheme will typically vary between 4 m and 8 m depending on the state of the tide and whether generation during both ebb and flood flow has been chosen. The peak spring tidal range of a good site may be greater than 8 m but for maximum electricity generation in a practical scheme this does not appear at the turbines (see the water level inside and outside the basin in Figure 7.7). With ebb-only generation the maximum head across the turbine will occur when there is low water outside the barrage and the basin level will then have dropped.

The turbine generator units for tidal range schemes are based on those used for low-head hydro schemes and have a high specific speed (see Section 3.6). However, there are significant differences between the details of the turbines used for low-head run-of-the-river hydro and those used for tidal range schemes. The seawater of a tidal range scheme leads to a requirement for additional corrosion protection either by the use of stainless steel in the turbines and/or by active cathodic protection. In contrast to a low-head hydro plant, the turbines of a tidal range scheme will experience a large number of starts and stops (at least twice per day) as well as continuous operation of the governing systems (the pitch of the runner blades and/or the position of the guide vanes of the wicket gate) due to the continuous changes in the head at the turbines through the tidal cycle.

In addition to their operation to generate power with a single direction of water flow, tidal range turbines may be designed for:

- generation during both directions of flow,
- pumping during both directions of flow,
- sluicing – the use of the turbine to empty or fill the basin but without generating or absorbing power.

An early design for the Severn Barrage scheme proposed using vertical-axis Kaplan turbines. The turbines proposed were double regulated with varying pitch blades on the runners and controllable guide vanes of the wicket gates. The advantage of the vertical-axis arrangement is that the electrical generators can then be located above the turbines in a dry plant room where it is easier to cool and maintain them. However, in this arrangement the flow of water has to turn twice through 90° and so hydraulic efficiency is reduced.

Most subsequent designs of tidal range schemes (including those that have been constructed) have used horizontal-axis turbines. These types can be:

- bulb,
- straight-flow (known as STRAFLO),
- tubular.

The La Rance scheme and Sihwa Lake both used bulb turbines with the electrical generator located in a sealed enclosure (Figure 7.8). The generator was positioned on the basin side of the turbine at La Rance and on the sea side at Sihwa Lake. The turbines were double regulated with variable pitch blades on the turbine runners and variable guide vanes forming the wicket gate. At La Rance the turbines were capable of bidirectional operation (in both ebb and flood flow) and could also be used in pumping mode. This flexibility allowed the operation of the scheme to be optimized in order to generate electricity at times of maximum price. The 10 MW generators had an operating voltage of 3.5 kV giving a current of 1.7 kA with local transformation up to 220 kV for connection to the transmission grid. At Sihwa Lake the turbines generate during the flood tide and are operated in sluicing mode (i.e. unpowered) to allow maximum flow of water out of the basin during ebb flow.

A disadvantage of the bulb turbine design is that the enclosure housing the generator obstructs the flow of water and so introduces an additional hydraulic head loss. To minimize this head loss, the diameter of the bulb is restricted, resulting in an electrical generator which is difficult to cool. The relatively long thin generator has a low inertia, which has undesirable consequences for the stability of the power system.

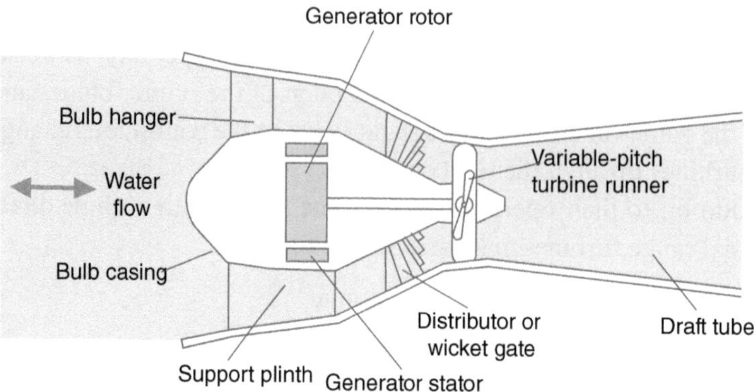

Figure 7.8 Bulb generator and double-regulated horizontal-axis Kaplan turbine.

An alternative design is the rim generator and straight-flow turbine. This was marketed under the trade name STRAFLO. The rotor of the generator is placed around the periphery of the turbine with the stator outside it. Rotating seals are used to retain the water within the turbine. The pitch angle of the blades of the turbine runner is fixed, so the static guide vanes in the distributor are used to regulate the turbine output power. The Annapolis Royal straight-flow turbine used a turbine runner diameter of 7.6 m to produce 20 MW.

Another conceptually simple approach is to use a long, inclined tubular shaft with the generator and possibly a gearbox located away from the turbine in a dry turbine hall. However, this tubular arrangement requires a diversion of the flow of water, and the long, high torque shaft that is required to connect the turbine and generator or gearbox can be subject to torsional instabilities.

7.1.6 Environmental Impact

The construction of a tidal range generating scheme (as with any major renewable energy project) will only be given permission to proceed by the civil authorities following a satisfactory, comprehensive environmental assessment and probably some form of public enquiry. A scheme that generates on the ebb tide will increase the mean water level in the basin above the mean sea level outside the barrage and reduce the tidal range within the basin by about 50%. This change in water level within the basin and reduction in the tidal range is likely to reduce the area of tidal mud flats that are exposed during the tidal cycle. The tidal mud flats provide valuable feeding grounds for wading birds and this potential loss of feeding habitat, particularly for migrating birds, was an important consideration during the evaluations of the Severn Barrage project.

Other environmental aspects that require careful attention include the following.

Salinity. The salinity of the water within the basin is likely to be reduced due to the restrictions of the exchange of water between the basin and the sea.

Turbidity. The level of suspended solids in the basin depends strongly on the velocity of the water flow, which is likely to be reduced by the barrage. Thus the turbidity in the basin will reduce, leading to deeper penetration of sunlight into the water, improving water quality and increasing marine life. However, care is needed to avoid silt build-up as solid particles drop out of the water.

Pollutants, both biodegradable and inorganic. The banks of suitable large estuaries near to cities, which provide the electrical load, are likely be highly populated and also have industrial facilities that discharge pollutants into the sea. Hence any change in the flows of the estuary must be modelled carefully.

Fish. The passage of migrating fish through the turbine at Annapolis Royal was a cause for concern and, as is usual practice with hydro stations, a fish ladder was installed. Fish mortality is likely to be the subject of detailed study for any tidal range scheme on an estuary through which fish migrate.

7.1.7 Tidal Lagoons

Any large tidal range development is likely to have significant environmental impact and attract opposition from environmental groups as well as from shipping interests and other users of the estuary. To avoid these objections, there is currently considerable interest in tidal lagoons that enclose an area of tidal water but do not restrict access to or change the environmental conditions of a tidal estuary. Rather than build a relatively short barrage across the mouth of an estuary to form the impounded area, a tidal lagoon is created by a much longer embankment that either encircles the lagoon completely or extends out from one shore of an estuary. Low-head turbines are used to generate electricity from the potential energy of the difference in water height inside and outside the lagoon as the tide changes in the same way as a tidal barrage scheme.

The supporters of tidal lagoons argue that they are attractive for the following reasons.

- They can be located in shallow water and so the cost per metre length of the barrage is reduced.
- They have much reduced environmental impact by not reducing the area of intertidal mud flats of the estuary or impeding the path of migrating fish.
- They interfere much less with established uses of an estuary.

The obvious disadvantage of a tidal lagoon is that the length of embankment for a given area of basin will be much longer than for a tidal barrage across the mouth of an estuary. Hence the cost of construction of the scheme is likely to be higher than for an equivalent traditional tidal range scheme that uses the banks of both sides of an estuary to create much of the impounded area of the basin. One rule of thumb that has been used when assessing the economic feasibility of sites for a tidal barrage generating scheme is that the ratio of the length of the barrage (in metres) to the impounded area (in km^2) should not exceed 80. This ratio for both the La Rance scheme and the Severn Barrage proposal is around 33. By not making use

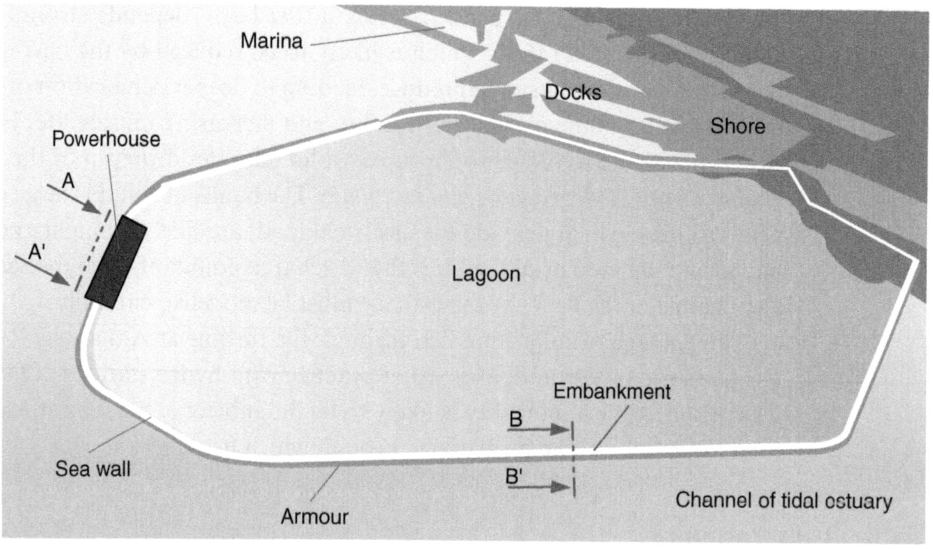

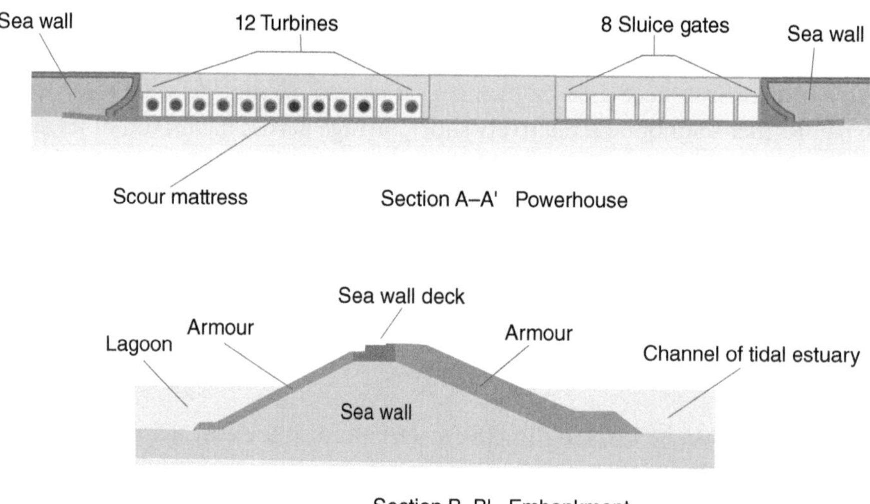

Figure 7.9 Illustration of a proposed tidal lagoon.

of the large impounded area that is created by the banks of the estuary, the ratio of barrage length to impounded area for a tidal lagoon greatly exceeds this ratio. So far, no tidal lagoon generating scheme has been built, although a number of studies have been undertaken.

Figure 7.9 shows an illustration of a proposed tidal lagoon. An embankment is built out from the shore to enclose part of the tidal estuary and form an enclosed basin. The embankment consists of a long sea wall and a much shorter powerhouse. The powerhouse contains multiple axial turbine generators as well as sluice gates.

The sea wall forming the embankment has rock armour protection and there is a scour mattress against erosion by turbulent flow near the powerhouse. Note that the marina and docks are outside the impounded area, avoiding the need for large locks. The sea wall deck and the lagoon may be used to site wind turbines or developed for leisure activities.

7.2 Tidal Stream Generation

The successful operation of the tidal range power plants at La Rance, Annapolis Royal and Sihwa Lake has demonstrated that the tides can be exploited effectively using tidal barrages to harness the potential energy of the difference in the water levels in an impounded basin and the sea. However, the barrage required for such a scheme is a major civil engineering structure that has a long construction period, so it is difficult to finance with loans from commercial sources. The change in water levels and flows within the basin can have a significant environmental impact, and tidal range schemes that alter the conditions in an estuary are likely to be opposed by some environmental groups. In spite of the technical success of these tidal range demonstration plants and numerous studies and proposals, no large tidal range generating scheme has been constructed in Europe or North America since 1984. Hence, attention has shifted to harnessing the kinetic energy of the marine currents that are caused by the tides. It is thought that tidal stream turbines will be easier to finance and can be installed with lower environmental impact than a tidal barrage. A number of countries have active programmes to support the development and demonstration of tidal stream energy devices, and the UK has a testing station at the European Marine Energy Centre on Orkney, where a variety of designs are demonstrated at ratings of up to ~1 MW. Architectures being investigated include horizontal and vertical axis, bottom mounted and floating, as well as shrouded rotors, but a preferred architecture has yet to emerge.

Figure 7.10 is an illustration of a horizontal-axis, bottom-mounted tidal stream turbine. Note the clearance to the seabed and the tripod support structure with tubular legs to resist the large thrust forces. The nacelle is orientated to face the direction of flow with a yaw drive and power is transmitted through flexible electricity cables.

7.2.1 The Tidal Stream Resource

The gravitational pull of the moon and the sun, together with the rotation of the earth, leads to a change in the height of the earth's oceans. The variation of the water height creates a long period wave that moves across the oceans. When the tidal wave reaches the shore, it creates the variation in the height of the water that is the tidal range (see Section 7.1).

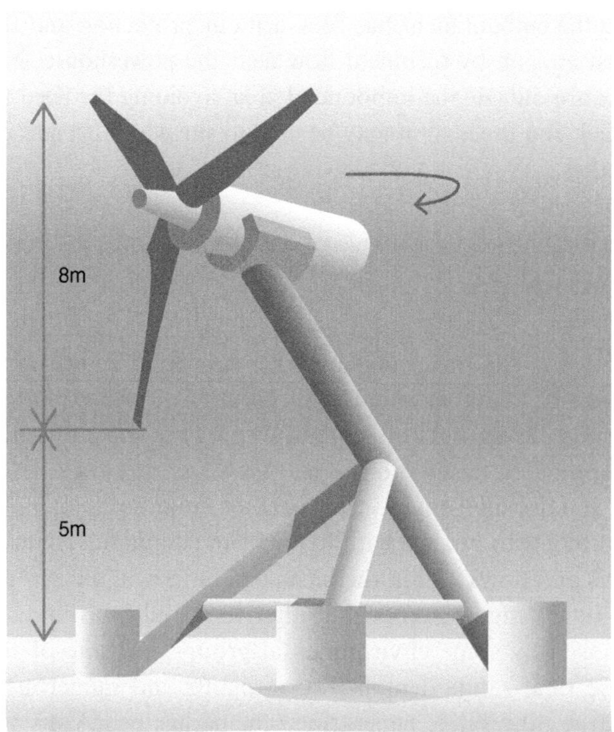

Figure 7.10 Illustration of a tidal stream turbine.

In the open ocean the speed of the tidal wave is low (0.02–0.05 m/s) but it is accelerated by underwater bathymetry and topography, particularly in narrow channels such as in the straits between islands and the mainland, at the entrances to sea locks and in shallow seas around headlands. In these locations during spring tides the peak velocities of the tidal stream currents can exceed 3 m/s. The speed of the tidal stream currents varies at the same frequency as the tidal range, but in estuaries and bays often with a different phase. It can be described using a similar technique of harmonic constituents that was used for predicting the tidal range (Section 7.1).

U, the speed of a tidal current, is described by:

$$U(t) = U_0 + \sum_{n=1}^{m} U_n \sin(\omega_n t + \alpha_n)$$

where

$U(t)$ is the speed at time t,
U_0 is the mean current speed,
U_n is the speed of the nth harmonic tidal current constituent,
ω_n is the angular velocity of the nth harmonic tidal current constituent,
α_n is the phase displacement of the nth harmonic tidal current constituent,
t is time,
m is the number of harmonic tidal current constituents.

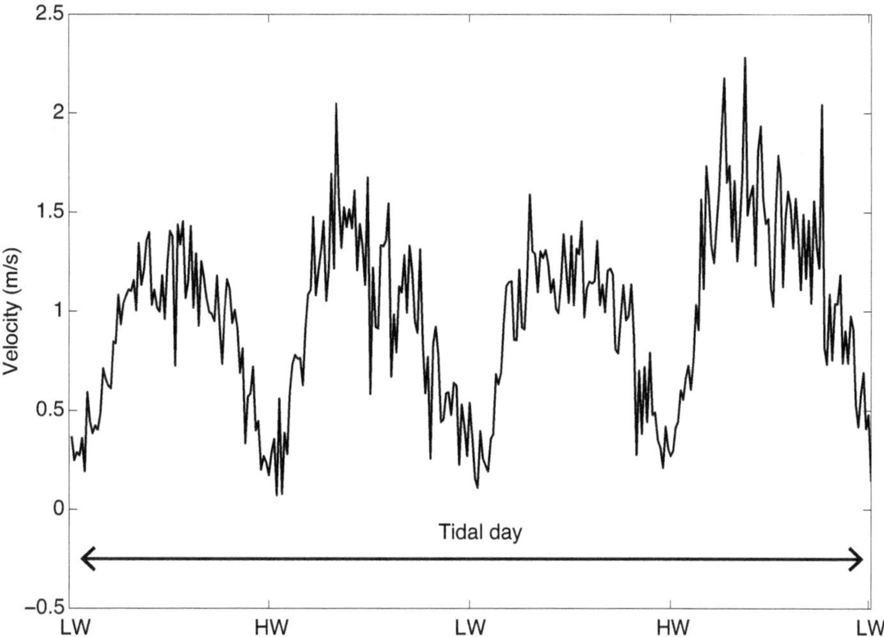

Figure 7.11 Example of the speed of a tidal stream over a lunar day (LW low water, HW high water). Note the ebb flow has been inverted to show scalar speed.
Source of data: Tidal Energy Ltd.

The speed of the tidal stream varies continuously with a reversal in the direction of the flow with each semi-diurnal cycle. Figure 7.11 is an example of the flow speed over a lunar (tidal) day. To obtain these data, a series of measurements of the magnitude and direction of the tidal stream flow were made every 5 minutes at a distance from the seabed of up to 30 m using an acoustic Doppler current profiler. At low and high water (LW and HW) the speed drops to near zero and reaches its maximum midway between these two points. The higher-frequency turbulent variations are superimposed on the daily cycle. It can be seen that these real measured tidal stream velocity data contain significant turbulence.

The turbulence of a tidal stream flow at depth is caused by obstructions and the roughness of the seabed. Nearer the surface, the turbulence is created by wave action. The level of turbulence varies with depth and depends on the wave climate, weather and local bathymetry and topography. As with wind energy, the longitudinal turbulence intensity I_u is defined as the ratio of the standard deviation of the speed of the flow σ to its mean $\overline{U}$. It is typically measured over, say, 10 minutes at a sampling rate of, say, 4 Hz.

$$I_u = \frac{\sigma}{\overline{U}}$$

Although flow conditions are site specific, the turbulence intensity of a tidal stream at a good site is likely to be around 10% compared with 15% for the turbulence of the wind at a good wind farm site. However, because of the much greater density of water than air, the forces generated by turbulence on a tidal stream device can be even more destructive than the effect of turbulent winds on a wind turbine. Hence it is usual to measure the turbulence of the flow when assessing a potential site for tidal stream generation, and also consider the impact of any adjacent turbines that might be sited in an array.

The direction of the tidal wave in the oceans is determined by a complex interaction of the gravitational pull of the moon and sun, the landmasses, local bathymetry and the rotation of the earth, including the Coriolis force. The tidal wave and the associated change in tidal range result in a tidal stream flow that varies in direction and magnitude. When represented as a polar plot of magnitude and direction the tidal stream describes an ellipse over the tidal day.

In some locations of good tidal stream resource, e.g. channels between islands or at the entrance to sea locks, the ellipse is compressed and the polar plot becomes bidirectional. Figure 7.12 shows a polar plot of 5 minutes averaged data at a site with strong bidirectional flows. The speed of the ebb flow (at 90°) is slightly greater

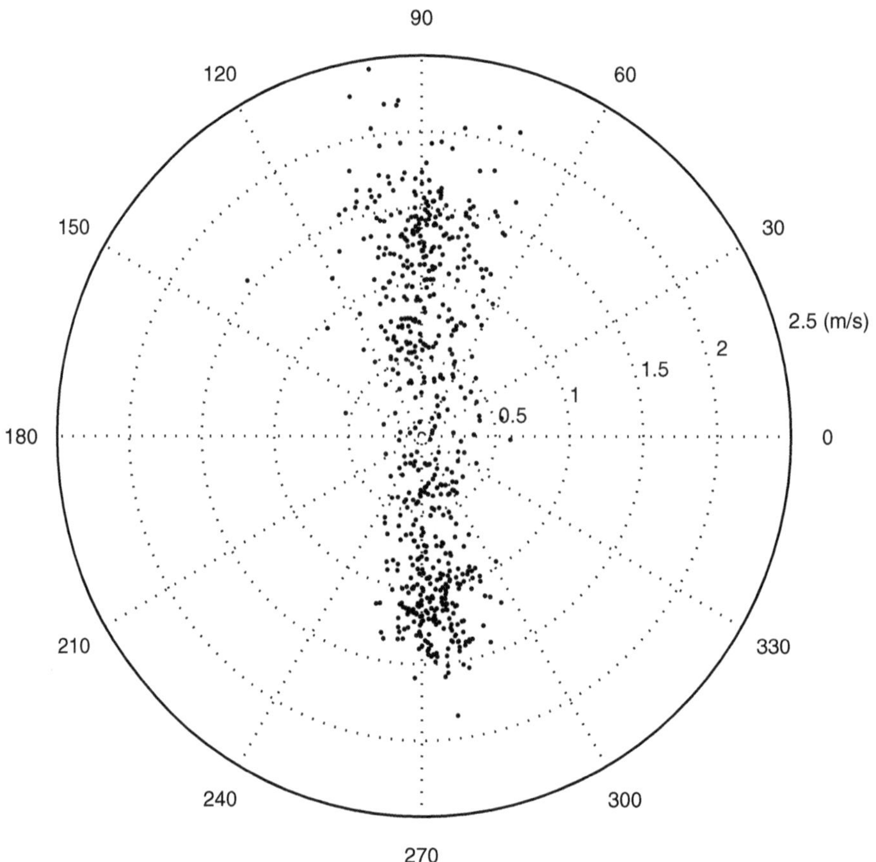

Figure 7.12 Polar plot of the magnitude and direction of a bidirectional tidal stream flow. Source of data: Tidal Energy Ltd.

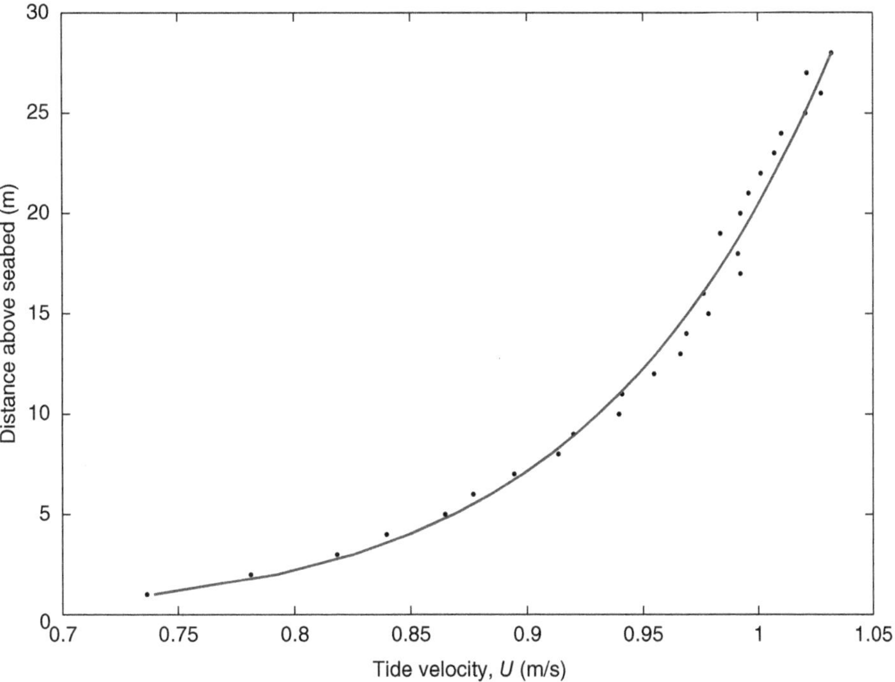

Figure 7.13 Variation of speed of a tidal stream with distance above the seabed. Curve fitted using the one-tenth power law between 1 m and 28 m.
Source of data: Tidal Energy Ltd.

than the flood flows (at 270°). Other sites, including those at which an array of multiple turbines is installed, may result in a more elliptical trace of the tidal stream when shown on a polar plot.

The mean velocity of the tidal stream flow varies with the height above the seabed (known as the shear) and the top and bottom of a rotor can experience significantly different flow speeds. Figure 7.13 shows the variation of mean flow speed with height above the seabed. Until detailed measurements are taken, it is usual to assume that the profile of water speed follows a one-seventh (IEC 62600-2) or one-tenth power law. This variation of flow speed with depth will create a cyclic loading at the rotational frequency on a horizontal axis rotor and an additional overturning moment on any tidal stream device.

Example 7.3 – Variation of Tidal Stream with Depth

Estimate the difference in the speed of the flow at the top and bottom of the rotor shown in Figure 7.14. Use the one-tenth power law to calculate the change of water velocity with depth.

Assuming a constant thrust coefficient C_T, estimate the difference in force at the top and bottom of the rotor.

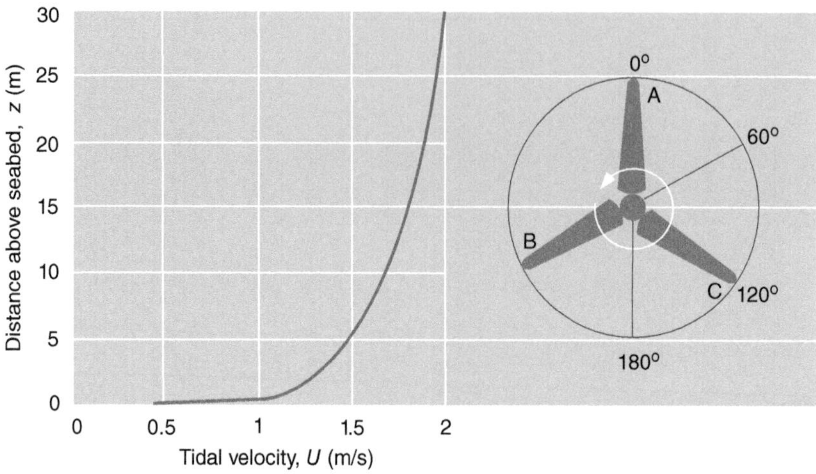

Figure 7.14 Variation of the velocity of the tidal stream flow with distance above seabed.

Solution

$$U = U_0 \left(\frac{z}{d} \right)^{1/x}$$

where

U is velocity at distance z above the seabed in m/s,
U_0 is surface velocity in m/s,
z is the distance above the seabed in m,
d is the depth of water in m,
x is the exponent, usually 7 or 10.

Using the one-tenth power law:

$$U_{25} = 2\left(\frac{25}{30} \right)^{1/10} = 2 \times 0.982 = 1.96 \text{ m/s}$$

$$U_5 = 2\left(\frac{5}{30} \right)^{1/10} = 2 \times 0.836 = 1.67 \text{ m/s}$$

The variation in water speed between the top and bottom of the rotor, ΔU, is

$$\Delta U = \frac{1.96 - 1.67}{1.96} = 14.8\%$$

The thrust force on a rotor is

$$F = \frac{1}{2} \rho A C_T U^2$$

where

C_T is the thrust coefficient.

Therefore, the variation in force between the top and bottom of the rotor, ΔF, is

$$\Delta F = \frac{1.96^2 - 1.67^2}{1.96^2} = 27\%$$

7.2.2 Tidal Stream Turbines

Over the past 30 years, the design of modern commercial wind turbines has converged to a common architecture of three-bladed, horizontal-axis, upwind-rotor, variable-pitch blades and variable-speed operation (see Chapter 2). In contrast to this established design for wind turbines, tidal stream turbines are at a much earlier stage, with a number of concepts under development, and no dominant architecture has yet emerged. For each of the major elements of a tidal stream turbine there remain several options that are being investigated and demonstrated. These are listed in Table 7.6.

The tidal stream of a semi-diurnal tide changes direction through approximately 180° every 6 hours. Vertical-axis crossflow turbines can naturally make use of flow from any direction while a horizontal-axis rotor must face the flow. Some horizontal-axis designs use an active yaw drive to orientate the nacelle so that the rotor is orthogonal to the flow. In a strongly bidirectional flow, a rotor of a fixed yaw, with horizontal-axis propeller turbine blades, can use symmetrical hydrofoils, or the blade pitch angle can be altered for operation on both the ebb and flood tides. All these approaches have been investigated and demonstrated. Two-bladed crossflow rotors generate significant cyclic torques as the blades experience different apparent water speeds over a rotation and if a vertical-axis rotor is mounted on the seabed the lower part of the blades will experience lower water speeds than the top of the rotor owing to the shear profile of the stream.

Some horizontal-axis architectures use ducts (sometimes known as shrouds) to accelerate the flow of water through a reduced-diameter turbine. Ducts were investigated for horizontal-axis wind turbines and the idea discarded, as they have to be orientated to face the wind and this proved to be prohibitively expensive.

Table 7.6 Options for main elements of a tidal stream turbine

Main element of tidal stream turbine	Options
Rotor	Propeller type Crossflow (horizontal or vertical axis) Shrouded or bare Rotating or oscillating hydrofoils
Support structure	Bottom mounted: tripod or mono-pile, gravity or piled Floating
Power regulation	Yaw Blade pitch Stall Overspeed
Generator	Fixed or variable speed Electrical high-speed generator through a gearbox Electrical multi-pole or rim generator Hydraulic transmission to shore

However, ducts are potentially useful for those tidal streams where the water flow is essentially bidirectional, reversing through 180° from the ebb to the flood flow. In such flows, a symmetrical duct can be used to accelerate the stream of water into the rotor and then expand the wake. The use of a duct allows a smaller-diameter turbine to be used for a given power output with a particular free stream velocity. The Betz limit still applies to ducted tidal stream turbines provided the calculation is based on the area of the duct and the flow is unconstrained by the surface of the water. Ducts accelerate the flow through a smaller diameter and reduce turbulence striking the rotor.

A rotating shaft is required to drive a conventional electrical generator but there have been attempts to use an oscillating hydrofoil to capture energy from the tidal stream and transmit the power using a high-pressure oil hydraulic system. The use of a linear electrical generator with an oscillating hydrofoil has been considered but not well developed. More speculative ideas to capture energy from the tidal stream include the use of tidal kites and Archimedes screw devices. Tidal kites are similar in concept to wind kites where either the tether is repeatedly extended and retracted to operate a fixed generator or alternatively the kite has a rotor and generator mounted on it and it generates electricity as it is 'flown' in the figure-of-eight pattern.

A variety of approaches have been used to locate the turbine in the tidal stream flow. Turbines have been mounted on a steel structure (often a tripod) that is fixed by gravity, by hydrofoils providing a down force or with piles into the seabed. In a similar manner to a large offshore wind turbine, a monopile may be driven into the seabed and one or more rotors supported from it. The monopile allows the turbine rotors to be yawed to face into the flow and raised above the surface of the water for maintenance. Floating turbines can be located using flexible tethers but then a flexible power cable must be used.

The speed of the tidal stream flow varies from zero at slack water to its maximum during peak spring tidal flows. It is unlikely to be cost-effective to design the generator and power take-off to accept all the power available in the tidal stream during peak spring tidal flows, so some means of rotor power control is required. The three usual techniques to regulate power are to yaw the entire rotor out of the flow, to pitch the blades by altering their angle with respect to the apparent flow or to allow the rotor to stall at high flow speeds. Stall control requires no moving parts but is likely to lead to high thrust forces. The requirement to control the overspeed of the rotor when a grid loss occurs favours the use of active pitch or yaw power control. At least one prototype design limits the power at high water speed by allowing the rotor to accelerate and so reduce the power coefficient by increasing the tip speed ratio.

Variable-speed operation of the generator and rotor is desirable – to maximize energy capture with the varying flow speeds, reduce mechanical loads and maintain good power quality. For variable-speed operation a power electronic converter is required to interface the generator to the local utility network. If this is not located beneath the water, then variable-frequency electricity is taken ashore. Fixed-speed operation is possible with a simple fixed-speed induction generator but this will lead to higher loads on the turbine.

Figure 7.15 A two-bladed, twin-rotor, horizontal-axis, variable-pitch tidal stream turbine mounted on a monopile.
Photo: RenewableUK. Turbine manufactured by Marine Current Turbines.
(A black and white version of this figure will appear in some formats. For the colour version, refer to the plate section.)

Figure 7.16 A three-bladed, horizontal-axis, fixed-pitch, bottom-mounted tidal stream turbine.
Photo: Tidal Energy Ltd.
(A black and white version of this figure will appear in some formats. For the colour version, refer to the plate section.)

Figure 7.17 A ducted horizontal-axis tidal stream turbine. Photo: RenewableUK. Turbine manufactured by Openhydro. (A black and white version of this figure will appear in some formats. For the colour version, refer to the plate section.)

In summary, a number of competing concepts are being investigated and demonstrated for tidal stream turbines and the most economically attractive architecture has yet to emerge. Examples of tidal stream turbine prototypes are shown in Figures 7.15–7.17.

7.2.3 Development of a Tidal Stream Project

The development of a tidal stream project follows a similar path to that of a wind farm, and indeed any renewable energy project (see Section 9.1). Initially a search is undertaken to identify a site with suitable tidal stream currents of high average flow speeds during both neap and spring tides and with low turbulence. The depth of water and seabed must be suitable for the type of tidal stream device being considered. It is usual to allow a depth of 5 m between the water surface level at the lowest tide and the highest point of any turbine to avoid small boats; a similar clearance is allowed at the seabed to ensure that the variation in water speed across the rotor disk is not too great and that debris does not strike the rotor.

An initial estimate of the speed of the tidal currents can be obtained from an atlas of the marine energy resource, in the same way that an initial indication of wind speeds for the development of a wind farm can be obtained from a wind energy atlas. A tidal stream atlas gives a useful general indication of the resource but, as the project is developed, progressively more detailed measured data are acquired and computer modelling undertaken. Suitable measured data in the form of harmonic constituents may be available from previous studies or they can be derived from site measurements taken using either a towed or fixed acoustic Doppler instrument. These site measurements of the time series are decomposed into the harmonic constituents, which are then used to predict the long-term tidal currents. Before construction of a scheme is finally sanctioned it is recommended that at least a three-month record of site data is obtained. This is used to calibrate a 2D or 3D numerical model of the site.

Figures 7.11–7.13 illustrated the complex nature of the flow in a tidal stream, and site measurements to calibrate a numerical model are required before the design of any scheme can be finalized and its energy output estimated with confidence. Access for installation and maintenance is particularly difficult for tidal stream turbines, which must be placed in fast tidal flows. Access may only be possible for a few hours around high and low water (Figure 7.11).

7.2.4 Comparison of a Tidal Stream Turbine with a Wind Turbine

The basic principle of operation of a tidal stream turbine is similar to that of a wind turbine. Energy is extracted from a flow of a fluid and the output power is proportional to the cube of the speed of the flow. If the flow of water is assumed to be unconstrained by the seabed and the water/air free surface, linear momentum theory allows a useful simple comparison of a tidal stream turbine with a wind turbine. In practice, the assumption of unconstrained water flow may not be realistic as tidal stream turbines are likely to be installed in dense arrays in flows that are constrained by limited depth and local topography. Hence, although a useful insight, this use of axial momentum theory is only approximate.

The major differences between a wind and tidal stream turbine are as follows.

- The density of water is around 837 times that of air.
- The rated flow speed of a tidal stream turbine is around ¼ that of a wind turbine,
- The mean flow speeds of a tidal stream can be predicted with a high degree of accuracy.
- The flow of a tidal stream is constrained by the free surface of the water/air interface, and turbulence is created by roughness of the seabed and wave action.
- A tidal stream turbine may be subject to cavitation.

Using linear momentum theory, the power generated by a tidal stream turbine is

$$P = C_P \frac{1}{2} \rho_w A U_\infty^3 \quad [\mathrm{W}]$$

where

C_P is the coefficient of performance of the rotor (typical maximum value around 0.45),
ρ_w is the density of water, 1025 kg/m^3,
A is the area of the rotor including any duct,
U_∞ is the free stream velocity of the water (typical rated value around 2–3 m/s).

The difference in the densities of water and air and in the speed of the flows of wind and the tidal stream leads to a tidal stream turbine needing a swept area around 1/10 of that of a wind turbine of similar rated power output, and blades of 1/3 the length for a horizontal-axis design (see Example 7.4). The smaller radius of the rotor reduces manufacturing and installation costs but leads to low inertia and makes control of the rotor more difficult.

Linear momentum or actuator disk theory (Section 2.4) considers the rotor of a turbine as an ideal element that extracts energy from a one-dimensional fluid flow. It ignores any variation of flow speed with height and applies equally to horizontal- and vertical-axis turbines. The theory shows that half of the reduction in the speed of the flow occurs upstream of the turbine and half downstream. For most effective operation the velocity of the downstream flow is 1/3 of its free upstream velocity and the axial induction factor a is 1/3. In this simple theory, the power extracted by the actuator disk is equal to the thrust force multiplied by the velocity of the flow through the rotor. Thus, the much lower velocity of the water across the tidal stream rotor leads to a much higher thrust force than for a wind turbine with a similar power rating. This is illustrated in Example 7.4.

Example 7.4 – Comparison of a Tidal Stream Turbine and a Wind Turbine

Use linear momentum theory to

(a) compare the radius of the rotor of a wind turbine and a tidal stream turbine with equal power rating of 1 MW,
(b) compare the thrust force on the rotor when operating with an axial induction factor, a, of 1/3.

Assume a rated free flow speed of 12 m/s for the wind turbine and 2.8 m/s for the tidal stream turbine, and that the two turbines have the same coefficient of performance, C_P, of 0.45. Take the density of seawater to be 1025 kg/m^3 and that of air to be 1.225 kg/m^3.

Solution

(a) *Comparison of rotors*

The power extracted by the rotor of a turbine in any fluid flow is

$$P = C_P \frac{1}{2} \rho \pi R^2 U_\infty^3 \quad [W]$$

where

C_P is the coefficient of performance of the rotor (typical maximum value around 0.45),
ρ is the density of the fluid in kg/m^3,
R is the radius of the swept area in m,
U_∞ is the free stream velocity of the fluid in m/s.

Therefore, for an equal power rating, the ratio of the areas of the rotor is

$$\frac{A_{wind\ turbine}}{A_{tidal\ stream\ turbine}} = \frac{\rho_{water}}{\rho_{air}} \times \frac{U_{water}^3}{U_{wind}^3} = \frac{1025}{1.225} \times \frac{2.8^3}{12^3} = 10.63 \simeq 10.6$$

and the ratio of the radius of the rotors (R) is

$$\frac{R_{wind\ turbine}}{R_{tidal\ stream\ turbine}} = \sqrt{\frac{A_{wind\ turbine}}{A_{tidal\ stream\ turbine}}} = \sqrt{10.63} = 3.26 \simeq 3.3$$

For 1 MW power and a C_p of 0.45 this leads to rotors of radius

$$R_{wind\ turbine} = 25.85 \simeq 26\,\text{m}$$

$$R_{tidal\ stream\ turbine} = 7.93 \simeq 8\,\text{m}$$

(b) *Comparison of thrust forces*

Linear momentum or actuator disk theory states that at the Betz limit when $a = 1/3$, the velocity of the flow at the disk, U_D, is 2/3 of the free stream velocity, U_∞. Also, remembering that the power extracted from the fluid flow is equal to the force on the disk multiplied by the speed of the fluid flow at the disk, we have

$$P_{turbine} = F \times U_D$$

$$F = \frac{P_{turbine}}{U_D} = \frac{P_{turbine}}{2/3 \times U_\infty}$$

The table shows the steady-state thrust force experienced during operation by an ideal 1 MW turbine. The thrust force of a tidal stream turbine compared with a wind turbine is increased by the inverse ratio of the flow speeds.

Comparison of the thrust force on an ideal 1 MW wind and tidal stream turbine rotor assuming $a = 1/3$

	Wind turbine	Tidal stream turbine
Free stream flow speed velocity (m/s)	12	2.8
Velocity at disk (m/s)	8	1.87
Thrust force (kN)	125	536

Electricity-generating systems use the spinning inertia of the generators to store the kinetic energy that is required for their stable operation. The rotating kinetic energy stored in the rotor of a tidal stream turbine is significantly less than that of a wind turbine. The moment of inertia about the rotational axis of a rotor is given by

$$I = \int_0^R mr^2\, dr$$

where

m is the mass of a blade element,
r is the effective radius of a blade element dr,
R is the length of the blade.

Therefore, for a reduction in the radius of the rotor by 1/3, one might expect a reduction in the inertia by 1/9. However, it is normally assumed that the mass of a turbine blade scales with its volume, which is proportional to the cube of the rotor radius. This would lead to a reduction in the inertia of a tidal stream turbine

Table 7.7 Comparison of the inertia and rotating kinetic energy in the rotor of a three-bladed, horizontal-axis 1 MW wind and tidal stream turbine assuming a tip speed ratio of 5 ($\lambda = 5$)

	Wind turbine	Tidal stream turbine
Mass of a blade (kg)	3000	1500
Centre of mass of a blade (from hub)	$0.4R$	$0.4R$
Radius of the rotor, R (m)	26	8
Inertia of three-bladed rotor (kgm^2 × 10^3)	973	46
Rotational speed (rad/s)	2.3	1.8
Rotating kinetic energy of rotor (kJ)	2573	74

compared with a wind turbine of similar rating by $1/3^5 = 1/243$. In practice, the blades of tidal stream turbines are heavier than those of wind turbines (not least in order to withstand the high thrust forces) but the inertia of a tidal stream turbine is still likely to be significantly lower than that of a wind turbine. This is illustrated in Table 7.7 using realistic assumed values for the mass of the blades and their centre of mass. This calculation indicates that the ratio of the inertia of a wind turbine to a tidal stream turbine rotor is approximately 20:1. For a similar tip speed ratio of 5, the tidal stream turbine rotates slightly slower than the wind turbine and the ratio of the rotating kinetic energy of the wind turbine to the tidal stream turbine is 35:1. Therefore the tidal stream rotor will appear very 'light' and difficult to control. As the turbulence intensity of a tidal stream is similar to that of a wind turbine site, the control of the speed of a tidal stream turbine rotor may be difficult. This is sometimes addressed by adding mass to the blades of a tidal stream turbine.

Example 7.5 illustrates the use of linear momentum theory to investigate the performance of a tidal stream turbine.

Readers may find it useful to refer to Section 2.4 for a revision of linear momentum theory.

Example 7.5 – Performance of a Tidal Stream Turbine

A tidal stream turbine has a rotor diameter of 16 m. It operates in a tidal stream of 2 m/s. The water speed at the rotor is 1.7 m/s.

Using linear momentum theory (see Section 2.4) calculate

(a) the axial induction factor a,
(b) the power and thrust of the rotor.

Solution

(a) Axial induction factor:

$$a = 1 - \frac{U_D}{U_\infty} = 1 - \frac{1.7}{2} = 0.15$$

(b) The power coefficient is

$$C_P = 4a(1-a)^2 = 4 \times 0.15 \times (1-0.15)^2 = 0.4335$$

The power in the flow of water is

$$P_{WATER} = \frac{1}{2}\rho A U_\infty^3 = \frac{1}{2} \times 1025 \times \pi \times 8^2 \times 2^3 = 842.3\,\text{kW}$$

The power on the turbine shaft is

$$P_{SHAFT} = P_{WATER} \times C_p = 824.3 \times 0.4335 = 357\ \text{kW}$$

The thrust coefficient is

$$C_T = 4a(1-a) = 4 \times 0.15 \times (1-0.15) = 0.51$$

Therefore the thrust on the rotor is

$$T = C_T \times \frac{1}{2}\rho A U_\infty^2 = 0.51 \times \frac{1}{2} \times 1025 \times \pi \times 8^2 \times 2^2 = 210\ \text{kN}$$

7.3 Wave Power Generation

The power in ocean waves is obvious to any observer, and attempts to harness wave energy have been made for more than two centuries. Following the oil crisis of 1973, the generation of electricity from wave power was investigated by several nations and for some years the UK Government funded a particularly active wave energy research programme to exploit the very large wave energy resource in the Atlantic. More than 50 years later, generation from wave energy remains the subject of research and development with different concepts under investigation and demonstration. No generally preferred technical architecture of the most cost-effective wave energy generator has emerged. Prototypes of different concepts have been demonstrated and evaluated, some at full scale, but so far none has achieved a breakthrough to commercial production and operation. This slow progress is not unusual in the development of a new generation technology but does underline the difficulties of harnessing the power of the waves. Difficulties of the conversion of wave power to electricity are due to the following.

- The location of the wave energy resource offshore in a very demanding environment where access is difficult. Many wave energy devices are placed in the particularly challenging environment of the splash zone either just above or below the water line.
- The slow oscillatory motion. Electrical generation is much easier from a high-speed, low-torque rotating shaft than from a slowly oscillating large force.
- The variable direction and magnitude of the waves.
- The need to generate in modest waves of power, ~10 kW/m, but to survive extreme storms of more than ~1000 kW/m. This very large difference between normal operation and extreme loading is less significant in tidal range and tidal stream generation.

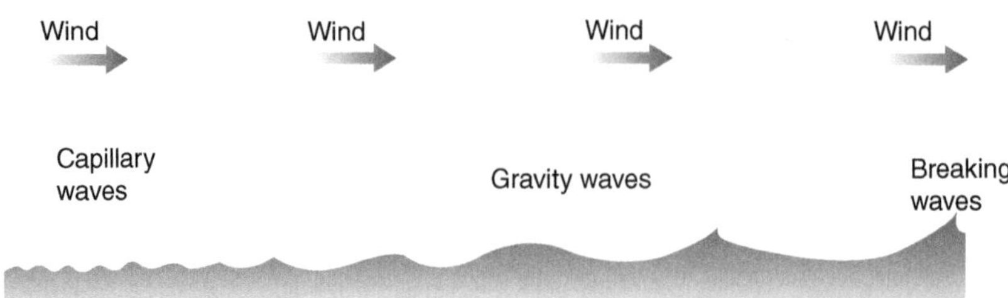

Figure 7.18 Formation of waves.

Wave energy is created by radiation from the sun heating the earth and creating winds that flow over the oceans. As illustrated in Figure 7.18, the wind deforms the water surface and initially creates small ripples, called capillary waves. The ripples grow as the wind causes a pressure difference across them and energy is transferred progressively from the wind into the water. As the wind continues to blow, the height, wavelength and speed of the waves increase. Thus the formation of waves can be thought of as a concentration of solar energy through wind.

The larger waves are known as gravity waves. The main restoring force of capillary waves is surface tension, while gravity is the main restoring force of the larger waves. As the wave height increases, the crests of the gravity waves become more pointed and the troughs more rounded. The energy of the waves increases with the wind speed, length of time the wind blows and the fetch. The fetch is the distance of water over which the wind blows.

The capillary waves have a wavelength of less than ~2 cm, which increases to ~10–50 m for typical gravity waves in the area near to their creation by the storm winds. As the wind continues to blow over a long fetch, the steepness of the waves (the ratio of wave height to wavelength) reaches around 1:7 before the waves break. The speed of the gravity waves can reach 1/3 of the wind speed.

When a storm occurs at sea, storm waves of different heights, direction and periods are created. As the waves travel away from the region of the storm, the higher-frequency, shorter-wavelength storm waves decay away, leaving only long-wavelength swell waves. The swell waves have a wavelength more than 100 m and relatively low height. They are particularly important for the generation of wave energy and can be analysed well with linear wave theory, which represents the waves as sinusoids.

Waves are surface phenomena, and in the deep oceans the swell waves can travel thousands of kilometres with little loss of energy. The waves striking the west coast of Europe may have originated in storms in the western Atlantic. The average power of the deep-water swell waves approaching the coastline of western Europe from the long fetch of the Atlantic is between 30 and 70 kW/m. Although a large amount of energy is transferred by the waves, there is limited net movement of water.

The worldwide wave energy resource has been estimated to be 2000–4000 TWh/year, equivalent to a continuous power of 230–450 GW. It is advantageous that the

Table 7.8 Estimates of the UK wave energy resource

	Offshore	Near-shore	Onshore
Distance to shore (km)	> 20	< 20	-
Depth (m)	> 50	20–50	-
Practical resource (TWh/y)	50–70	2.1–5.7	0.2–0.4
Equivalent continuous power (MW)	5710–8000	240–650	25–50

wave energy resource off the western coasts of Europe occurs mainly in the winter when the demand for electricity is high. The UK wave energy resource in October–March is twice that of April–September.

As a wave approaches the shore, it slows down, its wavelength decreases and it grows in height, which leads to increased steepness and the wave breaking. In shallow water, wave energy is lost through friction with the seabed and when the wave breaks. Hence only a fraction of the power of offshore deep-water waves reaches the shore. Near to the shore, the reduction in water depth leads to refraction and the waves becoming parallel to the contours of depth. The wave energy is then focused on certain locations, e.g. headlands.

Although some prototype wave power converters.have been located onshore or near-shore, where access for maintenance is easier, the energy resource is significantly less than offshore. Table 7.8 shows estimates of the UK wave energy resource. Note that most of the energy is more than 20 km offshore in water depths of more than 50 m and exploits the energy of deep-water swell waves. Locating wave energy devices offshore will produce maximum power but at greater costs of mooring in deep water, transmitting the power to shore and providing access for installation and maintenance.

7.3.1 Water Waves

Although the crests of fully developed ocean storm waves have pronounced curvature it is usual to represent waves as sinusoids for simple analysis. Figure 7.19 shows an idealized representation of two successive identical sine waves viewed at an instant in time. The wave height, H, is twice the amplitude, A, while the wavelength, L, is the distance between successive peaks or troughs.

The steepness, S, is given by

$$\text{Steepness}(S) = \frac{\text{Wave height}}{\text{Wave length}} = \frac{H}{L} = \frac{2A}{L}$$

Continuing with the idealized sinusoidal representation of linear wave theory, Figure 7.20 shows the vertical displacement of a progressive wave over time measured at a fixed location. The vertical displacement is:

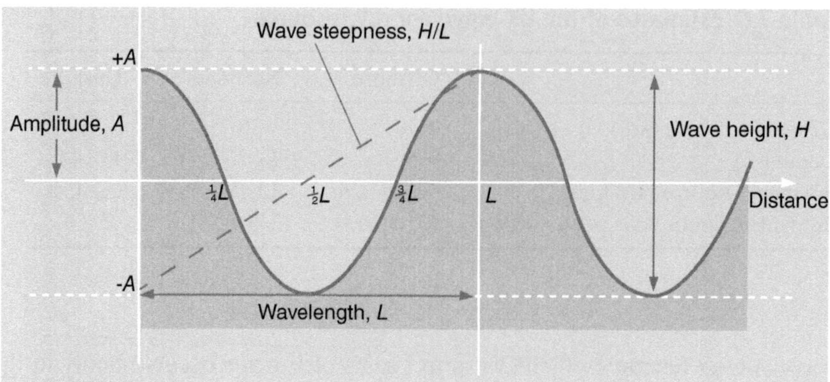

Figure 7.19 Profile of two successive ocean waves (viewed at an instant in time).
Source: *Wind, Waves and Shallow Water Phenomena*, 1999, Open University, with permission from Elsevier.

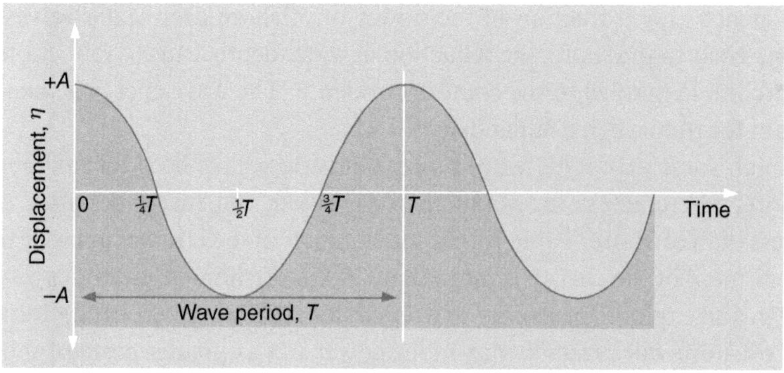

Figure 7.20 Displacement of a progressive wave (viewed from a fixed point).
Source: *Wind, Waves and Shallow Water Phenomena*, 1999, Open University, with permission from Elsevier.

$$\eta(t) = A\cos(\omega t)$$

where

$\eta(t)$ is the displacement in m,
A is the amplitude in m,
ω is the angular frequency in rad/s, $\omega = 2\pi f$,
f is the frequency in s^{-1} or Hz,
T is the wave period in seconds, $T = 1/f$.

Speed of Water Waves

The speed of a progressive wave, c, is the time taken for one wavelength to pass a fixed point thus:

$$c = \frac{L}{T} \quad [\text{m/s}] \tag{7.3}$$

The speed of a wave, c, in any depth of water is given by[1]

$$c = \sqrt{\frac{gL}{2\pi}\tanh\left(\frac{2\pi D}{L}\right)} \quad [\text{m/s}] \tag{7.4}$$

where

c is the speed of an individual wave (known as the celerity or phase velocity) in m/s,
D is the water depth in m,
L is the wavelength in m,
g is the acceleration due to gravity in m/s^2,
tanh is the hyperbolic tangent,
the wavenumber k is defined as $k = \frac{2\pi}{L}$ [m^{-1}],
the angular frequency is $\omega = 2\pi f = \frac{2\pi}{T}$ [rad/s].

Equation 7.4 can be rewritten as

$$\begin{aligned} c &= \frac{L}{T} = \frac{\omega}{k} = \sqrt{\frac{g}{k}\tanh(kD)} \\ c^2 &= \frac{\omega^2}{k^2} = \frac{g}{k}\tanh(kD) \\ \omega^2 &= gk\tanh(kD) \end{aligned} \tag{7.5}$$

Equations 7.4 and 7.5 are known as the dispersion equation and show how the speed (phase velocity) of a wave increases with its wavelength.

The behaviour of waves depends on the depth of water, so it is usual to consider three cases of waves at different depths: deep water, shallow water and at intermediate depths.

In water deeper than half the wavelength of the wave

$$D > \frac{L}{2}$$

so

$$\frac{2\pi D}{L} > \pi$$

tanh of an angle greater than π approximates to 1, so

$$\tanh\left(\frac{2\pi D}{L}\right) \approx 1$$

This results in Equation 7.4 simplifying to

$$c = \sqrt{\frac{gL}{2\pi}} \tag{7.6}$$

[1] Stated but not proved.

Thus in deep water the speed of a wave depends on the square root of the wavelength. Waves of longer wavelength travel faster.

Using Equations 7.3 and 7.6, and since

$$\frac{g}{2\pi} = 1.56 \text{ m/s}^2$$

for deep-water waves, we have

$$c = 1.56T \quad [\text{m/s}]$$
$$c = \sqrt{1.56L} \quad [\text{m/s}]$$
$$L = 1.56T^2 \quad [\text{m}]$$

In water much shallower than the wavelength of the wave

$$D < \frac{L}{20}$$

so

$$\frac{2\pi D}{L} < 0.3$$

tanh of a small angle approximates to the angle:

$$\tanh\left(\frac{2\pi D}{L}\right) \approx \left(\frac{2\pi D}{L}\right)$$

This results in Equation 7.4 simplifying to

$$c = \sqrt{gD} \tag{7.7}$$

Thus in shallow water the speed of a wave depends on the square root of depth of water.

At intermediate depths when

$$\frac{L}{2} > D > \frac{L}{20}$$

or the ratio of L/D is between 2 and 20, then the full form of Equation 7.4 must be used.

L appears twice in Equation 7.4, and it can be solved only graphically or by iteration, as shown in Example 7.6.

Example 7.6 – Waves at Intermediate Depths

Find the wavelength, L, and speed, c, of a wave of period 10 s in a water depth of 20 m. The wave period can be considered to be constant with depth.

Solution

Using the deep-water approximation of $L = 1.56T^2$, the ratio $L/D = 7.8$, so the full form of Equation 7.4 is required

$$c = \frac{L}{T} = \sqrt{\frac{gL}{2\pi}\tanh\left(\frac{2\pi D}{L}\right)}$$

which can be rearranged to give

$$L = \frac{gT^2}{2\pi}\tanh\left(\frac{2\pi D}{L}\right)$$

Forming the iterative equation, we have

$$L_{n+1} = \frac{gT^2}{2\pi}\tanh\left(\frac{2\pi D}{L_n}\right)$$

For deep water, $L = 1.56T^2 = 156$ m. Assume this for the start of the iteration, L_1.

n	L_n	L_{n+1}
1	156	104
2	104	130
3	130	116
4	116	124
5	124	120
6	120	122
7	122	121
8	121	121

Therefore, the wavelength, L, is 121 m and the speed of the wave, c, is 12.1 m/s.

Motion of Water at Depth

The particles of water in a deep-water wave move in an almost circular closed path. This is illustrated in Figure 7.21. Because of this circular motion, there is only a very small net movement of water. Near the surface, the diameter of the circular motion is equal to the wave height H. The radius of the motion decreases exponentially with the depth below the surface:

$$r_d = re^{\left(-2\pi d/L\right)} = \frac{H}{2}e^{\left(-2\pi d/L\right)}$$

where

r is the radius of circular motion in m; at the surface $r = H/2$,
r_d is the radius of circular motion at depth d in m,
d is the depth below the surface in m,
L is the wavelength in m.

At a depth below the surface of more than half the wavelength ($d > L/2$), the motion of the water particles is negligible. Some 95% of the energy is contained between the surface and a depth of a quarter of the wavelength ($L/4$).

As the wave enters shallower water, there is a progressive flattening of the ellipse. In shallow water, when $D < L/20$, the ellipse flattens further.

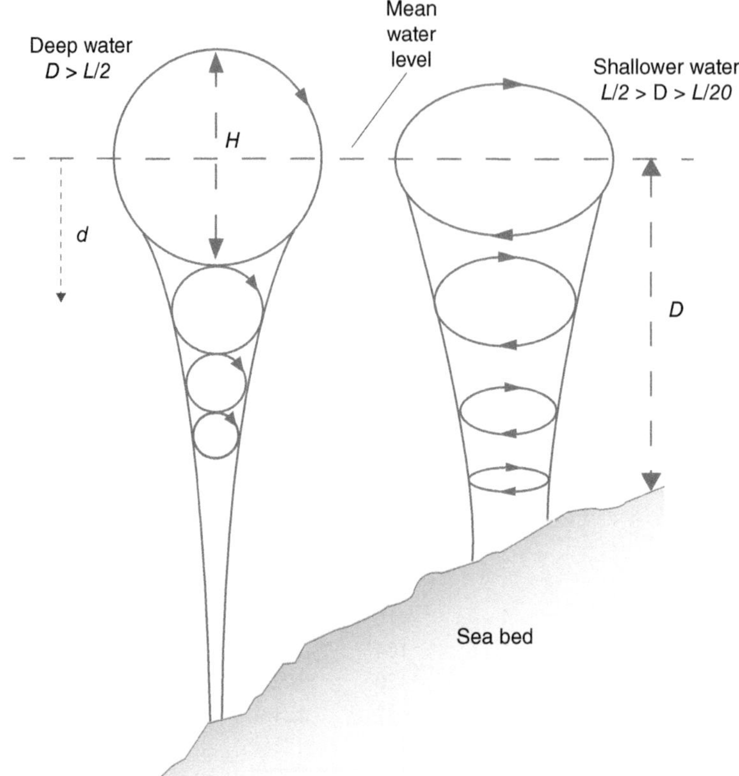

Figure 7.21 Water-particle orbits for different water depths. Note that in deep water the motion decreases with depth, while in shallow water the motion becomes elliptical.

Power of Monochromatic Waves

The total energy (E) per metre crest width averaged over a wavelength, due to kinetic and potential energy equally, is

$$E = \frac{1}{2}\rho g A^2 L = \frac{1}{8}\rho g H^2 L \quad [\text{J/m}] \tag{7.8}$$

The transfer of energy does not depend on the speed of individual waves (the phase velocity) but on the speed of the wave as a group (the group velocity, c_g). For regular waves (with a single period) in deep water

$$c_g \approx \frac{c}{2}$$

where

c_g is the speed of the wave group,
c is the speed of individual waves.

The power of the waves or energy flux (expressed as W/m of wave front) is the product of the average energy per unit area (E/L) and the group speed (c_g). Thus

$$P = \frac{1}{8}\left(\rho g H^2\right) c_g \quad [\text{W/m}]$$

As $c = gT/2\pi$ in deep water, the power may be rewritten

$$P = \frac{1}{32\pi}\rho g^2 H^2 T \quad [\text{W/m}] \tag{7.9}$$

where

H is the wave height (measured from peak to trough) in m,
T is the period of a monochromatic wave in s.

Hence monochromatic waves in deep water have a power of

$$P = 981 H^2 T \quad [\text{W/m}] \tag{7.10}$$

Note: in shallow water this equation does not apply as then $c_g \approx c$, but this is of less interest in assessing the wave energy resource as there is limited wave power in shallow water.

Example 7.7 – Power in Monochromatic Deep-Water Waves

What is the power in deep water of waves of amplitude A = 1.5 m and wavelength L = 50 m? Compare this power with that of extreme waves of amplitude A = 6 m and steepness S = 0.1.

Solution
For a deep-water wave of wavelength 50 m the speed is

$$c = \sqrt{1.56L}$$

$$c = \sqrt{1.56 \times 50} = 8.8 \text{ m/s}$$

and the speed of the wave group is $c/2$ = 4.4 m/s. Hence the power in the waves is

$$P = \frac{1}{8}(\rho g H^2)c_g = \frac{1}{8}(1025 \times 9.81 \times 3^2) \times 4.4 = 50 \text{ kW/m}$$

The wavelength of the extreme waves is given by

$$L = \frac{H}{S} = \frac{2 \times 6}{0.1} = 120 \text{ m}$$

The speed of the extreme waves is given by

$$c = \sqrt{1.56 \times 120} = 13.7 \text{ m/s}$$

Therefore the power of the extreme waves is

$$P = \frac{1}{8}(1025 \times 9.81 \times 12^2)\frac{13.7}{2} = 1.2 \text{ MW/m}$$

7.3.2 The Wave Energy Resource

Real ocean waves are not the simple monochromatic sine waves discussed in Section 7.3.1; the wave energy resource is highly complex. Figure 7.22 shows the heights of waves measured at the same point on three successive days. It can be seen

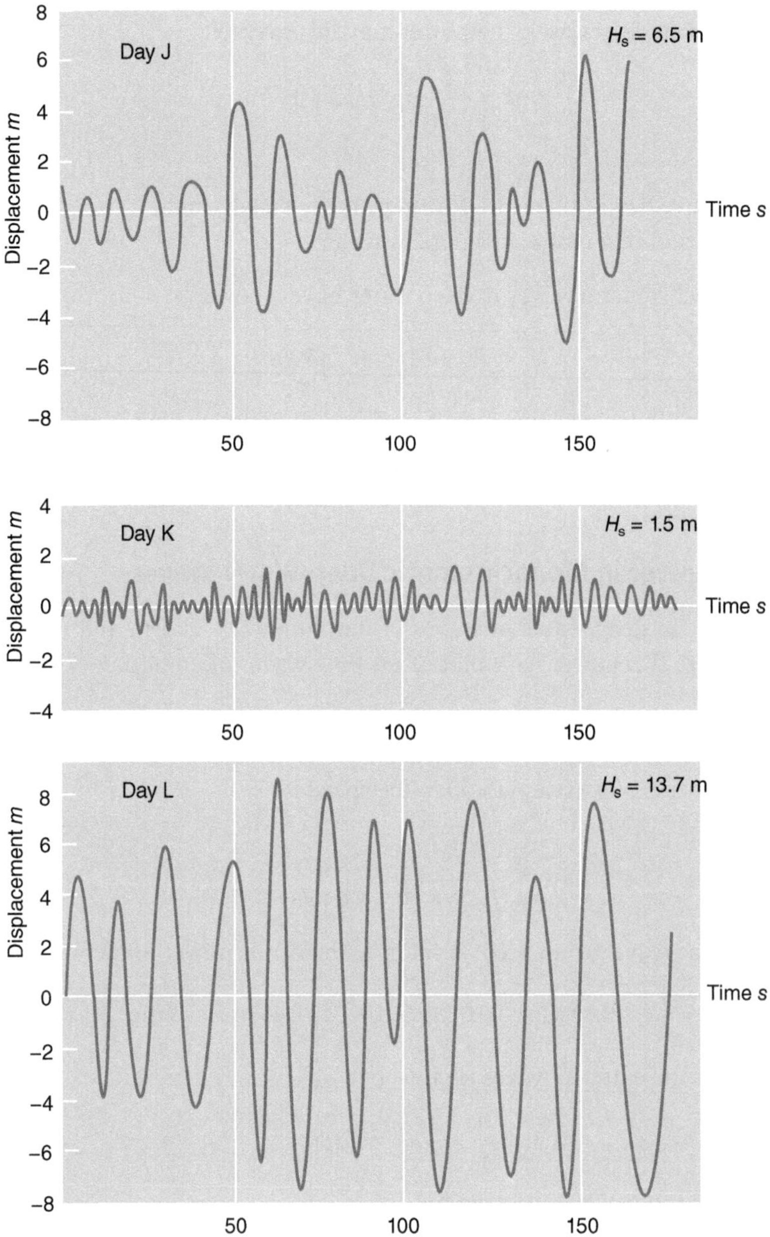

Figure 7.22 Example of waves at a location on three successive days.
Source: Godfrey Boyle (ed.), *Renewable Energy: Power for a Sustainable Future*, 1996.
By permission of Oxford University Press, USA.

that both the amplitude and period of the waves vary considerably from day to day. Waves created by a local wind have wavelengths of up to ~50 m. After travelling across the ocean, the swell waves that are caused by a storm many hundreds of kilometres away have wavelengths of more than ~100 m.

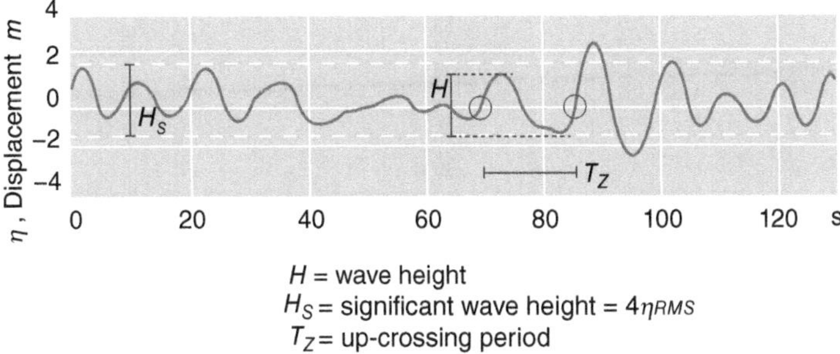

Figure 7.23 Displacement of a wave showing significant wave height and period.

In order to quantify the wave energy resource and predict the output of wave energy converters, there is a need for a simple description of the complex wave phenomenon. Figure 7.23 shows an irregular water wave and the conventional way of describing it. In former times when waves were evaluated by eye, the wave height was described by the average height of the highest 1/3 of the waves and defined as $H_{1/3}$. It is now taken from a series of automatic measurements and described as the significant wave height H_S. H_S is approximately equal to $H_{1/3}$ and is equal to four times the root mean square of the wave displacement, η_{RMS}, during the measuring period.

The zero-crossing period T_Z was originally found from a time series by counting the number of upward zero crossings in the measuring period, n, and dividing the measuring period by this number (as in Example 7.8). Now it is more common to define an energy period T_e from the power spectrum; T_e is the period of a wave with equal energy to that of the series of waves during the measuring period.

The extreme wave height (H) is the largest single wave height that can be expected for a given return period and can be approximated as:

$$H = 1.86 H_S$$

Before the widespread adoption of digital instrumentation and signal-processing techniques, observations of the height of waves were made by eye or with simple instruments such as a wave-staff to create a time series. $H_{1/3}$ and T_Z were found directly from the observed time series. Now it is common to use digital recording instruments and process the data in the frequency domain.

Example 7.8 – Wavelength of Deep-Water Waves

Compare the wavelengths of the deep-water waves shown in Figure 7.22.

Solution

By counting the number of upward zero crossings over the measuring period of 160 s the period of the waves can be estimated and hence the wavelength found.

	Day J	Day K	Day L
Number of upward crossings, n	15	44	11
Wave period (s), $T_Z = 160/n$	10.7	3.6	14.5
Wavelength (m), $L = 1.56T^2$	179	20	328

Power of Complex Waves

In shallow water, pressure sensors or acoustic Doppler profilers fixed to the seabed are used for measurement of the wave height. In deeper waters, accelerometer buoys and ship-borne wave recorders are used as well as satellite altimeters. An accelerometer buoy floats on the surface of the water and is anchored to the seabed with an elastic mooring. The vertical acceleration of the buoy is measured and integrated twice to give the vertical displacement of the water surface.

The resulting measurements of vertical displacement are used to create a power spectral density function $S(f)$ through a Fourier transform. An example of a power spectral density function from a series of waves is shown in Figure 7.24. The power spectral density function is then used to calculate the spectral moments and hence the parameters of the wave climate.

However, the use of measurements from digital instruments is expensive and time-consuming, and many of the estimates of the wave energy resource available today have been obtained from models linking historical wind speed data to wave climate.

For a sine wave the mean square value of displacement (surface elevation) is

$$\sigma^2 = \frac{A^2}{2}$$

and the root mean square (RMS) is

$$\sigma = \frac{A}{\sqrt{2}}$$

The significant wave height is defined for a typical spectrum as

$$H_s = 4\sigma$$

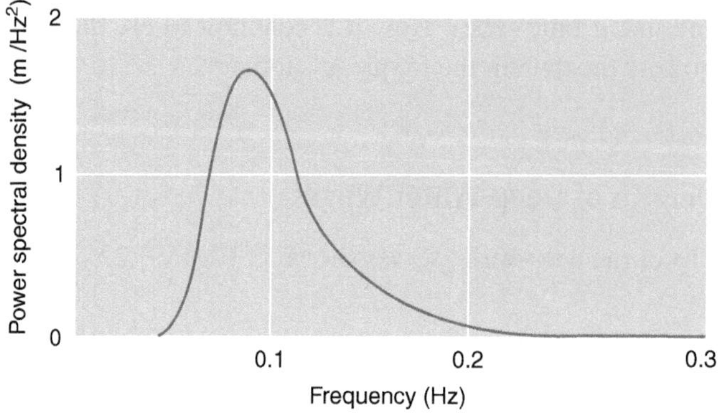

Figure 7.24 Power spectral density function of the amplitude of a series of waves.

and as $A = H/2$ we may write

$$P = \frac{\rho g^2}{32\pi} H^2 T = \frac{\rho g^2}{8\pi} A^2 T = \frac{\rho g^2}{4\pi} \sigma^2 T = \frac{\rho g^2}{64\pi} H_S^2 T \quad [\mathrm{W/m}]$$

T, the period of a monochromatic wave, is replaced by the energy period, T_e, of the sequence of waves to give

$$P = \frac{\rho g^2}{64\pi} H_S^2 T_e \quad [\mathrm{W/m}] \tag{7.11}$$

Assuming a density of seawater of 1025 kg/m^3 the power per unit width of wave front is

$$P = 491 H_S^2 T_e \quad [\mathrm{W/m}] \tag{7.12}$$

where T_e is the energy period.

For many seas $T_e \sim 1.12 T_Z$ and so

$$P = 549 H_S^2 T_Z \quad [\mathrm{W/m}] \tag{7.13}$$

where T_Z is the zero up-crossing period.

Note that:

H is measured from peak to trough, $H = 2A$,
H_S is an RMS measure,

$$H_S = 4\sigma = \frac{4A}{\sqrt{2}} = 2\sqrt{2}A$$

and so for a sine wave $H_S = \sqrt{2}H$.

Energy Output of a Wave Energy Device

If an accelerator buoy is used, measurements of its vertical displacement are typically taken every 0.5 seconds over 15 minutes once every 3 hours. The sea states at a location are assumed to remain statistically constant over several hours. These measurements are then used to form the wave power spectral density function and its moments, from which the significant wave height H_s and energy period T_e are calculated. The data for each combination of significant wave height and energy period are then plotted on a scatter diagram of T_e and H_s (Table 7.9) expressed either as hours or as occurrences per (typically) thousand records. It is desirable to acquire data over up to a year so that winter storms are represented. If the acceleration buoy also captures directional information it is plotted on a polar diagram.

To estimate the annual production of a wave energy device, the scatter diagram of the wave energy resource is combined with a scatter diagram defining the performance of the wave energy converter (Table 7.10). In a similar way to the calculation

Table 7.9 Example scatter diagram of a sea state (h), with hourly data over 4950 hours

	Wave period T_e (s)							
Wave height H_s (m)	5.5	6.5	7.5	8.5	9.5	10.5	11.5	>12
0.25	129	356	25	20	31	4	1	34
0.75	238	641	51	37	29	12	8	26
1.25	116	505	134	41	19	8	10	21
1.75	141	254	150	76	34	8	9	24
2.25	172	176	111	61	32	18	2	20
2.75	106	112	84	39	34	11	11	11
3.25	7	50	133	22	28	16	2	14
3.75		11	108	10	20	11	4	15
4.25		1	73	7	18	6		6
4.75			33	2	21	2		3
>5			17	7	57	22	2	31

of power from a wind turbine (Equation 2.3), but now in two dimensions, the energy generated is calculated from

$$\text{Energy} = \sum H(T_e, H_s) \times P(T_e, H_s)$$

where

$H(T_e, H_s)$ is the hours per year when the sea state has energy period T_e and significant wave height H_s,

$P(T_e, H_s)$ is the power generated by the wave energy converter with waves of energy period T_e and significant wave height H_s.

For some wave energy devices the direction of the waves is important and the calculation is made by direction.

7.3.3 Devices for Wave Power Generation

Using the energy of the waves to generate electricity is difficult and to date no design of a wave energy converter has emerged as being the most effective. Features that must be considered when designing a wave energy device include energy capture, cost, access for installation and maintenance and, above all, survivability in extreme storms.

A number of companies and research organizations are pursuing quite different devices and concepts. However, any wave energy generator must have a means of:

- collecting the wave energy,
- reacting the wave forces,
- controlling the device,
- locating the device,
- power take-off and export to the shore.

Table 7.10 Example scatter diagram of the performance of a wave energy converter (kW), with the device rated at 750 kW

	Wave period T_e (s)																
Wave height H_s (m)	5	5.5	6	6.5	7	7.5	8	8.5	9	9.5	10	10.5	11	11.5	12	12.5	13
1		22	29	34	37	38	38	37	35	32	29	26	23	21			
1.5	32	50	65	76	83	86	86	83	78	72	65	59	53	47	42	37	33
2	57	88	115	136	146	153	152	147	138	129	116	104	93	83	74	66	59
2.5	89	138	180	212	231	238	238	230	216	199	181	163	146	130	116	103	92
3	129	198	260	305	332	340	332	315	292	266	240	219	210	188	167	149	132
3.5		270	354	415	438	440	424	404	377	362	326	292	260	230	215	202	180
4			452	502	540	546	530	499	475	429	384	366	339	301	276	237	213
4.5			544	636	642	648	628	590	562	528	473	432	382	356	338	300	266
5				739	726	731	687	687	670	607	557	521	472	417	369	348	328
5.5				750	750	750	750	687	737	667	658	586	530	496	446	395	355
6					750	750	750	750	750	750	711	633	619	558	512	470	415
6.5					750	750	750	750	750	750	750	745	658	621	579	512	481
7						750	750	750	750	750	750	750	750	676	613	684	525
7.5							750	750	750	750	750	750	750	750	686	622	595
8								750	750	750	750	750	750	750	750	690	625

A simple classification of those wave energy devices that have been demonstrated so far is as follows:

- oscillating water columns (OWCs),
- overtopping devices,
- oscillating bodies.

An oscillating water column uses a partially submerged structure to enclose a column of water on top of which is a volume of air. Waves cause the water within the chamber to rise and fall, compressing and expanding the volume of air. The air flows to and from the atmosphere through a duct of reduced cross-section area in which a turbine generator is placed. Oscillating water column devices have been investigated extensively with several demonstration units installed with electrical outputs of up to 500 kW. The demonstration units have been onshore, near-shore supported from the seabed or near-shore floating. An advantage of the OWC is that the slow oscillatory wave motion of the water surface is transformed first into a reciprocating air flow and then into a high-speed unidirectional rotating shaft that is in air. OWCs are resonant devices with the mass and stiffness of the resonant system determined by the dimensions of the water column. The chamber is dimensioned for a particular wave height and period although the damping can be adjusted by control of the output power of the generator.

Figure 7.25 shows a shoreline oscillating water column. Waves enter the chamber, which is constructed on a steeply shelving shore. The air in the chamber is repeatedly compressed and expanded to drive an air turbine connected to a generator. In this example a Wells turbine is used, although a bidirectional impulse turbine has been proposed for use in other OWC devices. The Wells turbine has symmetrical aerofoils that generate a unidirectional torque from the air as its flow reverses.

Although OWCs have been shown to operate successfully, the early demonstration units experienced a number of difficulties. Shoreline OWCs are difficult to

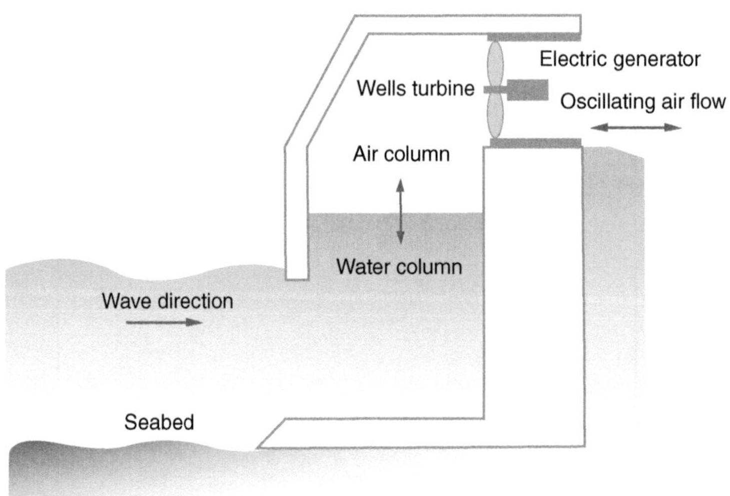

Figure 7.25 Shoreline oscillating water column.

construct, as civil work is required above and below the water line. The chamber needs to be constructed with a smooth floor to minimize losses, while extreme waves can lead to flooding of the turbine. Several OWCs that were installed as demonstration projects were damaged by extreme waves. One near-shore OWC was severely damaged during its installation.

Overtopping wave energy devices operate by channelling waves into an elevated storage reservoir (Figure 7.26). The kinetic energy of the waves is transformed into potential energy by raising the level of the water in the reservoir. The water then operates a conventional low-head hydro turbine. Overtopping devices can be onshore or near-shore, and both have been demonstrated. An early onshore overtopping device was known as the TAPCHAN after the tapered channel that was used to channel the waves into the reservoir. In one offshore overtopping design, parabolic collectors were used to focus the waves into the reservoir and a double ramp used to increase the water height.

Overtopping devices have the advantage that the power capture is less sensitive to wave period than resonant devices. They use low-head hydro turbines with an operating head of ~2 m. Floating devices have the advantage that they are not affected by the tidal range and the level of freeboard can be adjusted by controlling the outflow of the reservoir and air within flotation chambers.

Oscillating bodies extract energy directly from the motion of the waves. An oscillating body can move in a combination of six directions, three linear and three rotational (see Figure 7.27). It is usual to restrict the movement of the device to two or three directions as otherwise the power take-off becomes too complicated.

A large number of oscillating body devices have been proposed and some have been demonstrated to capture energy from the motion of bodies on the surface as the wave passes. Others use the motion of the water and pressure variations at depth. The advantage of oscillating bodies is that they can be located offshore and so capture energy from deep-water waves.

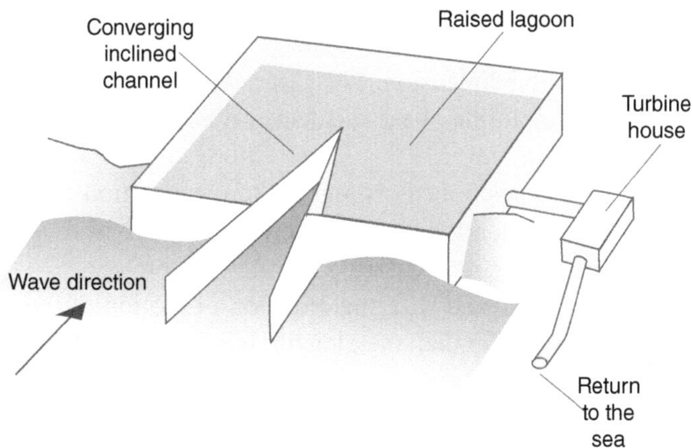

Figure 7.26 TAPCHAN overtopping wave energy converter.

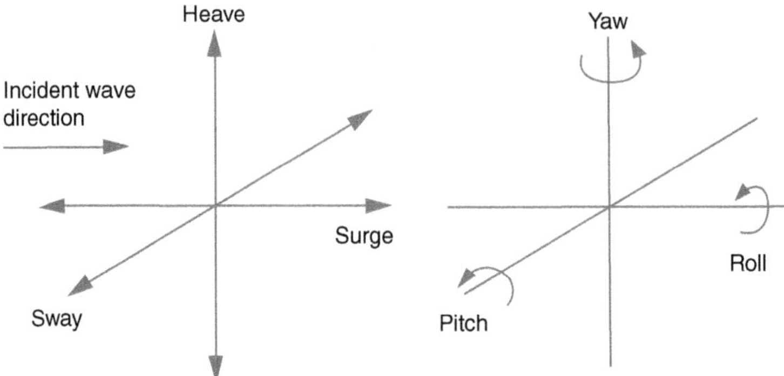

Figure 7.27 Directions of movement of a wave energy converter.

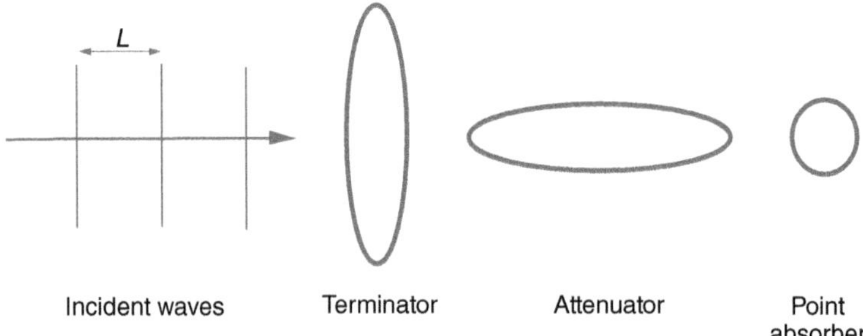

Figure 7.28 Classification of oscillating-body wave energy devices.

Oscillating bodies, shown diagrammatically in Figure 7.28, can be classified into three categories:

- point absorber,
- terminator,
- attenuator.

Point absorbers, typically moving in heave or pitch, are relatively small devices. They are capable of absorbing wave energy over a much greater width of wave front than their horizontal dimension. This large capture width implies oscillations much larger than the amplitude of the incident waves. Point absorbers can be surface or sub-surface devices.

Terminators are positioned orthogonal to the direction of motion of the incident waves and often rely on the surge of the waves. There have been several proposals for terminator devices, particularly in shallow water.

Attenuators are aligned with the direction of motion of the incident waves with their width much smaller than their length. It is usual for attenuators to be compliant or articulated structures with energy extracted through yaw and pitch movement.

All floating oscillating bodies are designed to capture energy from deep-water waves. Hence ensuring their survival in storms, their mooring and their power take-off are all difficult aspects of their design, construction and operation.

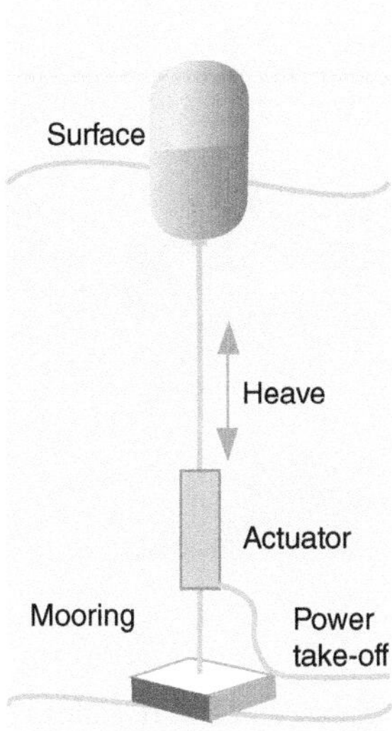

Figure 7.29 Point absorber.

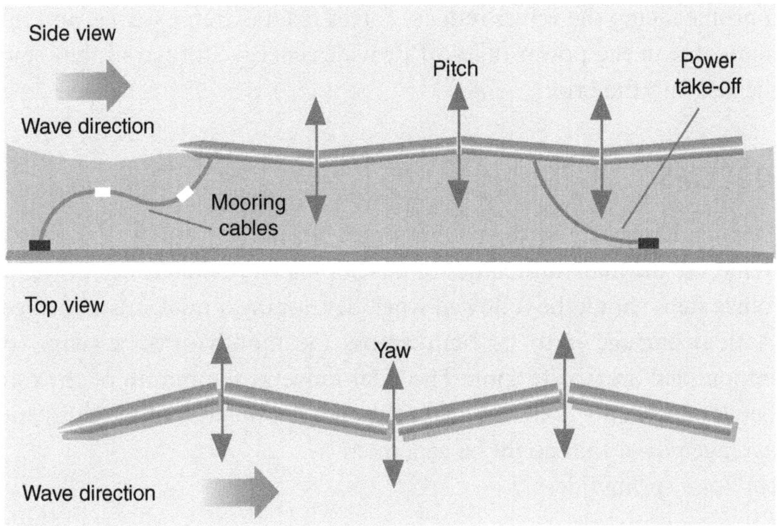

Figure 7.30 Surface attenuator (Pelamis).

Figure 7.29 shows the principle of operation of a bottom-mounted point absorber operating with a heave motion. In this example the device is anchored to the seabed and the float moves vertically either on or below the surface. The actuator transforms the movement either into electrical energy using a linear generator or into a high-pressure hydraulic fluid.

Figures 7.30 and 7.31 illustrate the operation of an attenuator wave energy device. The device is moored to the seabed and positioned parallel to the direction of the waves. The hinged sections are semi-submerged and follow the profile of the waves. The articulated joints allow movement in two directions that is resisted by hydraulic cylinders to create high-pressure oil. The high-pressure oil is stored in an

Figure 7.31 Pelamis surface attenuator wave energy converter.
Photo: Ashley Cooper pics/ Alamy Stock Photo. Rights Managed.
(A black and white version of this figure will appear in some formats. For the colour version, please refer to the plate section.)

accumulator and used to drive electrical generators. The power is transmitted to shore through a submarine electrical cable. Some demonstration units have been operated successfully. Advantages claimed for this design include that the long thin profile facing the waves reduces forces from extreme waves, and the hydraulic accumulator in the power take-off provides energy storage so that smooth power is injected into the grid.

PROBLEMS

1. Distinguish clearly between tidal range, tidal stream and wave power generation.
2. What are the main difficulties of developing large tidal range power schemes?
3. What steps should be followed when developing a tidal stream project?
4. A tidal barrage is to be built across the mouth of an estuary to create an impounded area of 15 km^2. The tidal range at the mouth of the estuary varies between 6 m and 12 m. Estimate the energy potential of the tides and hence the average power that might be generated:
   (a) for a spring tide,
   (b) for a neap tide.

   [**Answer:** (a) 486.5 MW; (b) 121.6 MW.]
5. A circular tidal lagoon that is 3 km in diameter is situated in an average tidal range of 6 m. Estimate the maximum average power if ebb-only generation with the performance shown in Figure 7.7 is used.

   [**Answer:** 16.3 MW.]
6. A tidal lagoon is constructed out from one bank of an estuary with a 9.5 km long embankment to give an impounded area of 11.5 km^2. It produces 420 GWh/year from turbines with a rated capacity of 320 MW.
   Calculate the capacity factor of the scheme and the ratio of length to area of the impounded area.

   [**Answer:** 15%, 826.]

7. Calculate the ratio of embankment length (in metres) to impounded area (in km^2) of a semi-circular tidal lagoon built out from the banks of an estuary. What are the implications of this calculation?
8. The tidal stream rotor shown in Figure 7.14 generates 400 kW. Calculate the power coefficient.

   [**Answer:** 0.42.]
9. A 10 m diameter horizontal-axis tidal stream turbine has a thrust coefficient C_T of 0.5 and a flow velocity at hub height of 2.8 m/s. Its hub is 15 m above the seabed. Calculate the thrust on the rotor and hence the overturning moment of a bottom-mounted support structure.

   [**Answer:** 157.8 kN, 2.37 MNm.]
10. Using suitable assumptions, compare the overturning moment of the tidal stream turbine described in Problem 9 with a wind turbine producing a similar rated power at a wind speed of 12 m/s. (Example 7.4 may be useful.)

    [**Answer:** wind turbine: 1.14 MNm, tidal stream turbine: 2.37 MNm.]
11. A three-bladed, 10 m diameter, horizontal-axis tidal stream turbine operates at a tip speed ratio (λ) of 5 in a flow velocity of 2.8 m/s. Each blade weighs 2000 kg with a centre of mass at $0.4R$. Calculate the rotational kinetic energy stored in the rotor. If the tidal stream turbine is rated at 800 kW calculate its inertia constant (H) and compare that value with the inertia constants commonly found in other generating plant. (See Section 10.4 for the definition of inertia constant.)

    [**Answer:** 94.1 kJ, 0.12 s.]
12. Calculate the wavelength and velocity of a wave of period 10 s in deep water.

    [**Answer:** 156 m, 15.6 m/s.]
13. Calculate the wavelength and velocity of a wave of period 10 s in shallow water of 1 m depth. Confirm that the shallow water approximation is appropriate.

    [**Answer:** 31.3 m, 3.13 m/s.]
14. Find the speed, c, of a wave of period 12 s in a depth of 25 m.

    [**Answer:** 13.8 m/s.]
15. Calculate the power in an ideal wave front of monochromatic waves of amplitude 1.5 m and wave period 10 s.

    [**Answer:** 88.3 kW/m.]

FURTHER READING

Baker A.C., *Tidal Power*, 1991, IET.
Classic text on tidal range generation based on studies of the Severn Barrage proposal.

Clark R.H., *Elements of Tidal-Electric Engineering*, 2007, IEEE–Wiley.
Important text on tidal range generation by an expert on the Bay of Fundy.

Cruz J. (ed.), *Ocean Wave Energy: Current Status and Future Perspectives*, 2008, Springer.
A very comprehensive text on wave energy.

Hardisty J., *The Analysis of Tidal Stream Power*, 2009, Wiley–Blackwell.
Accessible description of tidal stream resource and power conversion systems.

IEC TS 62600, 'Marine energy – Wave, tidal and other water current converters'.
A comprehensive set of standards for practitioners and researchers covering many aspects of marine energy.

McCormick M.E., *Ocean Wave Energy Conversion*, 2007, Dover.
A very useful description of wave power and conversion systems.

Open University, *Waves, Tides and Shallow-Water Processes*, 1999, Butterworth Heinemann.
A most useful explanation of waves and tides.

Shaw R., *Wave Energy: A Design Challenge*, 1982, Ellis Horwood.
An important early textbook on wave energy and conversion devices.

Website

UK Atlas of UK Marine Renewable Energy: ABPmer.www.renewables-atlas.info. Date accessed, 1 September 2022.

ONLINE RESOURCES

Exercise 7.1 Prediction of tidal range: tidal constituents are used to predict tidal range.

Exercise 7.2 Investigation of tidal stream data: 5-minute values of flow speed taken by an ADCP are used to visualize a time series of the speed of the flow, a polar plot of the flow and a shear profile.

Exercise 7.3 Calculation of the output energy of a tidal stream generator: 5-minute values of flow speed taken by an ADCP are combined with a power curve of a tidal stream generator to predict energy output.

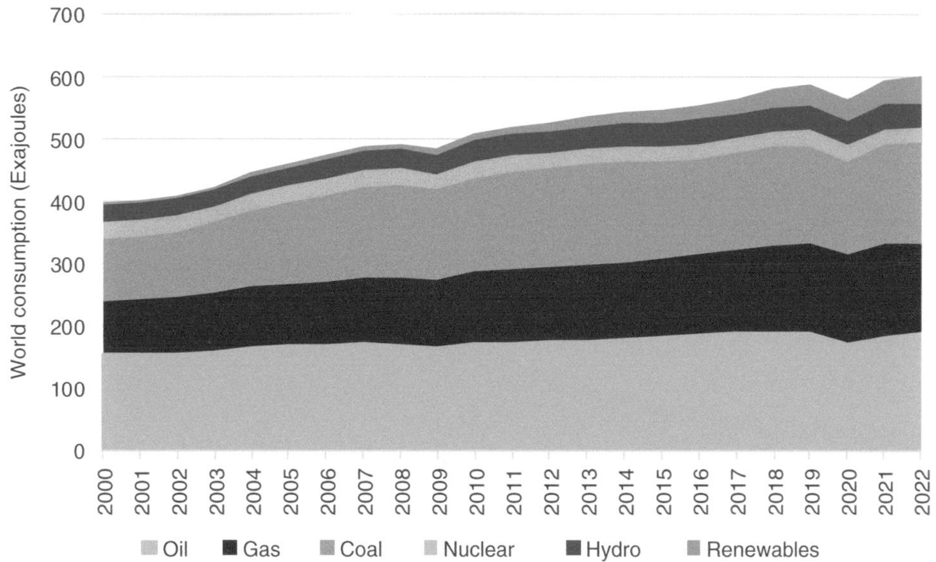

Figure 1.1 World consumption of primary energy.
Source: *Energy Institute Statistical Review of World Energy*, 2023.

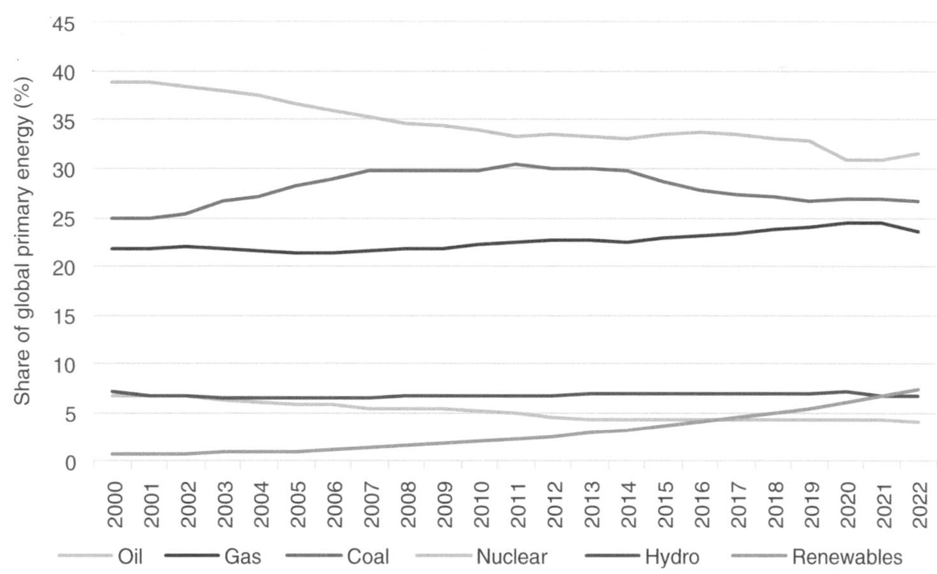

Figure 1.2 World share of primary energy.
Source: *Energy Institute Statistical Review of World Energy*, 2023.

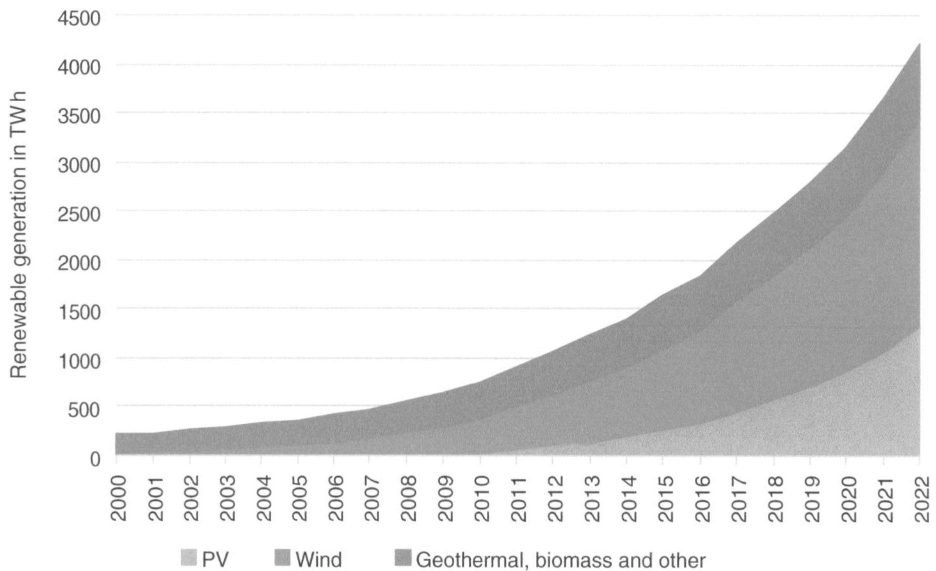

Figure 1.3 World annual generation of electricity from renewable sources.
Source: *Energy Institute Statistical Review of World Energy*, 2023.

Figure 1.6 Wind farm in South Wales.
Photo: RichardJones/BusinessVisual. Rights Managed.

Figure 1.7 Solar photovoltaic farm in South Wales.
Photo: RichardJones/BusinessVisual. Rights Managed.

Figure 2.2 Wind turbines on a site in the USA.
Photo: RWE Innogy UK.

Figure 2.11 Lifting the rotor of a wind turbine to attach it to the low-speed shaft. Photo: RES.

Figure 2.23 Barrow offshore wind farm showing turbines and offshore substation. Photo: Rob Arnold/Alamy Stock photo.

Figure 2.26 Construction of an offshore wind turbine using a jack-up barge to lift the final blade.
Photo: RES.

Figure 3.3 Small hydro scheme in Africa. The net head is 176 m with a flow rate of each penstock of 3000 litres/s. Each turbine generator unit is rated at 4.2 MW.
Photo: Gilkes.

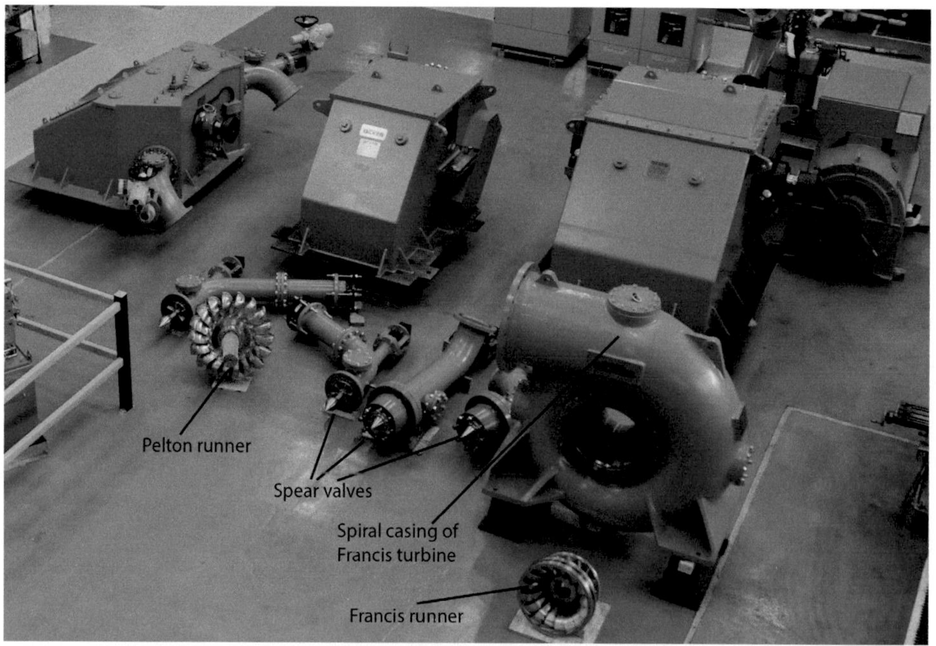

Figure 3.9 Small hydro turbines in the factory.
Photo: Gilkes. (Annotations by the author.)

Figure 3.11 Two Pelton runners.
Photo: Gilkes.

Figure 3.12 Horizontal-axis Pelton wheel in housing showing controls.
Photo: New Mills Engineering. (Annotations by the author.)

Figure 3.17 Turgo runner.
Photo: Gilkes.

Figure 3.21 Runner of a Francis turbine in the factory.
Photo: Gilkes.

Figure 3.23 Runner of a large Kaplan turbine. Note the adjustable angle blades.
Photo: sulaco229/Shutterstock.com.

Figure 3.25 Draft tube of 500 kW low-head axial turbine leaving the factory.
Photo: New Mills Engineering.

Figure 3.33 Archimedes screw generator being delivered to site. Note the coupling, gearbox and generator towards the front of the load and the upper and lower bearings of the screw.
Photo: Stephen Fleming/Alamy. Rights Managed.

Figure 5.1 Array of poly-crystalline silicon solar modules on a roof in a city centre.
Photo: RES.

Figure 5.2 A ground-mounted array in a rural solar farm in England. Photo: RES.

Figure 6.18 Roof-mounted domestic hot water heater using a thermo-syphon flat-plate solar collector. Photo: Roy Pedersen/Shutterstock.com.

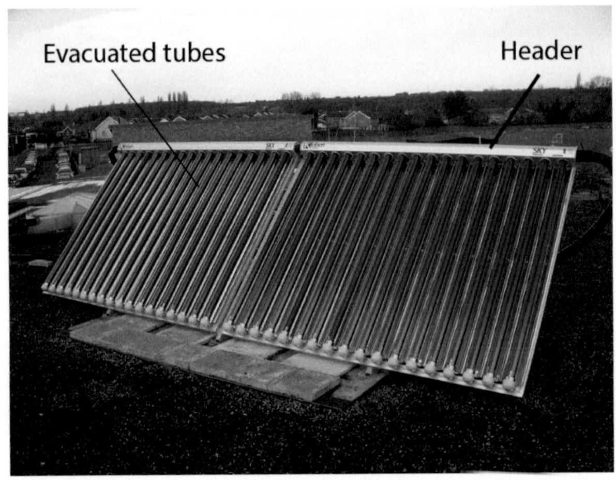

Figure 6.19 Evacuated-tube solar thermal collector on the roof of a school in England.
Photo: Rhico. (Annotated by the author.)

Figure 6.24 Solar thermal generating plant.
Photo: Tom Grundy/Shutterstock.com.

Figure 6.27 Solar thermal generating plant using Fresnel lenses.
Photo: Eunika Sopotnicka/Shutterstock.com.

Figure 6.28 Power tower solar generating plant.
Photo: raulbaenacasado/Shutterstock.com.

Figure 7.15 A two-bladed, twin-rotor, horizontal-axis, variable-pitch tidal stream turbine mounted on a monopile.
Photo: RenewableUK. Turbine manufactured by Marine Current Turbines.

Figure 7.16 A three-bladed, horizontal-axis, fixed-pitch, bottom-mounted tidal stream turbine.
Photo: Tidal Energy Ltd.

Figure 7.17 A ducted horizontal-axis tidal stream turbine.
Photo: RenewableUK. Turbine manufactured by Openhydro.

Figure 7.31 Pelamis surface attenuator wave energy converter.
Photo: Ashley Cooper pics/Alamy Stock Photo. Rights Managed.

Figure 8.5 Biomass feedstock at a 1 MW bioenergy plant.
Photo: Lanka Transformers Ltd.

Figure 8.6 View into the fixed-bed furnace.
Photo: Lanka Transformers Ltd.

Figure 8.7 Steam turbine coupled to a generator.
Photo: Lanka Transformers Ltd. (Annotated by the author.)

Figure 8.8 Wood pellet fuel used for large-scale generation.
Photo: Drax Power Ltd.

Figure 8.9 Biomass fuel-handling facilities at Drax power station. Note the large storage domes for wood pellets.
Photo: Drax Power Ltd.

Figure 8.12 A 35 kW_e reciprocating engine generator set (foreground) with gasifier (background). Photo: Lanka Transformers Ltd.

Figure 8.13 Biomass gasifier.
Photo: Lanka Transformers Ltd.

Figure 9.1 Delivery of a blade to a wind farm under construction.
Photo: RES.

Figure 9.2 Wind turbine concrete slab and positioning of the tower interface can.
Photo: RES.

Figure 10.5 Hydro power plant and substation.
Photo: Ceylon Electricity Board. (Annotated by the author.)

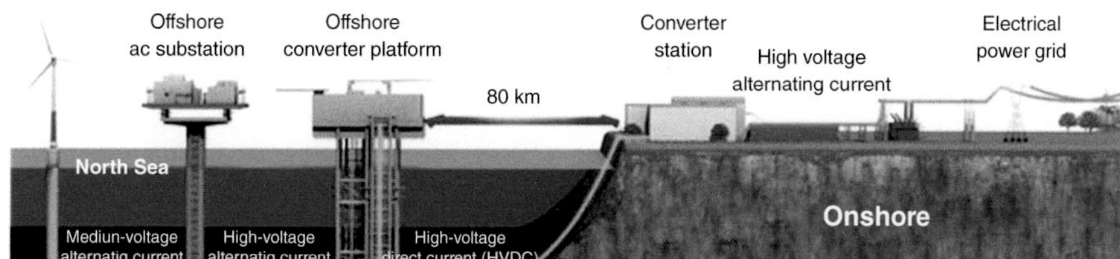

Figure 10.24 Offshore wind farm dc connection.
Source: © Alstom.

Figure 11.9 Computer image of a large utility flywheel unit.
Photo: Siemens Energy.

Figure 11.22 20 MW utility connected lithium-ion battery.
Photo: RES.

Figure 12.3 Photovoltaic array of a standalone system at a rural health centre in Zambia.
Source: Cardiff University project supported by the Mothers of Africa charity and Cardiff University. System design and photographs provided by D.J. Rogers, L.J. Thomas and J.M. Stevens.

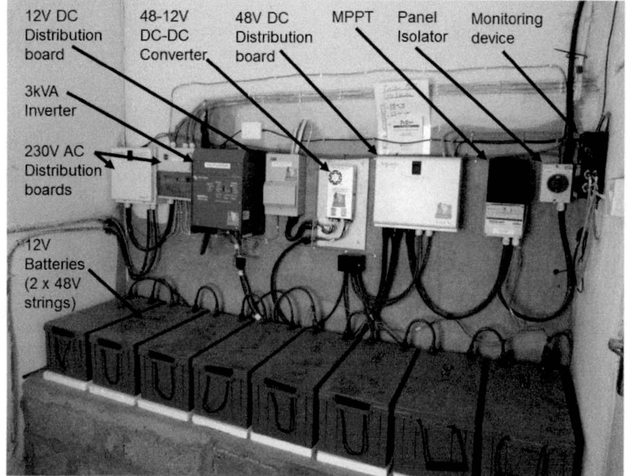

Figure 12.4 Battery bank, inverter and controller of a standalone PV system.
Source: Cardiff University project supported by the Mothers of Africa charity and Cardiff University. System design and photographs provided by D.J. Rogers, L.J. Thomas and J.M. Stevens.

Figure 12.9 Eluvaitivu Island installation. (a) Solar panel and microgrid assembly. (b) Solar inverters. (c) Wind turbines. (d) Li-ion batteries
Photos: Ceylon Electricity Board.

8 Bioenergy

INTRODUCTION

Biomass supplies some 14% of the world's primary energy, mostly from traditional fuels in developing countries. In these rural societies the traditional biomass fuels of wood, charcoal, crop residues or dried animal dung are widely used for cooking and heating. In the industrialized world, the use of bioenergy to generate electricity and as fuel for road vehicles is growing rapidly, although from a relatively low base.

There are a very large number of sources of biomass, and different processes by which it can be converted into useful energy, many of which are still being investigated and developed. This chapter does not deal exhaustively with each bioenergy feedstock and process, but rather describes the main routes by which biomass is converted into useful energy at commercial scale. Biomass processing techniques can be divided into thermochemical and biochemical processes as well as the extraction of oil from plants. Thermochemical processes use heat and catalysts to transform biomass into useful energy by combustion or gasification. Biochemical processes use enzymes and microorganisms in alcoholic fermentation or anaerobic digestion. Vegetable oils are extracted from plant seeds, processed and either used directly in compression ignition engines or converted into biodiesel.

Bioenergy has a number of very useful attributes. Biomass can be stored as a dry solid or converted into a gaseous or liquid biofuel. Hence, although bioenergy is a concentrated form of solar energy, it does not depend on the instantaneous irradiance of the sun and can be stored and used when needed. Biomass is often processed into biofuel in small units near where it is grown, contributing to rural employment. In some countries, the use of bioenergy can reduce the national requirement for importing fossil fuels so saving foreign exchange.

The other renewable energy technologies, once the equipment is manufactured and installed, generate electricity with very low emissions of greenhouse gases. With bioenergy, the same quantity of CO_2 that is absorbed from the atmosphere in photosynthesis is subsequently released when the biomass is converted into useful energy. The process can then be said to be CO_2 neutral. In practice, fossil energy is needed for the cultivation, fertilization, harvesting, transport and processing of biomass, so careful analysis is required to determine the lifetime environmental costs and benefits of any biomass scheme. Many of the existing bioenergy processes, particularly for the manufacture of vehicle fuels, use sugar, starch or oil crops that could otherwise be used for food. The use of food crops as a source of bioenergy and the consequential effect on world food prices is controversial.

Biomass is created by photosynthesis and requires a considerable area of land; its energy density is low in comparison to fossil fuels. Thus the costs of collection and transportation are high, and this limits the practical size of the majority of biomass or waste-fuelled electricity-generating plants to less than ~20 MW_e (megawatts of electric capacity). An exception to this is the very large electricity-generating units, up to 500 MW_e in GB, that are fuelled by imported wood pellets.

This chapter describes the main routes by which biomass is currently converted into useful energy. Section 8.1 provides an overview of the different sources and processes of bioenergy, including photosynthesis. Section 8.2 describes the properties of solid biomass; Section 8.3 gives an introduction to combustion. Section 8.4 reviews gasification and gasifiers. In Section 8.5, anaerobic digestion and the production of landfill gas are discussed. Finally, in Section 8.6, the conversion of biomass into fuel for road transport is addressed; the topics covered in this section include the fermentation of biomass into ethanol, extraction of natural vegetable oil and its conversion into biodiesel, as well as the social and environmental impacts of biomass vehicle fuel.

8.1 Bioenergy

Bioenergy is a broad term describing the conversion of solar radiation through photosynthesis into plant material, some of which may be eaten by animals. The organic matter from plants or animals is converted into heat and electricity or used to fuel transport. Biomass is any organic matter derived from plants or animals; it can be used directly to provide useful energy or processed into intermediate solid, liquid or gaseous biofuels.

Until the eighteenth century, biomass in the form of wood was the main source of energy for heating and the processing of metals. The wood was either burnt directly or first converted into charcoal. The industrial revolution led to a rapid increase in the use of coal, and subsequently oil and natural gas, but wood is still used in parts of Europe and North America for heating and as fuel for electrical generators. Methane from municipal waste in landfill sites (landfill biogas) is a useful fraction of the renewable energy used to generate electricity in the UK. Liquid biofuels, ethanol in Brazil and the USA, as well as biodiesel in Europe, are routinely used to supplement fossil fuels for road transport. In poor households of many developing countries, wood and animal dung are used directly for heating and cooking, often

Table 8.1 Main sources of bioenergy

Main sources of bioenergy	Details
Food (and fodder) crops	Edible parts of sugar, starch and oil plants that are traditionally grown as food for humans or animal fodder. Food crops currently being used for biofuels include wheat, maize, soya, palm oil, oilseed rape and sugar cane.
Agricultural residues	By-products from crops such as wheat straw, rice husks and sugar cane residue (bagasse). Also slurry and manure from animals.
Forestry and forest residues	Woody material from existing forests (which may be managed) plus residues from sawmills and tree pruning.
Municipal waste	Food and domestic waste, sewage and other biological waste.
Energy crops	Fast-growing trees and grasses grown for energy, e.g. miscanthus and willow; oil crops such as jatropha.

on open fires that are very inefficient and emit smoke that causes serious damage to human health.

Table 8.1 lists the main sources of bioenergy. These sources are very varied and a range of processing techniques are required to transform the biomass into useful energy. Compared with fossil fuels, most biomass has a limited energy density (expressed either as MJ/kg or MJ/m^3) and can be more expensive to store and handle. For example, natural gas (methane) has an energy density of 50 MJ/kg and can be compressed in pipes or canisters, while wood has an energy density of around 20 MJ/kg and has to be stored in dry conditions.

The creation of biomass and its subsequent conversion to bioenergy is more labour intensive than the exploitation of fossil fuels and so is an important source of employment in rural areas. Bioenergy is encouraged by many governments through financial support measures both for its environmental benefits and its contribution to rural economies.

In contrast to most renewable energy sources, dry biomass can be stored and used when needed. It is therefore well suited to being used as a fuel for electrical generators that balance supply and demand. Steam generator units that use processed wood as a fuel, either alone or mixed with pulverized coal in co-fired boilers, can provide the controlled generation that is needed for stable operation of an electric power system. However, the low density and hence large bulk of solid biomass leads to significant handling costs, and care is needed to ensure it is stored in dry conditions and any dust is controlled carefully.

As with all energy sources, the use of bioenergy has important environmental and social impacts that must be managed carefully. Brazil has manufactured bioethanol from sugar cane for many years, and in the USA and Europe it is common for 5–10% bioethanol and biodiesel, which are often derived from food crops, to be added to fossil fuels for road transport. This has led to concern that the use of valuable arable land for energy crops and the production of liquid biofuels has led to increases in the world price of food. Current research and trials are investigating how non-food crops grown on poor-quality land that is unsuitable for other uses can be converted into biofuel. It is obviously desirable to avoid competition for land with food crops. The environmental impact of large areas of monoculture energy crops is also controversial.

In principle, the cycle of growth of biomass and its subsequent use to create bioenergy can be carbon dioxide neutral, i.e. a similar quantity of CO_2 is absorbed from the atmosphere during photosynthesis as is released during the conversion of the biomass into energy. In practice, considerable fossil energy may be required during the growth and processing of biomass, e.g. for preparing the soil, harvesting the crop and for fertilizer. CO_2 emissions can be considerable if annual energy crops are planted, particularly if any change in land use results in the release of CO_2 that has been stored in the soil and vegetation for many years. Some biomass is traded internationally, and the costs and emissions incurred during its transport can be significant.

Careful evaluation of the entire life cycle of a bioenergy scheme is required in order to determine its environmental impact and the consequences for greenhouse gas emissions. The results of this life cycle analysis of the environmental impact of large-scale bioenergy schemes remain contentious, with some environmentalists arguing that the costs to the environment of using biofuels for road transport fuel and large-scale electricity generation exceed the benefits. For example, it can be argued that rather than burn wood for electricity generation, it is more effective to use it for the structure of buildings. Wood used for building material will last many years, retaining the CO_2 absorbed from the atmosphere, and will reduce the use of steel and cement, which are both manufactured using fossil fuels. Any assessment of the environmental costs and benefits of bioenergy depends critically on the boundaries of the calculation that are assumed, and the results remain contested.

It is anticipated that in the future it will become possible to capture CO_2 emitted from electricity-generating units or large industrial plants and sequester it underground. If CO_2 is captured from processes using biofuels, this will result in the bioenergy cycle reducing CO_2 in the atmosphere (i.e. overall negative CO_2 emissions). However, this technology, which is known as carbon capture and storage (CCS) has yet to be demonstrated to be cost-effective and deployed widely.

8.1.1 Bioenergy Processes

Bioenergy conversion processes can be divided into thermochemical processes, biochemical processes and extraction of fuel from plants. Thermochemical processes use heat and catalysts to transform biomass into bioenergy. Biochemical processes use enzymes and microorganisms for the same purpose. A more direct use is the extraction of vegetable oils from plants and the manufacture of biodiesel. Common examples of all three processes are as follows.

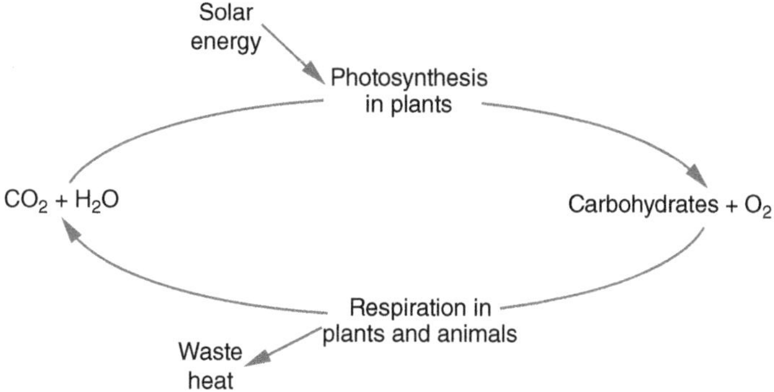

Figure 8.1 The CO_2 and O_2 cycle in the atmosphere.

Thermochemical processes
Combustion
Gasification
Biochemical processes
Anaerobic digestion
Alcoholic fermentation
Extraction of fuel from plants
Extraction of vegetable oil
Manufacture of biodiesel

8.1.2 Photosynthesis

All biomass is created initially through the process of photosynthesis, although there may be subsequent stages of processing to create biofuels, such as the fermentation of crops into ethanol or animals eating biomass and creating dung, which then produces methane through anaerobic digestion.

In photosynthesis, green plants (i.e. those containing chlorophyll) use energy from the sun to form carbohydrates from carbon dioxide and water, and so store energy (Figure 8.1). Chlorophyll pigments in the leaves of plants absorb photons from the red and blue parts of the visible solar spectrum and the captured energy is used to synthesize complex organic molecules that combine to form the structure of the plant. Oxygen (O_2) is created as a by-product of photosynthesis. In the process of respiration, which in plants may be thought of as the reverse of photosynthesis, the carbohydrates are broken down into carbon dioxide and water, releasing energy that eventually appears as low-temperature heat. In green plants under ideal conditions, the rate of photosynthesis is about 30 times the rate of respiration. The cycle of photosynthesis and respiration maintains the balance between CO_2 and O_2 in the atmosphere, and energy from the sun is captured and stored as biomass.

Photosynthesis can be represented by the general equation

$$CO_2 + H_2O + \text{light energy} \rightarrow [CH_2O] + O_2$$

where

$[CH_2O]$ is a general carbohydrate.

More specifically

$$6CO_2 + 6H_2O + \text{light energy} \rightarrow C_6H_{12}O_6 + 6O_2$$

where

$C_6H_{12}O_6$ is glucose.

Glucose within the plant is transformed into cellulose, hemicellulose and lignin. Cellulose is the main part of the plant cell wall with the formula $(C_6H_{10}O_5)_n$, where n = 50–4000. Hemicellulose is similar but has a more branched structure and is more susceptible to chemical degradation. Its formula is $(C_5H_8O_4)_n$, where n = 50–200. Lignin gives rigidity to the plant cell wall and is often bound to the cellulose and hemicellulose fibres to form a lignocellulose structure. A soft wood has the proportions of approximately 40% cellulose, 30% hemicellulose and 30% lignin.

A general formula for biomass that includes lignin is $CH_{1.4}O_{1.6}$.

The carbohydrates created by photosynthesis contain more energy than the CO_2 and H_2O that was used in their creation and so photosynthesis can be considered as a mechanism for concentrating solar energy. Only solar energy within the visible part of the spectrum (400–700 nm) is used in photosynthesis. Typically crops have an overall efficiency of conversion of solar insolation to stored energy of only around 0.5–1% but this can increase to 6–8% in sugar cane.

Photosynthesis occurs within the chloroplasts of green plants and proceeds in two stages: (1) light-dependent reactions that lead to the evolution of O_2 and the creation of energy storage molecules and (2) light-independent reactions that reduce CO_2 to carbohydrates by adding hydrogen. In plants, the complex light-independent reactions and formation of carbohydrates are known as the Calvin cycle. Most plants (including trees) fix carbon by a 3-carbon (C_3) cycle while some (e.g. sugar cane and miscanthus or elephant grass) obtain higher conversion efficiency by using 4-carbon (C_4) compounds. During the first stage of photosynthesis the C_3 plants form a pair of three-carbon-atom molecules while the C_4 plants initially form four-carbon-atom molecules.

At low levels of sunlight (irradiance < 100–$200\ W/m^2$), the rate of photosynthesis in C_3 plants is proportional to the intensity of the sunlight but then levels off and does not increase further at higher irradiance. Thus plants can use only 10–20% of the peak solar irradiance of 1000 W/m^2. In practice, this does not matter greatly, as the leaves within a crop are not oriented uniformly towards the sun. Higher rates of photosynthesis can be obtained by increasing the concentration of CO_2 in the air, as is done in some glasshouses.

Example 8.1 – Land Required for Bioenergy

Estimate the fraction of UK electricity demand that might be generated if an energy crop were grown on 10% of the arable land.

Solution

Assumptions	
Arable land in UK	65 000 km^2
Annual insolation	1000 kWh/m^2
Efficiency of biomass conversion	0.5%
Efficiency of electrical generation	30%
UK electricity demand	360 TWh/year

The energy in the crop is given by

Area of land × insolation × conversion efficiency

$$65\times10^9\times\frac{10}{100}\times1000\times10^3\times\frac{0.5}{100}=32.5\ \text{TWh}$$

The electrical energy generated (assuming an efficiency of 30%) is

$$32.5\times\frac{30}{100}=9.75\ \text{TWh}$$

This would supply

$$\frac{9.75}{360}=3\%$$

of annual UK electricity demand.

8.2 Properties of Solid Biomass

Biomass is more heterogeneous than most fossil fuels and exhibits a wide range of properties that influence its combustion. Table 8.2 shows typical characteristic properties of biomass fuel.

The moisture content of a fuel is the weight of water in it expressed as a percentage of the overall weight on a wet basis, on a dry basis or with respect to the weight of the dry and ash-free matter. This is illustrated in Figure 8.2 and Example 8.2. Biomass materials have a wide range of moisture content ranging from less than 10% for cereal grain straw to 50–70% for forest residues (on a wet basis). The maximum acceptable moisture content for efficient combustion of biomass is around 50–60% and for gasification is 20% (on a wet basis).

The inorganic ash content is usually expressed on a dry basis and ranges from 0.5% in wood to 20–30% in rice husks. High fractions of ash can lead to difficulties

Table 8.2 Typical characteristics of solid biomass fuels

Biomass material	Moisture content (wet basis) %	Ash content (dry basis) %	Lower heating value (MJ/kg)
Bagasse (sugar cane residue)	40–60	1.7–3.8	7.7–8
Peat	15	1–20	9–15
Rice husks	13–15	20–30	14
Straw	10	4.4	12
Wood	10–60	0.25–1.7	8.4–17
Charcoal	1–10	0.5–6	25–32

Source of data: P. Quaak, H. Knoef and H. Stassen, 'Energy from Biomass', 1999, World Bank Technology Paper No. 422.

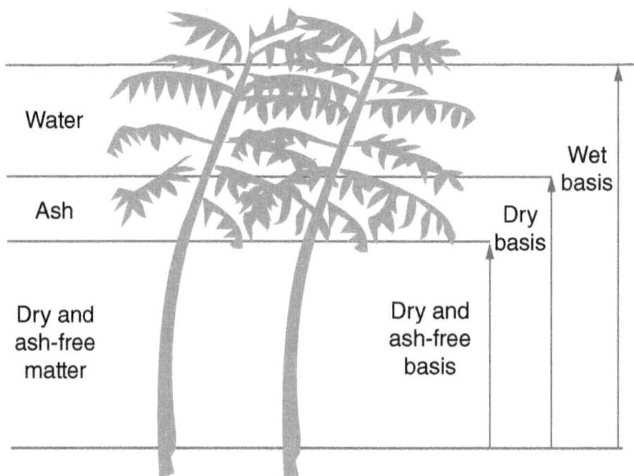

Figure 8.2 Definition of the basis of moisture and ash content.

with automatic ash-removal systems in furnaces and in the operation of fluidized bed combustors. Furthermore, at very high temperatures the ash will begin to melt and can transform into a fused product called slag, which causes blockage and damage to components such as heat exchangers.

The heating value of a fuel (usually expressed as MJ/kg) is measured with respect to a reference state of the products of combustion. For the lower heating value (LHV) the reference state is of water as a gas (i.e. the flue gases are not condensed) while the higher heating value (HHV) is with all the water in the flue gas condensed. The LHV is sometimes known as the net calorific value and the HHV as the gross calorific value.

Example 8.2 – Moisture Content of Biomass

1 kg of biomass (w_T) consists of:

0.8 kg of dry and ash-free matter (w_{daf}),
0.05 kg ash (w_{ash}),
0.15 kg water $\left(w_{H_2O}\right)$.

Calculate the moisture content on a wet, dry and dry and ash-free basis.

Solution
Moisture content on a wet basis, M_w, is

$$M_w = \frac{w_{H_2O}}{w_T}$$

$$M_w = \frac{0.15}{1} = 15\%$$

Moisture content on a dry basis, M_d, is

$$M_d = \frac{w_{H_2O}}{w_T - w_{H_2O}}$$

$$M_d = \frac{0.15}{1-0.15} = 17.6\%$$

Moisture content on a dry and ash-free basis, M_{daf}, is

$$M_{daf} = \frac{w_{H_2O}}{w_T - w_{ash} - w_{H_2O}}$$

$$M_{daf} = \frac{0.15}{1-0.05-0.15} = 18.75\%$$

Note also that

$$M_w = \frac{M_d}{1+M_d} = \frac{0.176}{1.176} = 15\%$$

$$M_d = \frac{M_w}{1-M_w} = \frac{0.15}{0.85} = 17.6\%$$

This section may be omitted at first reading.

Calculation of LHV from HHV

The energy content of a biomass material is often quoted as its higher heating value as this is independent of moisture content. The lower heating value (LHV) may be obtained from the higher heating value (HHV) by deducting the energy used in vaporizing the water in the biomass. This is shown in Equation 8.1, which includes terms representing the energy required to vaporize the moisture of the fuel as received as well as the additional water created during combustion.

(cont.)

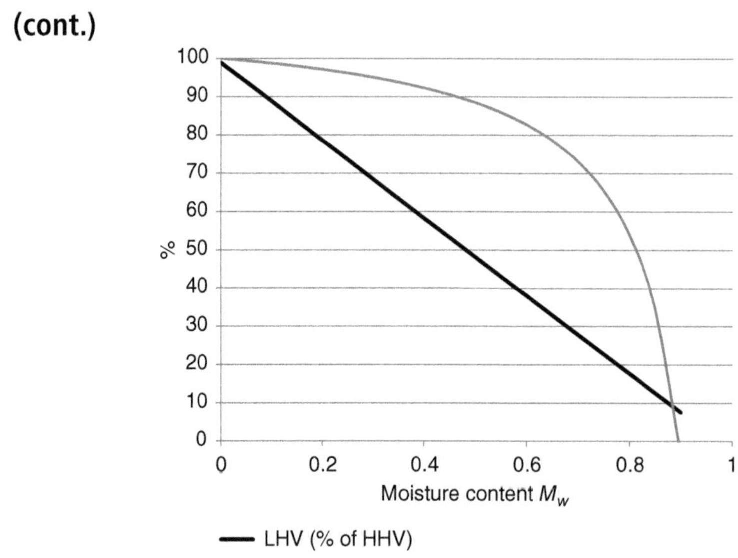

Figure 8.3 Effect of moisture content on the lower heating value and sensible energy available for combustion.

$$\begin{aligned} \text{LHV} &= \text{HHV}(1 - M_w) - 2.3M_w - 2.3X_H \times 9(1 - M_w) \\ &= (1 - M_w)\left[\text{HHV} - 2.3(M_d + 9X_H)\right] \quad [\text{MJ/kg}] \end{aligned} \tag{8.1}$$

where

HHV is the higher heating value,
M_w is the moisture content on a wet basis,
M_d is the moisture content on a dry basis,
2.3 is the energy of vaporization of water in MJ/kg,
9 is the ratio of molar mass between water and hydrogen,
X_H is the fraction of hydrogen in the biomass (by weight).

Useful Energy from Combustion of Wet Biomass

The moisture content of biomass has a major impact on the energy that is available for conversion into useful heat and it has been found that combustion cannot be maintained if the moisture content of the feedstock exceeds 60% (on a wet basis). Figure 8.3 shows the effect of moisture content on the lower heating value of a biomass feedstock with 5% hydrogen (by weight) calculated using Equation 8.1. The figure also shows the residual sensible energy available for combustion once the water in the feedstock has been evaporated. It was calculated using

$$Q_{residual}(\%) = 100\left[1 - \frac{2.3M_w}{(1 - M_w) \times \text{HHV}}\right]$$

where

$Q_{residual}$ (%) is the residual sensible energy after evaporation of moisture in feedstock,
HHV is the higher heating value on a dry basis, assumed in this example to be 20 MJ/kg.

Empirical Estimation of HHV

The higher heating value of biomass usually varies between 18 and 22 MJ/kg (on a dry basis) and based on experimental data, it can be estimated from:

$$\mathrm{HHV} = 34.91X_C + 117.83X_H + 10.05X_S - 10.34X_O - 1.51X_N - 2.11X_{Ash} \quad [\mathrm{MJ/kg}] \tag{8.2}$$

where

X_X is the fraction of constituent of the biomass (by weight),
subscripts are C carbon, H hydrogen, S sulphur, O oxygen, N nitrogen and Ash ash.

Estimates from this empirical model are not precise and may over-estimate the HHV of biomass by 2 MJ/kg.

8.3 Combustion

Combustion, or burning, is a complex sequence of chemical reactions between a fuel and oxygen (often air) accompanied by the production of heat and light in the form of a glow or flames. To achieve effective combustion a number of conditions must be met. Air must be supplied in excess (i.e. at a rate greater than the minimum air needed for complete combustion). An adequately high temperature must be maintained. Good mixing of the fuel gases and air must be assured and a sufficiently long dwell time in the reaction zone maintained. These conditions can be summarized as the three Ts of combustion:

- sufficient Time for complete chemical reaction (reaction time is often known as residence time),
- sufficient Temperature to heat the fuel through decomposition to ignition,
- sufficient Turbulence to mix the oxygen and fuel completely.

When combustion takes place, atoms are exchanged between colliding molecules to produce or absorb heat. In these reactions the atoms are preserved but the molecules are not. As the atoms are preserved the elements in the reaction must balance as can be seen in the simple reactions shown in Table 8.3.

The atomic and molar masses of the main elements and compounds found in simple combustion are given in Table 8.4 and may be used to calculate the mass of oxygen required for a combustion process and the products of combustion.

Table 8.3 Simple combustion reactions

Reactants → Products	Heat of combustion (kJ/mol)
$2H_2 + O_2 \rightarrow 2H_2O$	286
$C + O_2 \rightarrow CO_2$	394
$CH_4 + 2O_2 \rightarrow CO_2 + 2H_2O$	802

Table 8.4 Atomic and molar masses of elements and compounds

Substance	Symbol	Atomic mass	Molar mass	Nature – State
Carbon	C	12	–	Element – Solid
Hydrogen	H_2	1	2	Element – Gas
Oxygen	O_2	16	32	Element – Gas
Nitrogen	N_2	14	28	Element – Gas
Carbon dioxide	CO_2	$12 + 16 \times 2$	44	Compound – Gas
Methane	CH_4	$12 + 1 \times 4$	16	Compound – Gas

The supply of oxygen (or air) is critical for effective combustion. Stoichiometric conditions are when precisely the amount of air that is required for combustion is supplied and simple calculations usually start with this assumption.

Example 8.3 – Stoichiometric Combustion of Methane

Calculate the mass of oxygen required for the combustion of 1 kg of methane at stoichiometric conditions and the mass of carbon dioxide and water produced.

Solution

The combustion of methane in oxygen results in carbon dioxide and water

$$CH_4 + O_2 \rightarrow CO_2 + H_2O$$

Balancing the atoms on each side of this equation gives

$$CH_4 + 2O_2 = CO_2 + 2H_2O$$

Thus the combustion of 1 mole of methane requires 2 moles of oxygen and produces 1 mole of carbon dioxide and 2 moles of water.

The molar masses of the reactants and products are given in the following table.

	Substance	Molar mass
Reactants	CH_4	$12 + (4 \times 1) = 16$
	$2O_2$	$2(2 \times 16) = 64$
Products	CO_2	$12 + (2 \times 16) = 44$
	$2H_2O$	$2 \times ((2 \times 1) + 16) = 36$

Hence the combustion of 1 kg of methane requires 64/16 = 4 kg of oxygen.
It produces 44/16 = 2.75 kg of carbon dioxide.
The mass of water produced by burning 1 kg of methane is 36/16 = 2.25 kg.

It is usual to supply biomass furnaces with up to 40% excess air to ensure complete combustion and the minimum emission of un-burnt products. However, excess air has the effect of cooling the furnace and so reduces the temperature of combustion, which will affect the overall process efficiency and emission formation.

On a mass basis, the air/fuel ratio, AFR, of combustion is defined as

$$AFR = \frac{\text{mass of air}}{\text{mass of fuel}}$$

It is convenient to relate the air/fuel ratio to the special case at stoichiometric conditions AFR_{stoich} (where all the air and fuel reactants are used with no excess of either).

The air/fuel equivalence ratio, λ, is defined as

$$\lambda = \frac{AFR}{AFR_{stoich}}$$

The fuel/air equivalence ratio, ϕ, is defined as

$$\phi = \frac{AFR_{stoich}}{AFR} = \frac{1}{\lambda}$$

and

ϕ <1 means lean combustion,
ϕ >1 means rich combustion,
$\phi = 1$ is the stoichiometric condition, complete combustion.

The excess air, e, is

$$e = \frac{1}{\phi} - 1$$

Example 8.4 – Combustion of Biomass

Calculate the molar oxygen/fuel ratio for the combustion of biomass under stoichiometric conditions. Also calculate the mass of air required for the combustion of 1 kg of biomass with 40% excess air.

Solution

A general biomass material containing lignin can be represented as $CH_{1.4}O_{0.6}$ giving the process of combustion as

$$CH_{1.4}O_{0.6} + O_2 \rightarrow CO_2 + H_2O$$

Balancing hydrogen on both sides of the equation gives

$$CH_{1.4}O_{0.6} + O_2 \rightarrow CO_2 + 0.7H_2O$$

Balancing oxygen on both sides of the equation gives

$$CH_{1.4}O_{0.6} + 1.05O_2 = CO_2 + 0.7H_2O$$

Thus the molar fraction oxygen/fuel ratio is 1.05.

The composition of air by mass is 23% oxygen and 77% nitrogen, and so,

$$CH_{1.4}O_{0.6} + 1.05\left(O_2 + \frac{77}{23}N_2\right) = CO_2 + 0.7H_2O + 1.05\frac{77}{23}N_2$$

The molar mass of the biomass is

$$CH_{1.4}O_{0.6} \rightarrow 1 \times 12 + 1.4 \times 1 + 0.6 \times 16 = 23$$

and of the air is

$$1.05\left(O_2 + \frac{77}{23}N_2\right) \rightarrow 1.05\left(32 + \frac{77}{23}28\right) = 132$$

Therefore 1 kg of biomass requires 132/23 = 5.74 kg of air for stoichiometric combustion. With 40% excess air this increases to $1.4 \times 5.7 = 8$ kg.

8.3.1 Combustion of Solid Biomass

Combustion is the most direct use of biomass. The technology is well established, and equipment for electricity generation is offered by many manufacturers. However, the overall efficiency of electricity generation by combusting biomass is relatively low: only ~20% for older small generating plants rising to ~ 35% in larger modern units. Both the control of emissions and the disposal of ash from biomass combustion plants require careful attention.

Solid biofuels are burnt in devices that range in size and sophistication from small rural cooking stoves in developing countries to steam raising boilers for large electricity generators. Solid biomass is not easily flammable in its natural state and needs to be dried to reduce the water content before use. It then follows a complex thermochemical conversion process in order to burn satisfactorily. The stages of combustion of solid biomass are:

(1) heating and drying,
(2) pyrolytic decomposition to create volatile gases,
(3) combustion of the volatile gases above the fuel solid residues,
(4) combustion of the char in the fuel bed.

The solid biofuel must first be heated above its storage temperature of around 15–20 °C to drive off any water. From 225 °C to 700 °C the water-free fuel is pyrolysed and the volatile gases that emerge from the fuel bed are burnt. Long-chain molecules of volatile material are broken down into shorter-chain compounds releasing liquid tars and combustible carbon monoxide (CO), hydrogen (H_2) and gaseous hydrocarbons (C_mH_n). Above 225 °C these gases ignite, the process becomes exothermic and the water-free fuel decomposes under the influence of oxygen (O_2). From 500 °C to 700 °C the resulting solid carbon char is gasified in the presence of carbon dioxide (CO_2), water vapour and oxygen to form carbon monoxide and hydrogen that is subsequently burnt.

There are two basic types of furnaces commonly used for biomass combustion: fixed and fluidized bed systems. Fixed-bed systems include static and inclined grates (which themselves may move) with various approaches used for the supply of fuel and removal of ash including manual feed systems and arrangements using screw feeders. Temperatures in fixed-bed furnaces are typically between 850 °C and 1400 °C. Large fixed-bed furnaces achieve their desired operating conditions by using a separate air supply to the fire bed (primary air) and to the gas combustion zone (secondary air). The solid carbon char in the fuel bed is supplied with primary air and oxidizes to provide heat for the pyrolysis of newly added fuel. The secondary air ensures the more complete combustion of the volatile gases that are produced. The use of duplicate air streams leads to more even combustion and a low level of gaseous emissions. Modern fixed-bed furnaces maximize efficiency and minimize environmental impact by reducing the fraction of carbon monoxide, residual hydrocarbons and nitrogen oxide in the flue stack gases. The furnace is designed to minimize the excess air required for complete combustion as this also increases efficiency.

Fluidized-bed combustors have a bed of inert sand or limestone that is kept suspended in a turbulent state by a supply of pressurized air from fans. Fuel is burnt in and above the bed, which is typically kept at 700–1000 °C. There are two common types of fluidized-bed combustors; the bubbling fluidized bed and the circulating fluidized bed. A bubbling fluidized bed has a zone of suspended inert particles that is supported by the stream of air. There is a turbulent combustion zone above the bed into which the fuel is introduced. In a circulating fluidized bed, the air velocity is higher and some particles of both the inert bed and the burnt fuel flow upward with the gas stream, out of the combustor and are separated in an external cyclone and returned to the combustion zone. The turbulent conditions of a fluidized-bed combustor lead to intimate mixing of the products of combustion and the inert bed, providing good heat transfer. Fluidized-bed combustors are comparatively compact and they can accommodate a range of fuel types and shapes. They are also suitable for fuel with a high moisture and ash content.

Biomass can be burnt to provide energy for heating, industrial processes or the generation of electricity. Figure 8.4 shows the main elements of a steam turbine unit that uses the steam (Rankine) thermodynamic cycle for the generation of electricity from the combustion of solid biomass fuel. The fuel is combusted in the boiler, which consists of a furnace around which the boiler tubes are arranged. Purified water is injected into the tubes at high pressure from the powerful feed pump and circulates around the boiler, absorbing heat from the furnace. It then passes to the boiler drum where the water flashes off into steam. The steam is re-circulated to the superheater where it is heated further. The hot, high-pressure dry steam then leaves the boiler and drives the turbine. A large steam turbine consists of multiple stages of high-, medium- and low-pressure rotors and the steam may be reintroduced into the boiler between stages to reheat it. At the exit of the low-pressure stage the steam passes to the condenser where it is condensed back to water and returned to the boiler feed pump. The turbine drives a constant-speed synchronous generator.

The dashed lines in Figure 8.4 show some of the steam being taken from the turbine at an intermediate pressure and passing to a heat exchanger. The output of the heat exchanger can be used either for an industrial process or for district heating. Supplying both heat and electricity, known as combined heat and power,

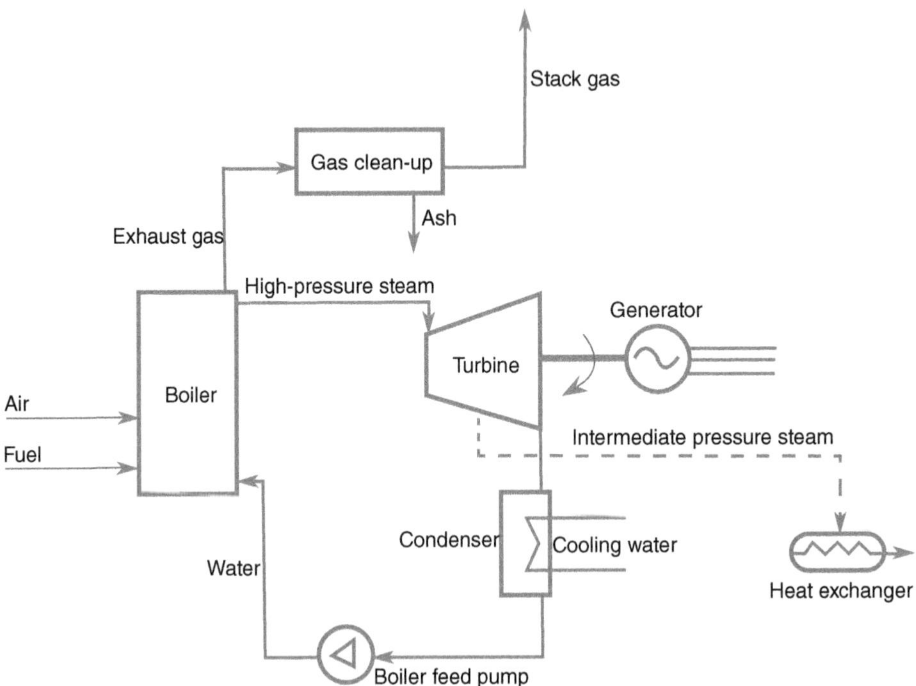

Figure 8.4 Steam turbine cycle.

can increase the thermal efficiency of the unit from around 30–35% to 65–70%. The particular arrangement shown is a pass-out condensing steam turbine that extracts steam from the turbine at intermediate pressure. An alternative arrangement is to use steam from the exhaust of the low-pressure turbine in a back pressure arrangement.

The hot gases from combustion are exhausted from the boiler and passed through a gas clean-up process that removes solid particles with an electrostatic precipitator and sometimes with fabric bag filters. Chemical cleaning of biomass flue gas can be complicated involving the injection of aqueous ammonia into the exhaust path to control NO_x, activated carbon to limit organic compounds such as dioxins and hydrated lime to control acidic compounds of nitrogen and chlorine. The low sulphur content of much biomass material reduces the need for flue gas de-sulphurization equipment, which is widely used in coal-fired plants to reduce emissions of SO_2 that cause acid rain. However, it may be necessary to remove nitrogen oxides (NO_x) that are formed from nitrogen in the air and bound into the organic matter when combustion takes place at high temperatures, 1200–1300 °C. Nitrogen oxides are the other main cause of acid rain. The cleaned gases are vented to the atmosphere through the stack.

The composition of the exhaust gas from all large combustion facilities is strictly controlled by legislation. Emissions from a waste-to-energy plant with its heterogeneous feedstock are likely to be a particular cause of public concern and lead to

Table 8.5 Details of a small biomass combustion plant in Sri Lanka

Fuel	Gliricidia wood
Fuel consumption	4 kg of wood per kWh_e
Steam	9000 kg/h at 400 °C
Turbine	1000 rpm coupled through a gearbox
Generator (synchronous)	1500 rpm, 1.6 MVA (de-rated for 1 MW plant)
Cycle	Rankine

Figure 8.5 Biomass feedstock at a 1 MW bioenergy plant. Photo: Lanka Transformers Ltd.
(A black and white version of this figure will appear in some formats. For the colour version, refer to the plate section.)

significant public opposition to the siting of any waste incineration plant during its planning and permitting.

Figures 8.5–8.7 show aspects of a small demonstration combustion biomass plant in Sri Lanka. Details of the plant are listed in Table 8.5.

Figure 8.5 shows the cut Gliricidia fuel wood being loaded on to a conveyer at the 1 MW plant. Note the covered storage area to reduce the moisture content of the feedstock. Gliricidia is a fast-growing wood that, because it is leguminous, fixes nitrogen in the soil.

Figure 8.6 is a view of the combustion in the fixed-bed furnace. Figure 8.7 shows the drive train of the turbine, gearbox and generator.

8.3.2 Analysis of the Combustion of Solid Biomass

Solid biomass fuels (in common with other fuels) are described using either proximate (thermochemical) or ultimate (elemental composition) analysis.

Figure 8.6 View into the fixed-bed furnace.
Photo: Lanka Transformers Ltd.
(A black and white version of this figure will appear in some formats. For the colour version, refer to the plate section.)

Figure 8.7 Steam turbine coupled to a generator.
Photo: Lanka Transformers Ltd. (Annotated by the author.)
(A black and white version of this figure will appear in some formats. For the colour version, refer to the plate section.)

Proximate analysis describes the fuel in terms of water content, volatile matter, fixed carbon and ash. The volatile matter consists of organic liquids and oils, plus permanent gases such as methane, carbon monoxide and carbon dioxide as well as vapours. Table 8.6 shows the proximate analysis of several biomass materials measured on a dry basis. Note the large difference in ash content between wood (2%) and rice husk (20%). This has important consequences for the design of furnaces and gasifiers that use rice residues as feedstock.

Ultimate analysis records the content of the biomass in terms of carbon (C), oxygen (O), hydrogen (H), nitrogen (N) and sulphur (S). For some crops, chlorine (Cl) is also recorded as it can lead to pollution and corrosion of the plant. Table 8.7 shows the

Table 8.6 Proximate analysis of biomass materials (percentage by weight on a dry basis)

	Wheat straw	Rice husk	Switchgrass	Sugar cane residue (bagasse)	Willow wood
Volatiles	75	64	77	86	82
Fixed carbon	18	16	14	12	16
Ash	7	20	9	2	2
HHV (MJ/kg)	18	16	18	19	19

Source of data: B.M. Jenkins *et al.*, Combustion properties of biomass, *Fuel Processing Technology*, **54** (1998) 17–46, with permission of Elsevier.

Table 8.7 Ultimate analysis of biomass materials (percentage by weight on a dry basis)

	Wheat straw	Rice husk	Switchgrass	Sugar cane residue (bagasse)	Willow wood	Coal (for comparison)
C	45	38	47	49	49	70
O	42	35	37	43	42	12
H	5	5	6	6	6	5
N	0.5	0.5	0.8	0.2	0.6	2
S	0.2	0.1	0.2	0	0.1	1
Cl	0.2	0.1	0.2	0	0.0	0

Source of data: B.M. Jenkins *et al.*, Combustion properties of biomass, *Fuel Processing Technology*, **54** (1998) 17–46, with permission of Elsevier.

ultimate analysis of some biomass materials. Biomass fuels have a high level of oxygen (typically 35–45%) compared with coal, which, depending on its source, has an oxygen content of less than 10%. They also have a low level of sulphur, which in coal may be up to 1–3%, so biomass boilers rarely require expensive flue gas de-sulphurization equipment.

Ultimate analysis of fuel can be used to calculate the quantity of air required as well as the products of combustion (see Example 8.5).

Example 8.5 – Combustion of Biomass Analysed Using Ultimate Analysis

Calculate the mass of air needed for the combustion of 1 kg of willow wood fuel.

If the fuel is burnt at an equivalent thermal rate of 1 MW and the higher heating value of the fuel is 19 MJ/kg, calculate the rate at which moisture will be produced.

Solution

The composition of willow wood by mass on a dry basis is given in Table 8.7:

carbon 49%, oxygen 42% and hydrogen 6%.

The combustion of carbon is described by the equation

$$C + O_2 = CO_2$$

The atomic weight of oxygen is 16, so the molar mass of O_2 is 32.

The atomic weight and molar mass of carbon C is 12. So the ratio of the mass of oxygen to carbon is 32/12 = 2.67.

But carbon makes up only 49% of wood, so the amount of oxygen needed to burn the carbon in 1 kg of wood is

$$\left(\frac{32}{12}\right) \times 0.49 = 1.31 \text{ kg}$$

The combustion of hydrogen is described by the equation

$$2H_2 + O_2 = 2H_2O$$

The atomic weight of hydrogen is 1 and so the molar mass of H_2 is 2.

Therefore the ratio of the mass of oxygen to hydrogen is 32/4 = 8.

But hydrogen constitutes only 6% of wood, so the amount of oxygen needed to burn the hydrogen in 1 kg of wood is

$$\left(\frac{32}{4}\right) \times 0.06 = 0.48 \text{ kg}$$

Oxygen makes up 42% of the wood and does not need to be supplied to the furnace through air. Thus 0.42 kg is deducted from the oxygen that has to be supplied.

Therefore, for each kg of wood fuel (on a dry basis) combusted at ideal stoichiometric conditions:

Oxygen required per kg of dry wood $= 1.31 + 0.48 - 0.42 = 1.37$ kg.

It is usual for 40% excess air to be needed for the combustion of biomass, increasing the oxygen required to $1.37 \times 1.4 = 1.92$ kg.

Air is 23% oxygen and 77% nitrogen by weight, so the mass of air needed is:

Mass of air $\left(1 + \frac{77}{23}\right) \times 1.92 = 8.35$ kg per kg of dry wood

Water is formed by the combustion of hydrogen:

$$2H_2 + O_2 = 2H_2O$$

Thus the mass of water produced per kg of fuel is

$$\frac{36}{4} \times 0.06 = 0.54 \text{ kg}$$

If the system operates at 1 MW_{th} (megawatt thermal) with an HHV of 19 MJ/kg

$$\frac{1 \times 10^6}{19 \times 10^6} = 0.053 \text{ kg/s}$$

Thus moisture will be produced at a rate of

$0.54 \times 0.053 = 0.029$ kg of water per second or 103 kg/h

It is convenient to set out the calculation in a table.

Constituent	Reaction	Mass of O_2 needed per kg of wood	Mass of product per kg of wood
C = 0.49	$C + O_2 \rightarrow CO_2$ 12 + 32 = 44 kg	$\frac{32}{12}(0.49) = 1.31$ kg	$\frac{44}{12}(0.49) = 1.79$ kg CO_2
H = 0.06	$2H_2 + O_2 \rightarrow 2H_2O$ 4 + 32 = 36 kg	$\frac{32}{4}(0.06) = 0.48$ kg	$\frac{36}{4}(0.06) = 0.54$ kg H_2O
O = 0.42	Less O_2 supplied by wood = −0.42 kg		None

8.3.3 Combustion of Wood

Solid biomass is processed before being burnt in order to ease handling and increase energy density and the efficiency of combustion. Typical products include wood pellets, wood chips, split logs, wood briquettes, pellets of refuse-derived fuel and straw bales. Wood pellets (Figure 8.8) are created by compressing dry sawdust or wood shavings with a binding agent at a pressure of around 100 kPa into pellets 6–8 mm diameter and up to 45 mm long. This produces uniform pellets with a density of 650 kg/m^3 and calorific value of around 5 kWh/kg (18 MJ/kg). The manufacture of wood pellets uses only a small fraction of the energy stored in the biomass but to this must be added the energy used for drying the wood and for the transportation of the pellets. Pellets are commonly used in furnaces up to 50 kW$_{th}$ rating with automatic fuel feed systems and for very large boilers. Wood chips that are directly cut from logs are also used. The chips are 1–10 cm long and 4 cm wide and are air-dried to a moisture content of ~20% before combustion.

8.3.4 Combustion of Biomass in Large Electricity Generating Stations

Large coal-fired boilers supplying steam to electricity-generating sets can be fired with biomass in order to reduce the use of fossil fuel and hence emissions of CO_2.

Figure 8.8 Wood pellet fuel used for large-scale generation.
Photo: Drax Power Ltd.
(A black and white version of this figure will appear in some formats. For the colour version, refer to the plate section.)

Figure 8.9 Biomass fuel-handling facilities at Drax power station. Note the large storage domes for wood pellets. Photo: Drax Power Ltd.
(A black and white version of this figure will appear in some formats. For the colour version, refer to the plate section.)

The dry biomass, often in the form of wood pellets, is introduced into the boiler either by co-milling with coal or by direct injection. With co-milling, limited quantities of biomass are blended with coal before the combined fuel is milled into a fine dust and blown into the boiler furnace. Up to 5–10% of the coal can be substituted by wood chips or pellets with limited modification to the boiler and fuel handling system. With direct injection, the fuel handling and milling of the biomass is separated from that of any coal and the milled biomass is blown directly into the furnace. Direct injection allows up to 100% biomass to be used, although the lower volumetric and energy density of the wood feedstock can result in a lower heat output of the boiler than if pulverized coal were used.

The Drax power station typically supplies around 6% of the electricity used in Great Britain. It was designed with coal-fired boilers using a Rankine cycle (Figure 8.4). Four of its six 660 MW generating units have been converted to operate only on biomass fuel giving a maximum generating capacity of the plant from biomass of 2600 MW. Wood pellets are manufactured in North America, typically from sawmill residues, forest thinnings and low-grade roundwood, and are shipped to the UK and delivered to the plant by train. The pellets are then stored in large domes (see Figure 8.9) and taken by a series of conveyers to ball mills where they are crushed into a fine powder and blown into the boilers in a similar manner to coal. The flue gas de-sulphurization plant that was installed to reduce SO_2 emissions from burning coal is not needed with the biomass fuel. The carbon intensity of electricity generation is quoted as 33 tonnes of CO_2/GWh.

Typical values of the proximate and ultimate analysis of the biomass fuel used at Drax are shown in Tables 8.8 and 8.9. The moisture content of the wood pellets is typically 4–8% rising to 10–12% for the other agricultural residues that are burnt in the plant.

Table 8.8 Proximate analysis of biomass wood pellets used at Drax power station (percentage by weight)

Volatiles	78.7
Fixed carbon (by difference)	20.4
Ash	0.9

Source of data: Drax Power Ltd.

Table 8.9 Ultimate analysis of biomass wood pellets used at Drax power station (percentage by weight on a dry basis)

C	47.8
O (by difference)	46.2
H	5.8
N	0.18
S	0.02
Cl	0.01

Source of data: Drax Power Ltd.

8.4 Gasification of Biomass

Gasification of biomass is the thermal conversion of organic material into combustible gases in the absence of oxygen or with only limited oxygen/air. Combustion of biomass requires air at 130–140% of stoichiometric conditions; this is reduced to ~25% for gasification.

Considerable heat energy is required for gasification, and in an indirectly heated gasifier this is supplied from an external source. More commonly the heat is provided by partial oxidation of some of the biomass, i.e. incomplete combustion at sub-stoichiometric conditions. Partial oxidation gasifiers use heat from the gasification process itself to maintain the required temperature. Water or steam can be injected into either an indirectly heated or a partial oxidation gasifier to increase the production of hydrogen.

Unfortunately the nomenclature used to describe the process of gasification and the product gases is not precise and often used in a loose manner. The terminology used in this book is described here.

The term *pyrolysis* is used to describe that part of the combustion and gasification process in which the biomass is heated and volatile gases formed that are then burnt. It is also used to describe the entire process by which solid biomass is transformed at sub-stoichiometric conditions into gases and char using either an external heat source or by partial oxidation. The manufacture of charcoal from wood is by pyrolysis. Fast pyrolysis is a high-temperature process that produces liquid bio-oils by rapidly heating small particles of biomass in the absence of oxygen.

Producer gas is the product of gasification and is a combination of combustible carbon monoxide (CO), hydrogen (H_2), methane (CH_4), traces of hydrocarbons and

Table 8.10 Approximate composition (percentage by volume) of producer (wood) gas made with air as the gasification medium

N_2	CO	H_2	CO_2	CH_4
50	20	15	10	5

inert carbon dioxide (CO_2). If air is used as the gasification medium, this results in a significant fraction of nitrogen (N_2). Producer gas has a rather low calorific value at ~5 MJ/kg, around 10% of that of methane. Other names are used for variants of producer gas depending on how it is manufactured including wood gas, coal gas and town gas. Table 8.10 shows the approximate composition of producer gas made from wood with air as the gasification medium. Note that only 40% of wood gas is combustible, as nitrogen and carbon dioxide are inert.

The strict definition of *syngas* or synthesis gas is that it consists of carbon monoxide (CO) and hydrogen (H_2) with small amounts of carbon dioxide (CO_2) and methane (CH_4). It is produced by gasification of carbon-rich material in oxygen and steam rather than in air, eliminating nitrogen in the product gas. Much of the carbon dioxide formed during gasification is removed. Syngas was developed originally for the synthesis of chemicals; it has about double the calorific value of producer gas and consists predominantly of combustible gases.

However, the term syngas has become widely used to describe any product gas of a gasification plant that uses a thermochemical process. Thus it is now generally accepted that the term syngas describes the product of all forms of thermochemical gasification and only needs to be distinguished from biogas, which is formed through the biochemical process of anaerobic digestion (see Section 8.6). This naming convention is followed in this book.

8.4.1 Gasification

Unprocessed biomass (e.g. straw, crop residues and wood) has a relatively low energy density with respect to both volume and mass compared with fossil fuels, so it is expensive to handle. It has to be stored carefully in dry conditions. The gasification of biomass changes a heterogeneous organic material into a homogeneous gaseous product that can be cleaned of impurities, and stored and transported more easily. Gas is a much more convenient fuel than solid biomass and can be burnt at higher temperatures, thereby increasing the thermodynamic efficiency of its use. Gas can also be compressed and pumped with relative ease compared with a solid material.

Heat is necessary for gasification. Partial oxidation gasifiers use the highly exothermic carbon-oxygen reaction to oxidize the carbon of part of the biomass feedstock and so provide heat. Air-blown partial-oxidization gasifiers operate with a fuel/air equivalence ratio, ϕ, of around 4, and the gas produced is around 50% nitrogen, reflecting the composition of the air that is used as the oxidant. If either steam or oxygen is used as the gasification medium rather than air, the resulting product gas contains higher proportions of hydrogen and carbon monoxide and hence more energy. Partial oxidation gasifiers operate over a wide temperature range of 600–1500 °C.

An alternative approach is that, rather than react air or oxygen with carbon to provide the heat required for gasification, an external heat source is used. Heat is generated externally, often from combustion of syngas or char, and transferred into the gasifier to provide an operating temperature of 600–800 °C. The gas produced by an indirectly heated gasifier has a high calorific value as no air or additional oxygen is used in its formation.

Gasification of biomass for electricity generation is still in the market entry phase with a number of demonstration plants in service and some manufacturers offering equipment. However, so far there has not been widespread adoption of the gasification of biomass. The technology holds great promise with a high efficiency of conversion of the chemical energy in the biomass feedstock to the energy of the gas produced. If the sensible heat of the gas exiting the gasifier is ignored (cold gas efficiency) then conversion efficiencies of 60–80% can be obtained rising to 80–95% if the heat of the gas leaving the gasifier is included in the calculation (hot gas efficiency). However, it has not been easy to develop reliable and cost-effective gasifiers that are fed automatically with biomass material of variable composition. A particular problem has been contamination of the syngas by tars.

The initial steps by which biomass is gasified are similar to those of combustion. The steps of gasification are:

(1) heating and drying,
(2) pyrolytic decomposition to create volatile gases,
(3) gas reactions with solid char,
(4) gas–gas reactions.

Heating and drying transforms biomass with a moisture content of 20–50% (on a wet basis) to dry matter at 300 °C and releases steam. Pyrolysis commences at a temperature of ~225 °C and becomes more rapid and complete as temperatures reach 400–500 °C. It produces volatile gases, permanent gases (gases that do not condense), tarry vapours and porous, carbonaceous solid char. Pyrolysis can convert up to 80% (by mass) of the biomass into vapours and gases. After pyrolysis, four key gas–solid reactions occur between the gases and the residual solid carbon char to produce additional carbon monoxide, hydrogen and methane.

The four important reactions that take place between the gases and solid carbon char are:

Carbon–oxygen reaction	$C+\frac{1}{2}O_2 \leftrightarrow CO$
Boudouard reaction	$C + CO_2 \leftrightarrow 2CO$
Carbon–water reaction	$C + H_2O \leftrightarrow H_2 + CO$
Hydrogenation reaction	$C+2H_2 \leftrightarrow CH_4$

The carbon–oxygen reaction is highly exothermic and provides heat for the endothermic Boudouard and carbon–water reactions. The hydrogenation reaction produces methane and is mildly exothermic.

The gas–gas reactions produce additional hydrogen and methane from the carbon monoxide. Gas–gas reactions are:

Water–gas shift reaction	$CO + H_2O \leftrightarrow H_2 + CO_2$
Methanation reaction	$CO + 3H_2 \leftrightarrow CH_4 + H_2O$

The formation of hydrogen can be increased by adding water to the process either by injecting steam with the air or by using wetter fuel. If chemical equilibrium were attained, then all the char would be converted to gas. In practice, about 10% (by mass) of char remains. When small particles of biomass are fed into a gasifier the reaction steps can occur in rapid succession, e.g. within a few seconds.

8.4.2 Gasifiers

Gasifiers that are offered commercially have either a fixed or fluidized bed with air partially oxidizing some of the biomass to provide heat. Figure 8.10a shows the principle of operation of a fixed-bed updraft gasifier. This is the simplest form of gasifier and is occasionally used in units of rating up to 1 MW. The biomass is fed into the top of the gasifier as a continuous feed, often through a rotary valve if wood pellets or chips are used. As the biomass descends against the flow of air, it is dried and pyrolysis takes place to release hydrogen, carbon monoxide and methane. Above the grate some of the char and tars that have been formed are combusted with air to provide heat for the gasification. Ash and any unburnt char exit at the bottom. Although the temperature in the combustion zone at the bottom of the gasifier can reach 1200 °C, the temperature at the gas exit is only around 100 °C. This low exit temperature leads to a high energy efficiency of conversion of the biomass to gas, but the syngas that is produced contains a significant quantity of tar, which has not passed through the high-temperature combustion zone at the bottom of the reactor and been burnt. Gas containing tar is not a suitable fuel for either reciprocating engines or gas turbines as it causes fouling. Hence it has been usual practice for the syngas from an updraft gasifier to be supplied directly to a furnace where the product gas and any tars are burnt. Some modern systems include a cleaning and filtration process where tar is extracted from the gas prior to it being fed into an engine.

Figure 8.10b shows a downdraft or concurrent-flow gasifier. In this design, air enters from the sides of the combustion zone. The biomass material is again fed into the top of the reactor and dried by the heat of the gasification process. The gases created by pyrolysis are burnt in the high-temperature combustion zone creating little tar. A constriction or throat made from ceramic to withstand high temperatures creates the reduction zone in which the tars and gases are broken down into carbon monoxide and hydrogen. The temperature of the product gas exiting the gasifier is ~700 °C.

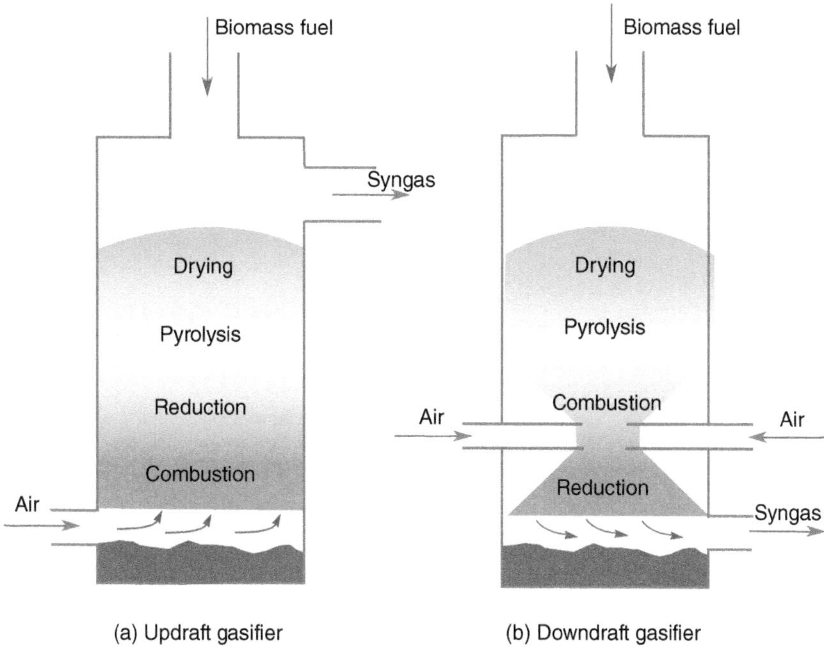

Figure 8.10 Fixed-bed gasifier: (a) updraft and (b) downdraft.

If the gas from a downdraft gasifier is used in a reciprocating engine it is first cooled then cleaned and filtered to remove particles of char, ash and any remaining tar. Cleaned gas from a downdraft gasifier can be used to power either a spark-ignition or compression-ignition reciprocating engine. Spark-ignition engines can operate on 100% syngas whereas compression-ignition engines are supplied with diesel fuel at a rate of 10–25% of full load consumption in addition to the gas. Downdraft gasifiers are more common than updraft gasifiers and were used widely in Sweden to power vehicles in the 1940s using wood as the biomass fuel when petrol was unavailable. However, the vehicles were rapidly converted back to fossil fuel once oil supplies were restored.

A fluidized-bed gasifier has many similarities to a fluidized-bed combustor. The gasification reaction takes place in and above a hot bed of inert granular material, such as sand, that is kept in a turbulent suspension by a stream of air. The bed behaves like a fluid, and fuel particles mix thoroughly with the sand. This forms a turbulent mixture with very good mixing and heat-transfer properties. The temperature of the fluidized bed is only 750–900 °C compared with 1200 °C at the hearth zone of a fixed-bed biomass gasifier.

The advantages of fluidized-bed over fixed-bed gasification are that it allows:

- a physically compact design as the turbulent conditions lead to high reaction and heat transfer rates,
- a wide variety of biomass feedstocks and moisture content can be used; material with high ash content such as rice husks can be utilized.

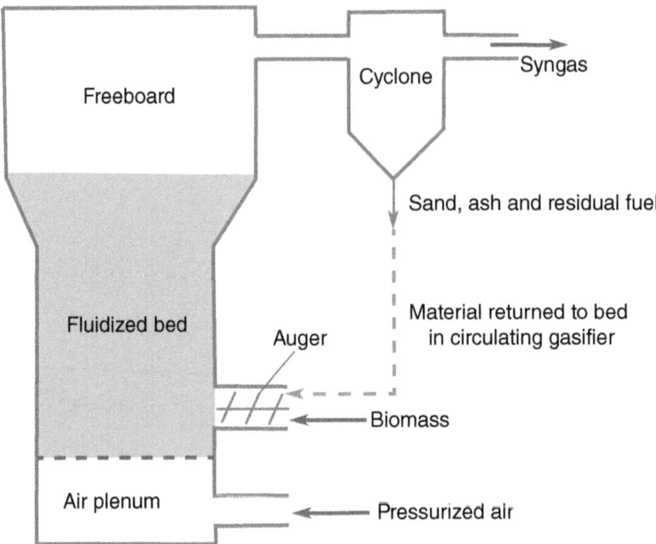

Figure 8.11 Fluidized-bed gasifier.

The disadvantages include:

- that a relatively small size of fuel is needed (< 20 mm),
- the need for power for the air compressor and the more complex control of fuel injection,
- the syngas is exhausted at high temperature and can contain significant tar and ash.

Figure 8.11 shows a fluidized-bed gasifier. There are two main forms of fluidized-bed gasifier: bubbling and circulating.

In a bubbling fluidized-bed gasifier, the biomass fuel is fed into a suspended hot-sand bed that creates a turbulent reaction zone. Biomass enters the gasifier either from the top or through the side of the bed with an auger. Injection of biomass from the side directly into the fluidized bed gives greater control of the residence time of the biomass feedstock, particularly of fine particles. Above the fluidized bed the cross-sectional area of the gasifier is increased to create a freeboard space that reduces the velocity of the gas. This limits the carry-over of bed material and residual fuel into the product gas. A cyclone is used to remove ash from the exhaust gas.

As the flow of oxidizing gas (air, oxygen or steam) is increased, the turbulent mixture of bed material and biomass occupies the entire volume of the gasifier and there is increased carry-over of this solid material into the product gas. It is then necessary to return the solids separated from the product gas by the cyclone, and the bed material and residual biomass solids are re-introduced into the gasifier. This is the principle of operation of the circulating fluidized-bed gasifier, which typically has air velocities some 3–5 times those of bubbling fluidized-bed gasifiers.

Figure 8.12 A 35 kW_e reciprocating engine generator set (foreground) with gasifier (background). Photo: Lanka Transformers Ltd.
(A black and white version of this figure will appear in some formats. For the colour version, refer to the plate section.)

Figure 8.13 Biomass gasifier. Photo: Lanka Transformers Ltd. (A black and white version of this figure will appear in some formats. For the colour version, refer to the plate section.)

Figures 8.12 and 8.13 show a 35 kW_e reciprocating engine generator powered by syngas from a biomass gasifier.

Biomass gasifiers have been the subject of active research and development since the 1990s but large units supplied from biomass have yet to be widely adopted.

8.5 Anaerobic Digestion

Anaerobic digestion is a natural biochemical process in which biomass is broken down by microorganisms in the absence of oxygen. Naturally occurring micro-organisms digest the biomass, and form biogas (a mixture of methane and carbon dioxide) and a solid/liquid residue. The biogas can be burnt for cooking, lighting and heating or used to power internal combustion engines to generate electricity. The methane can also be cleaned and injected into the gas supply grid. The residual solid matter from anaerobic digestion is known as digestate and, depending on the feedstock, can be a valuable fertilizer. However, particularly if the feedstock is classed by the environmental authorities as 'waste', disposal of the digestate is subject to very strict regulation. Anaerobic digestion has been used for many years and the technology is now mature.

A simplified general description for the anaerobic digestion of glucose is

$$C_6H_{12}O_6 \rightarrow 3CO_2 + 3CH_4$$

where

$C_6H_{12}O_6$ is glucose,
CH_4 is methane,
CO_2 is carbon dioxide.

The precise composition of biogas from anaerobic digestion depends on the feedstock and the particulars of the process. Typically biogas is made up of ~20%/80% methane (CH_4) and carbon dioxide (CO_2) to ~50%/50% CH_4 and CO_2 by volume. There are also likely to be traces (<1%), of nitrogen (N_2), hydrogen (H_2), ammonia (NH_3) and hydrogen sulphide (H_2S).

Anaerobic digestion occurs in four sequential stages, each of which is driven by different sets of anaerobic bacteria.

(1) Hydrolysis: large molecules of carbohydrates, fats and protein are broken down into simple sugars, lipids and amino acids that are readily available to other bacteria.
(2) Acidogenesis: the products of hydrolysis are converted into alcohols and fatty acids.
(3) Acetogenesis: the fatty acids and alcohols are converted to acetic acid.
(4) Methanogenesis: the intermediate products of the preceding stages are converted into methane and carbon dioxide to create biogas.

The calorific value of biogas is 17~25 MJ/Nm^3 (Nm^3 refers to normal cubic metres, volume at standard temperature and pressure). The theoretical efficiency of the anaerobic digestion of glucose to methane is around 90%; for practical anaerobic digestion systems using agricultural and municipal feedstocks, the efficiency of conversion drops to 40–60%.

The anaerobic digestion process can be either mesophilic, at a temperature of around 36 °C, or thermophilic with the sludge fermented in tanks at a temperature of 55 °C. Simple digesters, such as those used in rural villages of developing countries, use

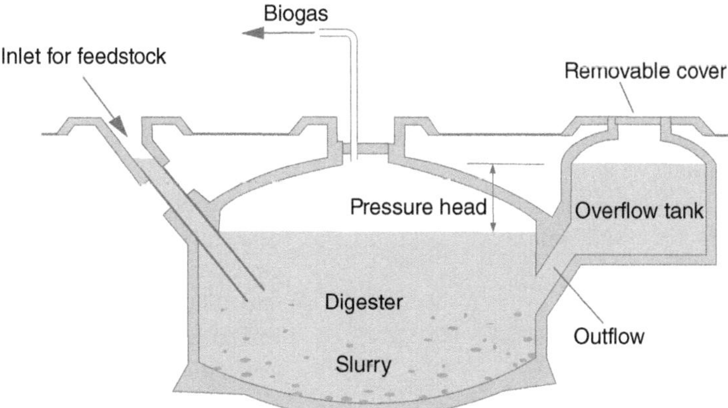

Figure 8.14 Single-stage anaerobic digester as used in homes in rural China.

mesophilic digestion. Supplementary heating of the digester and more sophisticated control of the process is required for thermophilic digestion, but the yield is increased at the higher temperature. It is usual for the contents of larger digesters to be stirred automatically to maintain an even temperature and mix in new biomass feedstock.

The elevated temperature within the digester acts to kill a large fraction of pathogens that are in the feedstock. This is particularly effective with thermophilic digestion. Digestate from animal manure slurry retains the nitrogen, phosphorous and potassium of the manure feedstock. The digestate formed from crop residues contains the lignin of the plant, which is unaffected by digestion, and when spread back on the land improves the structure of the soil.

Many forms of biomass are suitable for anaerobic digestion: suitable feedstocks, known as substrates, include food waste, farm slurry and animal manure, crops and crop residues. Woody biomass is rarely used because the microorganisms cannot break down the lignin of the woody residues without significant pre-processing. Typical yields of biogas from anaerobic digestion range from around 25 Nm^3/tonne for wet cow manure to 220 Nm^3/tonne if grass silage is used as the feedstock.

Biogas can be burnt to provide heat or used to generate electricity in spark-ignition reciprocating engines. Alternatively, it can be upgraded to biomethane through the removal of the carbon dioxide and the contaminant gases. If demanding standards of purity are met, biomethane can be injected into the national gas grid.

Anaerobic digestion is used with a variety of organic feedstocks and with different designs of digester ranging from small batch biogas digesters that provide energy for cooking and lighting in rural areas of India and China to large digesters using sewage and food waste that power gas engines either to generate electricity or in combined heat and power schemes.

Figure 8.14 shows a simple batch anaerobic digester of the constant-volume, variable-pressure type that has been widely used in rural China. Agricultural and domestic wastes are loaded into the fermentation compartment and create biogas in a single-stage digestion process. Fresh biomass is added daily at the inlet and the digestate is removed via the outflow. The resulting biogas is piped into the house for cooking and heating.

Table 8.11 Application of anaerobic digestion

Application	Feedstock	Use of biogas
Rural households in India and China	Agricultural and domestic waste	Cooking, lighting and heating
Large intensive livestock farms	Animal slurry and additional solid biomass	Gas engines for electricity generation, combined heat and power units. Biomethane blended into utility gas grids
Municipal sewage stations	Sewage sludge	Gas engines for electricity generation, combined heat and power units. Biomethane blended into utility gas grids
Municipal waste facilities	Organic fraction of municipal solid waste	Gas engines for electricity generation, combined heat and power units
Landfill gas sites	Municipal solid waste	Gas engines for electricity generation

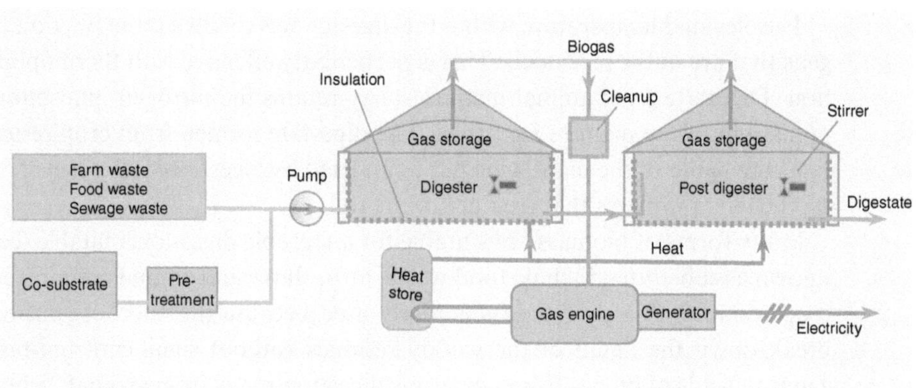

Figure 8.15 Multi-stage anaerobic digester supplying biogas to a combined heat and power generator.

Large farms in Europe and the USA use more complex multi-stage anaerobic digesters to dispose of animal waste. The slurry is supplemented by a co-substrate of solid agricultural waste such as crop residues to increase the production of biogas. Food waste is also used as a feedstock in municipal digesters (Figure 8.15). In some installations the biogas is cleaned and the CO_2 removed to produce biomethane for injection into the national natural gas grid, while in others a gas engine is used as a combined heat power plant to create heat for the anaerobic digestion and electricity. The digesters are insulated to maintain a high temperature and stirred.

The wide range of applications that use anaerobic digestion are summarized in Table 8.11.

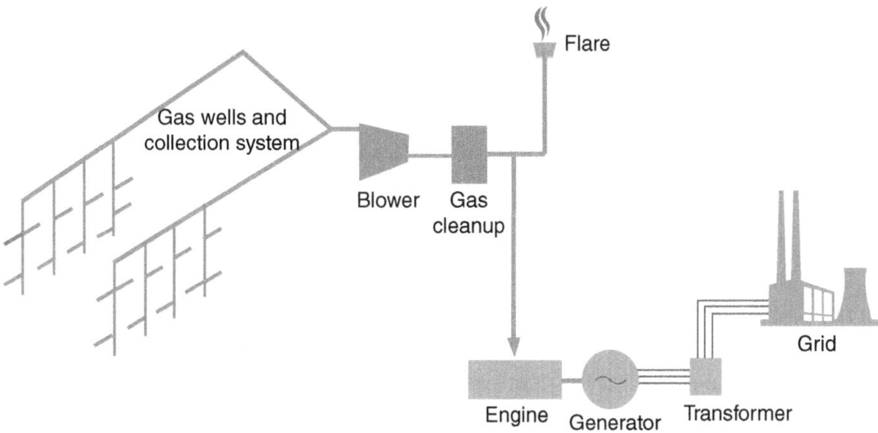

Figure 8.16 Collection of landfill gas and generation of electricity.

8.5.1 Landfill Gas

In the UK and USA it has been common to dispose of municipal waste in landfill sites. The municipal waste contains biological matter that over time decomposes in anaerobic conditions creating biogas. The emission of methane from a landfill site is subject to strict controls as it is a significant hazard both because of its action as a powerful greenhouse gas (25 times more powerful than CO_2) and because it can migrate along the outer surface of pipes and cables, enter dwellings and cause explosions.

In a landfill site, municipal waste is placed in a large excavation in the ground, compressed and buried. The excavation is lined with clay or plastic to stop liquid waste escaping and to create anaerobic conditions. When the site is full it is covered with a layer of soil, often clay, and then landscaped. The biogas that results from the anaerobic digestion of the biomass is captured through a series of gas wells located across the site, cleaned and either flared or used as fuel for electricity-generating engines. It can be difficult to use landfill gas for combined heat and power as any heat loads are usually remote from the landfill sites.

The UK has around 1.05 GW_e of electricity generation from landfill gas. When a landfill site is being filled, decomposition may be aerobic but, once the site is capped and sealed, anaerobic conditions exist, and biogas is produced. Depending on local conditions, peak biogas production is likely to occur up to ~5 years after the site is capped, and then the output of biogas declines over a further 20 years. The capacity factor of UK biomass generation peaked at 57% in 2012 but had dropped to 38% in 2020 as the biomass content of the waste decomposed. Figure 8.16 shows how the landfill gas is collected and used to power a spark-ignition engine to generate electricity.

UK and European policy is to reduce the quantity of municipal waste going to landfill and particularly limit the volume of the biodegradable waste that creates landfill gas. Thus increasingly, biodegradable material is being separated from the general municipal waste stream and treated in an anaerobic digester at a combined waste treatment and energy recovery facility.

8.6 Conversion of Biomass into Fuel for Road Transport

Biomass is converted into liquid fuel for vehicle engines in two quite different ways. Alcoholic fermentation is used to produce ethanol from sugar or starch crops. The ethanol is used as a partial or complete substitute for gasoline in spark-ignition engines. Alternatively, oil is extracted by pressing the seeds of crops such as rape, sunflowers or olives and either used directly in compression-ignition engines or processed by transesterification into biodiesel. Biodiesel is also produced from waste vegetable oils and animal fats, including waste cooking oil.

8.6.1 Fermentation of Biomass into Ethanol

The alcoholic fermentation of plant sugar is a well-established process to produce ethanol. The ethanol is either mixed with gasoline or used neat in spark-ignition engines. Bioethanol has been used in Brazil since the 1930s to fuel cars and motorcycles, reducing the quantity of fossil fuel that is imported. In Brazil, fuel for gasoline engines is sold as a mixture of 25% ethanol and 75% gasoline, while in the USA a mixture of 10% ethanol and 90% gasoline is common. In the UK, gasoline (petrol) for spark-ignition engines is sold with either 5% or 10% bioethanol. Most modern spark-ignition engines can accept up to 10% ethanol without adjustment. Flexible-fuel cars are sold in Brazil that are able to switch from using gasoline to ethanol, automatically adjusting their engine settings to operate on any combination of these fuels.

Bioethanol is produced most easily by the fermentation of sugar with yeast. In Brazil, sugar cane is processed to remove the sucrose for sale, leaving molasses. The molasses are fermented with yeast to produce carbon dioxide and ethanol. The carbon dioxide is often sold for the manufacture of soft drinks. The ethanol is then extracted from the water–ethanol mixture by distillation. The cane residue, known as bagasse, is burnt to provide heat for the distillation process. This integrated processing of sugar cane to make sugar, carbon dioxide and ethanol results in a very energy-efficient process.

The fermentation of glucose to produce ethanol and carbon dioxide is described by

$$C_6H_{12}O_6 \rightarrow 2C_2H_5OH + 2CO_2$$

where

$C_6H_{12}O_6$ is glucose,
C_2H_5OH is ethanol,
CO_2 is carbon dioxide.

Fermentation takes place in an aqueous mixture, but the yeast is poisoned by concentrations of ethanol greater than ~15% by volume. Hence fractional distillation is used to increase the concentration of ethanol to ~95%. The resulting product is known as hydrous ethanol. Hydrous ethanol can be used on its own in a spark-ignition engine but cannot be mixed with gasoline. For mixing with gasoline, it is necessary to remove the last 5% of water by a process of dehydration using vacuum distillation, distillation with benzene or molecular sieves. The final product is then anhydrous ethanol.

Starch from corn (maize) is widely used to manufacture bioethanol in the USA. Cobs of corn are milled and made into a slurry with water. The slurry is heated and enzymes are added to hydrolyse the starch into sugars. The subsequent process is then similar to that used with sugar cane. Yeast ferments the sugars into ethanol and carbon dioxide, and almost all the water is removed through a combination of distillation and molecular sieve dehydration to create anhydrous ethanol fuel.

There is great research interest in the fermentation of the lignocellulose material that is found in many non-food plants. This would allow bioethanol to be produced from a wide range of non-food crops that could be grown on much poorer land. However, at present, the fermentation of non-food plant material into alcohol is only possible with expensive pre-treatment and is not common commercial practice.

8.6.2 Extraction of Natural Vegetable Oil and Biodiesel

In a completely different process from the fermentation of sugar or starch into ethanol, oil-rich seeds are crushed and filtered to give straight vegetable oil (SVO). SVO is sometimes known as pure plant oil (PPO). Common crops used are oilseed rape, sunflowers and soya beans. The simple process of cold pressing followed by filtration has the advantage that it can easily be carried out on the farm. Alternatively warm pressing followed by extraction with organic solvents and purification can be used to give a higher yield. Waste cooking oil can also be recovered from commercial kitchens, filtered and purified and is then known as waste vegetable oil (WVO). Vegetable oil is considerably more viscous than fossil diesel fuel and so is usually heated before use. Unless the quality of vegetable oil is controlled carefully, its use can lead to the formation of wax in the engine and blocking of the fuel injectors. It is sometimes used in converted indirect-injection, compression-ignition engines, particularly in agricultural and forestry vehicles. The use of SVO and WVO for vehicle engines remains controversial and care is necessary not to void the vehicle manufacturer's warranty and to ensure that the appropriate tax is paid.

Biodiesel is a refined derivative of vegetable oil that has much reduced viscosity. Biodiesel is manufactured from SVO and other fat-based wastes using a process of transesterification. This involves a reaction with methanol, using sodium or potassium hydroxide as a catalyst, in which the triglycerides in the oil are replaced with mono-alkyl esters to produce biodiesel and glycerol. Biodiesel is sometimes referred to as fatty acid methyl ester (FAME). The glycerol by-product is refined and sold, and the crude biodiesel is filtered and refined. Biodiesel has substantially different properties from SVO and results in better engine performance. Biodiesel is commonly used as a blend with fossil diesel fuel. A blend of 5% biodiesel and 95% fossil diesel is used routinely in compression-ignition engines without modification, while some vehicle manufacturers approve the use of up to 20% biodiesel without compromising the warranty of the engine. In the UK, fuel for compression ignition engines is sold with 7% biodiesel.

8.6.3 Social and Environmental Impacts of Biomass Vehicle Fuel

The transport sector typically uses 20–30% of a nation's energy, with much of this coming from fossil fuels. Thus addressing the environmental impact of transport is a particularly important challenge. The available options for road transport fuel are as follows.

- Electric vehicles that use electricity from low-carbon sources. Electric propulsion is now common for battery/electric or hybrid passenger cars.
- Hydrogen-powered vehicles with hydrogen from low-carbon sources. At present, this is limited by the sources of hydrogen and the supply infrastructure.
- Fuel derived from biomass. This is limited by the social and environmental impact of the entire life cycle of growing, manufacturing and using biomass-derived vehicle fuels.

Using fuel derived from biomass could, in principle, significantly reduce the environmental impact of transport but an obvious issue is that many of the present sources of biomass transport fuel, e.g. sugar cane, corn, rapeseed and soya beans, are food crops grown on valuable arable land, sometimes using significant quantities of scarce water. The results of a number of studies have shown that the large-scale use of corn in the USA to produce bioethanol has had a measurable effect on the world price of food. More than 40% of the corn grown in the USA is used to produce bioethanol. Any increase in the price of food has particularly important implications for poor nations where expenditure on food can take half the family income. Precise quantification of the effect on the price of food of using food crops for vehicle fuel is difficult as the price of food depends on the weather, crop yields and market conditions. However, most studies indicate that diverting food biomass into vehicle fuel has contributed to an increase in world food prices.

An important indicator of the value of using crops for vehicle fuel is the ratio of energy in the fuel with respect to the fossil energy that is used to grow and process the crop. This is known as the fossil energy ratio (FER) or the fossil energy replacement ratio. The results of studies vary depending on assumptions and the boundaries that are drawn for the calculations but FER values of 1.4:1 for corn ethanol, 1.3–3.5:1 for biodiesel and 5–8:1 for sugar cane ethanol have been reported. If the fuel has a low FER it will make only a small contribution to reducing the emission of greenhouse gases but will reduce the need for a nation to import oil. A limitation of these studies is that they consider all energy sources on an equal footing. They ignore the increased value and convenience of energy-dense liquid fuel that can be easily stored and used in a wide variety of engines including in aviation gas turbines.

PROBLEMS

1. List the fundamental processes that are used to convert biomass into useful energy and give one example of each, with its product.
2. Distinguish clearly between biogas, syngas and producer gas. Discuss how each is formed and list their main constituent gases.
3. Describe the different ways that fuel is produced from biomass for spark- and compression-ignition engines.
4. What is the difference between the higher and lower heating value of a fuel?
5. Estimate the area of land that would be required to produce biomass to generate 1% of the GB average electricity demand of 40 GW.

   [**Answer:** 2335 km^2.]
6. Estimate the electrical energy that might be generated from 1 km^2 of land in England that can produce 10 dry tonnes of biomass per hectare each year.

   [**Answer:** 5.55 GWh or 190 kW continuous.]
7. An annual energy crop requires a fossil energy input of 35 GJ/hectare for cultivation and fertilizer. A further 8 GJ/tonne is required for harvesting and processing. The crop yield is 15 dry tonnes/hectare with an energy yield of 20 MJ/kg.
   (a) Calculate the fossil energy ratio.
   (b) Calculate the improved fossil energy ratio if the crop is coppiced annually over a three-year cycle.

   [**Answer:** (a) 1.94; (b) 2.28.]
8. A biomass fuel has a moisture content of 20% and an ash content of 5% on a wet basis. Calculate its moisture content on a dry and dry and ash-free basis.

   [**Answer:** 25%, 27%.]
9. The higher heating value (HHV) of an oven dried biomass material is 12 MJ/kg.
   (a) Write down the higher heating values if the material has a moisture content (on a wet basis) of 15% or 40%.
   (b) Calculate the sensible energy left for combustion once all the water has been evaporated.
   (c) If the biomass contains 5% hydrogen (by weight) calculate the lower heating values.

   [**Answer:** (a) 10.2 MJ/kg, 7.2 MJ/kg; (b) 9.86 MJ/kg, 6.28 MJ/kg; and (c) 9 MJ/kg, 5.7 MJ/kg.]
10. Write the stoichiometric reactions of hydrogen and sulphur in oxygen and air.
11. Write a balanced equation for the stoichiometric combustion of decane ($C_{10}H_{22}$) in oxygen and air, assuming that the only products are carbon dioxide, water vapour and nitrogen.
12. Write a balanced equation for the combustion of methane in air, with a fuel to air equivalence ratio of 0.8, and calculate the AFR (by mass).

    [**Answer:** 26.8.]

13. 1 kg of a biomass material requires 1.25 kg of oxygen to burn at stoichiometric conditions but is actually supplied with 6 kg of air per kg of biomass.

(a) Write down the mass of air required for combustion at stoichiometric conditions.

(b) Calculate the fuel/air equivalence ratio and the excess air.

[**Answer:** (a) 4.18 kg; (b) 0.7, 43%.]

14. A solid biomass fuel has an ultimate analysis of carbon 55%, hydrogen 15%, sulphur 1%, oxygen 9%, nitrogen 2%, ash 5% and moisture 13%.
Its calorific value is 1 MJ/kg (as received).

(a) For complete combustion with 28% excess air, determine the mass of air required per kg of fuel.

(b) If the fuel is burnt at an equivalent thermal rate of 1 MW, calculate the rate at which moisture will be produced, assuming all the water in the exhaust is condensed as liquid.

Atomic masses: C = 12, H = 1, S = 32, O = 16, H = 14 kg/kmol.

[**Answer:** (a) 14.3 kg; (b) 357 kg/h.]

15. A wood fuel has an ultimate analysis as shown in the table.

Element	Percentage dry matter (by weight)
C	50
H	6
S	0
O	40
N	1
Ash	3

Estimate the higher heating value.

[**Answer:** 20.3 MJ/kg.]

16. A biomass material has an ultimate analysis as shown in the table.

Element	Percentage weight on a dry basis
C	38
O	36
N	1
H	6
S	1
Ash	18

The fuel has a moisture content M_w of 30%, on a wet basis.

(a) Estimate the higher heating value (HHV).

(b) Calculate the lower heating value (LHV) for the fuel.

(c) The biomass will be combusted with 35% excess air. How much air is needed per kg of fuel for the combustion?

[**Answer:** (a) 16.32 MJ/kg; (b) 9.86 MJ/kg; (c) 4.7 kg.]

FURTHER READING

Brown R.C. (ed.), *Thermochemical Processing of Biomass*, 2019, Wiley.
Discusses a number of pathways for converting biomass into fuels, chemicals and power through thermochemical processes.

Caparada S.C., *Introduction to Biomass Energy Conversion*, 2014, CRC Press.
Comprehensive textbook covering all aspects of biomass energy conversion.

German Solar Energy Society and Ecofys, *Planning and Installing Bioenergy Systems: A Guide for Installers, Architects and Engineers*, 2004, Earthscan.
Practical guide to bioenergy systems.

Halford N. G., *An Introduction to Bioenergy*, 2015, Imperial College Press.
Entry-level textbook accessible to the non-specialist.

Hornung A. (ed.), *Transformation of Biomass: Theory to Practice*, 2014, Wiley.
Authoritative textbook covering a range of topics on biomass energy conversion.

Quaak P., Knoef H. and Stassen H., 1999, 'Energy from Biomass', *Technology Paper No. 422,* World Bank.
A review of combustion and gasification technology by the World Bank.

van Loo S. and Koppejan J. (eds), *The Handbook of Biomass Combustion and Co-firing*, 2010, Earthscan.
Authoritative textbook of biomass combustion.

ONLINE RESOURCES

Exercise 8.1 Biomass combustion: combustion of a solid biofuel is analysed using the ultimate analysis of the material.

Exercise 8.2 Combustion of wheat straw: combustion of wheat straw is analysed to determine the amount of air needed, the products of combustion and the heat produced.

9 Development and Appraisal of Renewable Energy Projects

INTRODUCTION

Compared with modern fossil fired power stations, many renewable energy plants to generate electricity are relatively small (typically less than 50 MW_e) with limited construction budgets (often less than £100 million). Hence many schemes are needed to meet a nation's electricity needs. Before contracts are finally awarded and any physical work started, the development phase of a project can take a significant fraction of the overall budget, and not all proposals will progress to construction. Therefore there is a great demand for engineers and other professionals to work on the project development and appraisal of renewable energy projects. This short chapter addresses three important aspects of the development and appraisal of a renewable energy project:

- project development and assessment of the renewable energy resource,
- economic appraisal,
- the Environmental Impact Assessment.

These topics are discussed in terms of a general renewable energy project because, although the characteristics of each form of renewable energy are different, the process of project development has many aspects that are common across a range of renewable energy resources and technologies. This chapter is written from the perspective of a project developer who would buy equipment and services to construct the project. The project developer can be either an independent company or the specialist arm of a larger organization.

Renewable energy projects differ from many other power projects in that a good understanding of the renewable resource is essential and the generating plants are often located in areas of great landscape and ecological value. Detailed and expensive investigations are needed to quantify the renewable energy resource and minimize potential impacts on the environment and local communities. During the feasibility phase and before planning permission is received, it is particularly important to strike a balance that limits expenditure (which might be wasted), while at the same time doing the work necessary to increase the chances of a successful outcome of the planning process, the project being commissioned on time and then working profitably over its life.

9.1 Project Development

The resource is the key factor in determining the energy output and hence income of any renewable energy project, while public acceptability will have a major influence over whether permission to construct and operate the plant is given by the planning authorities. Renewable energy projects usually require a large initial capital expenditure for the construction of the plant but have low operating costs. Hence the availability of project finance and the terms under which it is provided are important considerations. Although some technologies such as onshore wind power and photovoltaic generation using crystalline silicon modules are now well established, many of the newer renewable technologies carry significant technology risk. This will determine the sort of warranties that are available on the equipment, which in turn will influence how easy the project will be to finance. If the main equipment is supplied by a small newly formed company, then a guarantee of the performance of the technology and the contractor may have to be sought from a larger, longer-established organization. At present many new renewable technologies rely on some form of government support to compete with traditional energy sources as the environmental costs of burning fossil fuels are not fully reflected in the price of electricity from conventional power plants. Thus an additional risk of a renewable energy scheme is whether the government support will continue for the life of the project.

The site chosen for a renewable energy project is often in a sparsely populated area of the country where there has not previously been any reason to develop a strong electrical distribution network and the capacity of the grid to accept electrical power from a renewable energy scheme is limited. Reinforcing the grid often involves the construction of new overhead lines, which create their own environmental impact and can take a long time to obtain permission to build.

9.1.1 Phases of Project Development

The development of a renewable energy project is undertaken in phases of increasing complexity and expense. These are:

- initial site selection,
- project feasibility assessment, including costing, economic appraisal and identification of sources of funding,
- preparation of the Environmental Statement and submission of the planning application,
- *milestone:* planning permission obtained, project approved and financial close[1] achieved,
- construction,
- operation,
- decommissioning and land reinstatement.

[1] Financial close is when all agreements are in place and construction can commence.

The first step is to find a suitable site that has a good renewable energy resource and where a project will be welcomed by the landowner, planning authorities and the local population. The feasibility of the project is then assessed by estimating the costs of constructing and operating the plant, its energy output and future income. An economic appraisal is undertaken, and sources of funding identified. At the start of the feasibility assessment phase, the costs will be taken from budget estimates and are only known approximately. They are progressively refined until a firm price offer for construction of the plant is obtained, preferably from several equipment suppliers, and the preferred bidder chosen. An Environmental Impact Assessment is undertaken, and the resulting Environmental Statement forms an important part of the planning application.

The project planning and feasibility stage, up to the milestone when planning permission is obtained and the project approved for construction, can easily take several years, and consume 10% of the project budget. Until planning permission is obtained, finance is agreed and the project approved by the management of the developer, there is no certainty that it will proceed. Thus the project manager must balance expenditure during this preliminary phase, which may be wasted, with the necessity of undertaking the work that is required to pass the milestone and reduce risk in the project. Once the milestone has been passed successfully with all the necessary permissions granted, the actual construction phase can be quite rapid, particularly if only limited civil works are needed on-site. For example, the actual site installation work of a complete multi-megawatt onshore wind or solar farm takes only a few months, while the final erection of a single wind turbine and its tower on a previously prepared foundation can be completed in a day.

Figure 9.1 shows the blade of a wind turbine being delivered to an onshore wind farm. The length of the loads often requires improvements to minor public roads as well as the construction of on-site tracks.

Figure 9.2 shows a wind turbine slab foundation being constructed. The interface can is being lowered into position before more concrete is poured onto the reinforcing bar. The concrete turbine foundation takes several weeks to install and cure.

9.1.2 Assessment of the Renewable Energy Resource

The output of a renewable energy scheme, and hence the income, depends critically on the renewable energy resource. Therefore a careful evaluation of the future renewable energy resource is essential and will be scrutinized carefully by any investor in the project. Much of the initial assessment of the resource can be undertaken using published data sources and computer modelling, but at some point it often becomes necessary to conduct site measurements.

There are two objectives of the assessment of the renewable energy resource:

(1) to estimate the energy output and hence revenue over the life of the project,
(2) to characterize the site and the environmental conditions under which the equipment will operate.

Figure 9.1 Delivery of a blade to a wind farm under construction.
Photo: RES.
(A black and white version of this figure will appear in some formats. For the colour version, refer to the plate section.)

Figure 9.2 Wind turbine concrete slab and positioning of the tower interface can.
Photo: RES.
(A black and white version of this figure will appear in some formats. For the colour version, refer to the plate section.)

The estimation of the long-term energy output (and hence income) of a scheme is difficult as the project will have a life of at least 20 years, and the resource, such as wind speed, wave climate, river flows or even solar radiation, can vary significantly over a local area, throughout the year and from one year to the next. The variations of the resource over the local area can be determined by a site-measurement campaign, although these can be expensive, requiring the use of a considerable amount of equipment such as meteorological masts for wind measurements, buoys for wave assessment and weirs for river flows. Such measurements indicate the resource only at the time the measurements are taken and there are likely to be significant variations throughout the year and year-on-year. For example, the average wind speed at sites in the UK has been shown to vary by up to 20% from one year to the next, while the cloud cover of the monsoon can have a major influence on the solar resource in some tropical countries.

At many sites there will have been no reason to make measurements of the renewable energy resource at the site in the past and so no long-term records are available. For a site without long-term records, the only time for measurements to be made is during the feasibility study (which will last only one or two years). This difficulty of estimating the output over the next 20 years of a renewable energy plant at a new site, when only measurements of the period of the feasibility study are available, is addressed by using the general approach of measure–correlate–predict. The implementation of this technique for wind and hydro generation is described in detail in Chapters 2 and 3 but the approach can be used to increase confidence in estimates of solar power and even has applications for biomass where crop yields depend on the rainfall, irradiance and temperature of that particular year.

In general, measure–correlate–predict proceeds as follows. A set of measurements of the resource at the site is taken during the feasibility study for as long as possible. If a complete year of data cannot be acquired, it is important that measurements are taken over the time of year when the resource is high (e.g. in the winter for wind and hydro in UK). These measurements are then correlated with a series of similar measurements made at the same time by the national meteorological or environmental service at a nearby monitoring station, e.g. a local airport for wind speed or a gauging station fed by the same catchment for river flow. The monitoring station should have an environment similar to the project site. One or more scaling factors relating the site measurements to the monitoring station record are then calculated and applied to the long-term record that is held by the monitoring station. These factors are used to synthesize what would have been the long-term record of the resource at the site over the past 20–30 years. It is then assumed that the resource over the future life of the project will be similar to that in the past, and a prediction of the performance of the renewable energy scheme is made.

The use of measure–correlate–predict has the advantage that it is based on site measurements, but it also has several difficulties. There may not be a monitoring station near to the site or one with a similar environment. It is common for such data to be incomplete or partially corrupted and so require cleaning. An assumption is made that environmental conditions are stable and that the renewable resource in the future will be the same as in the past. In a period of climate change this assumption is open to question. The renewable resource is so important in determining the success of a project that it is prudent to reduce the anticipated energy output to a fraction (say 80–90%) of the value obtained from measure–correlate–predict to account for uncertainties. Another approach is to calculate the annual output of the project that is expected to be exceeded in 50% or 90% of the future years and use this as the estimate of the future energy yield.

It is also necessary for the project developer to characterize the site and inform the equipment supplier, through the purchase specification, of the conditions under which the plant will operate. Examples of the characteristics of the renewable resource that must be specified and made known to the plant manufacturer include: composition, moisture and ash content of biomass feedstock; mean values for extremes and turbulence of the wind speed and tidal stream resource; irradiation, temperature and wind speed at solar energy sites. For the well-established renewable energy technologies there are international standards that define how a site is characterized and broad classes of site conditions for which manufacturers design equipment. However, for the emerging renewable energy technologies particular site measurements may be necessary.

Levels of solar irradiance are more predictable than the wind and hydro resource, and a number of datasets are available based on ground or satellite measurements and modelling. Historically these datasets have been used directly to predict the output of solar installations but confirmation with direct measurement is desirable particularly for large schemes that rely heavily on diffuse irradiance.

9.1.3 Aspects of Project Development

An important part of the feasibility study is to define how the output energy of the scheme will be sold. For electrical energy there are two important aspects: the sale of the electrical energy, kWh, and the costs and time needed for connection of the generator to the network (determined by the rating of the plant, kW). In some countries the sale of energy and securing grid access will require agreements with two separate companies, while even with a single unified electric power utility these two aspects may be the responsibility of two departments.

For some renewable energy technologies, particularly for smaller schemes, the energy is purchased at a fixed rate, often decided by the government. Often this is through a feed-in-tariff that pays a fixed premium price per kWh for several years depending on the power rating of the equipment and generation technology. The feed-in-tariff provides certainty that, provided the plant operates as planned, the anticipated revenue will be received. Larger generation schemes must identify a buyer and sell the energy at a variable price determined by a market. This market risk is reduced by a contract-for-difference when the government compensates the project if the market price drops below an agreed strike price but receives payment from the project if the market price exceeds the strike price. It may not be possible to

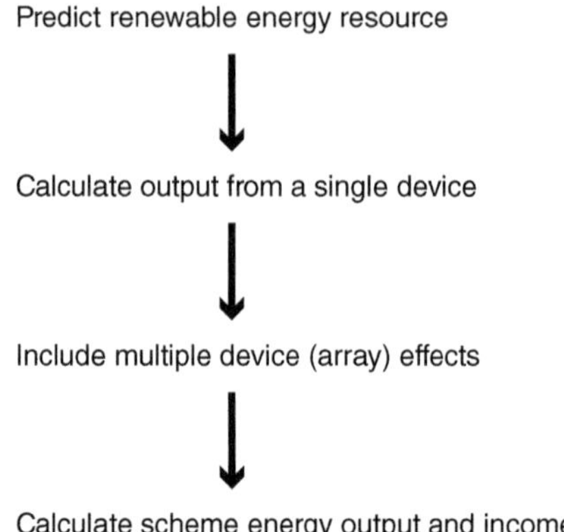

Figure 9.3 Prediction of the income of a renewable energy scheme.

obtain a firm price for the sale of electricity over the entire life of the plant, and the risk in finding a buyer for the output after the initial period must then be accepted by the project developer.

Access to the grid is arranged with the local electricity network operator and will depend on the electrical characteristics of the network in the area and how many other similar generators are already feeding into it. Clearly it is desirable for local loads to be supplied, limiting the power that needs to be transmitted a long distance. Any reinforcement of the grid is outside the direct control of the project developer and can cause long delays.

Prediction of the energy output and hence income of a renewable energy project is summarized in Figure 9.3. The renewable energy resource at the site is estimated over the future life of the project. Wind and hydro projects will certainly use some form of measure–correlate–predict while solar schemes may rely more on databases of satellite measurements and modelling results. The resource data are then passed through a mathematical model of the renewable energy device to obtain the electrical power output which, when integrated over time, gives the electrical energy generated. This model would typically be a simple power curve for a wind turbine that relates the output electrical power to the wind speed. For a photovoltaic (PV) array, where the output power of a module is specified under standard test conditions (STC), the model needs to include the change in module performance with increasing cell temperature as well as the effects of degradation over time and reduction in performance with pollution. Larger renewable energy schemes may consist of many devices connected together, and array effects are then considered. Array effects can include the reduction in the flow experienced by wind or tidal stream turbines that are positioned downstream of other devices and the electrical losses within the inverters and interconnections of a large PV installation. Finally the annual income from the scheme is predicted from the modelled farm output and the expected price of energy.

An important decision for the project developer is how to buy the equipment. A turn-key arrangement, whereby a single main contract for the engineering, procurement and construction of the entire plant is made with one supplier, moves maximum risk from the developer onto this main contractor. The main contractor undertakes detailed design, places contracts and assumes responsibility for the specialist equipment needed as well as for overall project management. After commissioning, a fully working plant is handed over to the developer, who may appoint a specialist operator. The project developer will probably still have to accept the risks of the renewable resource that differ from those anticipated and if the sale price of the product (e.g. electricity) is not fixed.

An alternative approach is to split the construction into a number of work packages, e.g. mechanical plant, civil works, electrical equipment, etc., and let contracts for each. This may result in a lower price of individual elements of the project but leaves the responsibility for the overall design and coordination of the work with the project developer or an engineer that is appointed for this purpose.

Even if the project developer is a large organization with strong financial resources, it is likely that a new company will be created to construct and operate each project. This company is known as a special purpose vehicle (SPV). The advantage of establishing an SPV for each project is that the parent company is protected if the project performs poorly, and one project is not affected by the performance of the parent organization or its other projects. The parent company will invest some equity in the project but most of the funds will be lent as debt by a bank or other financial institution. This debt is secured only on the project and so the bank (or the engineer and lawyer they appoint) will carefully examine all aspects of the project during the feasibility phase to ensure that all risks are adequately identified and controlled. There are likely to be a very large number of agreements and contracts that need to be scrutinized before funds are released at financial close (Table 9.1).

Renewable energy projects are sited to make best use of the resource and are often in valuable rural or marine environments. It is a mistake to think such places are not already being used (e.g. for agriculture or fishing) and it is particularly important to carry the support of the existing users of the environment and the local population.

Table 9.1 Agreements and contracts required for a renewable energy project

Planning permission
Land rights agreement
Grid connection agreement
Equipment supply contract
Energy yield assessment agreed
Balance of plant: civil and electrical contracts
Power purchase agreement
Operation and maintenance agreement
Insurance
Bank security
Corporate governance
Warranties and step-in agreement

This is sometimes achieved by a community group leading the development of the project or by local people having some financial interest in it. Another approach is to allocate a small percentage of the project revenue to be used for the benefit of the local community.

9.2 Economic Appraisal of Renewable Energy Schemes

Renewable energy schemes are expensive, and economic appraisal is used to decide how to allocate scarce financial resources to them. It has been conventional for many years to use techniques based on discounted cash flow (DCF). DCF appraisal was used in Section 1.2.2 of this book to evaluate if a proposed energy efficiency measure should proceed, and the same principles apply to the evaluation of generation projects. The approach is used by governments to decide which types of projects it should support and by firms to decide which projects to pursue. The key attribute of DCF analysis is that it recognizes the time value of money.

Renewable energy projects often require a relatively large expenditure of capital for their construction, while operating costs are low, and revenue is received over the life of the project. For example, hydro stations and wind farms are expensive to build but last many years. However, some schemes such as biomass plants have a significant fuel cost, and their equipment has a shorter life. Simple payback calculations that ignore the time value of money are of limited use in assessing if such projects should proceed. DCF analysis is a way to help choose between projects with different financial profiles. Of course, economic appraisal is only one aspect to be considered when comparing projects.

9.2.1 Simple DCF Appraisal

The basis of DCF appraisal is the time value of money and that humans prefer to receive money now rather than in the future. Hence, for the same benefit, a sum of money in *n* years' time has to be greater than the sum today.

This is expressed as:

$$V_n = V_p(1 + r_d)^n \tag{9.1}$$

where

V_n is the value of a sum in year n,
V_p is the present value of the sum,
r_d is the discount rate (%) reflecting our time preference for money.

It follows that the present value of a sum received or paid in the future is reduced and given by

$$V_p = \frac{V_n}{(1 + r_d)^n}$$

For a payment stream lasting m years, the present value is

$$V_p = \sum_{n=1}^{n=m} \frac{V_n}{\left(1 + r_d\right)^n}$$

This is a geometric series, which for equal payments (A) over m years, sums to

$$V_p = \frac{1 - \left(1 + r_d\right)^{-m}}{r_d} A \tag{9.2}$$

Equation 9.1 can be used for the calculation of the present value of any sum of money which is either paid or received at any time in the future (or in the past). The net present value (NPV) is simply the summation of all the present values of past or future income and expenditure.

General inflation is ignored in simple DCF calculations. The discount rate is then known as the real discount rate. Another way of expressing this assumption is that costs and benefits inflate at the same rate.

The choice of discount rate is a commercial decision made by the project developer. In a deregulated power system, discount rates of 8–12% would not be uncommon, reflecting the value placed on capital and the perceived level of risk. A high level of perceived risk in a project will increase the discount rate chosen by the developer. In economic terms the discount rate is an indication of the opportunity cost of capital. Opportunity cost is the return on the next best investment and so is the rate below which it is not worthwhile to invest in the project being assessed. The civil works of a hydro scheme can have a life of 80 years, while the main mechanical and electrical equipment can last 40 years. A comparison using DCF of a hydro scheme and a plant with a shorter lifetime but significant operating costs is particularly sensitive to the discount rate chosen.

The date to which all the costs and benefits are discounted (i.e. when $n = 0$) can be chosen as either the date of the initial capital expenditure at the start of the project or the date of commissioning of the plant. If the date of commissioning is chosen, then those costs incurred earlier before commissioning are negative.

Using the same discounting technique, other financial indicators can be calculated:

- the benefit–cost ratio is the present value of all benefits divided by the present value of all costs,
- the internal rate of return (IRR) is the value of the discount rate that gives a net present value of zero.

These indicators based on discounting (using Equation 9.1) should not be confused with indicators based only on immediate cash (non-discounted) sums such as:

- payback period, expressed in years, which is the capital cost of the project divided by the average annual return (i.e. income less operating costs),
- return on investment, which is the average annual return divided by the capital cost, expressed as a percentage.

For economic appraisal of any significant project, some form of discounted cash flow is normally required, and the non-discounted indicators are rarely used as the primary indicators of whether a project should proceed. Financial DCF calculations for projects can be further complicated by considerations of tax and how much of the capital is borrowed.

DCF analysis is based on our assumed time preference for money and is the established approach to economic appraisal of projects. However, the assumption that a benefit (or cost) now is more valuable than a similar benefit (or cost) in the future is an assumption that is in principle open to challenge.

Example 9.1 – Economic Appraisal using Discounted Cash Flow

A small hydro scheme with details as shown in the table is proposed.

Rated power of hydro scheme	5 MW
Annual capacity factor	30%
Capital cost	£6 500 000
Annual operating cost	£30 000
Revenue from sale of energy	8.5p/kWh
Lifetime	10 years of operation
Discount rate	9%

(a) Calculate the simple payback period (in years) of the project.
(b) Calculate the net present value of the project.
(c) Should the project proceed?

Solution

The annual energy production is

$$E = 5000 \times 0.3 \times 8760 = 13\,140\,000 \text{ kWh}$$

(a) The annual income is

$$13.14 \times 10^6 \times \frac{8.5}{100} = £1\,116\,900$$

Annual net revenue (return) is £1 116 690 – £30 000 = £1 086 900.

Therefore the simple (undiscounted) payback is

$$\frac{6\,500\,000}{1\,086\,900} = 6 \text{ years}$$

(b) The DCF appraisal calculation is set out in the table.

Discounted cash flow calculation

Capital cost	£6 500 000	
Annual energy	13 140 000 kWh	
Net annual income	£1 086 900	
Payback period	6.0 years	
	Discount factor	Present value
Expenditure Year 0	1.000	–£6 500 000
Income Year 1	0.917	£997 156
Income Year 2	0.842	£914 822
Income Year 3	0.772	£839 286
Income Year 4	0.708	£769 987
Income Year 5	0.650	£706 410
Income Year 6	0.596	£648 083
Income Year 7	0.547	£594 572
Income Year 8	0.502	£545 478
Income Year 9	0.460	£500 439
Income Year 10	0.422	£459 118
NPV		£475 352

The discount factor is given by

$$\frac{1}{\left(1 + r_d\right)^n}$$

From the table it can be seen that the NPV at the end of year 10 is £475 352.

(c) With the 9% discount rate and 10 years of operation the NPV is positive and so the project should proceed.

Note that the present (discounted) value of the annual income has decreased from £997 156 to £459 118 (42%) over the 10 years.

9.3 Environmental Impact Assessment of Renewable Energy Projects

Renewable energy projects can make a major contribution to improving the environment, particularly by generating electrical energy without the emission of greenhouse gases and other pollutants. However, the equipment is often spread over a wide area to capture the diffuse renewable energy resource and sited in areas of considerable ecological and landscape value. Thus a formal process is used to determine whether the benefits to society offered by a scheme outweigh the environmental, social and other costs. A final decision as to whether a project should proceed is made by elected office holders (e.g. the local planning committee or local mayor and councillors) but they are advised by officials on the technical aspects of the proposal and how it fits into wider environmental and societal policy objectives.

The officials in turn are informed about the site and the project by the results of the Environmental Impact Assessment and by the Environmental Statement document prepared by the project developer.

An Environmental Impact Assessment (*EIA*) is a well-defined process that is used to assess the environmental consequences of a project and is an essential step in securing planning (permitting) permission from the civil authorities. The requirement for an Environmental Impact Assessment is set out in national planning legislation.

The Environmental Statement (*ES*) is the formal document that is based on the work undertaken during the Environmental Impact Assessment. The ES is submitted to the planning authorities as part of the application for planning permission to construct the plant. Without planning permission, a project cannot proceed.

The scope of the Environmental Impact Assessment and the extent of the Environmental Statement depend on the technology, size and location of the project. Smaller projects using equipment with a limited environmental impact, e.g. a small crystalline silicon photovoltaic installation, say 10–100 kW, on an industrial building that is not visible from dwellings or publicly accessible places, may require only a limited environmental assessment. However, in most cases an Environmental Impact Assessment and the preparation of an Environmental Statement will be required for any renewable energy project of reasonable (MW) size. The Environmental Statement required for a large wind farm, hydro scheme or biomass combustion plant is a very wide ranging and expensive document. The extent of the Environment Impact Assessment and the investigations that are required is agreed with the planning authorities early in the project development process.

9.3.1 Uses of an Environmental Statement

The prime purpose of an Environmental Statement is to inform the officials and decision makers of the planning authority of the proposed scheme, the extent of the environmental impact, how possible alternative schemes were considered and any environmental impact mitigation measures that will be adopted. An underlying philosophy is that it is always preferable to avoid environmental damage by finding alternatives rather than relying on mitigation measures to limit impact.

A most important secondary role of the document is to inform the local population and other stakeholders of the project. Thus it always includes a non-technical summary that is used during the public consultation.

The detailed objectives of the Environmental Statement may be listed as follows:

- to explain the need for the project and describe the physical characteristics, scale and design of the scheme,
- to examine the environmental character of the site and the area surrounding it that may be affected by the scheme,
- to predict the likely environmental impacts,
- to describe measures that will be taken to avoid, offset or reduce any adverse environmental impacts,

- to describe measures that will be taken to improve the environment in the area of the scheme,
- to provide the planning authorities, other stakeholders and the public with the information they need to assist them in making a decision as to whether planning permission should be granted for the scheme to go ahead.

9.3.2 Contents of a Typical Environmental Statement

The contents of the Environmental Statement and the studies undertaken during the Environmental Impact Assessment are specific to the scheme being proposed but are likely to include:

- an introduction describing the applicant, site location, nature of the scheme and structure of the document,
- a description of national energy and local planning policies and how the scheme fits into these,
- a discussion of the need for the scheme and why this particular site is preferable to any other,
- a description of the scope of the assessment and any studies that have been undertaken,
- a description of the project and how the design has been developed including any changes to avoid adverse impacts,
- a series of detailed technical chapters that will vary according to the technology of the project and the site, but which are likely to include:
  - landscape and visual impact,
  - air quality (particularly important for biomass generation plants),
  - noise,
  - cultural heritage and archaeology (are there any important archaeological sites that will be affected?),
  - access to the site and impacts of traffic during both construction and operation on the plant biodiversity and natural heritage (a baseline site survey to determine the flora and fauna before the project may be required),
  - water environment (particularly for hydro developments but also addressing impact of other schemes on wetlands and catchments),
  - disposal of waste (particularly for biomass schemes),
  - socio-economic aspects.

As can be seen, the list of possible studies required is long and many require specialist expertise from consultants.

PROBLEMS

1. Describe how the energy output of the following types of generating scheme can be estimated:

   a wind farm,
   a solar PV project,
   a biomass plant.

2. List and discuss the main risks to the projects of the types listed in Problem 1.
3. What is meant by a turn-key contract? Discuss the advantages and disadvantages of this type of procurement.
4. What is the impact of a high discount rate on renewable energy projects?
5. List and discuss the main issues that should be addressed in the Environmental Statement for:
   a small hydro scheme,
   an onshore wind farm,
   a biomass generation project.
6. A small hydro scheme with details shown in the table is proposed.

Head	40 metres
Peak flow rate	3 m^3s
Annual capacity factor	45%
Capital cost	£800 000
Annual operating cost	£5500
Revenue	4p/kWh
Lifetime	10 years
Discount rate	11%
Turbine efficiency	85%
Generator efficiency	97%

   (a) Calculate the simple payback period (in years) of this project.
   (b) Calculate the net present value of the project.

   [**Answer:** (a) 5.4 years; (b) £68 454.]

7. A small wind farm with details shown in the table is proposed.
   (a) Calculate the simple payback period (in years) of the project.
   (b) Calculate the net present value of the project. Assume the capital cost of the wind farm is incurred when construction starts and revenue starts to flow after one year.
   (c) Calculate the internal rate of return.
   (d) List and discuss the main financial risks to the project.

Rated power of wind farm	5 MW
Annual capacity factor	28%
Capital cost	£3 000 000
Annual operating cost	£10 000
Revenue from energy (kWh)	5.5p/kWh
Lifetime	7 years of operation
Discount rate	12%

   [**Answer:** (a) 4.5 years; (b) £32 000; (c) 12.3%.]

FURTHER READING

Burton A., Jenkins N., Sharpe D. and Bossanyi E., *Wind Energy Handbook*, 3rd ed., 2021, Wiley. Addresses the environmental impact of onshore and offshore wind farms.

National Research Council. *Environmental Impacts of Wind-Energy Projects*, 2007, National Academies Press. https://doi.org/10.17226/11935.

Describes the Environmental Impact Assessment of wind energy projects, particularly in the USA.

Glasson J. and Therivel R., *Introduction to Environmental Impact Assessment*, 4th ed., 2012, Routledge.

A comprehensive textbook introducing Environmental Impact Assessment.

Khatib H., *Financial and Economic Evaluation of Projects in the Electricity Supply Industry*, 1997, Institution of Electrical Engineers.

An accessible book describing the economic evaluation of projects in the electricity sector.

10 Electrical Energy Systems

INTRODUCTION

The world's reliance on electricity to meet its needs for heat, light, motive power and the exchange of information continues to increase. Reducing emissions of CO_2 to limit climate change will require a transition from fossil fuels to electricity for many services, including for heating (e.g. from gas boilers to heat pumps) and for transport (e.g. from internal combustion engines to electric vehicles). This will further increase the demand for electricity.

A traditional power system is supplied by large fossil-fired or hydro units with synchronous generators. The operation of these units is well understood and, because of the energy stored in the fossil fuel or in large reservoirs, they are relatively easy to control. In contrast, many renewable generators, e.g. wind and solar power units, respond to the instantaneous renewable resource and are connected to the network through static power electronic interfaces which have quite different operating characteristics from those of large rotating generators.

The power generated by a large renewable generator such as a wind farm or hydro generator must be transported through transmission and distribution networks to the load. Smaller generators, e.g. domestic photovoltaic (PV) systems, are connected close to the customers but rely on the central power system to provide a stable reference of voltage and frequency. Depending on their location and power output with respect to the local load, renewable generators that are distributed over the network will have a greater or lesser impact on the voltages and the frequency of the power system.

With an increasing penetration of renewables, the operation of the power system becomes more difficult. However, many national electricity systems now operate routinely with more than 50% of their electrical power being supplied from renewables. Traditionally a power system has attempted to supply any load when electricity is demanded. With a high fraction of renewable generation operating on the system it becomes necessary to take control of the load, particularly in the event of an unexpected failure of a generator or an error in forecasting the renewable resource.

This chapter discusses the factors governing the connection and operation of renewable generators and some of the issues that can be anticipated in operating a future electricity system that has a large proportion of renewable generators. After a general discussion of energy vectors, it presents a review of how an ac electric power system is configured and operated, before moving on to address how renewable generators can affect voltages and frequency. Then several particular characteristics of renewable generators, e.g. low inertia and the need for increased reserve generation, are discussed.

The chapter is supported by Tutorial I on Electrical Engineering, which provides a simple description of electrical generators and inverters together with a revision of single- and three-phase electrical fundamentals.

10.1 Energy Systems

Energy is transported to the location at which it is needed through a number of pathways (sometimes known as energy vectors). These include electricity, heat or combustible gases. In many countries, electrical energy accounts for around one-third of the total final energy consumed with the remainder being supplied directly from fossil fuels: natural gas, oil products or coal. Steam or hot water thermal systems are used to transmit energy in industrial plants and urban areas where there is a dense heat load. Although electricity is a very flexible energy vector that is essential for modern societies, in most countries it transports only a fraction of the energy that is used.

One key attribute of fossil fuels is their high energy density (Table 10.1). They can be transported easily by truck or pipeline and converted in devices occupying a small volume. Fossil fuels can be converted into useful energy centrally, such as by the generation of electricity in large power stations, or locally by the creation of heat in domestic natural gas boilers as well as into heat and electricity in combined heat and power units. In contrast, renewable sources of energy are significantly less dense and the resource must be exploited where it occurs. Wind and solar energy are particularly diffuse, although wave and tidal energy are more concentrated (Table 10.2). The values of Table 10.2 are for normal operating conditions and do not reflect the often destructive energy density during abnormal events such as storms.

Table 10.1 Approximate energy densities of fuels

Fossil fuel	Energy density (MJ/litre)	Energy density (MJ/kg)
Diesel oil	36	45
Coal, bituminous	20	25
Wood	2	15
Hydrogen gas (room temperature and pressure)	0.01	140
Natural gas (room temperature and pressure)	0.04	50

Table 10.2 Approximate power density of renewable energy resources

Renewable energy resource	Power density	Conditions
Wind	625 W/m^2	Wind speed of 10 m/s
Solar	1000 W/m^2	Bright sunlight
Tidal stream	8 kW/m^2	Flow speed of 2.5 m/s
Wave	50 kW/m	Average linear power in an exposed location

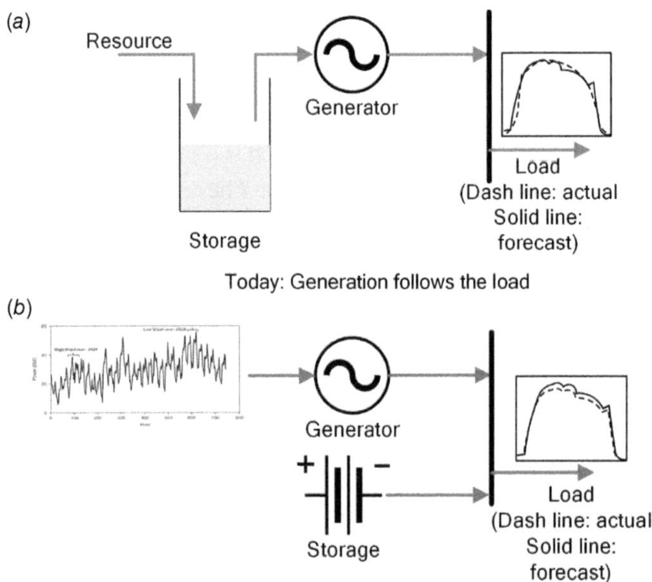

Figure 10.1 (a) Conventional and (b) future power system.

A second important characteristic of fossil fuels, which is shared by large reservoir hydro schemes, is that energy can be stored for long periods with little degradation. Fossil fuels store energy in a dense form, and this most useful attribute has allowed the conventional energy system to be developed in its present architecture, where energy demand is unconstrained and energy supply is arranged always to meet it. It is unlikely that a future mainly renewable energy system can be built and operated cost-effectively without at least some of the energy demand being controlled in response to variations of the renewable resource. Unless expensive energy storage is used, renewable sources have to be converted into useful energy when the resource is available. Figure 10.1a depicts the operation of the conventional power system and Figure 10.1b the future power system with a large penetration of renewables.

Large electricity systems benefit greatly from the diversity of their loads. For example, an individual modern domestic house may have a once-a-year instantaneous maximum electric load of 10 kW when, by chance, all the appliances (e.g.

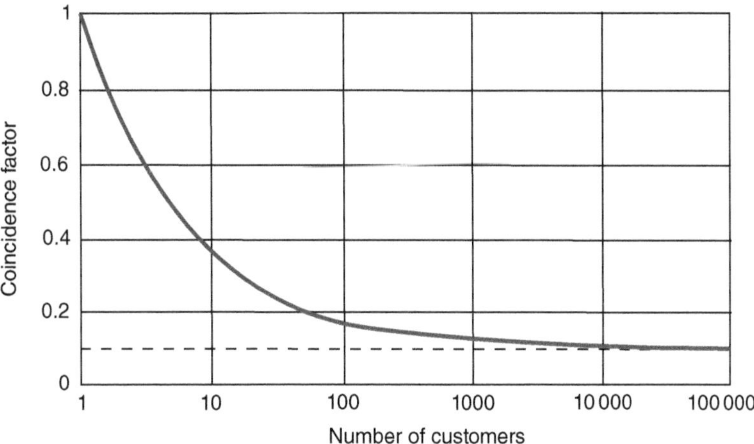

Figure 10.2 Load coincidence factor of typical households.

dishwasher, washing machine, tumble dryer, fridge, iron) happen to be working simultaneously. However, over the year, the average load of many houses is only around 1 kW as appliances are used rarely and at different times. Thus the final supply cable to a single house must be rated at 10 kW but the connection of many dwellings to the grid is only sized at 1 kW per house. Figure 10.2 shows the coincidence factor of domestic electricity loads and it can be seen that once more than 10 000 houses are supplied from an item of equipment (e.g. a generator or high-voltage transmission circuit) the summation of all the individual peak loads can be reduced dramatically and the power rating of the items of plant chosen accordingly.

10.2 Alternating Current (ac) Power Systems

Modern electrical power systems mainly follow the arrangement shown in Figure 10.3. Large central power stations have generating units with ratings up to 1000 MW driven by high-pressure steam from fossil-fuel boilers and nuclear reactors or by hydro power. The electrical power is fed through generator transformers to a high-voltage interconnected transmission network, known as a grid. The transmission system is used to transport the bulk electrical power, sometimes over considerable distances, and it is then extracted and passed down through a series of distribution transformers to radial distribution circuits for delivery to the customers. Some large renewable energy generators such as wind farms and hydro sets are connected to the transmission or high-voltage distribution network, while smaller units such as PV systems are connected at lower voltages.

Transformers are used to connect the various voltage levels of the three-phase alternating current (ac) power system. For example, 400 kV is used for bulk transmission in Great Britain (GB) while consumers need 230 V single-phase for their domestic equipment. Typical voltage levels used for the various circuits of the GB power system are shown in Figure 10.4.

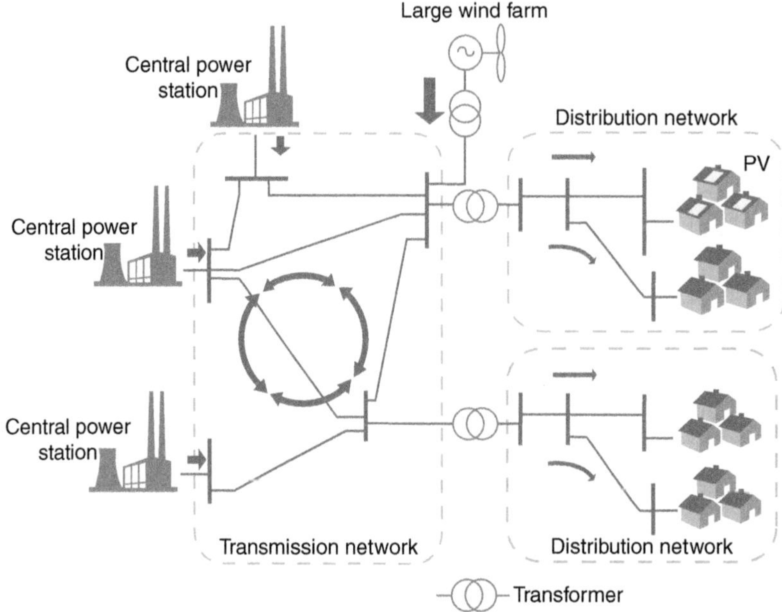

Figure 10.3 Conventional large electric power system.

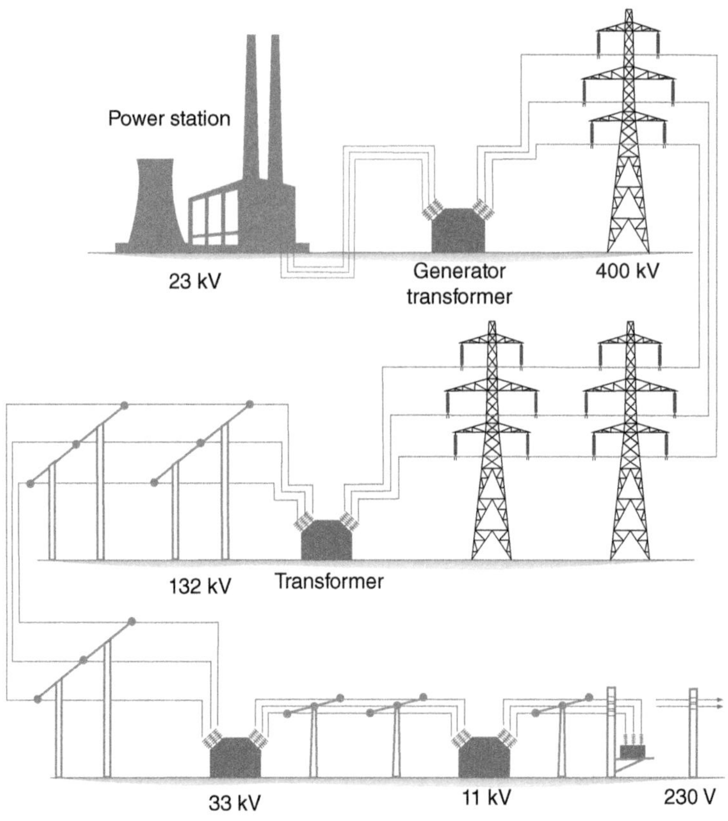

Figure 10.4 Voltage transformations of the GB power system.

Figure 10.5 Hydro power plant and substation.
Photo: Ceylon Electricity Board. (Annotated by the author.)
(A black and white version of this figure will appear in some formats. For the colour version, please refer to the plate section.)

Figure 10.5 shows a 60 MW hydro power plant with its substation in Sri Lanka. Inside the building shown, two 30 MW generators are in operation. The rating of each generator is 37.5 MVA, 12.5 kV, 500 rpm. They are vertical-axis synchronous generators driven by Francis turbines. The output of the generators is connected to the 132 kV power grid through 38 MVA, 12.5/132 kV transformers. The photo shows the high-voltage switchyard, two generator transformers connected to the transmission line and a spare transformer. The two large pipes running up the hill behind the power station are the water supply penstocks for the turbines.

High voltages are used for transmitting large amounts of power over long distances as (considering only direct current for simplicity) the power transmitted in a circuit is

$$P_{\text{transmitted}} = VI \quad [\text{W}] \tag{10.1}$$

where

V is the voltage of the circuit,
I is the current through the circuit.

For high-power overhead transmission, it is cheaper to build high-voltage rather than high-current equipment as air can be used as the insulation. Hence high voltages are used for bulk power transmission. In Europe transmission voltages of up to 400 kV are common (750 kV in America and China) but with the normal operating currents of each circuit limited to less than 2 or 3 kA.

At high voltage a large quantity of power can be transmitted with only limited electrical losses. The resistive losses in a circuit are

$$P_{\text{losses}} = I^2 R \quad [\text{W}] \tag{10.2}$$

where

R is the resistance of the circuit.

Equation 10.1 shows that when a high voltage is used the current required to transmit a large amount of power is limited. Thus the resistive losses in a transmission line can be kept low.

National electrical power systems operate at a constant frequency. The operating frequency that was chosen by engineers at the beginning of the twentieth century is 50 Hz in many parts of the world and 60 Hz in North America. In Japan either 50 Hz or 60 Hz is used in different parts of the country.

10.3 Real and Reactive Power

Figure I.12 of Tutorial I shows the active and reactive components of the instantaneous power. The active power has a positive average value given by $P = VI\cos\phi$ and the reactive power has a zero average but a peak value of $Q = VI\sin\phi$.

Figure 10.6a shows a generator connected to a load of impedance **Z**. The sign convention of the current is that positive current always flows into the component (generator or load). The load impedance **Z** can have three forms: resistive, capacitive or inductive. Consider the case of an inductive load as shown in Figure I.11. The active power consumed by **Z** is represented by a current in phase with the voltage phasor and the reactive power is represented by a current in quadrature with the voltage phasor (Figure I.11b). Figure 10.6b shows the phasor diagram and Figure 10.6c shows the *PQ* quadrant diagram (Figure I.11c) of this. The phasors of load current ($\mathbf{I_L}$) and generator current ($\mathbf{I_G}$) are shown by the dashed line in Figure 10.6b. $\mathbf{I_L} = -\mathbf{I_G}$ and the negative sign is due to the direction of the current marked in Figure 10.6a. P_L and Q_L lie somewhere within the first quadrant of the *PQ* diagram depending on the values of resistance and inductance. In this case the load consumes both active and reactive power and the power consumption is $P + jQ$.

If $\mathbf{Z} = R$, then Q is zero and only real power is consumed.

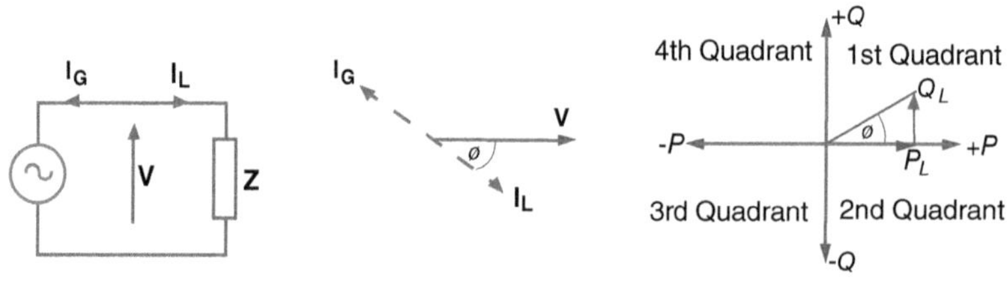

Figure 10.6 (a) A simple generator–load circuit, with (b) its phasor diagram and (c) its active reactive power diagram.

When **Z** has a resistive and a capacitive component, the load consumes active power and generates reactive power and the power consumption is $P - jQ$. In this case P and Q lie in the second quadrant.

A generator always generates active power and then P is negative. A synchronous generator can consume or generate reactive power (so P and Q can be in the third or fourth quadrant). A fixed-speed induction generator always consumes reactive power thus P and Q are in the fourth quadrant. Renewable energy sources that are connected through an inverter (wind or hydro plant connected through a Full Power Converter (FPC) or a PV panel connected through an inverter) can consume or generate reactive power (so P and Q are in the third or fourth quadrant).

In a real power system a number of loads are connected to a distribution circuit. These can be a combination of resistive, inductive and capacitive loads. A part of an industrial load is shown in Figure 10.7. The low-voltage (LV) loads are lighting and heating in buildings as well as small motors used for fans and pumps. The total power consumed by this factory is the power consumed by the large induction motor, power consumed by all the LV loads and power losses in wires and the 1 MVA transformer. That is:

$$P_{\text{supplied}} - P_{\text{motor}} - P_{\text{LV loads}} - P_{\text{losses}} = 0 \tag{10.3}$$

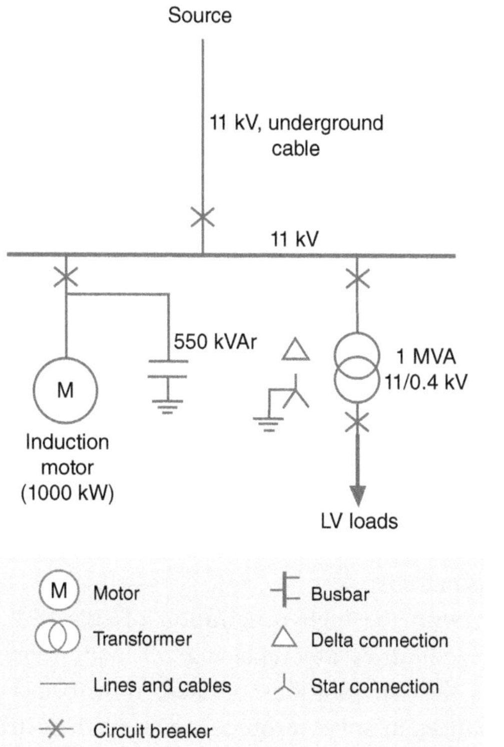

Figure 10.7 Typical industrial load.

The flows of reactive power can be considered in a similar way. The induction motor and LV loads consume reactive power whereas the capacitor generates reactive power. Therefore the total reactive power drawn from the supply line is the reactive power consumed by loads and losses minus the reactive power supplied by the capacitor. That is:

$$Q_{\text{supplied}} + Q_{\text{capacitor}} - Q_{\text{motor}} - Q_{\text{LV loads}} - Q_{\text{losses}} = 0 \quad (10.4)$$

In general, at a node or a boundary $\sum P = 0$ and $\sum Q = 0$. The P and Q flows at a circuit node or at a boundary can be treated independently.

The real power losses (P_{losses}) create heat and so energy is lost. In contrast, the reactive power losses (Q_{losses}) are caused by energy oscillating in and out of the reactive components (capacitors and inductors) and so no energy is lost.

In this conceptual framework the flows of active power ($\pm P$) and reactive power ($\pm Q$) are considered independently. This useful concept is widely used when considering high-voltage transmission networks but is also a helpful representation for describing the effect of renewable generators on distribution systems.

10.4 Voltage of the Power System

As shown in Figure 10.4, different parts of the power systems use different voltage levels. These voltages are maintained within limits defined by national guidelines. For example, in the UK, the voltages on the 275 kV and 132 kV parts of the transmission system will be kept within ±10% of the nominal value unless abnormal conditions prevail. Active and reactive power flows in the transmission and distribution lines cause voltage drops, therefore means are employed to maintain the voltages at different parts of the network within the limits.

10.4.1 Transformer Tap Changing

A common technique used to control voltage is to vary the ratio of the windings of the transformers. A mechanical switch that alters the connections on one of the transformer windings is used to change the turns ratio. Transformers in high-voltage networks are capable of changing their turns ratio while in operation. Such transformers are called on-load tap-changing transformers. Even though a transformer connected to the lower-voltage networks has a tap-changing mechanism, the transformer must be taken out of service before changing the taps. These tap changers are called either off-load (when no current flows) or off-circuit tap changers (when no voltage is present).

However, with the high penetration of solar PV into the lower-voltage networks some countries are replacing off-load transformers with on-load transformers. As the tap positions of these transformers change many times a day with the changes in solar resource, some of these transformers employ a hybrid tap-changing arrangement as shown in Figure 10.8. Initially S1 is ON, S2 is OFF

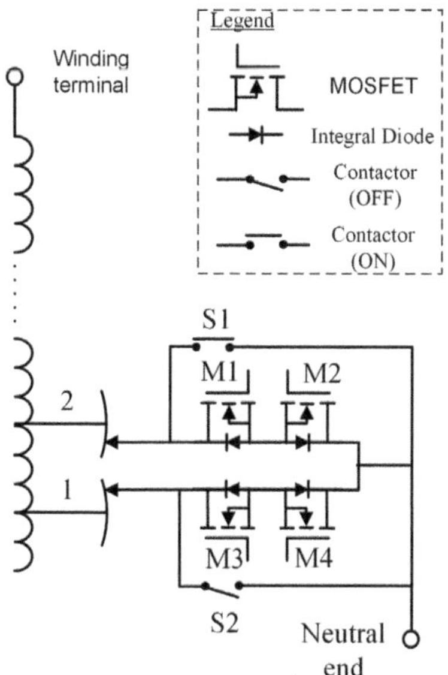

Figure 10.8 Hybrid tap changer.

and all metal–oxide–semiconductor field-effect transistors (MOSFETs) are OFF. If the tap position should be changed from 2 to 1, initially MOSFETs M1 and M2 are turned ON. Then S1 is turned OFF. Then current is commutated from switch pair M1 and M2 to M3 and M4 using a mechanical switch. Then S2 is turned ON and finally M3 and M4 are switched OFF.

10.4.2 Voltage Drop and Power Flows

Figure 10.6a showed a generator connected to a load through an ideal circuit. A more realistic circuit is modelled as a resistor and an inductor connected in series. Figure 10.9 shows the connection of a load of impedance **Z** to the generator through a circuit with a resistance R and reactance X.

As defined in Tutorial I, the apparent power taken by the load is:

$$\mathbf{V}_\mathrm{L}\mathbf{I}^* = P + jQ$$

$$\mathbf{I} = \frac{P - jQ}{\mathbf{V}_\mathrm{L}^*}$$

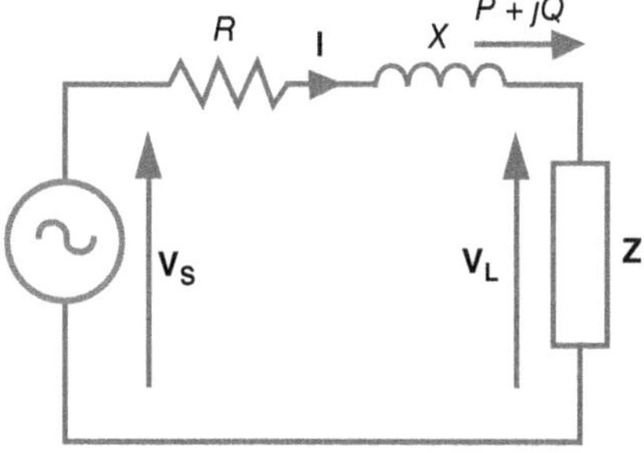

Figure 10.9 Part of a transmission circuit.

The voltages at the generator or the sending end ($\mathbf{V}_S$) and the load ($\mathbf{V}_L$) are related by:

$$\begin{aligned}\mathbf{V}_S &= \mathbf{V}_L + (R + jX)\mathbf{I} \\ &= \mathbf{V}_L + (R + jX)\left[\frac{P - jQ}{\mathbf{V}_L^*}\right]\end{aligned} \tag{10.5}$$

If $\mathbf{V}_L$ is defined as the reference phasor, $\mathbf{V}_L = \mathbf{V}_L^* = V_L$, and the angle between the phasors $\mathbf{V}_S$ and $\mathbf{V}_L$ is δ, then

$$\begin{aligned}V_S(\cos\delta + j\sin\delta) &= V_L + (R + jX)\left[\frac{P - jQ}{V_L}\right] \\ &= \left[V_L + \frac{RP + XQ}{V_L}\right] + j\left[\frac{XP - RQ}{V_L}\right]\end{aligned} \tag{10.6}$$

Equation 10.6 shows that the difference between the sending and load voltages is determined by the active and reactive power flows through the impedance of the circuit.

It is usual to describe voltage levels above 150 kV as transmission and voltage levels below 150 kV as distribution. Although governed by the same physical laws, in practice, transmission and distribution circuits show different characteristics. Transmission circuits have very low resistance, while distribution circuits have a very small angle between the voltages of the sending and load end.

Since δ is small for distribution circuits, $\cos\delta \approx 1$ and $\sin\delta \approx 0$. Then Equation 10.6 simplifies to

$$\Delta V \approx V_S - V_L = \frac{RP + XQ}{V_L} \tag{10.7}$$

For transmission circuits, since resistance is small, Equation 10.6 can be reduced to

$$\Delta V \approx \frac{XQ}{V_L} + j\frac{XP}{V_L} \tag{10.8}$$

Equation 10.8 shows that the active power or P flow influences the imaginary term and therefore the voltage angle, whereas the reactive power or Q flow influences the magnitude of the voltage drop. Therefore the voltages of transmission circuits are usually controlled by varying the flows of reactive power.

10.4.3 Changes of Local Voltage with *P* and *Q* Flows

Figure 10.10 shows a section of a distribution network to which a small wind farm is connected. Busbar A is a stiff source of the main power system represented as a voltage behind a small impedance. Busbar A is effectively an infinite busbar as its voltage is more or less constant irrespective of variation of the loads and generation in the downstream distribution network. In power system studies, voltages are normally described as a per unit voltage based on a base voltage. For busbar A with a base voltage of 33 kV the per unit voltage is 1.0 pu (33 kV is a line quantity as explained in Tutorial I). Figure 10.10 also shows the locations of circuit breakers and fuses that are used to isolate a part of the network if that part develops a short-circuit fault.

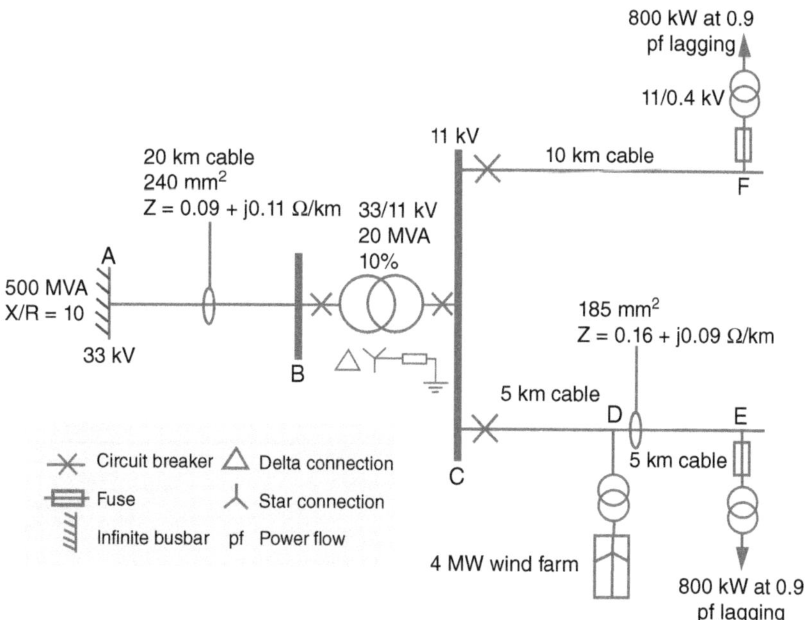

Figure 10.10 Part of a distribution network to which a small wind farm is connected.

As the voltage of busbar A is known, the voltage of busbar B can be calculated using Equation 10.6. However, this equation needs the active and reactive power flows at busbar B as well as the voltage at busbar B (which is unknown). The active and reactive power flowing out of B is the addition of the active and reactive power of the loads and the wind farm, together with the losses in transformers and 11 kV circuits. As shown in Equation 10.2, the calculation of losses requires the current in the lines. However, in order to calculate the currents, the voltages in the 11 kV network must be known. Therefore it is not possible to calculate the voltages of different parts of the network and currents (and hence power flows) directly. In power system studies iterative techniques are used for such calculations and they are commonly known as load flow programs.

A load flow program was used to obtain the voltages at each busbar and power flow in the network shown in Figure 10.10. Three cases were examined: (a) when the wind farm connected to busbar D is not producing any output, (b) when it is producing full real-power output and exporting reactive power, (c) when it is producing full real-power output and importing reactive power. The results are shown in Table 10.3. These voltages are given as per unit quantities based on the base voltage shown in the first column of Table 10.3.

Figures 10.11a and b (Figure 10.10 is rotated by 90° and all the circuit breaker symbols are removed) show the active and reactive power flows in different parts of the network when the wind farm exports and imports reactive power. Losses in the lines and transformers are neglected. As can be seen, at every node the addition of incoming active and reactive power is equal to the addition of active and reactive

Table 10.3 Voltage of each busbar of Figure 10.10

		Voltage at each busbar (pu)		
			When the wind farm output is 4 MW	
Base voltage	Busbar	When the wind farm output is zero	Exporting reactive power	Importing reactive power
33 kV	A	1.0	1.0	1.0
	B	0.996	1.006	0.998
11 kV	C	0.992	1.012	0.983
	D	0.985	1.037	0.986
	E	0.978	1.031	0.989
	F	0.978	0.998	0.969

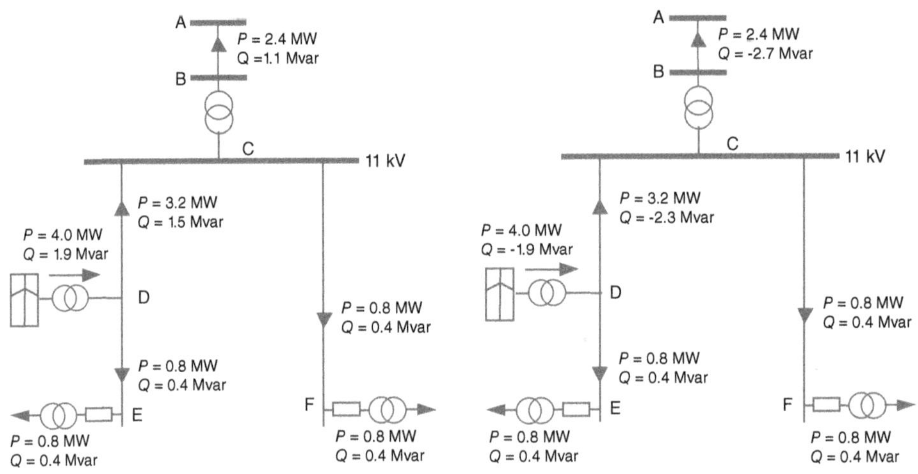

Figure 10.11 Load flows with reactive power export and import.

power leaving that node. For example, when the wind farm exports reactive power, it produces active power of 4 MW and reactive power of 1.9 Mvar. Active and reactive power leaving busbar D is the power consumed by the load connected to busbar E ($P = 0.8$ MW and $Q = 0.4$ Mvar) and the power flow into busbar C ($P = 3.2$ MW and $Q = 1.5$ Mvar). Therefore the total power entering and leaving busbar D is $P = 0.8 + 3.2 = 4$ MW and $Q = 0.4 + 1.5 = 1.9$ Mvar.

Compare the two cases when the wind turbine is either exporting or importing reactive power; the active power flow in line AB is the same for both cases (2.4 MW from B to A). When the wind turbine is exporting reactive power, reactive power flows from B to A and the voltage at busbar B is greater than that at busbar A (see Table 10.3). On the other hand, when the wind turbine is importing reactive power, reactive power flows from A to B and the voltage at busbar B is less than that of busbar A.

Table 10.3 with Figure 10.11 show that the voltage variations are local and depend on the active and reactive power flows in the neighbouring circuits.

10.4.4 Voltage Control by Reactive Power

In addition to the use of tap changers on the transformers, varying the reactive power flows can control the network voltage.

Some small wind farms and hydro units use fixed-speed induction generators (FSIGs). As discussed in Tutorial I, an FSIG absorbs reactive power to establish its magnetic field. The amount of reactive power absorbed by an FSIG depends on the active power generation (which changes with the wind speed or water flow). This is shown in Figure 10.12 where the real power output varies between 0 and −1 pu and the reactive power drawn between 0 and 0.55 pu. If the generator is rated at 1 MVA, P varies between 0 and −1 MW and Q between 0 and 550 kvar.

If an FSIG is connected to the system through a circuit with a resistance R and reactance X, then the voltage rise in the circuit caused by the export of active power and import of reactive power is given by:

$$\Delta V \approx \frac{RP - XQ}{V_S} \tag{10.9}$$

Note that as the FSIG always absorbs reactive power, the XQ term in Equation 10.9 is negative. Further, as the FSIG is situated at the sending end, the denominator of Equation 10.9 is the voltage at the sending end.

A capacitor bank that supplies reactive power to compensate the no-load reactive power requirement of FSIG is normally connected at the terminals of the FSIG. If the capacitor supplies reactive power equal to Q_c locally then Equation 10.9 is modified to:

$$\Delta V \approx \frac{RP - XQ + XQ_c}{V_S} \tag{10.10}$$

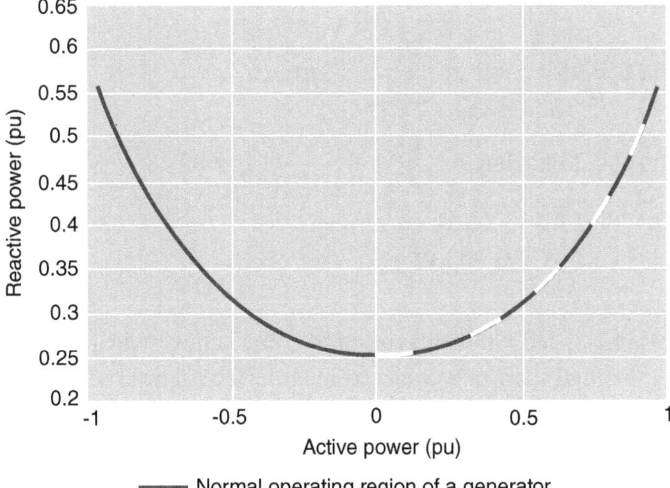

Figure 10.12 Reactive power demanded by a fixed-speed induction generator.

Example 10.1 – Voltage Rise at the Connection of a Renewable Generator

A wind farm with FSIGs is shown in Figure 10.13. The 8 km long, 33 kV collector cable has a resistance of 0.127 Ω/km and reactance of 0.114 Ω/km. The voltage at the point of connection is held at 33 kV by the power system. The active and reactive power, measured at the point of connection, are 15 MW and 1 Mvar (importing).

(a) What is the voltage at the wind farm end of the collector cable?
(b) Calculate the losses in the 33 kV cable due to the active (in-phase) current.

Solution

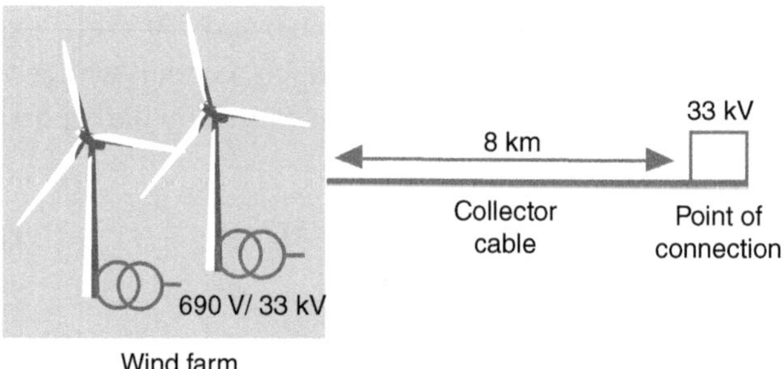

Figure 10.13 Typical wind farm connection.

$$R = 0.127 \times 8 = 1.016\,\Omega$$
$$X = 0.114 \times 8 = 0.912\,\Omega$$

(a) Per phase active power at the point of connection $= 15/3$ MW

Per phase reactive power at the point of connection $= 1/3$ Mvar
Per phase voltage at the point of connection $= 33/\sqrt{3}$ kV $= 19.05$ kV
From Equation 10.9

$$\Delta V \approx \left[\frac{1.016 \times 15/3 \times 10^6 - 0.912 \times 1/3 \times 10^6}{33 \times 10^3/\sqrt{3}}\right]$$

$$\Delta V \approx 144 \text{ V}$$

The voltage of the collector cable at the wind farm is:
$\left(19.05 \times 10^3 + 144\right) \times \sqrt{3} = 33.25$ kV

(b) Active current in the 33 kV line $= I_P = \left[\dfrac{15/3 \times 10^6}{\sqrt{3} \times 33 \times 10^3}\right] = 87.48$ A

Losses in the 33 kV line due to active current is:
$3I_P^2 \times R = 3 \times 87.48^2 \times 1.016 = 23.33$ kW

Modern variable-speed wind turbines and some small hydro units have reactive power control through their power electronic interface. The converter connected to the network can be controlled to produce a variable amount of reactive power. The reactive power capability of the converter depends on its rating. If the rating is S in MVA, then its reactive power capability Q in Mvar is given by

$$Q = \sqrt{S^2 - P_{\text{Generated}}^2}$$

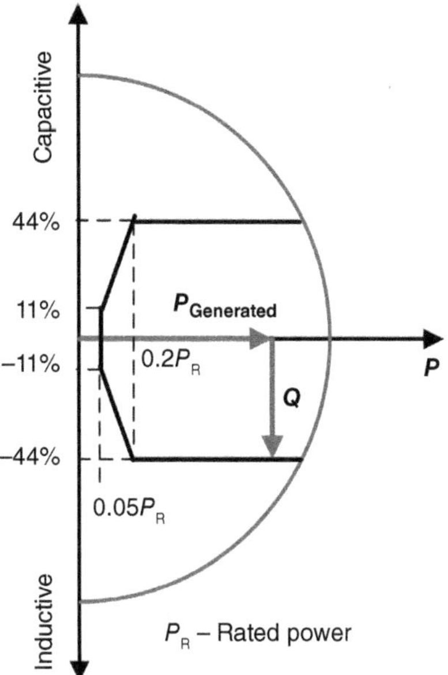

Figure 10.14 Reactive power requirement of a power electronic interface.
Source: reprinted from Appendix H of *IEEE 1547: Standard for Interconnection and Interoperability of Distributed Energy Resources with Associated Electric Interfaces*, © 2018 IEEE, with permission.

According to IEEE standard 1547:2018, the power electronic interface should have a capacity to provide reactive power as shown in Figure 10.14.

As shown in Table 10.3, by controlling the reactive power generated by a wind turbine (or other generator connected through power electronics) the voltage in the network around the wind turbine can be controlled.

10.5 Frequency

Conventional power systems use fossil-fuelled units or hydro turbines fed from a reservoir to control frequency. Both these types of generation have energy stored either as fossil fuel or in the potential energy of water. The turbines are connected to synchronous generators as shown in Figure 10.15.

If a single generator supplies a load (this is the case when the power system starts from a total blackout), the frequency depends on the balance between the mechanical power from the turbine and the electrical power drawn by the load. The frequency is controlled by measuring the rotational speed of the generator shaft and admitting more steam or water to the turbine when its speed drops and less when its speed rises.

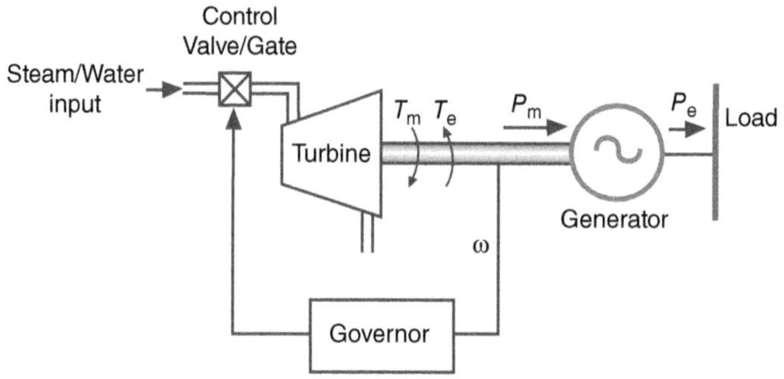

Figure 10.15 Conventional generator showing frequency control. T_m is the torque developed by the mechanical prime mover, T_e is the retarding torque developed by the generator. P_m is the power input by the turbine to the shaft, P_e is the power output of the generator.

In an interconnected power system with many generators, the frequency is determined by the balance between the power output of all the generators and the total load connected to the power system. Under normal operating conditions the frequency of an interconnected electric power system is the same throughout the system. Thus the frequency of the GB system is the same from Scotland to the south of England while across much of Europe the power system has the same frequency. The rotor of a synchronous generator is locked to the rotating stator field by magnetic coupling (see Tutorial I, Section I.3.1). The speed of rotation of the stator field, and hence the speed of the generator rotor, is determined by the frequency of the power system. All the rotors of generators connected to a power system operate at the same rotational speed (synchronous speed, ω_s).

As shown in Figure I.15 of Tutorial I, when the rotor of a two-pole machine rotates by one turn, the induced voltage across the coil changes by one period. With any synchronous machine there is a fixed relationship between the angular velocity (ω) and frequency (1/period). For a two-pole machine in a 50 Hz power system this is:

$$\omega = 2\pi f = 2\pi \times 50 = 314.15 \text{ rad/s} = 3000 \text{ rpm}$$

To share the output power of generators running in parallel, proportional control of the governors based on a droop characteristic is used (Figure 10.16). When the system frequency changes the governors of the prime movers of large generating units change their power output following their droop characteristics. For example, if the power system operates at 50 Hz, the generating unit having the droop characteristic shown in Figure 10.16 operates with the output power of P_1 at point A. If the load on the power system increases, then the frequency drops by Δf to point B and the generator increases its output to P_2 by admitting more steam or water to the turbine.

Increasing the output power of large steam generators is possible using the heat energy stored in the boilers and, more slowly, by burning more fuel. Hydro units with reservoirs to store water can also vary their output as the system frequency changes. Other renewable energy generators do not have these energy stores and so do not take part in the control of power system frequency. Rather, they inject energy into the electrical power system as the renewable energy resource is available.

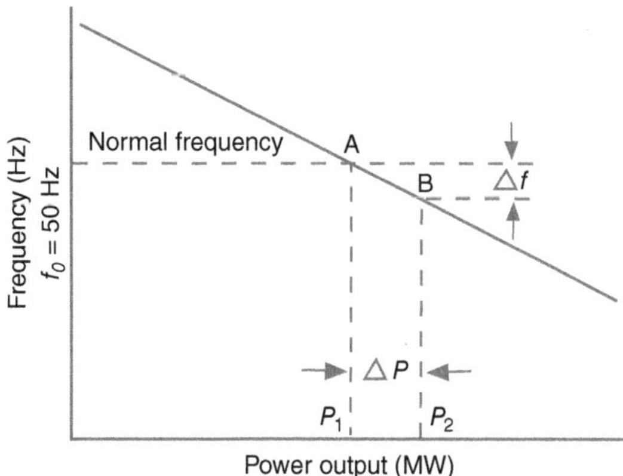

Figure 10.16 Droop characteristics of a governor.

In terms of frequency, an electric power system can be viewed simply as a single rotating mass of equivalent inertia I_{eq}. All the turbine generators and spinning loads operate as a single coherent generator with one rotational speed (ω) and a single system frequency (f). The equivalent inertia, I_{eq}, is the sum of the inertias of all the turbines and generator rotors that are connected to the power system.

The energy stored in all the rotating masses is:

$$KE = \frac{1}{2} I_{eq} \omega^2 \tag{10.11}$$

where

KE is the kinetic energy in J,
I_{eq} is the equivalent moment of inertia in kgm^2,
ω is the rotational speed in rad/s.

Now the torque balance of any generator determines its rotational speed:

$$\frac{d\omega}{dt} = \frac{1}{I}(T_m - T_e) \tag{10.12}$$

where

T_m is the torque developed by the mechanical prime mover in Nm,
T_e is the retarding torque developed by the electrical generator in Nm,
I is the moment of inertia of the generator.

The inertia constant (H) of a turbine-generator unit is defined as the ratio of the kinetic energy stored at synchronous speed to the generator kVA or MVA rating.

$$H = \frac{\frac{1}{2} I \omega^2}{S} \tag{10.13}$$

where

H is the inertia constant in seconds,
S is the rating of the turbine generator in MVA.

Substituting I in Equation 10.12 gives

$$\frac{d\omega}{dt} = \frac{1}{2H} \frac{T_m - T_e}{S} \omega^2 \tag{10.14}$$

Since the relationship between power (P), torque (T) and angular velocity (ω) is given by $P = T\omega$, and P/S is power in pu with respect to the turbine-generator base, Equation 10.14 can be simplified to:

$$\frac{d\omega}{dt} = \frac{1}{2H}\left(\frac{T_m\omega}{S} - \frac{T_e\omega}{S}\right)\omega = \frac{1}{2H}\left(P_m - P_e\right)\omega \tag{10.15}$$

where

P_m is the power input by the turbine to the shaft,
P_e is the power output of the generator,

(Both P_m and P_e are expressed as per unit quantities based on the power rating of the generator set, S.)

Since $\omega = 2\pi f$,

$$2\pi\frac{df}{dt} = \frac{1}{2H}\left(P_m - P_e\right)2\pi f$$

$$\frac{df}{dt} = \frac{1}{2H}\left(P_m - P_e\right)f \tag{10.16}$$

If H is now the inertia constant of all spinning mass connected to the power system, that is,

$$H = \frac{\frac{1}{2}I_{eq}\omega^2}{S_{sys}}$$

and $\left(P_m - P_e\right)$ is the mismatch between all the generation and demand in a system, then Equation 10.16 applies to the entire power system.

Equation 10.16 shows that the smaller the value H, the greater is the rate of change of frequency. Large conventional generator sets have a large spinning mass and increase the system inertia, H, whereas renewable energy generators normally do not contribute to H. The power electronic interface of a variable-speed wind turbine decouples the generator and mechanical drive train from the network, thus making no contribution to system inertia, H. A photovoltaic system is static, has no spinning mass and again makes no contribution to system inertia.

When renewable energy generators are connected to the system, they displace conventional power plants and the system inertia drops, resulting in more rapid frequency excursions for changes in generation or load. This reduction in the inertia of the power system caused by increasing operation of renewable generators is a major topical concern of power system operators.

Example 10.2 – Effect of PV Generation on System Inertia

A power system operates with 1000 MW of coal, 500 MW of oil and 1000 MW of hydro power generation. The inertia constants of the generator types are shown in the table.

(a) If 750 MW of coal and 250 MW of oil generation are replaced by 1000 MW of PV generation, calculate the percentage reduction in the inertia of the system.

Generator type	Inertia constant, H (s)
Coal	4.5
Oil	6
Hydro	4.5

(b) The power system operates at 50 Hz with these generators. A large generator of 250 MW is switched out due to a fault. Assuming that the frequency drops linearly, calculate the new frequency after 1 second for the system without and with PV generation.

Solution

(a) The equivalent inertia constant, H_{eq}, on the system base is calculated using:

$$H_{eq} = \sum_{i=\text{coal,gas,...}} H_i \times \frac{S_i}{S_{sys}}$$

where

H_i is the inertia constant and S_i is the MVA rating of the different types of power plant.

Before the PV generation is connected

$$H_{eq} = \frac{4.5 \times 1000 + 6 \times 500 + 4.5 \times 1000}{[1000 + 500 + 1000]} = 4.8 \text{ s}$$

After the PV generation is connected

$$H_{eq} = \frac{4.5 \times 250 + 6 \times 250 + 4.5 \times 1000 + 0 \times 1000}{[250 + 250 + 1000 + 1000]} = 2.85 \text{ s}$$

The percentage reduction in system inertia $= (4.8 - 2.85) \times 100/4.8 = 40.63\%$

(b) With a change in generation, $P_m - P_e = 250$ MW.

From Equation 10.16

$$\Delta f = \frac{1}{2H}(P_m - P_e) f \Delta t$$

Without PV generation:

$$\Delta f = \frac{1}{2 \times 4.8} \times \frac{250}{2500} \times 50 \times 1 = 0.52 \text{ Hz}$$

$$\text{New frequency} = 50 - 0.52 = 49.48 \text{ Hz}$$

(Note that the 250 MW generation loss is expressed as a pu value with respect to the total generation of 2500 MW.)

With PV generation:

$$\Delta f = \frac{1}{2 \times 2.85} \times \frac{250}{2500} \times 50 \times 1 = 0.88 \text{ Hz}$$

$$\text{New frequency} = 50 - 0.88 = 49.12 \text{ Hz}$$

Typically a transmission system operator (TSO) tries to maintain the frequency within ±1% (0.5 Hz). With the PV generation displacing fossil-fuel plant, the inertia of the system is reduced and the frequency falls below this limit for this contingency.

10.6 Operating the Power System

10.6.1 Generation Scheduling

The electrical demand of a power system changes continuously. At any time only some generators are in use to maintain the balance between the demand and generation. Other units are held in reserve to meet increasing load and be ready in case a generator breaks down. Generators are called on to operate depending on their cost of operation and maintenance, fuel and CO_2 emissions as well as the need for reserve capacity. The arrangement through which this control of generators, known as scheduling, is administered depends on how the power system is managed. In some countries market mechanisms are used, while in others a central organization determines the optimum operation of the power system.

The costs of a power plant can be split into capital and operating costs. Table 10.4 shows typical values of these costs for different-generation technologies as well as their rate of emission of CO_2.

When scheduling power plants only the operating cost is considered as the capital costs of building the plant will already have been incurred. The operating cost includes the fixed operating and maintenance (O&M) cost and the variable costs. The fixed O&M cost (C_F) depends on the plant size (capacity) and type. It includes the wages for the operating staff, the cost of routine maintenance and other costs of plant operation. The variable costs (C_V) are determined by the fuel cost (C_{FU}), the power output and the efficiency of each generator.

The input–output characteristic of a turbine influences the economic operation of a power plant. For a fossil-fuelled generator, the heat rate input (MJ/h) against output of the generator (MW) is approximately a no-load value with a reasonably straight line over the operating range. Therefore the heat rate (HR) can be approximated to

$$\mathrm{HR} = r + sP \quad [\mathrm{MJ/MWh}]$$

where

P is the power output in MW,
r and s are constants.

Therefore $C_V = C_{FU} \times \mathrm{HR} = C_{FU}(r + sP) \quad [£/\mathrm{MWh}]$

The total operating cost per hour, C, is then

$$\begin{aligned} C &= C_F + C_V P = C_F + C_{FU}(r + sP)P \\ &= C_F + rC_{FU}P + sC_{FU}P^2 \quad [£/\mathrm{h}] \end{aligned}$$

which may be rewritten as

$$C = a + bP + cP^2 \quad [£/\mathrm{h}] \tag{10.17}$$

where a,b,c are constants.

Table 10.4 Costs and CO_2 emissions of different generation technologies

		Operating costs		
	Capital cost (£/kW)	Fixed O & M, C_F (£/kW per year)	Fuel, C_{FU} (£/MWh)	Emissions (kg CO_2/MWh)
Coal	2500	44	30	972
Gas (CCGT)	750	25	48	411
Biomass	4000	120	40	15
Nuclear	3500	100	5	12
Wind – onshore	1400	100	0	9
Wind – offshore	4000	280	0	22
Solar PV	2700	220	0	51

Example 10.3 – Cost Function of Renewable Generators

Using the data given in Table 10.4, draw the operating cost functions of a 100 MW biomass and 100 MW onshore wind power plant. For the biomass power plant assume the rate at which fuel is used is given by a heat rate of $\eta = 0.3 + 0.001P$ MJ/MWh.

Solution

For the biomass power plant:

$$\begin{aligned}\text{Fixed O \& M cost}\left(C_F\right) &= 120 \text{ £/kWy} \\ &= 120 \times 1000/8760 \text{ £/MWh} \\ &= 13.7 \text{ £/MWh}\end{aligned}$$

Therefore for a 100 MW plant

$$a = 13.7 \times 100 \text{ £/h} = 1370 \text{ £/h}$$

$$\begin{aligned}\text{Variable cost} &= rC_{FU}P + sC_{FU}P^2 \\ &= 0.3 \times 40P + 0.001 \times 40P^2 \\ &= 12P + 0.04P^2 \text{ £/h}\end{aligned}$$

$$\text{Total cost} = C_{Biomass} = 1370 + 12P + 0.04P^2 \text{ £/h}$$

For the wind power plant:

$$\begin{aligned}\text{Fixed O \& M cost} &= 100 \text{ £/kWy} \\ &= 100 \times 1000/8760 \text{ £/MWh} \\ &= 11.4 \text{ £/MWh}\end{aligned}$$

Therefore for a 100 MW plant

$$\text{Total cost } C_{Wind} = 11.4 \times 100 \text{ £/h} = 1140 \text{ £/h}$$

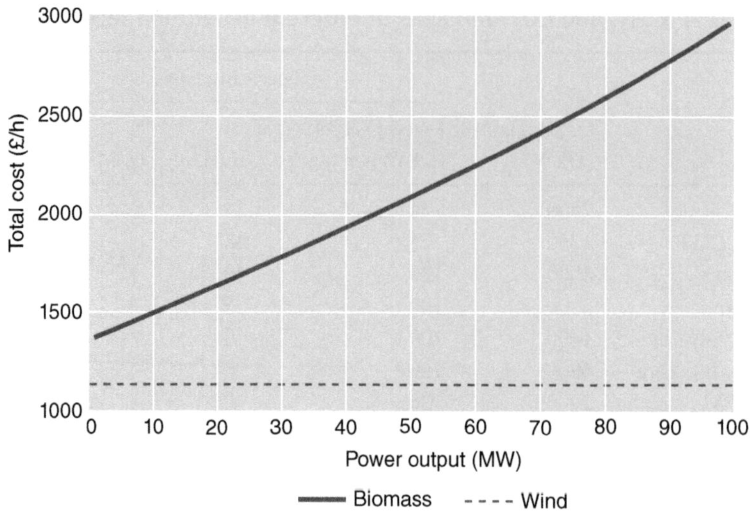

Figure 10.17 Total cost of operation of a biomass and wind power plant.

Costs $C_{Biomass}$ and C_{Wind} $C_{Biomass}$ with respect to power output are shown in Figure 10.17.

Now consider the scheduling of power plants. The task of allocating generation is an optimization problem, with the generation operating cost to be minimized while ensuring that the total power generated is equal to the system demand.

If two generating plants have cost curves of $C_1 = a_1 = b_1 P_1 + c_1 P_1^2$ [£/h] and $C_2 = a_2 + b_2 P_2 + c_2 P_2^2$ [£/h] then the optimization can be described by

$$\min \quad [C_1(P_1) + C_2(P_2)] \tag{10.18}$$

The demand must be met at all times, so

$$P_1 + P_2 = P_D \tag{10.19}$$

where P_D is the demand.

Equation 10.18 can be manipulated by eliminating one of the variables using Equation 10.19, giving $\min \quad [C_1(P_1) + C_2(P_D - P_1)]$.

At the optimum, the derivative of the objective function is equal to zero:

$$\therefore \frac{d}{dP_1}[C_1(P_1) + C_2(P_D - P_1)] = 0 \tag{10.20}$$

Example 10.4 – Generator Scheduling

Consider two generating plants with the following cost functions (in the form shown in Equation 10.17):

$$C_1 = 1000 + 11P_1 + 0.03 \times P_1^2 \quad [\text{£/h}]$$
$$C_2 = 1200 + 14P_2 + 0.04 \times P_2^2 \quad [\text{£/h}]$$

where P is in MW.

If the total demand is 200 MW, calculate the most economic division of load between the generators.

Solution

From Equation 10.20

$$\frac{d}{dP_1}\left[\left(1000+11P_1+0.03\times P_1^2\right)+\left(1200+14(200-P_1)+0.04\times(200-P_1)^2\right)\right]=0$$
$$\therefore 11+0.06P_1-14-0.08(200-P_1)=0$$

Solving this linear equation gives $P_1 = 135.7$ MW. Given that the total generation of the units must be equal to 200 MW (demand), the generation of the second unit is $P_2 = 64.3$ MW.

When finding the solution it was assumed that the two generating plants can operate at any output. However, in practice, there is a minimum stable generation and a maximum possible generation for each power plant and the solution is constrained by these limits.

When there are more than two generators, the optimization defined by Equation 10.20 can be obtained considering the marginal or incremental cost of each generator $MC = dC/dP$. Assume two generators have incremental costs of MC_1 and MC_2 with $MC_1 > MC_2$. If the load on the generator with higher MC is reduced, that would reduce the total cost. Obviously, with the reduction of the load on the first generator, the load taken by the other generator should increase to meet the demand. This transfer will be beneficial until the values of MC for both generators are equal, after which the machine with the previously higher MC now becomes the one with the lower value, and vice versa. There is no economic advantage in further transfer of load, and the condition when $MC_1 = MC_2$ therefore gives optimum economy.

Consider the marginal costs of the two power plants of Example 10.4:

$$MC_1 = 11+0.06\times P_1 \quad [\text{£/h}]$$
$$MC_2 = 14+0.08\times P_2 \quad [\text{£/h}]$$

Figure 10.18 shows these marginal costs. Suppose that initially Plant 1 generates 150 MW (point S_1) and Plant 2 generates 50 MW (point S_2). At these operating points, the marginal cost of Plant 1 is 20 £/MWh and that of Plant 2 is 18 £/MWh. The first row of Table 10.5 shows the operating costs of Plants 1, 2 and the total cost. As can be seen, the total operating cost is 5325 £/h. On the other hand, if Plant 1 generates 125 MW (point U_1) and Plant 2 generates 75 MW (point U_2), the marginal cost of Plant 1 is 18.5 £/MWh and that of Plant 2 is 20 £/MWh. As can be seen from the third row of Table 10.5, the total operating cost is 5319 £/h. The minimum cost of generation, i.e. 5310 £/h, is when the marginal costs of the generators are equal and the two plants are operating at points T_1 and T_2. At this operating point $MC_1 = MC_2 = 19.14$ $[\text{£/MWh}]$ and load on each generator is equal to the values obtained in Example 10.4.

Table 10.5 Operating cost of generating plants in Example 10.4

Loading condition	P_1(MW)	P_2(MW)	C_1(£/h)	C_2(£/h)	$C_1 + C_2$(£/h)	MC_1(£/MWh)	MC_2(£/MWh)
S	150	50	3325	2000	5325	20	18
T	135.7	64.3	3045	2265	5310	19.14	19.14
U	125	75	2844	2475	5319	18.5	20

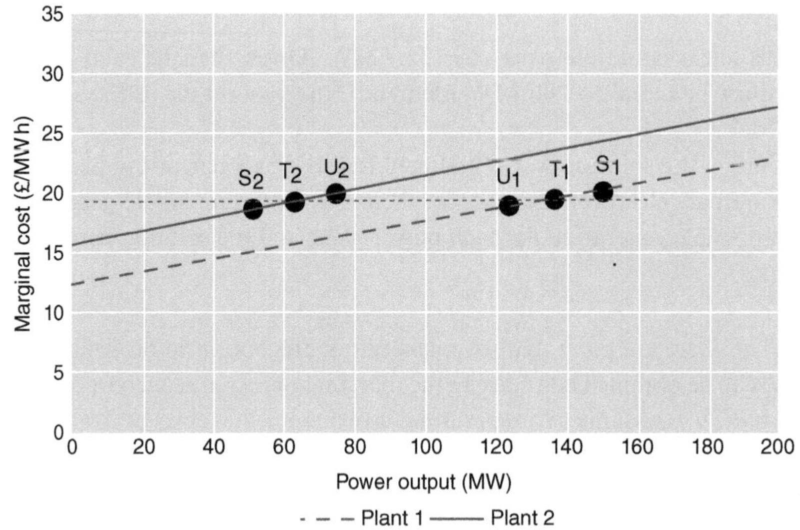

Figure 10.18 Marginal cost curves of the two generators of Example 10.4.

In many countries (and across the European Union) a cost is applied to the emission of CO_2 from power stations to represent the damage caused to the environment and to encourage the use of renewable energy. Generation scheduling with CO_2 emission costs is demonstrated in Example 10.5.

Example 10.5 – Generator Scheduling with CO_2 Cost

It is required to schedule a combined cycle gas turbine (CCGT) power plant of rating 250 MW and a biomass plant of 100 MW to meet a demand of 200 MW. Calculate the most economic division of load between the generators (a) without considering the cost of CO_2 and (b) with a CO_2 cost of £50 per tonne.

The costs of operation of the two plants, ignoring the cost of CO_2, are

$$C_{CCGT} = 200 + 5P_{CCGT} + 0.03 \times P_{CCGT}^2 \quad [\text{£/h}]$$
$$C_{Biomass} = 1200 + 14P_{Biomass} + 0.04 \times P_{Biomass}^2 \quad [\text{£/h}]$$

where P is in MW.

Solution

Without considering the cost of the CO_2 emission, by equating marginal costs, the following equation can be obtained:

$$\frac{dC_{CCGT}}{dP_{CCGT}} = 5 + 0.06 \times P_{CCGT} = \frac{dC_{Biomass}}{dP_{Biomass}} = 14 + 0.08 \times P_{Biomass} \quad [\text{£/MWh}]$$

$$0.06 \times P_{CCGT} - 0.08 \times P_{Biomass} = 9$$

From the demand–supply balance

$$P_{CCGT} + P_{Biomass} = 200$$

Solving these equations gives the most economic division of generation as $P_{CCGT} = 178.6$ MW and $P_{Biomass} = 21.4$ MW.

If the cost of CO_2 emission is included, the cost function becomes (emission figures from Table 10.2)

$$C_{CCGT} = 200 + 5P_{CCGT} + 0.03 \times P_{CCGT}^2 + \frac{411 \times 50}{1000} P_{CCGT}$$
$$= 200 + 25.55P_{CCGT} + 0.03 \times P_{CCGT}^2 \quad [\text{£/h}]$$

$$C_{Biomass} = 1200 + 14P_{Biomass} + 0.04 \times P_{Biomass}^2 + \frac{15 \times 50}{1000} P_{Biomass}$$
$$= 1200 + 14.75P_{Biomass} + 0.04 \times P_{Biomass}^2 \quad [\text{£/h}]$$

By equating the marginal costs:

$$\frac{dC_{CCGT}}{dP_1} = 25.55 + 0.06 \times P_{CCGT} = \frac{dC_{Biomass}}{dP_2} = 14.75 + 0.08 \times P_{Biomass} \quad [\text{£/MWh}]$$

$$0.08 \times P_{Biomass} - 0.06 \times P_{CCGT} = 10.8$$

Again, from the demand–supply balance

$$P_{CCGT} + P_{Biomass} = 200$$

Solving these equations gives the most economic division of load as $P_{CCGT} = 37.1$ MW and $P_{Biomass} = 162.9$ MW.

Note that the inclusion of the cost of CO_2 has resulted in an increased output of the biomass generator.

Depending on the technology, it may take several hours for a power plant to provide the output required. Therefore scheduling is always done ahead of real time, often several hours ahead. It may not be possible to predict accurately the renewable energy resource and hence the output of renewable generators. This uncertainty brings extra complexity to the economic dispatch of generators.

10.6.2 Mismatches Between the Generation and Load

Even in a power system without a significant proportion of renewable generation, mismatches between the scheduled generation and demand are inevitable. These are due to (a) variations in customer demand (minor deviation), (b) sudden connection or disconnection of large loads (moderate deviation), (c) sudden loss of part of the transmission network due to a fault (moderate deviation) and (d) sudden loss of a large generator (major deviation).

Any mismatch between the generation and load results in a frequency change that is compensated initially by the kinetic energy of rotating generators. As shown in Section 10.4, the speed of the deviation of frequency during the initial few seconds after a disturbance depends on the inertia of the power system.

Minor mismatches between customer demand and generation are compensated automatically by the action of the governors of large generators. To accommodate moderate mismatches, some generators are operated part loaded so that they can increase their output when needed. These generators are sometimes known as 'spinning reserve'.

Significant mismatches between demand and generation caused by a sudden loss of generation or a large part of the customer load may cause large changes in frequency. To control the frequency change, different automatic frequency response services are used by the system operator. These include response, reserve and load shedding. If the mismatch cannot be arrested with these automatic measures, the power system will become unstable, resulting in a partial or complete blackout.

When renewable generators are in operation, mismatches between generation and loads are also caused by the variability of the generation. Figure 10.19 shows the variation of the power output of a 10 MW wind power plant over one day in June. As can be seen, rapid variations of wind power are experienced over the day.

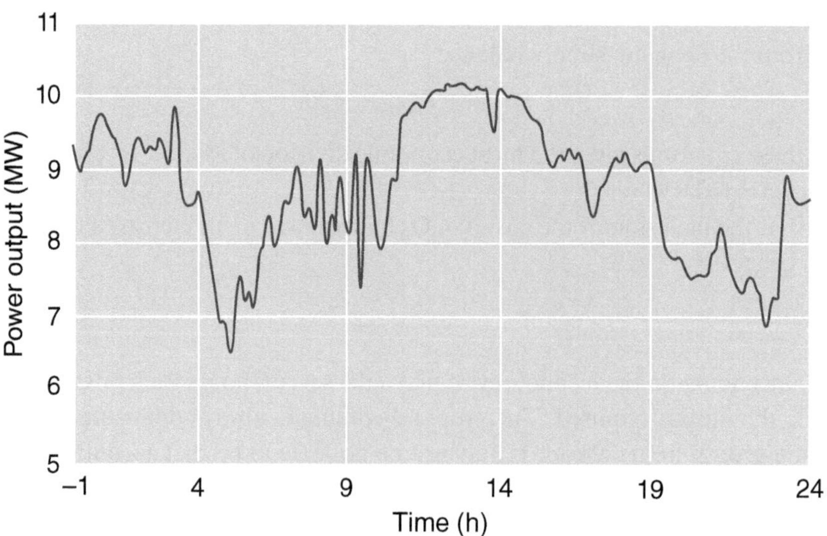

Figure 10.19 Output variations of a 10 MW wind power plant in Sri Lanka.

10.6.3 Reserve Generation Requirements

Traditionally the need for reserve generation was determined by considering the sudden failure of the largest generator on the system as well as the anticipated deviation of the actual demand away from its forecast value. However, with the introduction of renewable generation, the difference between the forecast and realized output from renewable generators must be added to the reserve requirement.

The uncertainties that determine how much reserve generation is required are characterized as follows:

(a) *Uncertainty in demand.* Electricity demand forecasting is a mature process and the standard deviation of the difference between actual and forecast demand is usually taken as 1% of the peak demand.
(b) *Uncertainty in renewable generation.* The difference between the forecast and realized output from renewable generation depends on the forecast horizon. Figure 10.20 shows the standard deviation of the forecast error vs forecast horizon for the wind generation of a power system. If the forecast horizon is 4 hours, then the standard deviation of forecast error is approximately 8.5% of the total installed wind capacity.

The reserve allocated to cover the imbalances caused by errors in load and wind predictions is usually taken as 3 times the standard deviation of forecast error.

(c) *Generation outages.* To compensate for generation outages, the system holds reserve at least equal to the size of the largest generator in operation (P_{MAX}).

Based on these uncertainties the total reserve requirement is estimated by adding the rating of the largest generator to 3 times the standard deviation of the expected differences between actual and forecast demand and the output of renewable generation.

$$\text{Reserve} = P_{MAX} + 3\sqrt{(\sigma_{demand})^2 + (\sigma_{renewables})^2} \tag{10.22}$$

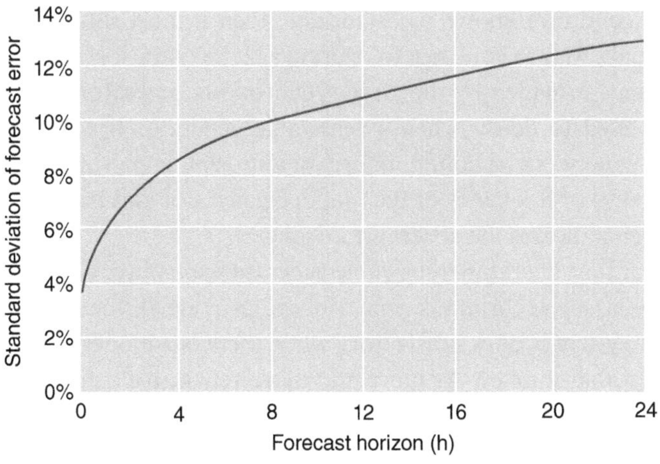

Figure 10.20 Standard deviation of wind energy forecast error.

Example 10.6 – Reserve Requirement

Calculate the total reserve requirement assuming the peak demand of a power system is 3000 MW, the largest generator on the system is 500 MW and there is 375 MW of wind generation. Also assume that four-hourly wind predictions are used.

Solution

$$P_{MAX} = 500 \text{ MW}$$
$$\sigma_{demand} = 3000 \times 0.01 = 30 \text{ MW}$$

As the wind forecast horizon is 4 hours, the standard deviation is 8.5% of the total wind capacity:

$$\sigma_{renewables} = 375 \times 8.5/100 = 31.88 \text{ MW}$$

From Equation 10.22, we have

$$\text{Reserve} = 500 + 3 \times \sqrt{(30)^2 + (31.88)^2} = 631.3 \text{ MW}$$

Without wind generation the reserve capacity would have been 590 MW.

10.6.4 Stability

The stability of the conventional power system is provided by the large synchronous generators. These units adjust their output to ensure there is a balance between system demand and supply at all times. They also provide the majority of the system inertia. Most renewable generators do not take part in the balancing of the system but merely supply energy from the wind or sun when the resource is available. These renewable generators can operate only when there is a strong voltage and frequency reference provided by the large synchronous generators.

If a small renewable energy generator becomes disconnected from the main power system, a condition known as 'islanding', then it must shut down immediately. It cannot be allowed to continue to operate as it has now lost the frequency and voltage reference provided by the large synchronous generators. 'Anti-islanding' protection is used to detect when a renewable generator becomes isolated from the main power network. A common form of anti-islanding protection is to measure the rate of change of frequency of the islanded generator and trip the unit if it exceeds a value which indicates the generator is isolated.

It is clear that operating renewable generators as sources of energy only and relying on the large synchronous generators to provide all the control and stability of the power system is only satisfactory when there is a modest fraction of generation from renewable sources. As more and more renewable generation is connected to the power system, this simple way of operating will have to change and the renewable generators must evolve to contribute fully to the control and stability of the power system.

10.7 Demand-Side Participation

Power system operators maintain reserve capacity to allow for the uncertainty in predicting demand and the output of renewable generation as well as for the inevitable breakdowns of generators. In many countries, this reserve is provided by thermal generators either part-loaded or on standby. When a thermal generator operates at less than its rated output, or is kept warm but not generating, its fuel efficiency reduces. Keeping generators on standby is expensive as the cost of their construction has been incurred and they are not being used.

Much renewable generation responds immediately to the resource (i.e. the wind, waves, sun) and so has a variable output that cannot be predicted with certainty. As more and more renewable generators are connected, the requirement for reserve capacity increases. Maintaining a large reserve of thermal generators to overcome uncertainties in predicting the output of renewable generation is expensive and increases CO_2 emissions.

Rather than trying to meet demand by changing the output of generators, another approach is to make the demand flexible so that it can follow the generation. This is called demand-side participation. As shown in Figure 10.21, demand-side participation can be divided into two categories: incentive based and price based.

In incentive-based demand-side participation, customers are rewarded financially for reducing their electricity consumption at times when the power system is short of generation. These rewards are distinct from the price paid for energy.

With direct load control, the power system operator remotely reduces or disconnects the load of domestic and small commercial customers. The loads that are controlled are usually heating, ventilation and air conditioning (HVAC) systems.

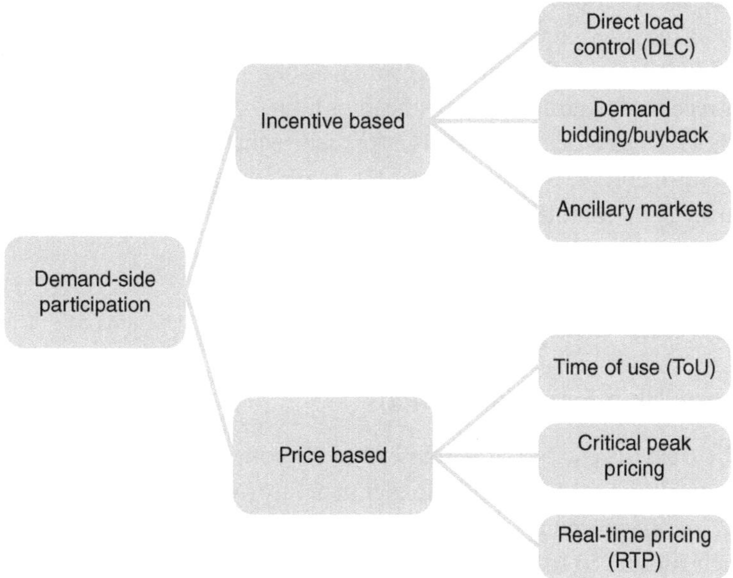

Figure 10.21 Demand-side participation categories.

A customer enters into a contract with the operator to hand over the control of the load (e.g. an air conditioner) for a number of hours per year. When a generation shortage occurs, a signal is broadcast to reduce the participants' loads. Upon receiving the signal, the controllable loads such as air conditioners modify their operation. For example, the temperature of the thermostat can be raised by a few degrees or the air conditioner switched off.

In demand bidding/buyback programmes the participating customers can bid in a market to provide demand reduction. The bid is for day-ahead or same-day events. There is a minimum duration for the demand reduction. The customer is penalized if the reduction in demand is not delivered when required.

System operators use another ancillary market to procure reserve from large electricity users (e.g. aluminium or steel manufacturing plants). The participant must commit to reduce load when the system operator decides this is needed.

In price-based demand-side participation the price of electricity to customers is changed in real time to reflect the value of electrical energy at that time.

Time-of-use tariffs incentivize customers to shift their demand from high-price to low-price periods. They have a minimum of two rates: peak and off-peak. The rates can also vary depending on the season. One of the most successful time-of-use tariffs was implemented in France, called 'Heures Pleines / Heures Creuses' (Full Hours / Hollow Hours) with 10 million participant customers. The electricity rates are reduced by 30% for 8 Hollow Hours each day. In the UK approximately 3 million residential customers pay for their electrical energy through time-of-use tariffs that reduce the price of electricity during the night when demand is low.

Critical peak pricing is applied at hours where the demand is exceptionally high. The price of electrical energy is then increased by three to five times the standard rate. Participants are notified the day before if it will be enabled. An average of 14% peak demand reduction for residential customers during the critical peak pricing days was recorded in California's state-wide pricing pilot.

The electricity rate for a customer with real-time pricing varies hour by hour. The customer receives the tariff on a day-ahead or hours-ahead basis. The hourly rate follows the wholesale electricity cost. This approach requires a communication link between the operator and the meter as well as the meter's capability of recording energy use on an hourly or 30-minute basis. Meters with such capabilities are called smart meters.

10.8 Renewable Energy Connections

10.8.1 Onshore Wind Farm Connections

Wind turbines generate at a relatively low voltage, usually below 1000 V. Some larger turbines (> 3 MW) use a higher generator voltage, up to around 3–5 kV, but at this increased voltage the safety requirements are more onerous: it is necessary for each turbine to have an individual transformer either within or adjacent to the tower. This increases the voltage for connection to the power collection network,

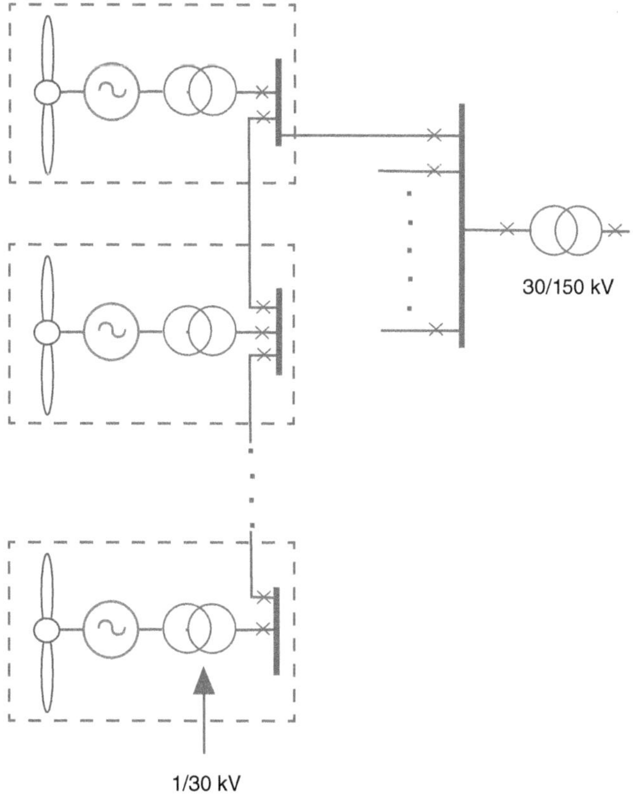

Figure 10.22 Typical onshore wind farm connection.

e.g. to 30 kV. Each turbine also requires some form of switchgear (circuit breaker, isolators and perhaps fuses). Many wind turbine manufacturers now include the transformer and switchgear as part of the turbine supply.

A typical onshore wind farm connection is shown in Figure 10.22. A number of strings of wind turbines are connected to the collector busbar at 30 kV. A central wind farm transformer steps up the voltage to 150 kV for connection to the public network.

10.8.2 Offshore Wind Farm Connections

An offshore wind farm can be connected to the onshore network through an ac or a dc submarine cable. The choice of ac or dc depends on the distance to shore and the power of the wind farm. For cable route lengths above 80 km, dc transmission becomes cost-effective. Above around 100 km and 50 MW only dc transmission is possible owing to the high reactive power generated by the capacitance of ac cables. For wind farms larger than 200 MW, dc can be attractive as fewer cables are needed to connect the wind farm to shore and so multiple cables crossing a beach can be avoided.

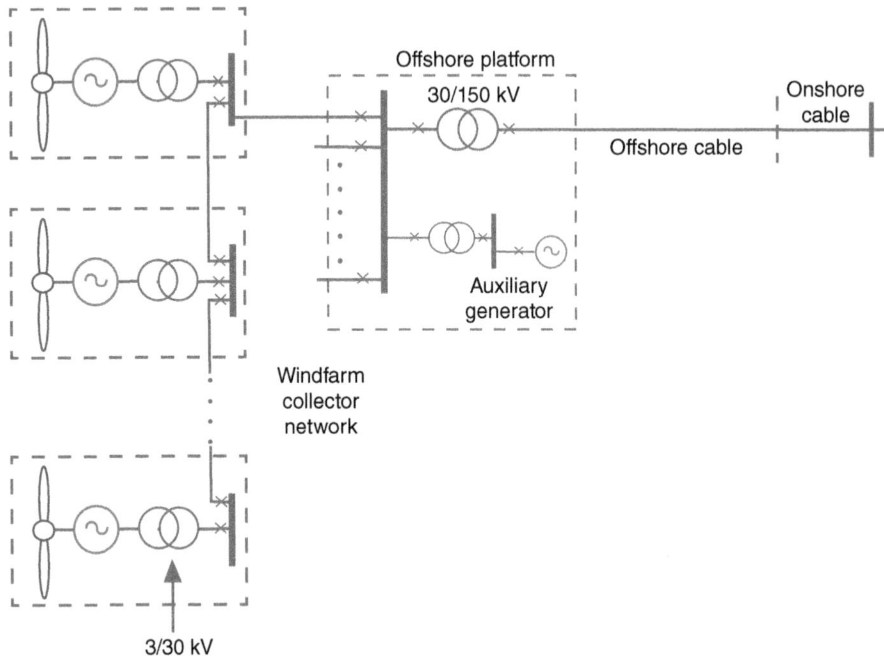

Figure 10.23 Offshore wind farm ac connection.

Figure 10.23 shows the connection of an offshore wind farm using ac. The wind farm collector network voltage is again 30 kV. The cables from the wind turbine array are connected to a switchboard situated on the offshore platform. This platform carries the main step-up transformer, auxiliary/earthing transformer, auxiliary generator, switchgear, as well as metering and protection equipment.

A wind turbine has a number of motors and pumps, heaters and navigation lights. When the connection to the shore is lost a diesel auxiliary generator located on the offshore platform supplies these auxiliary loads.

Wind farms that are located more than 80 km offshore may use dc transmission. This is shown in Figure 10.24 and schematically in Figure 10.25. The power from the wind turbines is collected by an ac medium voltage (~30 kV) network. Large wind farms use an offshore ac substation to increase the voltage to around 150 kV, as shown in Figure 10.24. The power is then taken to another offshore platform on which a power electronic converter is located. This converts the generated power from ac to dc. The ac to dc converters use a form of pulse width modulation (PWM) power electronic converter, shown in Figure I.24 of Tutorial I. The power is then transmitted through a dc submarine cable to an onshore converter station, where it is converted back to ac for onward transmission to the main electrical power grid. Smaller offshore wind farms may omit the offshore ac substation and connect at ~30 kV to the offshore converter platform, as shown in Figure 10.25.

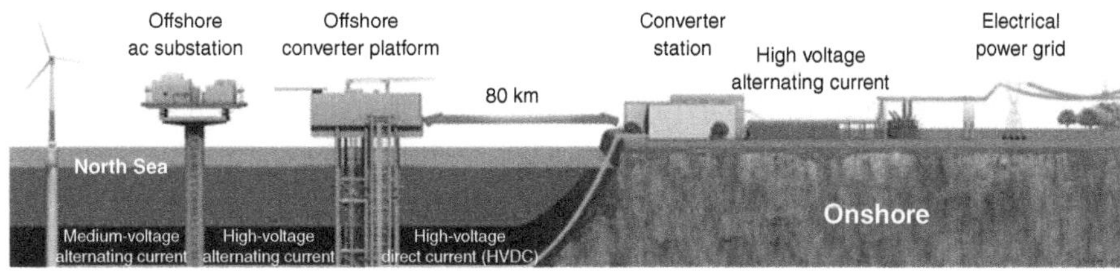

Figure 10.24 Offshore wind farm dc connection.
Source: © Alstom.
(A black and white version of this figure will appear in some formats. For the colour version, please refer to the plate section.)

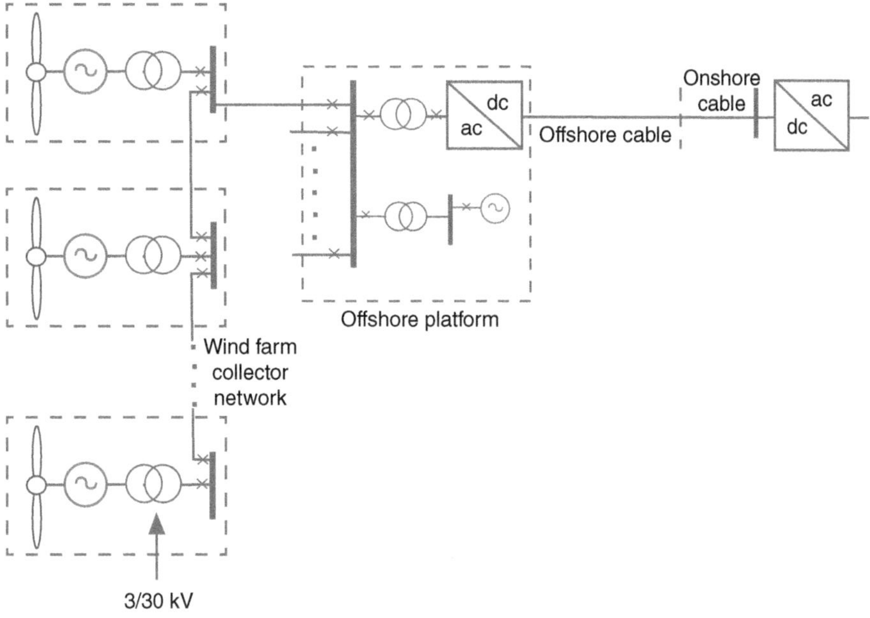

Figure 10.25 Wind farm dc connection using a single offshore platform.

10.8.3 PV Connection

PV cells within a module are connected in series and parallel to increase the power. In a similar way, large PV arrays are formed by connecting strings of modules in different formations. Commonly used topologies are: strings connected in parallel to a central inverter, strings connected through multiple inverters and individual modules connected through micro-inverters.

The connection of a PV array into a central inverter is shown in Figure 10.26. In this example, the PV module strings are shown connected in parallel through string overcurrent protective devices (OCPD). OCPDs are also used to protect the array cables. An OCPD is a series-connected fuse or a circuit breaker that protects the string or array from reverse current flow in the event of a short-circuit. As the fuses used in the strings are subjected to constant sun irradiance and high temperatures, they should be de-rated by applying the temperature correction factor specified by the manufacturer. A disconnector is connected in series with each OCPD for manual disconnection. Combiner boxes are used to mount the OCPDs and isolators, and combine the string cables into the higher-power array cables.

Even though the central inverter connection can be cost-effective, the lifetime of the central inverter is often less than that of the PV modules, thus requiring its replacement at least once over the life of the solar array, adding significant cost.

Figure 10.27 shows the use of string inverters. Each string of modules is connected through a dc/ac converter, which allows the maximum power point of each PV string to be optimized. The PV system can be expanded easily by installing additional strings of modules and their own inverters.

PV modules can also be connected as shown in Figure 10.28. Each PV panel has a micro-inverter of 150–300 W. The micro-inverters convert dc to ac from each module and their output is connected in parallel to form an array. Even though this scheme is attractive in terms of performance (unlike the other scheme, the performance does not depend on the poorest module), the main disadvantage of this scheme is high cost.

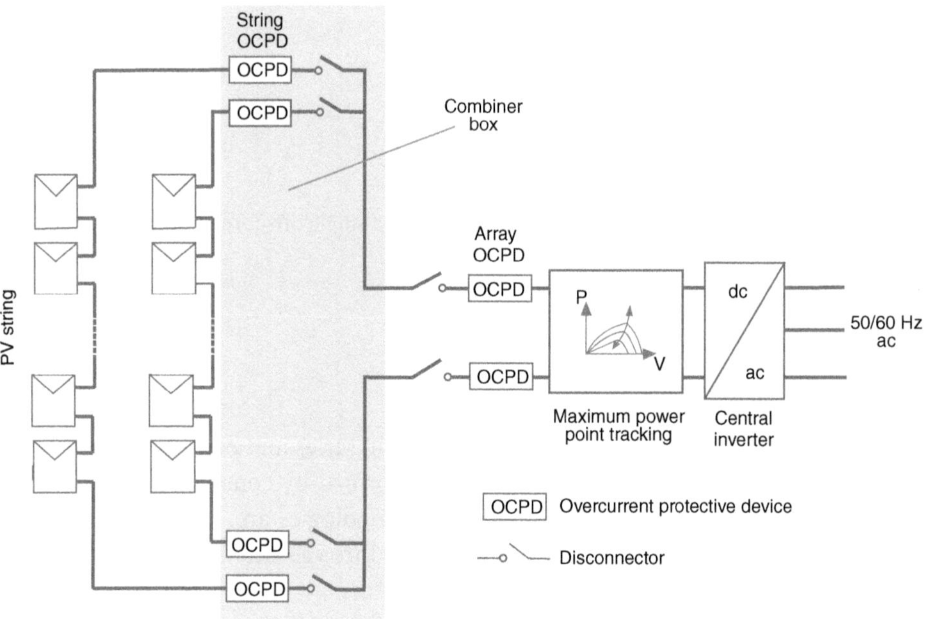

Figure 10.26 PV array connected to a central inverter.

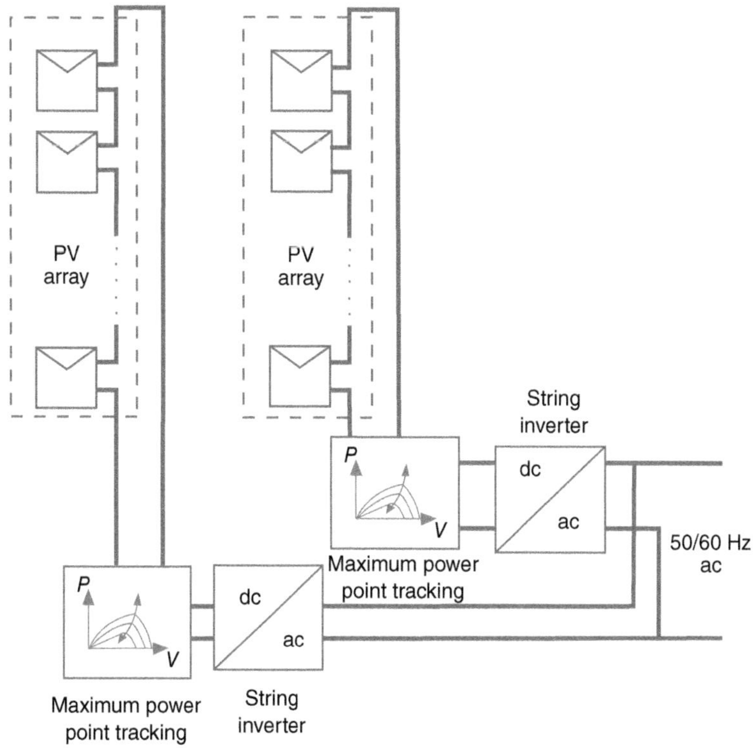

Figure 10.27 String of PV panels connected to a number of string inverters.

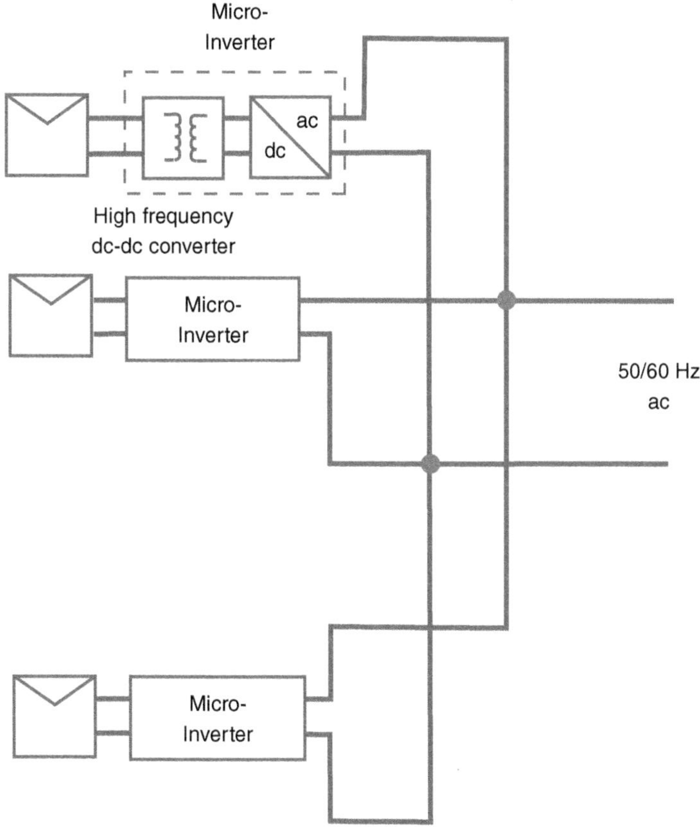

Figure 10.28 Each solar panel is connected to a micro-inverter.

PROBLEMS

1. Compare fossil fuel with renewable energy as an energy source.
2. Why are high voltages used for transmitting large amounts of power over long distances?
3. After the connection of a large load, the frequency decreases rapidly in a power system that has a large penetration of renewables. Describe the reason for this.
4. A transmission line has the following data:

   length 100 km
   line resistance 38 mΩ/km
   power flow in the line 100 MW at unity power factor

   Calculate the losses in the line, when the line voltage is (a) 132 kV and (b) 220 kV.

   [**Answer:** (a) 2.2 MW; (b) 785 kW.]

5. A factory has a dedicated connection through an 11 kV line of length 4 km as shown in Figure 10.29. The 11 kV line (AB) has an impedance of $0.25 + j0.3\ \Omega/\text{km}$. The following equipment is connected to busbar B:

   1 MW induction motor operating at full load with a power factor of 0.9 lagging,
   300 kvar power factor correction capacitor bank,
   1 MVA, 11/0.4 kV transformer with a peak load of 500 kW at 0.85 power factor lagging,
   500 kW PV array operating at unity power factor.

   If the peak 400 V load, full-load operation of the motor and full output of the PV panel coincide, calculate:

   (a) the active and reactive power drawn by all the equipment connected to Busbar B,
   (b) the power losses in line AB.

   [**Answer:** (a) 1000 kW and 494.2 kvar; (b) 8.3 kW.]

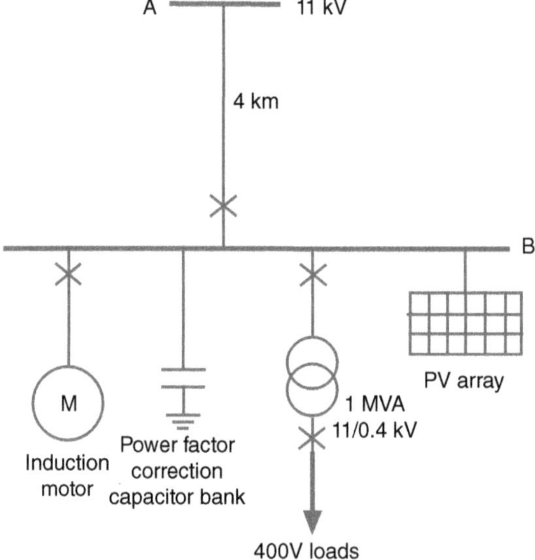

Figure 10.29 Network for Problem 5.

6. Part of a distribution network with a 2 MW biomass power plant is shown in Figure 10.30. The 11 kV line has an impedance of $0.25 + j0.3\ \Omega$/km. The 11 kV busbar voltage varies by ±1% owing to the size of the transformer tap step. A 1 MVA, 0.9 power factor load is connected to point B. Determine:

(a) the losses in the line section BC when the biomass power plant produces 2 MW (assume the voltage of busbar C is 11 kV),

(b) the maximum and minimum voltage of the load.

[**Answer:** (a) 41.3 kW, (b) 11.15 kV, 10.8 kV.]

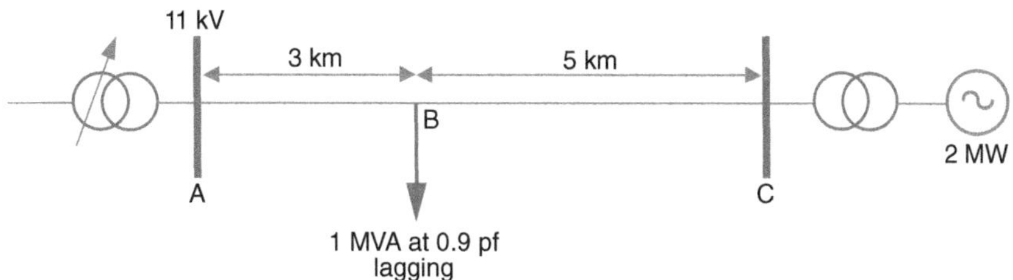

Figure 10.30 Network for Problem 6.

7. A 50 MW wind farm is to be connected to the grid through a 10 km collector circuit. The following connections are considered during the planning of the circuit:

132 kV connection through a single-circuit 400 mm^2 cable,
33 kV connection through a double-circuit 300 mm^2 cable.

The cable data are given in the following table.

	R (Ω/km)	X (Ω/km)
132 kV, 400 mm^2 cable	0.05	0.12
33 kV, 300 mm^2 cable	0.06	0.11

The wind farm can operate with 15 Mvar importing or exporting reactive power. Calculate the maximum voltage rise at the wind farm as a percentage of the nominal voltage and the losses in the cable for the two sizes of conductor considered.

[**Answer:** For the 132 kV connection, 0.25%, 78.2 kW; for the 33 kV connection, 2.13%, 750.7 kW.]

8. A power system with a total installed capacity of 3000 MW has an equivalent inertia constant H of 4.5 s and operates at 50 Hz. A large generator of 350 MW is switched out owing to a fault. Calculate the rate of change of frequency. If the equivalent inertia drops to 3.5 s, due to the connection of wind generators, what is the new rate of change of frequency?

[**Answer:** 0.65 Hz/s, 0.83 Hz/s.]

9. An island is supplied from two 100 MW biomass power plants and a 50 MW wind power plant. The cost functions are:

$$C_{\text{Biomass\_1}} = 1000 + 11P_{\text{Biomass\_1}} + 0.03 \times P^2_{\text{Biomass\_1}} \text{ £/h}$$
$$C_{\text{Biomass\_2}} = 1200 + 14P_{\text{Biomass\_2}} + 0.04 \times P^2_{\text{Biomass\_2}} \text{ £/h}$$
$$C_{\text{Wind}} = 1140 \text{ £/h}$$

where P is in MW.

(a) If the peak demand is 200 MW, how should the generators be despatched?

(b) If the minimum generation of the biomass power plants is 50 MW, what is the new economic dispatch for the same demand?

(c) A demand-side participation programme was introduced in the island. This resulted in shifting 25 MW of load from peak to off-peak. What is now the economic dispatch without considering the minimum generation limit?

[**Answer:** (a) Biomass 1: 107.1, Biomass 2: 42.8 MW; (b) Biomass 1: 100, Biomass 2: 50 MW; (c) Biomass 1: 92.85, Biomass 2: 32.2 MW; for (a), (b) and (c) wind 50 MW.]

10. A power system has the following characteristics:
peak demand: 5000 MW,
largest generator on the system: 500 MW,
wind generation: 500 MW.

Hourly wind predictions are used to determine the system reserve.

(a) Using the wind energy forecast error curve shown in Figure 10.20, calculate the reserve requirement. Assume that the uncertainty in the demand is 1%.

(b) This reserve is supplied by open cycle gas turbines (OCGT). For the OCGT, the fixed O&M cost is £50/kWy, the variable O&M cost is £2.5/MWh and the fuel cost is £60/MWh. What is the total cost of providing reserve in £/h?

[**Answer:** (a) 667.7 MW, (b) 45 543 £/h.]

FURTHER READING

Jenkins N., Ekanayake J. and Strbac G., *Distributed Generation*, 2010, IET.
Addresses the electrical integration of small generators to distribution systems.

Grainger J.J. and Stevenson W.D., *Power System Analysis*, 1994, McGraw Hill.
Covers both basic and advanced topics in electrical power systems.

Weedy B.M., Cory B.J., Jenkins N., Ekanayake J.B. and Strbac G., *Electric Power Systems*, 2012, Wiley.
A comprehensive treatment of electrical power systems for undergraduate students

Wood A.J. and Wallenberg B.F., *Power Generation Operation and Control*, 2013, Wiley.
Discusses the characteristics of power generating units, scheduling and control of generators.

ONLINE RESOURCES

Exercise 10.1 Wind farm Q and V control: a load flow is used to demonstrate the effect of Q and V control of a wind farm. The Gauss Seidel method was used to develop the load flow programme and MATLAB code is provided.

Exercise 10.2 Power generation constraints: this exercise demonstrates the maximum capacity additions of wind power possible with Q control.

11 Storage of Renewable Energy

INTRODUCTION

The need for energy often occurs at times that do not coincide with the renewable energy resource. Electricity is needed for lighting and to cook food after dark when there is no sun, and the periods of high demand in a national electricity system do not always coincide with times of strong winds or bright sunlight. In a temperate climate, the main need for heating occurs in the winter when the solar resource is low. Thus an energy supply system relying heavily on the intermittent renewable resource of the sun and wind needs energy storage.

Renewable energy is often used as one component of an energy system that includes some fossil fuels. In these mixed fossil–renewable systems, the balance between the energy needed and supplied is maintained by adjusting the rate at which fossil fuel is burnt and the chemical energy stored within the fossil fuel is released. Fossil fuels store energy very conveniently but releasing the energy results in emissions of greenhouse gases and other pollutants. Fully de-carbonized electricity and energy systems will need to use the energy storage technologies discussed in this chapter, as well as new technologies still under development.

Some renewable energy systems store energy naturally. These have been discussed in earlier chapters and are listed in Table 11.1. The renewable technologies that naturally store energy are particularly useful for operating electrical power systems and providing electricity when it is needed.

Table 11.1 Typical duration of energy stored in renewable energy technologies

Renewable energy technologies	Energy store	Chapter of this book	Typical duration of intrinsic energy storage
Hydro generation	Water reservoir at height above turbine	2	Hours/weeks
Solar thermal energy	Fabric of buildings, hot water, high temperature salts	6	Daily
Tidal range generation	Impounded area of water	7	Hours
Bioenergy	Biogas, liquid biofuel, dry biomass	8	Days/months

11.1 Storage of Renewable Energy

Figure 11.1 shows the main ways in which renewable energy is stored in significant quantities at present. They are divided into the storage of electricity and the storage of heat.

11.1.1 Storage of Electrical Energy

Electricity is the most useful form of energy but is difficult and expensive to store in large quantities. The bulk storage of electrical energy has traditionally been through mechanical systems, particularly pumped hydro storage. Table 11.2 summarizes the characteristics of mechanical systems and devices that are used to store renewable electrical energy.

Pumped hydro is the dominant technology for storing large quantities of electrical energy and is estimated to provide 98% of all electricity storage worldwide. Pumped hydro energy storage works by pumping water from a lower reservoir

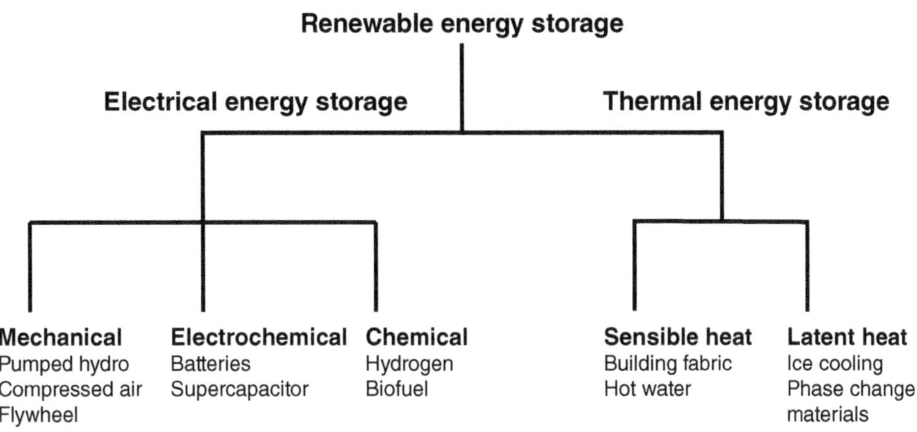

Figure 11.1 Storage of renewable energy.

Table 11.2 Mechanical storage of renewable electrical energy (typical values)

Technology	Power capacity (MW)	Energy capacity (GWh)	Efficiency (%)	Response time	Lifetime of scheme (years)
Pumped hydro storage	100–1000	Up to 100	75–85	Seconds to minutes	40–60 years
Compressed air energy storage (CAES)	50–500	0.1–10	70–89	1–15 minutes	20–40 years
Flywheels	0.01–20 (up to ~100 for increasing grid stability)	10–100	70–95	100 milliseconds	>15 years, 100 000+ cycles

or water course to a higher reservoir and storing the water (and its potential energy) until it is needed. Depending on the height difference between the two reservoirs and the volume of water, large pumped storage schemes can supply more than ~1000 MW. Many pumped hydro schemes store energy for a few hours in a daily cycle by pumping water to an upper reservoir at a time of the day when electricity is plentiful and cheap. They then generate when electricity is scarce and expensive. Other schemes with large reservoirs store energy for weeks or even months.

Compressed-air energy storage (CAES) systems compress air and store it until electricity is needed, at which stage the air is released and passed through a turbine to drive a generator. Fuel is burnt in the air as it is released from the store to compensate for the temperature drop that occurs as the air expands. There have been two large demonstration CAES schemes with power ratings of 290 MW (Huntorf, Germany, 1978) and 110 MW (McIntosh, USA, 1991). The Huntorf plant was originally intended to provide black start[1] and McIntosh was intended for load shifting. Several similar large underground CAES projects have been proposed but have not progressed to construction. There have also been smaller designs demonstrated that store compressed air in pressure vessels on the surface.

Flywheels store rotating kinetic energy. They have been used to store energy to smooth the rotational speed of piston engines for many years, and large, low-speed flywheels can be seen on the sides of historical steam engines. The flywheel effect is essential in an ac electrical power system to limit the rate of change of frequency when the load and generation are not equal. Kinetic energy is stored naturally in rotating electricity generators and loads but sometimes additional flywheels coupled to motor–generator sets are added to the power system to increase the kinetic energy stored in it. Flywheel systems store energy with few losses and have a short response time. Both high- and low-speed flywheels have been used for storing electricity generated from renewable energy in various demonstration projects. Flywheels are most effective when they provide large injections of power for a relatively short time (typically minutes).

Electrochemical batteries are widely used for domestic, commercial and industrial applications and several different chemistries are common. The rechargeable lead–acid battery was invented in the nineteenth century; this type of battery is produced in very large numbers for starting, and providing the auxiliary power for, petrol and diesel vehicles. Lithium batteries now provide the motive power for most electric vehicles. Stationary batteries are essential in microgrids that are not connected to a strong ac electricity network, and batteries are used increasingly to help manage the frequency of large electricity grids.

Domestic-scale battery energy-storage systems can enhance the performance of domestic and small commercial rooftop PV systems. Many combined PV–battery

[1] Black start is the process of restarting an electrical power system that has shut down.

systems are connected on the consumer side of the meter and maximize the consumption of solar energy within a building by storing any excess solar electricity rather than exporting it to the grid. Around midday on sunny days when there is a lot of PV generation, the value of grid electricity is low so the battery stores energy from the PV array and power is not exported. The stored energy is then used to reduce the amount of energy drawn from the grid in the evening when the value of grid electricity is high. These domestic battery systems use either lead–acid or lithium-ion batteries. The systems are typically less than 5 kW in power rating and can provide this level of power for 2–3 hours before recharging. The power electronic converters that connect the battery to the grid may not allow independent operation to continue to supply the building if there is a grid power outage. There is no fundamental technical reason for this limitation; it is a design feature to reduce the costs of the inverters and to meet national electricity supply regulations.

In those countries with a very good solar resource, such as in parts of Australia, some consumers are choosing to disconnect completely from the distribution network and operate an autonomous PV system supported by battery energy storage within their dwelling. This can create commercial difficulties for the utility power company as it reduces the volume of electricity flowing through the distribution network and the revenue available to the network company to operate and maintain the network.

There is an increasing use of large batteries in utility electric power systems that have a high fraction of renewable generation. These are typically lithium-ion or sodium–sulphur and are up to 50–100 MW in power capacity. They are usually able to provide their full output energy for a few hours. Some large batteries are located on the sites of wind power or solar PV farms where there is physical space and a suitable high-voltage connection to the electrical network.

There have been several demonstrations of flow batteries that store their electrolytes in separate tanks. This allows the energy capacity of the flow battery to be increased by storing a large amount of charged electrolyte without increasing the size and output power of the battery cells and hence the expense of the battery.

Renewable electricity can be stored as chemical energy, particularly in a gas. Hydrogen produces only water when it is burnt and has been considered as a possible means of storing and transmitting energy for over 100 years. The term 'green hydrogen' is used to describe hydrogen that is created by the electrolysis of water using electricity that has been generated without the emission of greenhouse gases. The hydrogen is stored and re-converted to electricity through a fuel cell.

Biofuels and biomass store energy in chemical form and, although they produce CO_2 at the time of their use, it is argued that the complete cycle of biomass production through photosynthesis (which absorbs CO_2), transport and storage of the biomass and subsequent oxidation results in reduced emissions of greenhouse gases.

The typical capacities of the technologies currently used for storing electricity produced by renewable energy are summarized in Figure 11.2. Other ideas for storing bulk electricity continue to be investigated but are not commercially significant at present.

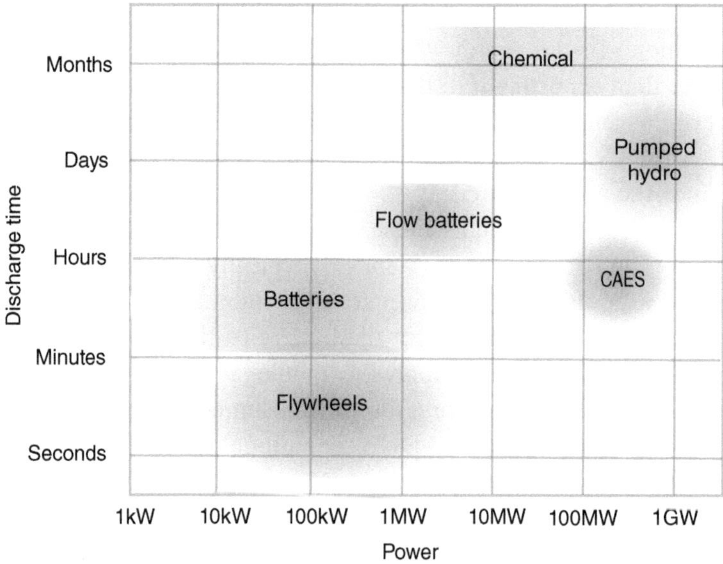

Figure 11.2 Typical power capacity and discharge time of technologies for storing electrical energy.

Use of Energy Storage for Frequency Control of a National Electricity System

Figure 11.3 is an illustration of how the frequency of a national electricity system is maintained by the constant balancing of the net load by adjusting the controllable generation or energy storage. The net load is the load drawn by consumers minus the generation from renewable generators that do not take part in frequency control such as wind turbines and photovoltaic systems, which have no natural energy storage and generate whenever the wind or sun conditions are suitable. Therefore the net load changes all the time as more or less electricity is used, and the outputs of wind farms or solar PV systems change with the strength of the wind and sun. All controllable generation has some form of energy storage either in fossil fuel or in the elevated water reservoirs of hydro schemes. Separate energy storage systems such as batteries or flywheels are controlled to release energy when the system frequency drops and absorb energy when the frequency rises.

The frequency of an ac power system is maintained close to 50 Hz (60 Hz in some countries) so that the generators and transformers are not damaged. If the load exceeds the generation, then the frequency drops and if the generation exceeds the load the frequency rises. Typically a large power system will be controlled to maintain its frequency to within ±0.5 Hz and will shut down if variations of more than ±2 Hz are experienced. The speed at which the frequency changes depends on the magnitude of the mismatch between generation and loads, but also on the inertia of all the spinning mechanical elements of the power system (generators and motors). These spinning generators and motors are of course storing kinetic energy.

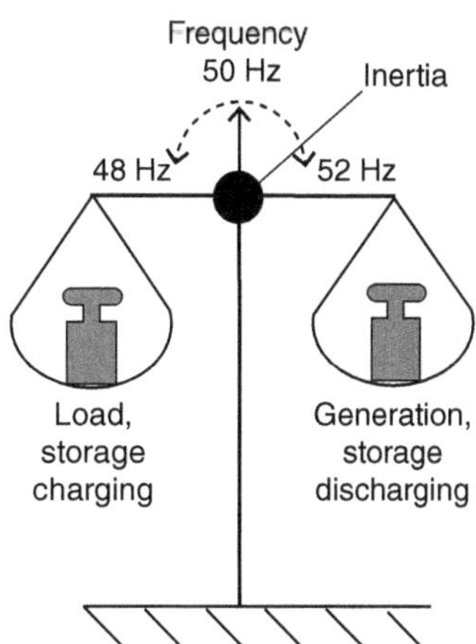

Figure 11.3 Illustration of the control of frequency of an electric power system.

The control of frequency in an electrical power system is a continuous process but for convenience can be thought of in two time frames. Slow frequency changes are balanced by generation or controllable demand being adjusted in response to instructions from the system operator. Rapid frequency variations are managed by automatic controls and ensuring that the system has adequate inertia.

(For a detailed explanation of frequency control of an electric power system, see Section 10.5.)

11.1.2 Storage of Thermal Energy

Many countries are making good progress in increasing the fraction of electricity that is generated from renewable energy, reducing the use of fossil fuels. However, in a temperate climate significantly more energy is used for heating than is currently used as electricity. The need for energy for heating and cooling varies throughout the day and over the year, so the challenge of storing heat energy in a renewable energy system is even greater than that of storing electricity.

Heat energy can be stored as either sensible or latent heat. Sensible heat is stored by changing the temperature of a material, while latent heat storage involves a change of its phase. Buildings store heat as sensible energy in their fabric and therefore smooth the effect of the daily variations in solar radiation and outdoor temperature to provide a comfortable indoor environment. The Trombe wall in a low-energy house (see Section 6.2) stores sensible heat in its large mass of concrete to maintain a constant temperature over the daily fluctuations of the solar resource. Domestic solar water heating systems have a thermally insulated tank to store the water at an increased temperature using sensible heat energy. The temperature of the liquid water is typically increased from 15 °C to around 60 °C.

Greater quantities of heat can be stored as latent heat energy, as the phase of the material changes. Before electric refrigerators became available, ice was used to cool food in some rich households. The ice was brought from a glacier or mountain top, stored in a highly insulated outhouse, and then allowed to melt over several months absorbing heat energy and keeping the temperature of the food low. A number of new phase-change materials and their delivery systems are being developed to store thermal energy using latent heat for both heating and cooling.

11.2 Pumped Hydro Energy Storage

Hydro generation (discussed in Chapter 3) stores potential energy in the water of the reservoir which is at a height above the turbines. This stored energy, and the ability of a turbine generator to convert the flow of water rapidly to electricity whenever it is needed, makes hydro power plants extremely useful in the operation of electric power systems. In a conventional hydro scheme, the reservoir is filled from rivers that are fed by rainfall over the area of the catchment.

If the local topography and geology are suitable, this water supply can be supplemented or replaced by pumping from a lower reservoir or water source. When the load on the electric power system is low, typically at night, water is pumped into the upper reservoir and stored for use when additional electricity is needed.

In the past, pumped hydro energy storage was used mainly to smooth variations in electricity demand and provide a constant load for nuclear generators and large coal-fired power stations. It is desirable to operate both nuclear and large thermal generators at constant output to maintain stable temperatures and pressures in the reactors and boilers. Pumped hydro energy storage is now used increasingly to accommodate the variations in output of renewable power generation.

Reservoir hydro and pumped hydro storage are the main ways to store electricity at present. In 2020, pumped hydro power stored approximately 10 times the energy that was stored in batteries, while the energy stored in reservoir hydro was another two orders of magnitude greater (see Figure 11.4).

Figure 11.5 shows the principle of pumped hydro energy storage. An upper reservoir stores potential energy in the elevated water and releases it to a lower reservoir while generating electricity. When there is a surplus of electricity from other generators in the power system, water is pumped from the lower reservoir to be stored in the upper reservoir. In the early days of pumped storage, the pump and turbine were separate hydraulic machines on a common shaft but now they are usually integrated into a single hydraulic machine that can act as either a turbine or a pump. The motor–generator is usually a single electrical machine capable of generating or motoring. It can reverse its direction of rotation and may be able to vary its speed. Pumped hydro storage schemes require an upper reservoir and a lower source of water at the appropriate heights and these sites are often difficult to find and are frequently in environmentally sensitive areas. Any pumped hydro project that is

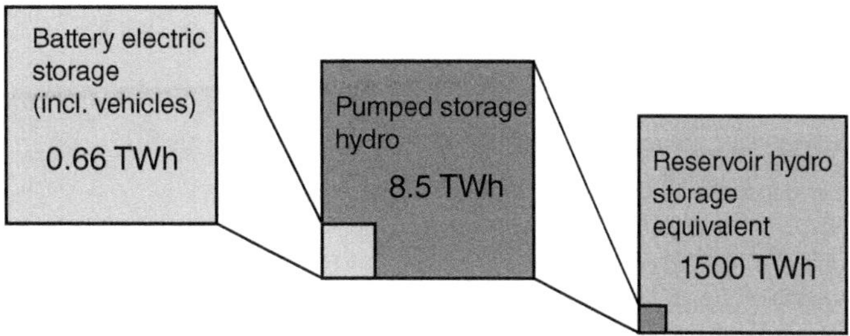

Figure 11.4 Comparison of worldwide energy stored in batteries, pumped hydro and reservoir hydro schemes.
Source of data: IEA Special Reports on Hydro Power, July 2021.

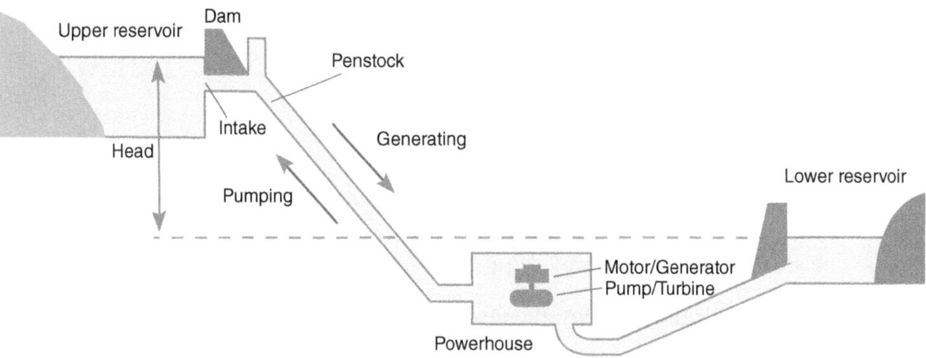

Figure 11.5 Pumped hydro energy storage.

proposed will require the same attention to the Environmental Impact Assessment as is needed for a large hydro scheme.

Simple or pure pumped hydro schemes exchange water between the upper and lower reservoirs and only use additional water from the catchment area to make up evaporation and seepage losses. These schemes generate less electricity than is used for pumping and may have a combined round trip electrical efficiency of 70–80%. Other schemes, known as mixed pumped storage plants, integrate some pumping into conventional hydro schemes that mainly use water from the catchment area. These schemes generate more energy than they pump.

Pumped hydro storage schemes with large turbines and directly connected synchronous motor–generators provide useful inertia to limit the rate of change of frequency of the power system (see Section 10.5). Some modern pumped storage hydro schemes can operate with the turbines spinning in air and then ramp from zero to full generation in only a few seconds. The ability to increase power output rapidly and increase system inertia are both extremely useful in controlling system frequency.

Example 11.1 – Pumped Hydro Energy Storage Scheme

A pumped storage hydro scheme has an average net head of 250 m and a flow rate of 25 m^3/s. It has an efficiency of 85% when generating and 90% when pumping.

Using the equations of Section 3.2 of this book:

(a) calculate the electrical power that is generated,
(b) calculate the electrical power required for pumping,
(c) calculate the useable volume of the upper reservoir required to generate 50 MW of electricity for 4 hours,
(d) estimate the diameter of the circular surface area of the upper reservoir if the water level varies by 10 m.

Solution

(a) Electrical power that is generated:

$$P_{generating} = \rho g h Q \eta_G = 1000 \times 9.81 \times 250 \times 25 \times 0.85 = 52.12\ \text{MW}$$

(b) Electrical power required for pumping:

$$P_{pumping} = \frac{\rho g h Q}{\eta_P} = 1000 \times 9.81 \times 250 \times \frac{25}{0.9} = 68.12\ \text{MW}$$

(c) The stored electrical energy required is 4 × 50 = 200 MWh.

The stored hydraulic potential energy required is

$$PE = \frac{200}{0.85} = 235.3\ \text{MWh}$$
$$= 235.29 \times 10^6 \times 3600 = 847.06 \times 10^9\ \text{J}$$

The potential energy of elevated water is

$$PE = \rho V g h \quad [\text{J}]$$

Therefore the useable volume of water required is

$$V = \frac{PE}{\rho g h} = \frac{847.06 \times 10^9}{1000 \times 9.81 \times 250} = 345.4 \times 10^3\ \text{m}^3$$

(d) If the water level in the upper reservoir varies by 10 m, the approximate diameter of the circular surface of the reservoir is given by

$$\text{Area of reservoir} = \frac{345.39 \times 10^3}{10} = 34.539 \times 10^3\ \text{m}^2$$

$$\text{Diameter} = \sqrt{34\,539 \times \frac{4}{\pi}} = 210\ \text{m}$$

11.3 Compressed-Air Energy Storage

Figure 11.6 shows how compressed-air energy storage (CAES) can be used to store energy in electrical power systems. Air is compressed and stored at high pressure when there is a surplus of electrical generating capacity. Then, when additional electrical power is needed, the pressurised air is released, heated and used to drive a turbine in a similar manner to a gas turbine. A motor/generator is connected by clutches either to drive the compressor or to generate electricity when driven by the turbine. Advantages of the system include that much of the electromechanical equipment is conventional, the system can start within minutes and there is limited environmental impact. If air is simply compressed, stored and used to drive a turbine to generate electricity, the efficiency will be low. A useful round-trip efficiency can be obtained only by heating the air that is discharged from the store with fuel such as natural gas (methane) or biofuel.

For bulk electricity storage in a large utility system the compressed air is stored underground. This may be in large caverns that have been formed by the extraction of salt or in porous rock formations that are surrounded by impermeable rock barriers or aquifers. Effective sealing of the underground store is essential. Geological surveys indicate that locations suitable for CAES are distributed more widely than sites that that are suitable for pumped hydro energy storage. Smaller compressed-air energy storage systems that store air or other gases in pressurized tanks on the surface are common in chemical plants and factories and their use has been demonstrated for energy storage.

A key challenge of CAES is that when air is compressed its temperature increases, and when it is expanded its temperature drops. If this heat is lost to

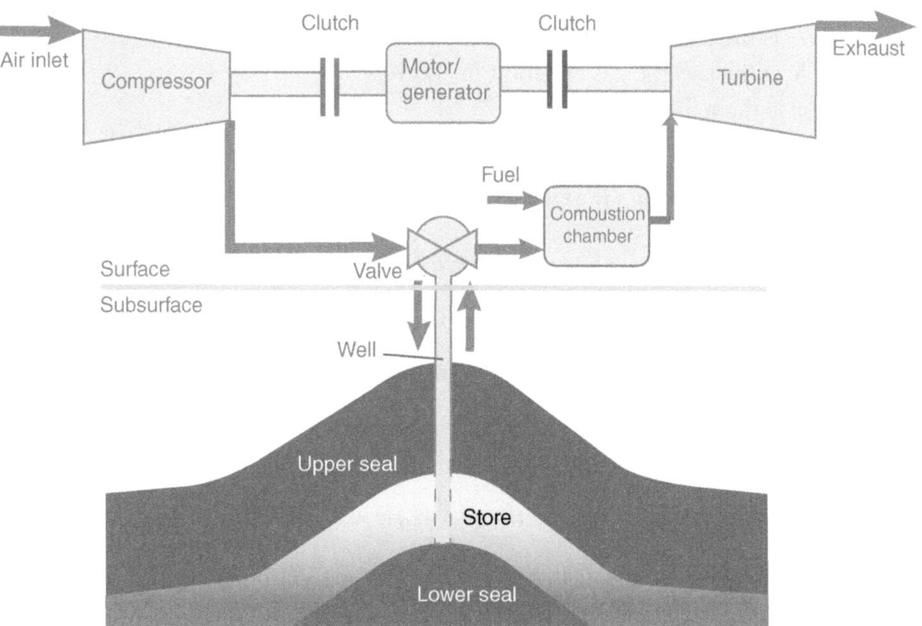

Figure 11.6 Compressed-air energy storage (CAES) with underground store.

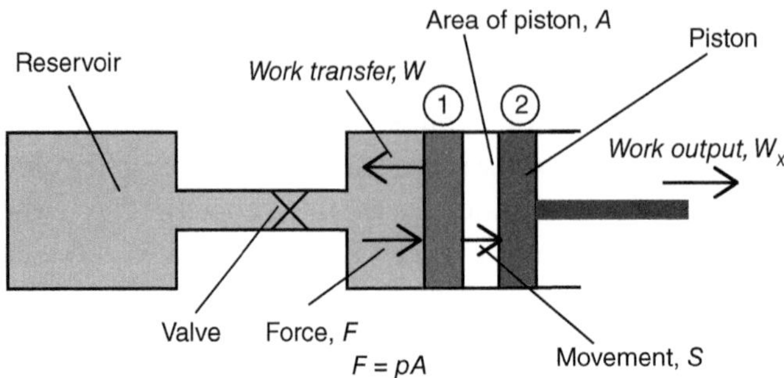

Figure 11.7 Operation of gas piston.

the surroundings then the round-trip efficiency of the energy storage will be low. There have been considerable research and development efforts to retain the heat created during compression for use during expansion (an adiabatic process) or to keep the air temperature constant (an isothermal process) and so improve the efficiency of CAES. So far, only compressed-air systems that use additional fuel to heat the air during discharging through a turbine have been implemented at utility scale.

An understanding of the energy that can be stored in compressed air can be gained by considering a gas piston, shown in Figure 11.7, and the ideal gas law that states:

$$pV = mRT$$

where

p is the absolute pressure in Pa,
V is the volume in m^3,
T is the temperature in K,
R is the specific gas constant in J/kgK,
m is the mass of gas in kg.

Consider the piston shown in Figure 11.7 moving from Position 1 to Position 2. The *work output* (W_X) when the piston moves from Position 1 to Position 2 is

$$W_X = \int F dS = \int pA dS$$

The *work transfer*[2] to the gas from the movement of the piston (W) is given by

$$W = -\int_{V_1}^{V_2} p dV$$

If the pressure of the gas is held constant with the valve to the large reservoir open in an isobaric process, the work transfer to the gas is

$$W = -\int_{V_1}^{V_2} p dV = -p(V_2 - V_1)$$

[2] Work transfer (W) is positive when work is done on the control mass or volume and has the opposite sign to work output (W_X). $W \equiv -W_X$.

Rather than the case of constant pressure, consider the case of the valve being shut and a mass of gas held at constant temperature in an isothermal process. The ideal gas law then gives

$$pV = mRT = constant$$

$$pV = p_1V_1 = p_2V_2$$

and

$$p = \frac{p_1V_1}{V} \text{ and } \frac{V_2}{V_1} = \frac{p_1}{p_2}$$

Then

$$W = -\int_{V1}^{V2} pdV = -\int_{V1}^{V2} \frac{p_1V_1}{V} dV = -p_1V_1 \int_{V1}^{V2} \frac{1}{V} dV$$

$$W = -p_1V_1 \left[\ln V\right]_{V_1}^{V_2} = -p_1V_1\left(\ln V_2 - \ln V_1\right) = -p_1V_1 \ln\left(\frac{V_2}{V_1}\right)$$

and expressed in terms of pressure

$$W = -\, p_1V_1 \times \ln\frac{p_1}{p_2} = -\, p_1V_1 \times \ln\frac{p_1}{p_2}$$

To maintain a constant temperature, heat energy (Q) must be transferred to the environment.

$$Q = -W$$

If no heat is transferred to or from the gas in an adiabatic process, then the temperature of the air will change as:[3]

$$\frac{T_1}{T_2} = \left(\frac{p_2}{p_1}\right)^{\gamma-1/\gamma}$$

where

γ is the ratio of specific thermal capacity at constant pressure and volume (c_p/c_v) and for air is ~1.4.

Example 11.2 – Energy Stored in Compressed Air

A compressed-air energy store has an effective volume of 1000 m^3 and operates at 1–70 bar.

(a) Assuming ideal isothermal operation, estimate the maximum energy that can be stored.
(b) Assuming ideal adiabatic operation, estimate the maximum temperature of the air.
(c) In a CAES project of the type shown in Figure 11.6, 1 kWh of electrical output requires 0.69 kWh of electrical input to compress the air and 1.17 kWh of fuel to heat the air. Calculate the energy efficiency of the cycle.

[3] Stated but not proved.

Solution

(a) For a volume of 1000 m^3 and isothermal expansion of 70 to 1 bar

$$W = -p_1 V \times \ln\frac{p_1}{p_2} = 70 \times 10^5 \times 1000 \times \ln\left(\frac{70}{1}\right)$$
$$= -29.74 \times 10^9 \text{ J}$$
$$= -8.26 \times 10^6 \text{ Wh} = -8.26 \text{ MWh}$$

The energy stored is 8.26 MWh.

(b) The maximum temperature of the gas is given as follows. Assume atmospheric temperature is 300 K.

$$T_1 = T_2\left(\frac{p_1}{p_2}\right)^{\gamma-1/\gamma} = 300 \times \left(\frac{70}{1}\right)^{0.4/1.4} = 1010 \text{ K}$$

This is unrealistically high for conventional materials; cooling stages would be introduced in any practical system.

(c) Energy efficiency of a practical CAES cycle:

$$\eta = \frac{Elec_{out}}{Elec_{in} + Fuel_{in}} = \frac{1}{0.69 + 1.17} = 0.537 \text{ or } 54\%$$

11.4 Flywheel Energy Storage

Flywheels are widely used to store rotating kinetic energy and smooth variations in the speed of many rotating mechanical systems. They have also been demonstrated to provide motive power to bus and rail vehicles either onboard or by storing energy at the stations; these flywheels absorbed energy as the vehicle decelerated and released it to provide a power injection for re-acceleration. Flywheel energy storage systems are most effective in providing a large injection of power for a short time. They are used in some uninterruptible power supplies to provide energy during a power interruption or a sudden transient drop in voltage and to allow time for a standby generator to start. Electric power systems rely on the flywheel effect (the energy stored in the inertia of spinning generators and loads) to slow the speed of the change in frequency caused by any mismatch between generated power and the load.

The kinetic energy of a spinning mass such as a flywheel is given by

$$E = \frac{1}{2} I\omega^2 \quad [\text{J}] \tag{11.1}$$

where

I is the moment of inertia in kgm^2,
ω is the radial speed in rad/s.

For a disk with its mass concentrated at the rim, the moment of inertia is given by

$$I = \text{mr}^2 \quad \left[\text{kgm}^2\right]$$

while for a solid disk with uniformly distributed mass, the moment of inertia is given by

$$I = \frac{1}{2} m r^2 \quad \left[\text{kgm}^2\right] \tag{11.2}$$

where

m is the mass of disk in kg,
r is the radius of the disk in m.

Example 11.3 – Energy Stored in a Flywheel

A solid disk flywheel has a mass of 50 kg and a diameter of 1 m. It operates at between 1000 and 3000 rpm. Calculate the energy stored.

Solution
Inertia of flywheel:

$$I = \frac{1}{2} m r^2 = \frac{1}{2} \times 50 \times 0.5^2 = 6.25 \text{ kgm}^2$$

Rotational speeds:

$$\omega_1 = \frac{3000 \times 2\pi}{60} = 314.16 \text{ rad/s}$$
$$\omega_2 = \frac{1000 \times 2\pi}{60} = 104.72 \text{ rad/s}$$

The energy that is stored is

$$\begin{aligned} E &= \frac{1}{2} I \times \left(\omega_1^2 - \omega_2^2\right) \\ &= \frac{1}{2} \times 6.25 \times \left(314.16^2 - 104.72^2\right) \\ &= 274.16 \text{ kJ} = 76.15 \text{ Wh} \end{aligned}$$

11.4.1 Kinetic Energy Storage Systems (KESS)

Traditional low-speed flywheels are heavy metal devices rotating at up to several thousand rpm, and often at much lower speeds. Because of their low speed, even heavy, high-inertia flywheels are able to store only a limited amount of energy. Since the 1970s composite materials have allowed the development of lighter, very-high-speed flywheels with speeds of rotation typically up to 50 000 rpm. These materials have sufficient strength to resist the high stresses caused by very fast rotational speeds.

The modern high-speed flywheel systems used to store energy are often described as kinetic energy storage systems (KESS). KESS have a long cycle life, high power and energy density, rapid charge/discharge rates, high round-trip efficiency and limited environmental impact. However, self-discharge can be significant, and

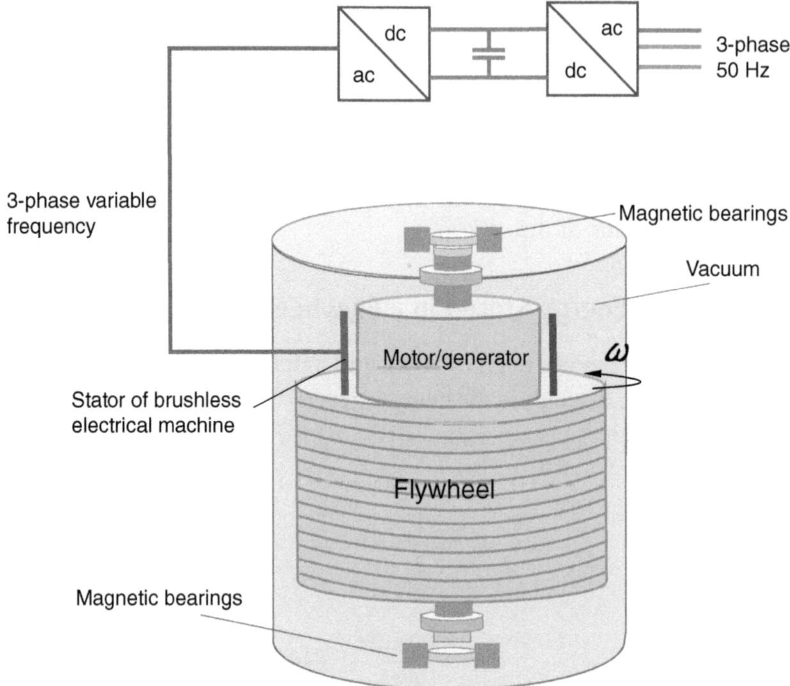

Figure 11.8 Kinetic Energy Storage System.

precautions need to be taken against dangerous overspeed and bursting. For a number of years KESS have been demonstrated to inject a relatively large amount of electrical power (up to 1 MW) for a short period of time (seconds or minutes). They have been used on several small autonomous renewable energy systems but the technology has yet to achieve commercial breakthrough.

Figure 11.8 shows the main elements of a KESS. It consists of a high-speed flywheel, an electrical machine acting either as a motor or as a generator, and a bidirectional power electronic interface connected to the electricity supply. The flywheel is a cylinder of composite material rotating at high speeds within a robust housing that forms a protective enclosure in case of overspeed and mechanical failure. The flywheel operates in a vacuum to limit windage losses, and magnetic bearings are used to minimize friction losses.

The brushless motor–generator may be a separate electrical machine or integrated into the flywheel. It acts as a motor to accelerate the flywheel and charge the energy store. When the energy is needed, the flywheel is decelerated by the generator and the energy injected into the electricity network. One design used a synchronous machine with its permanent magnet rotor bonded into the flywheel to withstand high rotation forces. The power electronic interface was two voltage source converters connected through a dc link which allowed the flywheel to rotate over a wide range of rotational speeds.

Combining Equations 11.1 and 11.2 for a disk with its mass concentrated at the rim gives the stored energy as

Table 11.3 Properties of materials used for flywheels

Material	Density, ρ (kg/m^3)	Ultimate strength, σ (MN/m^2)	Theoretical specific energy (kJ/kg)	(Wh/kg)
Cast iron	7150	150	19	5
High strength steel	8000	1900	237	66
Glass reinforced plastic (unidirectional)	2500	1300	520	144
Kevlar reinforced plastic (unidirectional)	1500	1200	860	240

$$E = \frac{1}{2} I\omega^2 = \frac{1}{2} mr^2\omega^2 = \frac{1}{2}\rho V r^2\omega^2 \quad [\mathrm{J}] \tag{11.3}$$

where V is the volume of the disk and ρ is the density of the material.

Dividing Equation 11.3 by the mass gives the energy density as

$$\frac{E}{m} = \frac{1}{2} r^2\omega^2 \quad [\mathrm{J/kg}]$$

and dividing Equation 11.3 by volume gives the volume energy density as

$$\frac{E}{V} = \frac{1}{2}\rho r^2\omega^2 \quad [\mathrm{J/m^3}]$$

The speed of rotation of a flywheel is limited by the strength of the material to resist the circumferential stress. The maximum circumferential, or hoop, stress of a rotating ring is given by

$$\sigma_{max} = \rho r^2\omega^2 \quad [\mathrm{N/m^2}]$$

The maximum mass energy density is then

$$\left(\frac{E}{m}\right)_{max} = \frac{1}{2}\frac{\sigma_{max}}{\rho}$$

and the maximum volume energy density is

$$\left(\frac{E}{V}\right)_{max} = \frac{1}{2}\sigma_{max}$$

These equations, with the properties of materials shown in Table 11.3, show that a high-speed flywheel should use a composite material with high tensile strength and low density. However, it should be noted that the strength of composite materials varies according to the direction of the fibres, so the design of a composite flywheel is complex.

11.4.2 Flywheels to Stabilize the Power System

Figure 11.9 is a computer image of one of the large flywheel systems that are being added to increase the inertia of the electrical power system in GB, which has a

Figure 11.9 Computer image of a large utility flywheel unit.
Photo: Siemens Energy.
(A black and white version of this figure will appear in some formats. For the colour version, please refer to the plate section.)

high proportion of renewable energy generation. The need for these flywheels can be understood by recalling the box entitled 'Use of Energy Storage for Frequency Control of a National Electricity System': the supply and demand of the power system must always balance and, if these are not equal, the frequency changes at a rate that depends on the system inertia. When renewable generation increases, the conventional thermal generators must generate less power, and when their output drops too low, they must be disconnected. Each conventional fossil fuel turbine generator unit is a large mass spinning at 3000 rpm and adds significant inertia to the power system. In contrast, large wind turbines and PV generators are connected to the power system through static power electronic converters and do not add inertia to the system. Therefore, as the fraction of generation from renewable energy grows, the inertia of the power system decreases and the speed of the changes in frequency increases. This is sometimes described as the power system becoming 'lighter' and these flywheel systems are being installed to increase the system inertia as well as provide extra voltage support through control of the motor–generator.

The flywheel shown in Figure 11.9 stores 1100 MWs driven by a 3000 rpm, 15 kV synchronous machine rated at 125 MVA. A variable-speed pony motor is used to accelerate the flywheel and motor–generator up to synchronous speed when the unit is synchronized to the 50 Hz network. The motor–generator is connected through a transformer to the 66 kV network. The motor–generator is air cooled and can provide or absorb real and reactive power. The flywheel is air cooled and is in a reduced pressure to limit windage losses. There is a water-cooling system to transfer the heat from the unit to outside the building. The motor–generator can provide short-circuit current of 10 times its rating, but for short-circuit faults on the network this is reduced by the impedance of the transformer.

Example 11.4 – Power Injected by a Grid-Connected Flywheel

The flywheel illustrated in Figure 11.9 rotates at 3000 rpm and is connected by a synchronous motor–generator to a power system that normally operates at 50 Hz. At this speed it stores 1100 MWs (MJ) of energy. What is the energy injected from the flywheel? What is the power that will be supplied if the power system frequency drops to 48 Hz over three seconds?

Solution

The speed of rotation of the flywheel is

$$\omega = \frac{N \times 2\pi}{60} = \frac{3000 \times 2\pi}{60} = 314.16 \text{ rad/s}$$

The inertia of the flywheel, from $E = \frac{1}{2} I \omega^2$, is

$$I = \frac{2E}{\omega^2} = \frac{2 \times 1100 \times 10^6}{314.16^2} = 22\ 291 \text{ kgm}$$

The energy injected is

$$E_{Inj} = \frac{1}{2} I \omega^2 = \frac{1}{2} \times 22.29 \times 10^3 \times 314.16^2 \left(1 - \frac{48}{50}\right) = 44 \text{ MJ}$$

The power injected over 3 seconds $= \frac{44}{3} = 14.67$ MW.

11.5 Batteries

Batteries store chemical energy and deliver it as electricity. Primary batteries use the chemical energy in the materials used during manufacture and cannot be recharged. Secondary batteries take electrical energy and store it in chemical form for later reconversion to electricity. For many years batteries have ensured a secure supply of electrical power to high-value loads such as telecommunications and medical equipment. They have been used with solar photovoltaic modules to create small systems that provide limited amounts of power where there is no mains electricity. In practical applications, batteries have losses of up to ~15% in round-trip efficiency and are too expensive for widespread use in bulk utility electricity supply. However, their price is dropping rapidly with the development of new battery technologies and the increasing volumes of production of batteries for electric vehicles. Batteries are beginning to be used in power systems with a large fraction of electricity generated from variable renewable energy to provide energy storage and an injection of power for a few hours.

Batteries use direct current (dc) electricity, so a power electronic converter is needed either to change the dc voltage to the level needed by the load or to connect the battery to an alternating current (ac) supply. A battery consists of cells that are connected into modules, which are then arranged to form the battery pack. Each cell develops only several volts, and electric vehicles often use lithium-ion cells rated at only a few amps, so a lithium-ion vehicle battery may consist of several thousand cells. However, cells of other chemistries with large-area electrodes can deliver several hundred amps.

Example 11.5 – Connection of Cells to Form a Battery

Each cell of a battery has a useable capacity of 150 Ah at a constant voltage of 2 V. How many cells are required to store 90 kWh? Sketch the arrangement of the cells if it is to provide a voltage of 100 V. If the battery is discharged over 10 hours, what current would flow in the connecting circuit?

Solution

The energy capacity of each cell is:

$$E = 150 \times 2 = 300 \text{ Wh}$$

To store 90 kWh requires:

$$\frac{90\,000}{300} = 300 \text{ cells}$$

To supply 100 V, each series string must have

$$\frac{100}{2} = 50 \text{ cells connected in series}$$

Each series string would store

$$300 \times 50 = 15 \text{ kWh}$$

To store 90 kWh would require

$$\frac{90}{15} = 6 \text{ strings in parallel.}$$

Discharged over 10 hours the current would be

$$I = \frac{90\,000}{10 \times 100} = 90 \text{ A}$$

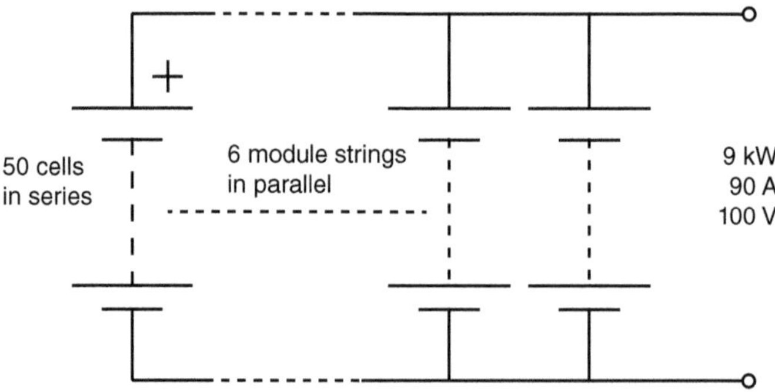

Figure 11.10 Arrangement of cells to form a battery.

Table 11.4 shows the battery technologies that have been used for storing renewable energy, while Figure 11.11 shows the energy densities of batteries using these chemistries. At present, most stationary batteries for storing renewable energy are based on lithium-ion chemistry with either lithium iron phosphate (LFP) or lithium manganese cobalt (LMC) as the material of the positive electrode. For many

Table 11.4 Typical performance of batteries that have been used to store renewable energy

Technology	Power capacity	Energy capacity	Round-trip efficiency (%)	Lifetime	Nominal voltage of one cell (V)
Lead–acid	1 kW–10 MW	Up to 10 MWh	75–90	3–15 years, 500–3000 cycles	2.1
Nickel–cadmium	1 kW–50 MW	Up to 15 MWh	60–65	10–20 years, 2000–3500 cycles	1.2
Sodium–sulphur	50 kW–30 MW	300 kWh–50 MWh	80–90	10–15 years, 2500–4500 cycles	2.0
Lithium-ion	1 kW–300 MW	Up to 450 MWh	80–90	5–15 years, 1000–10 000 cycles	3.2–3.7

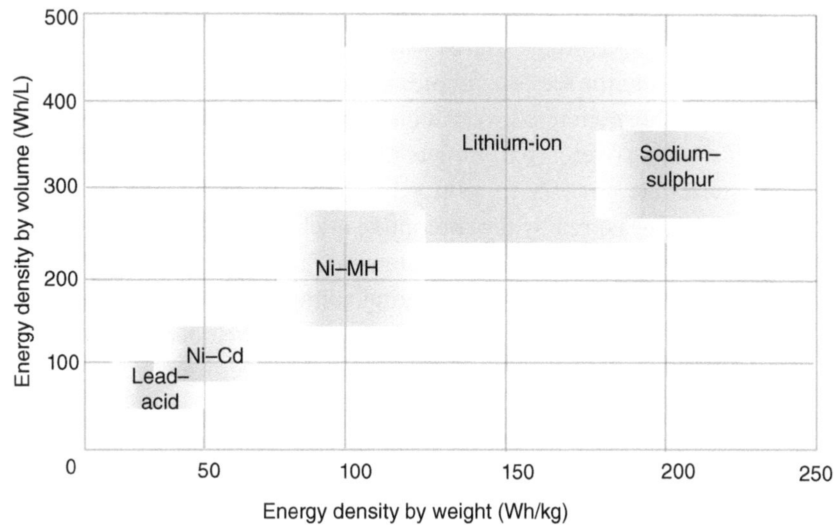

Figure 11.11 Energy density of batteries used to store renewable energy.

applications, lead–acid and nickel–cadmium batteries are being superseded by lithium-ion. Several demonstration projects have used sodium–sulphur batteries but these operate at 300–350 °C. In the past electric vehicles have used nickel metal hydride cells (NiMH) but this chemistry is now being superseded by lithium-ion.

Figure 11.12 shows a simple model of a battery. It consists of the open-circuit voltage (U) in series with an equivalent series resistance (R). In this simple model, the open-circuit voltage is assumed to be constant. The equivalent series resistance is a simple representation of the complex internal structure and chemistry of the battery. More sophisticated equivalent-circuit models vary the open-circuit voltage with the state of charge and the temperature of the battery, and use a circuit of multiple resistors and capacitors to represent dynamic characteristics.

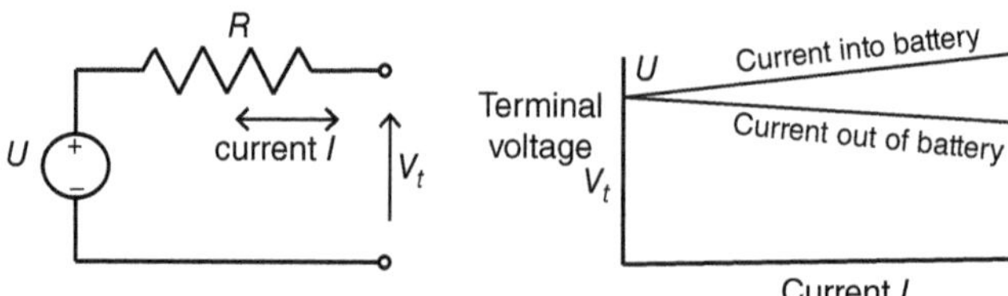

Figure 11.12 Simple equivalent-circuit model of a battery cell, and the variation of terminal voltage with current.

The model illustrates several important characteristics of a battery. When current is drawn from the battery during discharge, the terminal voltage (V_t) drops, whereas to charge the battery, the terminal voltage must be higher than the open-circuit voltage.

The battery is represented by an ideal voltage source (U) and the equivalent series resistance (R). A voltage source will always try to supply its voltage irrespective of the current, and the series resistance is small. Therefore if by accident the battery terminals are short-circuited, very high currents will flow. Thus it is important to provide fast-acting overcurrent protection, such as a fuse, close to the battery terminals.

Batteries are not able to return the entire charge that is put into them, and this efficiency is measured by the coulombic or charge (in)efficiency. The coulombic efficiency is calculated from Ah measurements at a defined rate of charge and discharge to a final voltage. Also, as the charging voltage is higher than the discharge voltage, due to the internal resistance of the battery, and the open-circuit voltage varies with state of charge, there is an additional voltage (in)efficiency. The overall round-trip efficiency of the battery is the product of the voltage and coulombic efficiencies.

All batteries lose their charge if left charged but unused. The rate at which charge is lost is known as the self-discharge of the battery and depends on its construction and chemistry as well as the temperature. If a large battery bank is used with a standalone PV system to provide energy storage for many days of low solar insolation, then careful attention must be paid to its self-discharge as this can use a significant fraction of the output of the photovoltaic array.

Example 11.6 – Performance of a Battery

Each cell of the battery shown in Figure 11.10 has an open-circuit voltage of 2 V and an equivalent series resistance of 5 mΩ. Calculate the terminal voltage when the battery is being charged and discharged with a current of 90 A. Calculate the current if the battery is short-circuited.

Solution

As there are six strings in parallel, one string of the battery passes $90/6 = 15\text{A}$ either charging or discharging (Figure 11.13).

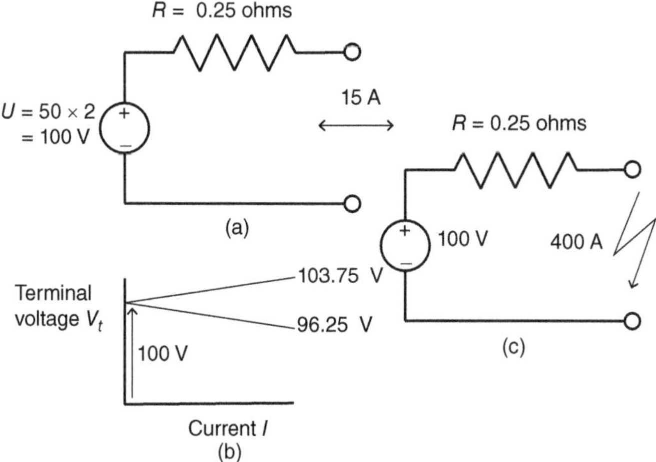

Figure 11.13 (a) Model of one module. (b) Terminal voltage of modules. (c) Short-circuit current of one module.

The open-circuit voltage of a string is $OCV = 50 \times 2 = 100$ V.

The equivalent series resistance of a string is $R = 50 \times \frac{5}{1000} = 0.25\ \Omega$.

Therefore, with a current of 15 A, the terminal voltage is

$$V = 100 \pm 15 \times 0.25 = 103.75 \text{ V (charging) or } 96.25 \text{ V (discharging)}$$

If the terminals are short-circuited, the short-circuit current from one string is

$$I_{SC} = \frac{OCV_s}{R_s} = \frac{100}{0.25} = 400 \text{ A}$$

The battery has six modules connected in parallel, so the short-circuit current of the battery is 2400 A.

11.5.1 Battery Electrochemistry

Figure 11.14 shows the principle of operation of one cell of an electrochemical battery. Each cell of the battery consists of a positive and negative electrode, electrolyte and a separator. The electrochemistry of the battery creates a potential difference between the two electrodes.

When the electrodes of a charged battery are connected by an external circuit, current flows from the positive to the negative electrode and the battery discharges energy into the load. The flow of current is caused by electrons, which are negatively charged and so flow in the opposite direction to the current. When the battery is being discharged, the direction of the current in the external circuit is from the positive to the negative electrode, and the flow of electrons is from the negative to the positive electrode. The electrons flowing away from the negative electrode in the

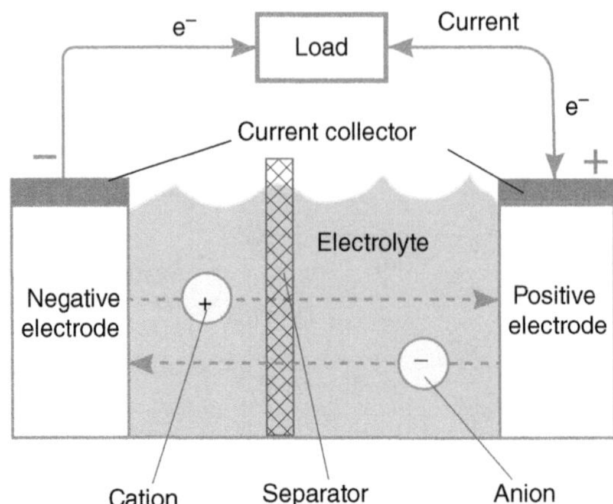

Figure 11.14 Electrochemical battery during discharge. (The direction of the movement of electrons, current and ions is reversed when the battery is being charged.)

external circuit cause positively charged ions (known as cations) to flow from the negative to the positive electrode through the electrolyte. At the positive terminal, the charge on the cations is balanced by the electrons flowing into the positive electrode from the external circuit. With some battery chemistries, negatively charged ions (known as anions) are formed at the positive electrode and flow towards the negative electrode.

To charge the battery, the load is replaced by a battery charger that applies a voltage higher than the potential of the electrodes of the battery. When the battery is being charged, current flows in the external circuit from the negative to the positive electrode and electrons flow in the external circuit from the positive to the negative electrode. The positively charged cations in the electrolyte then flow through the electrolyte towards the negative electrode.

The polarity of the voltages of the electrodes remains unchanged whether the battery is being charged or discharged, but the direction of the current flowing in the external circuit and the ions flowing in the electrolyte are reversed.

In an electrochemical battery, the material of the negative electrode is often a metal and that of the positive electrode a metallic oxide, e.g. in a lead–acid cell the negative electrode is lead and the positive electrode is lead dioxide. The electrolyte provides a path for the ions to flow between the electrodes and is often an acid or salt in a solvent. Aqueous sulphuric acid is used as the electrolyte in a lead–acid cell. The electrolyte and separator inhibit the flow of electrons between the electrodes and limit self-discharge of the cell. The separator ensures there is no electrical short-circuit between the electrodes while allowing the flow of ions through the electrolyte.

Terminology of Batteries

The anode is defined as the terminal at which current enters the battery. During discharge, current flows into the negative terminal, which acts as the anode, while during charging, current flows into the positive terminal, which then becomes the anode. The cathode is the other terminal at which current exits the battery.

This traditional scientific definition, which identifies a battery terminal as the anode or cathode depending on whether it is being charged or discharged, is confusing. Therefore many documents describing the engineering applications of batteries always define the anode of a battery during discharge, i.e. the negative terminal is always referred to as the anode irrespective of whether the battery is charging or discharging. This possible confusion can be avoided by describing the electrodes as positive and negative (which does not change depending on whether the battery is charging or discharging) rather than as anode and cathode, which reverses.

In an electrochemical battery, oxidation, or the loss of electrons from the electrode material to the external circuit, occurs at the anode. Reduction, or the addition of electrons to the electrode material from the external circuit, occurs at the cathode. This leads to the name redox battery.

An ion is an atom or molecule with a net electric charge caused by the loss or gain of one or more electrons. An anion has a negative charge and always moves towards the anode. A cation has a positive charge and always moves towards the cathode.

The rate of discharge of a battery is described by C-rate or the number of hours taken to fully discharge the battery. The 1C rate will discharge the battery in 1 hour while the C/10 rate (sometimes written 0.1C) will discharge the battery over 10 hours. For example, the C/10 rate of a 120 Ah battery is 12 A. The capacity of a battery is not constant but depends on the rate at which it is discharged. At high rates of discharge only active material near the surface of the electrodes is effective and so the capacity is reduced. Figure 11.15 shows how the capacity of a lead–acid battery varies with rate of discharge.

The performance of batteries is often specified at the 10- or 20-hour rate (C/10 or C/20) but the rate of discharge and hence the capacity depends on the application. Batteries that are grid connected and used to inject power for 1 or 2 hours at times of high load and low solar radiation will operate at a 1C or 0.5C rate. Standalone photovoltaic systems will operate with five days of no-sun storage, implying operation at a 120-hour rate (C/120), or even at a 300-hour rate (C/300).

11.5.2 Lead–Acid Batteries

Lead–acid batteries have been common for small, isolated dc renewable energy systems and continue to be used where cost is a major concern. They are widely

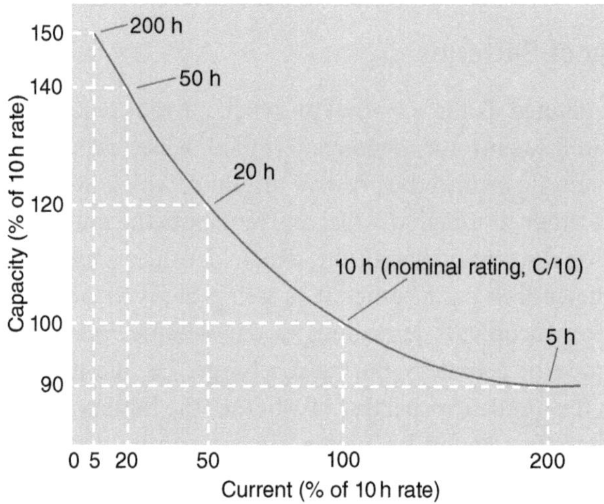

Figure 11.15 Variation of available capacity of a lead–acid battery with rate of discharge.
Source of data: M.S. Imamura, P. Helm and W. Palz, *Photovoltaic System Technology*, 1992. Published by H.S. Stephens and Associates on behalf of the European Commission.

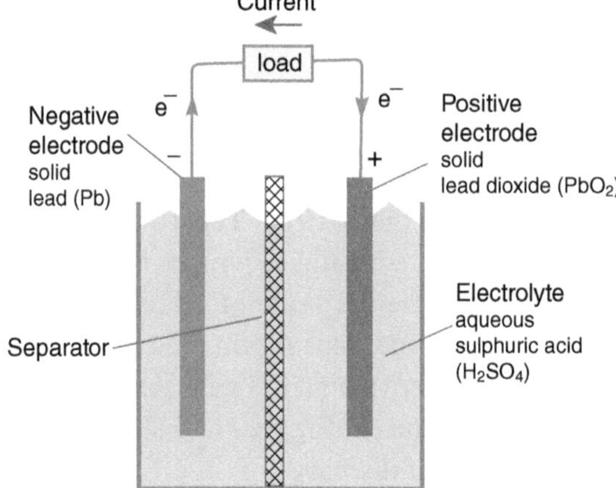

Figure 11.16 Cell of a lead–acid battery during discharge.

available and are relatively cheap. There are well-established routes for recycling lead–acid batteries at the end of their lives and this results in much lower raw material requirements than for some modern batteries. Several large (up to 10 MW), grid-scale energy storage schemes have been developed using lead–acid batteries, but lithium-ion chemistry is now preferred in new large stationary installations.

Figure 11.16 shows a schematic diagram of a cell of a lead–acid battery. A lead–acid battery cell has a negative electrode of porous, metallic lead (Pb) and a positive

Figure 11.17 Operation of a lead–acid battery cell.

electrode of lead dioxide (PbO_2); both are immersed in an electrolyte of aqueous sulphuric acid (H_2SO_4 and H_2O). During discharge, lead sulphate ($PbSO_4$) is formed at both electrodes and the electrolyte is diluted by water formed at the positive electrode. During charging, the electrodes and electrolyte are restored to their original state.

The overall operation of a lead–acid battery cell is described in Figure 11.17. Lead–acid batteries are large, heavy and appear to be robust. However, they are very sensitive to temperature and method of operation. They can easily be damaged by excessive discharging, and the electrolyte can be boiled off by overcharging. Experience shows that the battery is usually the most sensitive piece of equipment in an autonomous PV system.

The battery in a solar PV system must withstand a daily cycle of charge and discharge and so the battery for this application should be suitable for this and sized to have a large enough capacity to limit the depth of discharge. Lead–acid batteries suitable for deep cycling have electrode plates to which antimony is added. They can withstand up to 1000 cycles to 80% depth of discharge (20% residual charge) and up to 3500 cycles to 20% depth of discharge, i.e. 10 years of daily charge/discharge. However, adding antimony causes a higher rate of self-discharge, which increases at high temperatures. The battery is sized to supply electricity if there are several consecutive days with little sun; this is known as the days of no-sun storage. The high rates of self-discharge of deep-cycle lead–acid batteries mean that if the size of the battery is increased to give many days of no-sun storage, too much of the output of the PV array may be taken only to supply its self-discharge current.

If a lead–acid battery is charged too rapidly or a fully charged battery continues to be charged, hydrogen gas is evolved at the negative electrode and oxygen at the positive electrode. This leads to a loss of electrolyte. Flooded lead–acid batteries for autonomous photovoltaic systems have an increased capacity of electrolyte, to withstand some loss due to gassing. Valve-regulated lead–acid batteries are sealed and recombine the hydrogen and oxygen that is evolved during charging to form water and so conserve the electrolyte. An alternative is to use electrolyte in the form of a gel. Both gelled and valve-regulated lead–acid batteries require careful control of charging for a long life.

Lead–acid batteries of the type used in vehicles, known as starting, lighting and ignition (SLI) batteries, are not suitable for photovoltaic systems as their design is optimized to provide a high current to the vehicle starter motor to crank the engine and then the battery is recharged rapidly by the vehicle generator. SLI batteries have thin plates strengthened with calcium and can tolerate depths of discharge of around only 20% (80% residual charge) before they experience a significant loss of life. If they are used in an autonomous PV system and experience a 20% daily depth of discharge, an SLI battery will have a life of around 500 cycles or less than two years. In high temperatures their life may be reduced further. SLI batteries can sustain deep discharges to 80–90% depth of discharge on only around 10 occasions before their life is reduced.

In small autonomous solar PV systems, it is common to use the terminal voltage of a lead–acid battery as an approximate indication of its state of charge. If the terminal voltage of the battery drops too low, indicating the battery is discharged, the load is disconnected, while if it rises too high the PV array is disconnected to prevent overcharging.

11.5.3 Nickel–Cadmium Batteries

Nickel–cadmium (Ni–Cd) batteries are sometimes used for autonomous photovoltaic systems that power high-value loads, such as radio repeater stations. They have a number of attractive features:

- they can be overcharged without damage,
- they can be fully discharged and so there is no need for oversizing,
- they maintain a uniform voltage during discharge,
- they have good cycle life,
- they can be operated over a wide temperature range.

However, Ni–Cd batteries are significantly more expensive than lead–acid batteries, cadmium is highly toxic and disposal of the battery at the end of its life can be difficult. In many countries, the use of Ni–Cd batteries is restricted by environmental regulations, and they have been replaced by nickel–metal hydride (Ni–MH) or lithium-ion batteries.

11.5.4 Sodium–Sulphur Batteries

A sodium–sulphur battery has electrodes of molten sodium and sulphur and a solid electrolyte. It operates at a temperature of 300–350 °C and requires heating. A sodium–sulphur battery has high energy density, a very low rate of self-discharge and a life of 4500 cycles (or 15 years' daily cycling) at 90% depth of discharge (10% residual charge). Both sodium and sulphur are low-cost elements that are widely distributed throughout the world, and the use of liquid electrodes improves the reliability and extends the life of the battery. The round-trip efficiency of the charge–discharge cycle is around 80–85% but heating may take 10% of the energy rating of the battery a day when the battery is not in use. Molten sulphur is highly corrosive, and accidental contact between liquid sulphur and liquid sodium can be dangerous. A sodium–sulphur battery is typically constructed of 50 kW modules with a capacity of 375 kWh made from 320 cells. There is a 34 MW, 245 MWh sodium–sulphur battery associated with a wind farm in Japan.

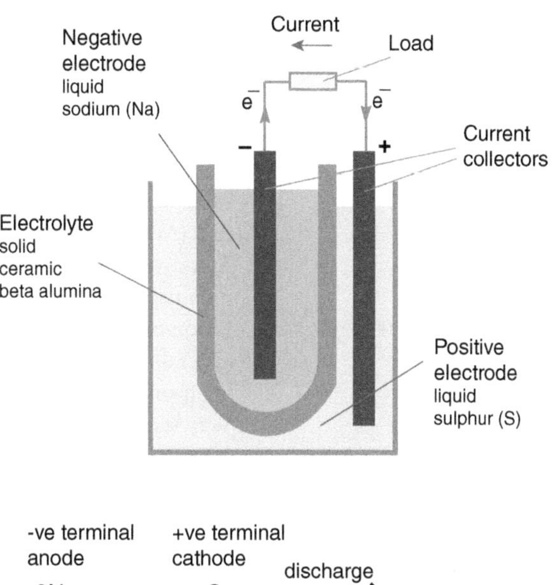

Figure 11.18 Sodium–sulphur battery during discharge.

-ve terminal anode | +ve terminal cathode

$$2Na + xS \underset{\text{charge}}{\overset{\text{discharge}}{\rightleftharpoons}} Na_2S_x$$

$3.3 \leq x \leq 5$ | Na_2S_x: +ve terminal anode

Figure 11.19 Operation of a sodium–sulphur battery cell.

Figure 11.18 shows the operation of a sodium–sulphur battery during discharge. The negative electrode is molten liquid sodium, and the positive electrode is molten liquid sulphur. The electrolyte is solid beta-alumina ceramic that allows sodium ions to flow through it. The liquid electrodes require current collectors to make the connections to the external circuit. During discharge, electrons flow in the external circuit from the negative electrode through the load to the positive electrode, and sodium cations (Na^+) from the negative electrode pass through the solid electrolyte to form sodium polysulphide (Na_2S_x) at the positive electrode.

The overall operation of a sodium–sulphur battery cell is described by Figure 11.19.

11.5.5 Lithium-Ion Batteries

Lithium-ion batteries are now the dominant battery technology and are almost always used in electric vehicles. They are being manufactured in very large volumes and the consequent cost reduction together with their high energy densities (see Figure 11.11) have resulted in lithium-ion batteries also becoming widely used for the storage of electricity produced from renewable energy.

Table 11.5 lists some of the advantages and disadvantages of lithium-ion batteries. Lithium is a highly reactive element, and this leads to high energy densities but creates the need for a battery management system to ensure safe operation.

Table 11.5 Advantages and disadvantages of lithium-ion batteries

Advantages	Disadvantages
High volumetric and gravimetric energy density	The raw material supply chains are not yet well developed and need careful management.
High cell voltage	Chemicals are highly reactive and, unless managed effectively, a battery can become a safety hazard. An advanced battery management system is often required.
Long cycle life (typically 6000 cycles to 80% Depth of Discharge)	Li-ion batteries suffer from ageing over time (calendar ageing).
High efficiency	High cost
Low self-discharge	Very sensitive to overcharging

A battery management system (BMS) typically monitors voltage, current, state of charge, state of health and temperature to protect against unsafe or damaging operation. The sophistication and expense of the BMS depends on the size of the battery and the consequences of any maloperation. Fire-suppression systems are required for large batteries. Lithium-ion batteries have a long cycle life but suffer from degradation over time. They are very sensitive to overcharging, which must be avoided, and their operation and rate of degradation are sensitive to temperature, leading to a need for an active temperature management system in many applications. The supply chains of some material used in lithium-ion batteries, particularly cobalt, are not well developed and require careful management to avoid environmental damage, exploitation of child labour and other poor working practices.

Lead–acid, nickel–cadmium and sodium–sulphur batteries (redox batteries) rely on chemical reactions that oxidize or reduce the material of the electrodes as electrons pass through the external circuit. Lithium-ion batteries work rather differently and rely on the movement of lithium ions between two electrodes that have an open crystal structure into or from which lithium is inserted or removed. This process is known as intercalation and the electrodes are known as insertion electrodes.

Figure 11.20 shows a lithium-ion cell during discharge. The negative electrode is usually graphite, which accepts lithium atoms between its layers of carbon atoms. There is sufficient space between the layers of graphite for the lithium atoms to move freely. A number of different materials are used for the positive electrode, usually some form of lithium metal oxide.

During discharge the negative electrode emits electrons into the external circuit and lithium ions into the electrolyte. These ions pass through the electrolyte and separator to the positive electrode where they combine with electrons from the external circuit to form lithium that is intercalated into the electrode. There is no free lithium metal in the cell. When the battery is being charged, the flow of electrons and ions is reversed.

Table 11.6 Materials used for positive electrode

Material of positive electrode	Acronym	Common application
Lithium cobalt oxide	LCO	Consumer electronics
Lithium manganese oxide	LMO	Electric vehicles, consumer electronics
Lithium iron phosphate	LFP	Grid energy storage, electric vehicles, domestic batteries, consumer electronics
Lithium nickel manganese cobalt oxide	NMC	Electric vehicles, grid energy storage

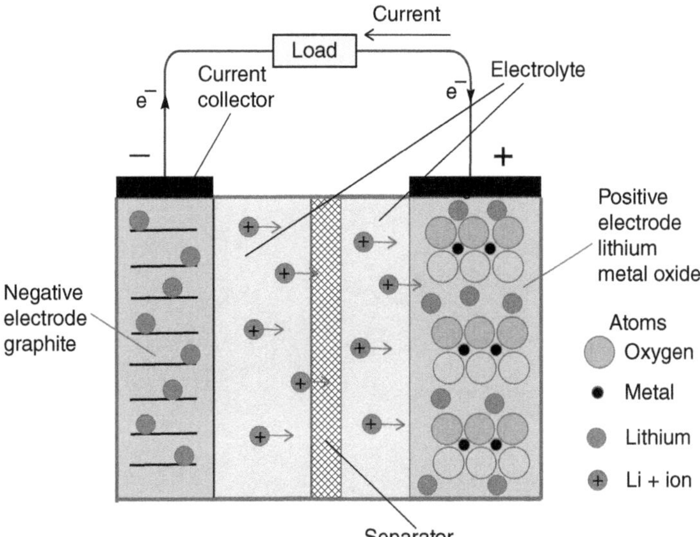

Figure 11.20 Lithium-ion battery during discharge.

Table 11.6 shows some of the materials currently used for the positive terminal of lithium-ion batteries. The material of the positive electrode determines the characteristic of the battery, including its open-circuit voltage. The electrolyte is often a lithium salt in a non-aqueous solvent and the separator a fibrous membrane. At the time of writing, the most common technologies for stationary batteries for the storage of renewable energy are LFP and NMC. LFP cells tend to be cheaper but with lower energy density than NMC cells hence requiring more space for a given rating.

Figure 11.21 is an illustration of a containerized lithium-ion battery and inverter installed at a 2 MW solar farm. The battery is rated at 500 kWh and the inverter is rated at 250 kVA and connected to the utility three-phase 400 V network. The battery consists of a number of modules placed in racks in the battery room. The battery has a sophisticated BMS and active temperature management. The high-energy density of the lithium-ion cells requires careful assessment of the safety of the battery and a fire suppression system.

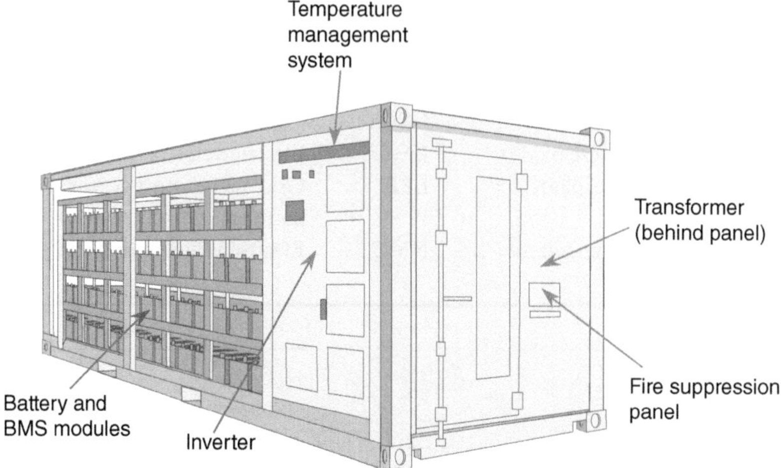

Figure 11.21 Utility lithium-ion battery and inverter.

Figure 11.22 20 MW utility connected lithium-ion battery.
Photo: RES.
(A black and white version of this figure will appear in some formats. For the colour version, please refer to the plate section.)

The battery/inverter system can be used for a number of purposes including (1) to take energy from the solar farm, store it and export it to the grid at times of high electricity price, (2) to take electricity from the grid at times of low price, store it and return it to the grid at times of high price, (3) to provide rapid injections of power in the event of a sudden shortage of power and drop in grid frequency.

Figure 11.22 shows 6 Lithium-ion battery blocks and 11 inverter units forming a 20 MW battery/inverter to provide balancing services to the GB power system.

11.5.6 Flow Batteries

Flow batteries work in a similar way to other batteries with ions being exchanged between electrodes as electrons circulate in an external circuit creating a current. However, in a flow battery, liquid electrolytes are stored in tanks and pumped into the cells when the battery is operated. This allows the power and energy capacities of the battery to be designed independently. The maximum current and hence power that is available from a flow battery is determined by the area of the electrodes, while the energy capacity depends on the volume of the electrolytes stored in the tanks. Figure 11.23 is a diagram of an experimental redox flow battery that used an electrolyte of sodium polysulphide as the negative electrode and sodium bromide as the positive electrode. A cation selective membrane separated the two electrolytes to avoid direct chemical reaction. During discharge, sodium cations (Na^+) flowed from the negative electrode through the ion selective membrane as electrons circulated in the external circuit. This design was constructed but did not enter commercial service.

Advantages claimed for flow batteries include high energy ratings and long life. Disadvantages can include the use of large volumes of sometimes toxic chemicals; relatively low energy density; and complex system requirements (including sensors, pumps, flow and battery management systems). Flow batteries have been investigated for many years and no preferred chemistry has yet emerged. Some designs are currently in demonstration and early commercialization, but flow batteries are not in common commercial use for storing renewable energy.

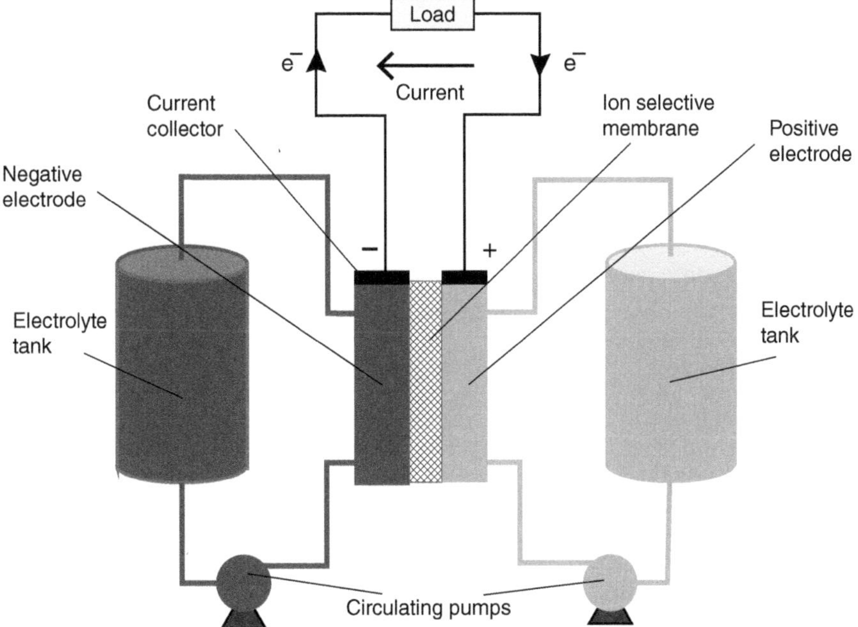

Figure 11.23 Redox flow battery during discharge.

11.5.7 Lifetime of Batteries

The intercalation process of a lithium-ion battery does not cause fundamental chemical changes in the material of the electrodes. This is in contrast to the redox process of electrochemical batteries; lithium-ion batteries have a much longer cycle life than electrochemical batteries.

Figure 11.24 shows a comparison of the cycles to failure (CF) for different depths of discharge (DoD) of typical lithium-ion and lead–acid batteries. These equations were derived from experimental test results of several batteries. Note that the cycle life of a battery depends on cell construction, temperature management, rates of charge and discharge, etc., and the constants of these formulas apply to these example batteries only.

$$CF_{Lithium\text{-}ion} = e^{\left(7.27 - \left[\frac{\ln(DoD)}{0.685}\right]\right)}$$

$$CF_{Lead-acid} = e^{\left(3.826 - \left[\frac{\ln(DoD)}{0.767}\right]\right)}$$

The models show that the lithium-ion battery is capable of operating for 15 000 cycles with a 20% depth of discharge, while the lead–acid battery can operate for fewer than 400 cycles at this depth of discharge. With the daily cycling of a solar system this corresponds to just over one year of operation.

The reduction to the life of a battery from cycling at different depths of discharge can be evaluated simply using Miner's rule. Miner's rule applied to battery cycle degradation states that the damage caused by the cycles at one depth of discharge can be added to the damage caused by cycles at other depths of discharge. Miner's

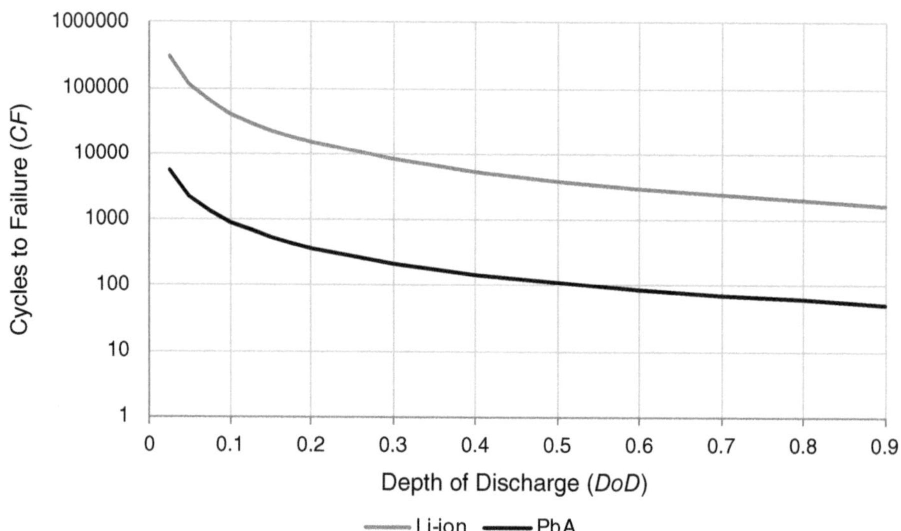

Figure 11.24 Cycles to failure versus depth of discharge of example lead–acid and lithium-ion batteries.

rule considers only the number of cycles at a particular depth of discharge and ignores the order in which the cycles occur.

$$D^{cycle} = \sum_{i=1}^{M} \frac{N_i}{CF_i}$$

where

D^{cycle} is the damage due to cycling,
M is the number of levels of discharge, e.g. 0–10%, 10–20%, etc.,
CF_i is the Cycles to Failure at DoD_i,
N_i is number of cycles that occurred at a depth of discharge i.

The damage (D) ranges from 0 when the battery is new to 1 when the battery is worn out.

For the lithium-ion battery data of Figure 11.24 this gives

$$D^{cycle} = \sum_{i=1}^{M} \frac{N_i}{e^{\left(7.27 - \left[\frac{\ln(DoD)}{0.685}\right]\right)}}$$

Lithium-ion batteries also experience degradation over time, known as calendar ageing. A common model is that batteries age linearly over time, and the battery reaches its end-of-life at a percentage of its energy capacity. Damage due to calendar ageing can be estimated as:

$$D^{age} = \frac{\text{Years of use}}{\text{Years of life of battery}}$$

where

D^{age} is the damage due to calendar ageing.

Adding the damage due to calendar ageing to the damage caused by cycling gives the approximate total damage to the battery:

$$D^{total} \cong D^{age} + D^{cycle}$$

The end of life of a battery is normally taken to be when its capacity is reduced to 70–80% of its original energy capacity, although this reduction is sometimes increased to 50% for stationary applications if the batteries are reused from another application.

The capacity of a lithium-ion battery reduces, and the internal resistance increases, as the damage increases over time. If the capacity at the end of life of the battery is assumed to be 80% and the series resistance to be 120% of their initial values, then the capacity and series resistance at any time can be estimated from:

$$\text{Capacity} = 100 - \left(20 \times D^{total}\right) \quad [\%]$$

$$\text{Series resistance} = 100 + \left(20 \times D^{total}\right) \quad [\%]$$

Example 11.7 – Cycle Degradation of a Battery

A battery stores solar energy during the day and releases it at night. The photovoltaic system and the solar resource cause the battery to experience equal numbers of daily cycles with depths of discharge of 10%, 15%, 20%. Compare the damage that a lead–acid and a lithium-ion battery (with the characteristics of Figure 11.24) will experience over three years due to cycling. If the lithium-ion battery has a life of 10 years, what is the capacity of the battery after three years?

Solution

Apply Miner's rule to sum the damage that occurs at these three depths of discharge over three years.

	Lithium-ion battery	Lead–acid battery
Cycles to Failure @ 10% DoD	41 417	923
Damage from 365 cycles @ 10% DoD	0.009	0.395
Cycles to Failure @ 15% DoD	22 914	544
Damage from 365 cycles @ 15% DoD	0.16	0.671
Cycles to Failure @ 20% DoD	15 056	374
Damage from 365 cycles @ 20% DoD	0.24	0.976
Total cycle damage over 3 years	0.049	2.042
Cycle lifetime (years)	61.22	1.47

This shows that a lead–acid battery used for this application would be worn out due to cycling within 1½ years. A lithium-ion battery would not fail due to cycling for 61 years and its life would be limited by calendar ageing.

After three years the capacity of the lithium-ion battery would be

$$\text{Capacity} = 100 - 20 \times \left(\frac{3}{10} + 0.049 \right) = 93\%$$

11.6 Storage of Heat

Fossil fuels store energy cost-effectively for long periods of time, with little degradation and few losses. Coal is stored in stockpiles in the open air, gas is stored in high-pressure pipes and bullets and oil is stored in tanks. These fuels are burnt to create high-temperature heat for industrial processes and low-temperature heat for space and water heating. Many renewable energy technologies do not naturally store heat and, as the use of fossil fuels decreases to limit the emissions of greenhouse gases, finding cost-effective ways to store heat is becoming increasingly important.

A good heat store should have the following attributes. It should

(a) store a high density of heat energy (with respect to its volume and mass),
(b) be easy to get heat in and out of the store, and the heat should be well distributed internally,

(c) have a high level of external insulation giving a good round-trip efficiency,
(d) have a long service life and be low cost with limited environmental impact.

11.6.1 Sensible and Latent Heat Storage

The two main mechanisms for storing heat from renewable energy systems are sensible and latent heat storage. Sensible heat storage is based on the change of temperature of a material without changing its phase. The material may be in a solid, liquid or gas phase. Sensible heat storage is described by

$$Q_{sensible} = mc\Delta\theta \quad [\mathrm{J}]$$

where

m is the mass in kg,
c is the sensible specific heat in J/kgK,
$\Delta\theta$ is the change in temperature in K.

Latent heat storage uses the energy associated with a change of phase, which is typically melting or evaporation. Latent heat storage is described by

$$Q_{latent} = lm \quad [\mathrm{J}]$$

where

m is the mass in kg,
l is the latent heat of fusion or evaporation in J/kg.

Figure 11.25 shows the operation of sensible and latent heat energy storage. This simple example shows the ice/water packs that are used to maintain a low temperature within an insulated box. The icepacks are cooled to a low temperature, around −20 °C, in a freezer and placed inside the insulated box, usually

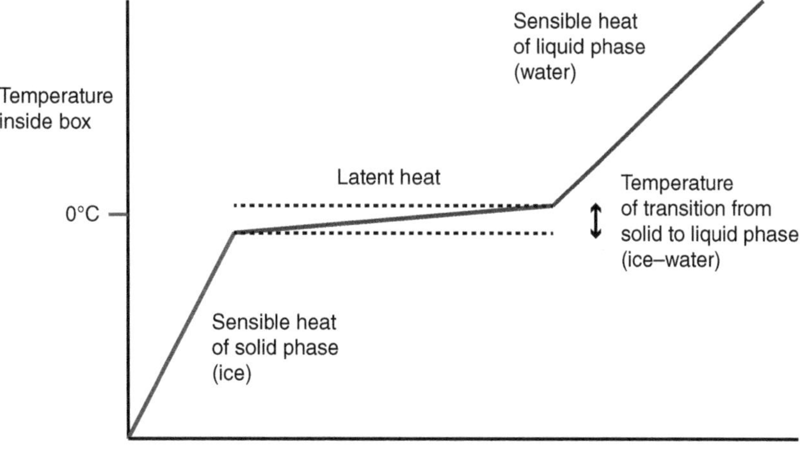

Figure 11.25 Example of sensible and latent heat storage.

around its sides. The thermal insulation of the box is not perfect, and the box absorbs heat from its environment, increasing the temperature inside the box. Initially the icepacks remain frozen, but they reduce the speed of the temperature increase inside the box by absorbing the heat that leaks into the box as sensible heat energy. When the temperature of the icepacks reaches melting point, the ice starts to melt, absorbing latent heat energy. The temperature of the box is then maintained at around 0 °C. Finally, when the ice is fully melted, the temperature inside the box starts to increase as the water continues to absorb sensible heat. There may be a small variation in temperature around 0 °C if the icepacks do not melt uniformly.

Example 11.8 – Performance of Icepacks

Icepacks containing 2 litres of water are used to control the temperature inside an insulated box that is used to transport vaccines in a warm climate. Assume the icepacks are frozen at a temperature of −18 °C and the insulated box absorbs heat energy from its environment at 15 W. The vaccines will be damaged by a temperature greater than 5 °C. How long can the vaccines be stored in the box?

Data:
Mass of water, $m = 2$ kg
Sensible heat of ice, $c_{ice} = 2108$ J/kgK
Sensible heat of water, $c_{water} = 4200$ J/kgK
Latent heat of fusion of water, $l_{water-ice} = 334$ kJ/kg

Solution

Assume that the three stages of the heating process occur slowly, and the temperature is uniform throughout the box.

$$Q_{stored} = \int_{-18}^{0} mc_{ice}d\theta + lm + \int_{0}^{5} mc_{water}d\theta$$

The energy absorbed by raising the temperature of the icepacks from −18 °C to 0 °C is

$$Q_{sensible} = mc_{ice}\Delta\theta = 2 \times 2108 \times 18 = 75.9 \text{ kJ}$$

The energy absorbed by melting the icepacks at 0 °C is

$$Q_{latent} = ml = 2 \times 334 = 668 \text{ kJ}$$

The energy absorbed by raising the temperature of the icepacks from 0 °C to 5 °C is

$$Q_{sensible} = mc_{water}\Delta\theta = 2 \times 4200 \times 5 = 42 \text{ kJ}$$

The length of time that the temperature can be kept below 5 °C is therefore given by

$$\text{Time} = \frac{(75.9 + 668 + 42) \times 10^3}{15 \times 3600} = 14.6 \text{ h}$$

11.6.2 Storage of Heat in Buildings

In hot climates, many old buildings maintain a comfortable internal temperature in the middle of the day even with a high external temperature and bright sunlight. At night, these buildings retain some warmth even when the ambient temperature falls. This is because their thick stone walls have a large heat capacity and low thermal conductivity. This is the basis of the Trombe wall where a thick concrete wall is used to capture energy from solar radiation during the day, store it and transfer it to the building in the evening.

A simple model of the heat flows of a building is shown in Figure 6.10. This uses thermal resistance to create an electrical circuit analogue to represent the path of heat to/from the building. Temperature is taken to be analogous to voltage, and heat power to be analogous to current. This model is now extended to include a capacitor to represent the heat stored in a building (thermal mass) and its change of temperature over time.

In Figure 11.26:

T_{amb} is the external ambient temperature in K,
T_{int} is the internal temperature of the building in K,
A is the area of the fabric in m^2,
U is the U-value of the fabric in W/m^2K,
Q represents the solar gain, and the thermal energy from any internal heating or cooling, in W, where the reference temperature is 0 °C,
R represents the thermal resistance of the fabric

$$R = \frac{1}{UA} \quad \left[\frac{\mathrm{K}}{\mathrm{W}}\right]$$

C represents the thermal mass of the building,

$$C = \text{mass of material} \times \text{specific heat of material} \quad \left[\frac{\mathrm{J}}{\mathrm{K}}\right]$$

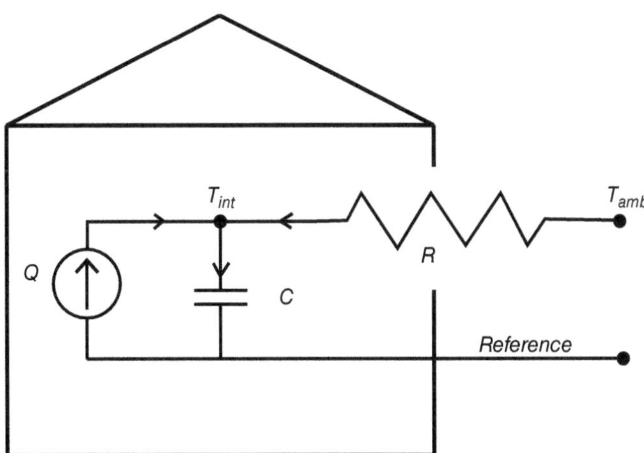

Figure 11.26 Simple dynamic thermal model of a building.

Considering the heat flows as analogous to currents in an electrical circuit and summing the heat flows at the T_{int} node, we have

$$Q+\frac{T_{amb}-T_{int}}{R}=C\frac{dT_{int}}{dt}$$

This first-order differential equation may be rearranged as

$$\frac{dT_{int}}{T_{\text{int}}-\left(T_{amb}+QR\right)}=\frac{dt}{-RC}$$

Integrating both sides, we have

$$\int\frac{dT_{int}}{T_{int}-\left(T_{amb}+QR\right)}=-\frac{1}{RC}\int dt$$

$$\ln\left(T_{int}-\left(T_{amb}+QR\right)\right)=-\frac{t}{RC}+K$$

where K is the integration constant.

Taking the exponential of both sides, we have

$$T_{int}-\left(T_{amb}+QR\right)=e^{\left[\frac{-t}{RC}+K\right]}=e^{K}e^{\left[\frac{-t}{RC}\right]}$$

with $K_1=e^K$. This equation may be rearranged as

$$T_{int}-\left(T_{amb}+QR\right)=K_1e^{-\frac{t}{\tau}}$$

where the time constant $\tau=RC$ and $T_{int}\left(t\right)$ is the internal temperature of the building at time t.

At $t=0$, with $T_{int}=T_{int,t=0}$, $K_1=T_{int,t=0}-\left(T_{amb}+QR\right)$.

Therefore the internal temperature of the building at time t is

$$T_{int}\left(t\right)=\left(T_{int,t=0}-\left(T_{amb}+QR\right)\right)e^{\frac{-t}{\tau}}+\left(T_{amb}+QR\right)$$

A large thermal mass of the building (C) with a high thermal resistance (R) of the fabric will lead to a slow rate of change of internal temperatures.

Example 11.9 – Time Constant of a Building

A building is represented by the model of Figure 11.26. The building has a mass of 25 tonnes of concrete, a total surface area of 600 m^2 and a composite U value of 1.5 W/m^{2}K. The specific heat of concrete is 880 J/kgK.

(a) Calculate the thermal time constant of the building.

(b) If the ambient temperature is 10 °C and the initial temperature inside the building is 12 °C, what would be the temperature of the building 2 hours after turning on a 6 kW heater? Assume that the ambient temperature remains constant.

Solution

(a) Equivalent thermal resistance:

$$R = \frac{1}{(U \times area)} = \frac{1}{(1.5 \times 600)} = 1.11 \times 10^{-3}\ \text{K/W}$$

Equivalent thermal capacitance:

$$C = \text{specific heat of concrete} \times \text{mass} = 880 \times 25\,000 = 22 \times 10^{6}\ \text{J/K}$$

The building has a time constant

$$\tau = R \times C = \frac{1.11 \times 22 \times 10^{6}}{60 \times 60} = 6.8\ \text{h}$$

(b) From $T_{int}(t) = \left(T_{int,t=0} - \left(T_{amb} + QR\right)\right)e^{\frac{-t}{\tau}} + \left(T_{amb} + QR\right)$ with $T_{int,t=0} = 285$ K and $Q = 6000$ W, we have

$$\begin{aligned} T_{\text{int}}(t) &= (285 - (283 + 6000 \times 1.11 \times 10^{-3}))e^{\frac{-2}{6.8}} + (283 + 6000 \times 1.11 \times 10^{-3}) \\ &= 286.2\ \text{K} \end{aligned}$$

The temperature inside the building after 2 hours is 13.2 °C.

11.6.3 Storage of Hot Water

Heating domestic hot water from solar thermal energy was described in Section 6.7 and is shown in Figure 11.27. Heat is taken from the solar collector in a closed hydraulic circuit whenever the strength of the sun allows. The input of the hydraulic circuit from the solar panel is positioned at the bottom of the tank so that the temperature of the heat transfer fluid is kept as low as possible to reduce heat losses in the solar panel. A second hydraulic circuit is positioned higher up the cylinder to provide auxiliary heat from a boiler when it is needed. The hot water is stored in a well-insulated tank before being delivered to the building.

A simple model of the heat flows and energy stored in the hot water can be used. Consider the heat flows entering and exiting the tank and the change in temperature of the mass of water. The equation of the balance of heat flows is

$$\Sigma Q + \frac{T_{amb} - T_{water}}{R} = mc_{water}\frac{dT_{water}}{dt}$$

which may be rearranged as

$$\frac{dT_{water}}{dt} = \frac{1}{mc_{water}}\left(\Sigma Q - \frac{\left(T_{water} - T_{amb}\right)}{R}\right)$$

where

T_{water} is the water temperature in K,
T_{amb} is the ambient temperature around the cylinder in K,

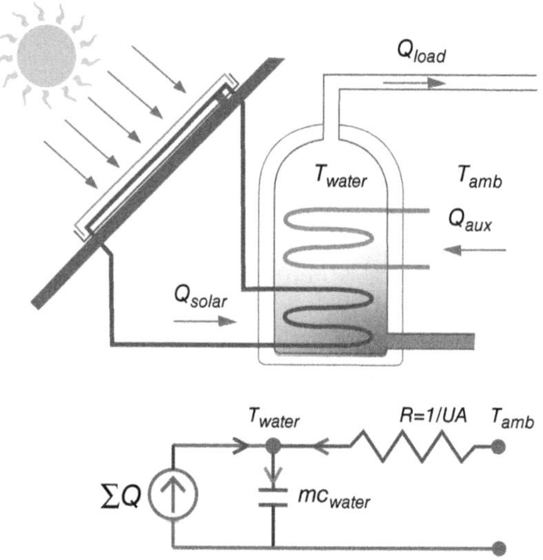

Figure 11.27 Solar thermal water heating and model.

m is the mass of water in kg,
c_{water} is the specific heat of water in J/kgK,

$$\Sigma Q = Q_{solar} + Q_{aux} - Q_{load}$$

where
Q_{solar} is the heat gain from the solar panels in W,
Q_{aux} is the heat gain from the auxiliary heater in W,
Q_{load} is the water heat load in W.

$$R = \frac{1}{UA}$$

where
U is the U value of the hot water cylinder in W/m^{2}K,
A is the surface area of the water cylinder in m^2.

This equation can be solved approximately with a time-step of Δt seconds assuming that temperatures and heat flows do not change within a time-step.

$$T_{water}^{n+1} = T_{water}^{n} + \frac{\Delta t}{mc_{water}}\left\{Q_{solar} + Q_{aux} - Q_{load} - UA\left(T_{water}^{n} - T_{amb}\right)\right\}$$

Example 11.10 – Solar Water Heating

Water is heated by a solar thermal water heater and stored as shown in Figure 11.27. Table 11.7 shows the solar irradiance, heat load and ambient temperature over morning hours. Assuming that the temperature of water at 5 am is 40 °C, obtain the water temperature at the end of each hour. Assume the efficiency of the solar collector is 100% and that the water temperature is uniform throughout the tank.

Assume:

Solar thermal rated power at irradiance of 1000 W/m^2 is 3 kW.
Volume of hot water cylinder is 120 litres.

Table 11.7

Hour	6	7	8	9	10	11	12
Solar irradiance (W/m^2)	29	116	318	537	731	874	950
Q_{load} (W)	500	1000	1500	2000	1000	1000	1000
Ambient temperature	17	19	20	20	20	21	22

Loss coefficient of cylinder, UA, is 2 W/K.
Specific heat of water = 4190 J/kgK.

Solution

The step-by-step solution may be written as:

$$T_{water}^{n+1} = T_{water}^{n} + \frac{3600}{120 \times 4190}\left\{Q_{solar} - Q_{load} - 2\left(T_{water}^{n} - T_{amb}\right)\right\}$$

$$Q_{solar} = \text{irradiance at each hour} \times 3000/1000$$

With the water temperature of 40 °C at 5 am, the temperature at 6 am is:

$$T_{water}^{6} = 40 + \frac{3600}{120 \times 4190}\left\{29 \times 3000/1000 - 500 - 2(40 - 17)\right\}$$
$$= 36.71\,°\text{C}$$

Now if we take $T_{water}^{n} = 36.71\,°\text{C}$, the table can be completed.

Hour	6	7	8	9	10	11	12
Solar irradiance (W/m^2)	29	116	318	537	731	874	950
Q_{load} (W)	500	1000	1500	2000	1000	1000	1000
Ambient temperature	17	19	20	20	20	21	22
Q_{solar} (W)	87	348	954	1611	2193	2622	2850
T_{water}^{n+1} (°C)	36.71	31.79	27.71	24.82	33.29	44.73	57.65

11.6.4 Seasonal Storage of Heat

Figure 6.2 illustrated the difficulty of using solar energy for space heating in a temperate climate. The solar resource peaks in the summer when space heating is not required but is very low in the winter when heat is needed. Heat can be stored in a domestic hot water tank or in the fabric of a building for hours, but storing heat for longer requires larger stores, often underground. A similar problem can arise in the summer when there can be a sustained requirement for cooling.

Four methods for the seasonal storage of heat are commonly recognized. These are: tank, pit, borehole and aquifer thermal energy stores, shown in Figures 11.28–11.31. Larger heat stores are more efficient than smaller stores as the energy stored is proportional to the volume of the store, while heat losses are proportional to the surface area.

Tank Thermal Storage (Figure 11.28)

A large tank containing water can be used to create a long-duration heat store. Water has a high specific heat and the rate at which heat can be transferred is controlled by the speed of pumping of water in and out of the store. When the tank is mounted on the surface it is often made from steel; tanks made from concrete are often placed underground. The inlets and outlet flows are located and controlled to encourage the water to stratify and form temperature layers to reduce heat losses. The tank has a polymer or steel liner and thermal insulation is applied to the external surfaces. The temperature of the water can be up to 98 °C. The typical store size is 1000 m^3 with energy density of 60 to 80 kWh/m^3.

Pit Thermal Storage (Figure 11.29)

An alternative, lower-cost method is to use an excavated pit filled either with water or with a gravel and water mixture. The pit is lined with a watertight foil and insulated. A roof limits heat losses to the atmosphere. A pit has lower construction costs than a tank and can be made larger but needs a suitable, sometimes floating, roof. A gravel–water mixture has the advantage of providing support for the roof but has a lower heat capacity than water. A typical energy density of a pit thermal energy storage system is 30 to 50 kWh/m^3.

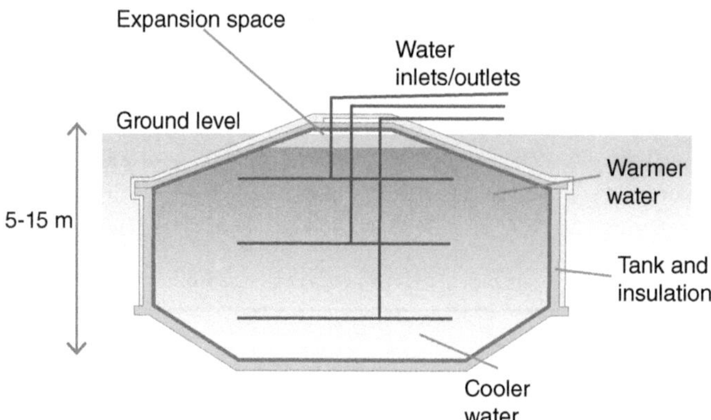

Figure 11.28 Underground tank seasonal heat energy store.

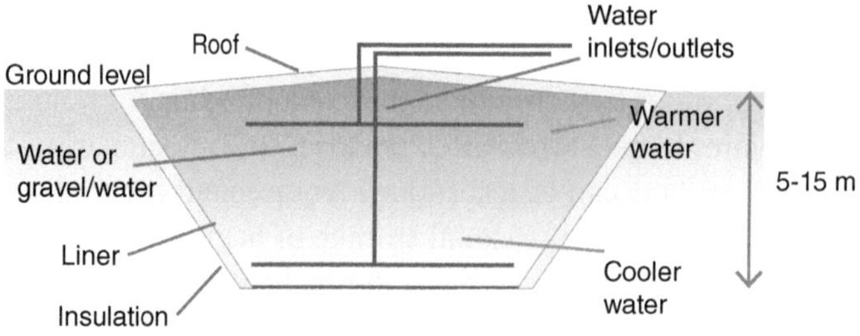

Figure 11.29 Pit heat energy store.

Borehole Thermal Storage (Figure 11.30)

Heat can be stored directly in the ground if geotechnical conditions are suitable, i.e. the rock or soils have high thermal capacity and little movement of groundwater. Boreholes are drilled and a vertical U-shaped duct is placed in thermal grout. A heat transfer fluid, usually water, is pumped through the ducts in a closed circuit and passed through a heat exchanger on the surface. Multiple boreholes are connected in a pattern so that heat is injected or extracted from the centre of the borehole pattern and the temperature gradient reduces towards the periphery. The surface of the ground above the boreholes is insulated to reduce heat loss to the atmosphere. The ground can be heated up to 80 °C but borehole storage can alternatively be used to store cold. A typical energy density of a borehole thermal energy storage system is 15 to 30 kWh/m^3 of ground.

Aquifer Thermal Storage (Figure 11.31)

Subterranean aquifers are naturally occurring layers of groundwater. They can be used to store large quantities of heat over a prolonged period if the ground conditions are suitable, i.e. the ground has high porosity and limited mixing between layers of water. Heat is stored in the water and in the surrounding porous ground. Heated or cooled water is pumped to and from the aquifers through wells and passed through a heat exchanger on the surface. Hot and cold underground reservoirs are separated using two wells or groups of wells spaced horizontally. Aquifer heat stores can be used for cooling large commercial premises as well as for heating, sometimes with the additional use of heat pumps. If suitable conditions can be found, aquifer energy

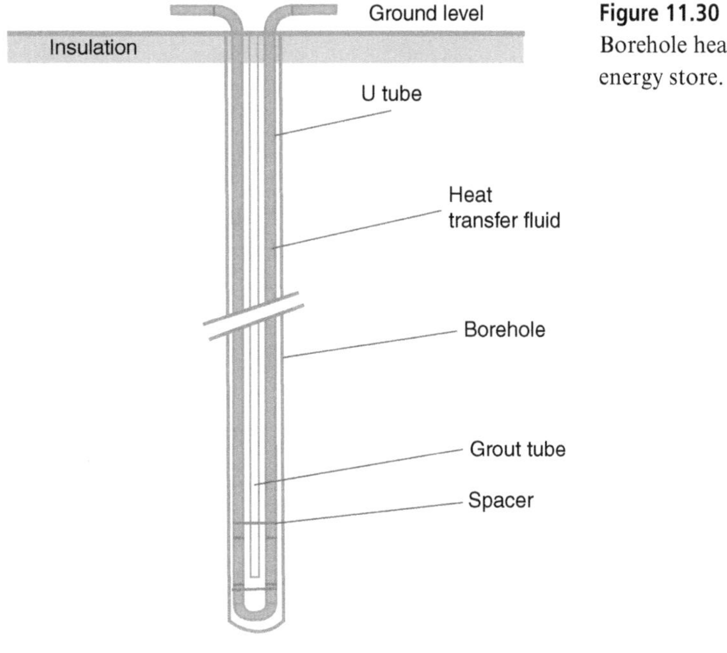

Figure 11.30 Borehole heat energy store.

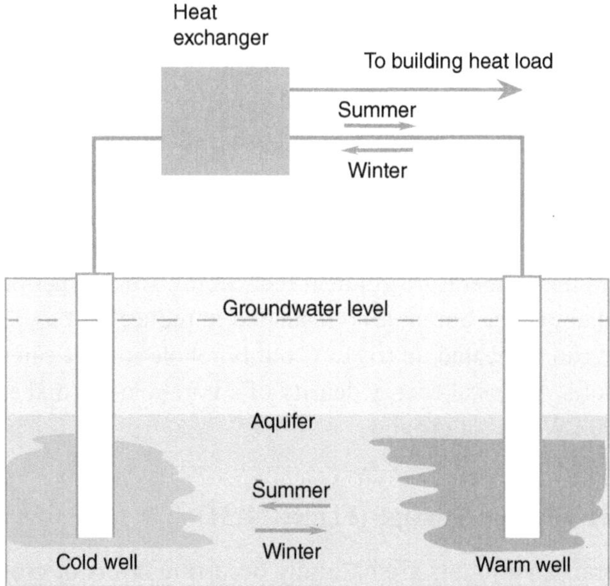

Figure 11.31 Aquifer thermal energy store.

stores are cheaper than borehole energy stores. A typical energy density of an aquifer thermal energy storage system is 30–40 kWh/m^3.

The operating strategy shown in Figure 11.31 is to take cold water from the cold well in summer to cool a building and inject the heated water into the warm well. Heat is stored in the water and in the surrounding ground. In the winter the flow is reversed: warm water is taken from the warm well, used to heat the building and the cooled water is injected into the cold well.

11.6.5 Phase Change Material (PCM)

Phase change material (PCM) absorbs or releases large quantities of latent heat at an almost constant temperature as the phase of the material changes, usually liquid–solid or solid–liquid. In Example 11.10, more than 5 times as much energy was absorbed as latent heat when the ice melted than as sensible heat when the ice and water increased in temperature. PCM with a phase change temperature of 20–30 °C can be used to reduce the variation in the temperature of the fabric of buildings, while PCM with a transition temperature of 40–70 °C can be integrated into solar hot water systems.

The use of latent energy through changes in phase increases the density of thermal energy stores and can lead to smaller equipment than sensible energy storage units that use small temperature differences. The main disadvantages are the higher cost compared to sensible heat storage and the limited transfer of thermal power, which is often restricted by the thermal conductivity of the PCM.

A good PCM has:

(a) a suitable temperature of phase change,
(b) a large transfer of latent heat energy during the phase transition,
(c) high thermal conductivity in the solid and liquid phase,
(d) long cycle life,
(e) repeatable phase change temperature,
(f) high specific heat to provide additional sensible heat storage,
(g) small volume change,
(h) chemical stability and compatibility with other materials,
(i) low cost, limited environmental impact and good ability to be recycled.

The most common phase change material that is used with renewable energy is water-ice, but many other materials with different transition temperatures are available. Paraffins derived from petrochemicals are widely used and the melting temperature can be chosen to be between 0 °C and 70 °C by altering the number of carbon atoms in the organic material. A similar range of transition temperatures can be obtained with fatty acids. Some modern commercial PCM is described as bioparaffin wax and consists of fatty acids, alcohols and esters derived from recycled coconut or palm oil. The latent heat of fusion of commercially available organic PCM is around 230 kJ/kg and the thermal conductivity around 0.2 W/mK.

Inorganic PCM is usually a mixture of salt hydrates; it has a wider range of transition temperatures and higher latent heat energy than organic PCM. However, it experiences larger volume changes during changes of phase, can be corrosive, and can suffer from long-term instability and low thermal conductivity. Additives are needed to ensure it changes phase reliably at a repeatable temperature, and crystallization can be delayed below the melting point, an effect known as subcooling.

PCM can be used to improve the performance of low-energy buildings. Microparticles of PCM several micrometres in diameter are encapsulated into polymer shells and combined with building materials. This increases the thermal mass of a building and slows the rate of change of temperature.

PCM can also be enclosed in plastic spheres or in larger slabs. The PCM is contained in plastic and heat is transmitted using a reversible flow of a heat transfer fluid, which is typically air or a water-based fluid. Figure 11.32 shows a small thermal energy store using PCM contained in plastic spheres.

Alternatively, the phase change material can be placed directly in a heat exchanger. Figure 11.33 shows a shell and tube heat exchanger with a liquid heat transfer fluid and the PCM in direct contact with the tubes and fins. The fins increase the area of contact between the heat exchanger and PCM. An alternative design to improve the transfer of heat into and out of the store is for the tubes to be connected in a serpentine arrangement to increase the area of contact between the tubes and the PCM. Because of the low thermal conductivity of PCM and its complex behaviour during the phase change, it can be difficult to rapidly charge or discharge an energy store if there is a limited temperature difference between the heat transfer fluid and the PCM.

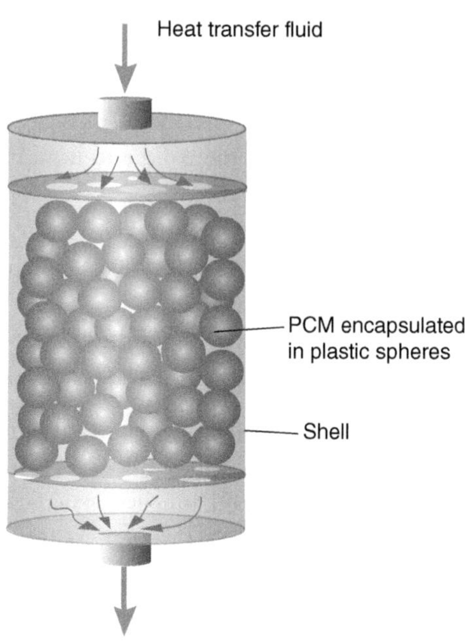

Figure 11.32
Thermal energy store using phase change material contained in plastic spheres.

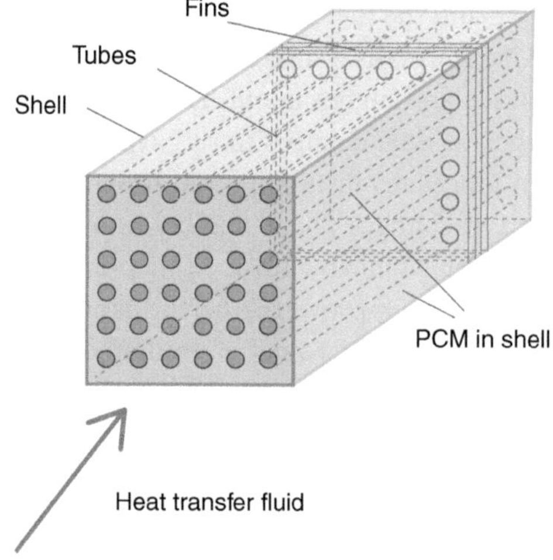

Figure 11.33 Shell and tube heat exchanger containing phase change material.

Typical Values of the Sensible Heat of Common Materials

Material	Sensible heat (J/kgK)	Density (kg/m^3)	Thermal conductivity (W/mK)
Water	4190	1000	0.6 (still water)
Ice	2108	920	2.2
Concrete	880	2240	1.5
Stone	820–880	2569	1.0
Cast iron	837	7900	37

Typical Values of the Latent Heat of Common Materials

Material	Latent heat (kJ/kg)	Melting point (°C)
Water – ice	334	0
Water – steam	2250	100
Paraffins	147–190	22–68
Commercial bioparaffin	235	58
Commercial salt hydrate	259	58

PROBLEMS

1. Compare how electrical energy is stored using high-speed flywheels and pumped storage hydro. Which technology would be more suited for storing 50 kW for 1 minute and which for 100 MW for 5 hours? Identify important risks that would be considered in the impact assessments of these projects.
2. How are sensible heat and latent heat storage making your indoor environment more comfortable?
3. Compare the principle of operation of a lead–acid and a lithium-ion battery cell.
4. How can heat be stored between summer and winter? Why are larger heat stores more efficient than smaller ones?
5. Explain what is meant by stratification of water in a heat storage tank and why this reduces heat losses.
6. A pumped storage hydro scheme has an average net head of 50 m and a flow rate of 3 m^3/s when pumping or generating. The efficiency with these water conditions when generating is 80% and the efficiency when pumping is 85%. The upper reservoir has a useable volume of water of 10^5 m^3.
   (a) Calculate the electrical power that can be generated.
   (b) Calculate the electrical power required for pumping.
   (c) How long can the generator supply its rated power?
   (d) Calculate the cost of electricity used for pumping if electricity costs £40/MWh.

   [**Answer:** (a) 117 kW; (b) 1731 kW; (c) 9.26 hours; (d) £436.]
7. A 10 m^3 pressure vessel stores air at 50 bar.
   (a) If this air is used to operate a piston of area 0.25 m^2, what force will it exert?
   (b) Assuming constant temperature, what is the initial volume of the air at atmospheric pressure?
   (c) Estimate how much energy is used to compress the air assuming constant temperature.

   [**Answer:** (a) 1.25 MN; (b) 500 m^3; (c) 54.3 kWh.]

8. A CAES uses 0.69 kWh of electrical energy to compress the air and 1.17 kWh of methane (CH_4) to heat the air for each 1 kWh of electrical output. The electrical energy for compression is supplied from zero carbon generation. What is the carbon intensity of generation of the stored energy?

   Data:

   The lower heating value of methane is 13.9 kWh/kg.
   The combustion of 1 kg of methane produces 2.75 kg of carbon dioxide (see Example 8.3).

   [**Answer:** 231 gCO_2/kWh.]

9. An underground store has an effective volume of 310 000 m^3 and operates between 45 bar and 70 bar. To produce 1 kWh of electrical output requires 0.8 kWh of electrical energy to compress the air and 1.6 kWh of fuel to heat the air.
   (a) Calculate the energy efficiency of the cycle.
   (b) How much electrical energy can be produced?
   (c) How much fuel energy is needed?

   [**Answer**: (a) 0.417; (b) 333 MWh; (c) 533 MWh.]

10. The flywheel of Example 11.4 is connected to the electrical power system through a variable speed drive that can operate down to 10% of nominal speed. What is
    (a) the total energy that can be injected into the power system,
    (b) the power that could be injected over 10 seconds?

    [**Answer:** (a) 275 kWh; (b) 99 MW.]

11. A battery provides 10 kWh at 100 V using cells that each develop 3.7 V and have a capacity of 10 Ah.
    (a) How many cells are in series and parallel?
    (b) If the battery discharges at a constant rate over 5 hours what is the current in each cell?

    [**Answer:** (a) 27 in series, 10 in parallel; (b) 2 A.]

12. A battery has an open-circuit voltage of 48 V and a series resistance of 50 mΩ. Calculate:
    (a) the terminal voltage when it is charged or discharged at 20 A,
    (b) the current that would flow if there was a short-circuit at the battery terminals.

    [**Answer:** (a) 49 V, 42 V; (b) 960 A.]

13. A domestic solar collector heats a 100-litre storage tank to 85 °C. Domestic hot water at above 40 °C is needed during the night.
    (a) How much useful heat energy can be stored in the tank?
    (b) How much could the weight of the tank be reduced if a phase change material was used?

Data on phase change material:

Latent heat (kJ/kg)	Sensible heat (solid and liquid) (J/kgK)
235	2000

[**Answer:** (a) 18 855 kJ; (b) 42 kg.]

14. How much energy is required to convert 1 kg of ice at −18 °C into liquid water at 40 °C?

Data:

Sensible heat of ice is $c_{ice} = 2108$ J/kgK,
Sensible heat of water is $c_{water} = 4200$ J/kgK,
Latent heat of fusion of water is $l_{water-ice} = 334$ kJ/kg.

[**Answer:** 540 kJ.]

15. A building is represented by the model of Figure 11.26. The building has a mass of 40 tonnes of concrete, a total surface area of 1000 m^2 and a composite U value of 1.75 W/m^{2}K. The specific heat of concrete is 880 J/kgK.
(a) Calculate the thermal time constant of the building.
(b) If the initial temperature inside the building is 12 °C, what would be the temperature of the building 3 hours after turning on a 5 kW or 20 kW heater? Assume the following ambient temperature values after each hour:
from 0 to 1 h, 10 °C,
from 1 to 2 h, 12 °C,
from 2 to 3 h, 13 °C.

[**Answer:** (a) 5.58 h; (b) 13.1 °C, 16.7 °C.]

16. Water is heated by a solar thermal heater and an auxiliary heater. The temperature at 5am is 35 °C and the expected load at 6am is 500 W. What is the amount of heat that should be supplied by the auxiliary heater to raise the water temperature to 40 °C at 6 am? Assume that the ambient temperature is 15 °C. Also, assume the efficiency of the solar collector is 100% and the water temperature is uniform throughout the tank.

Data:

Volume of hot water cylinder is 150 litres,
Loss coefficient of cylinder, UA, is 1.5 W/K,
Specific heat of water = 4190 J/kgK.

[**Answer:** 1.4 kW.]

17. In Problem 16, the expected load at 7am is 1000 W. Assume that the ambient temperature is 16 °C and the solar heater produces 100 W. What amount of heat should be supplied by the auxiliary heater to maintain the water temperature at 40 °C?

[**Answer:** 936 W.]

18. In an industrial plant, it is required to maintain the temperature of a water tank always between 50 °C and 60 °C. Water is heated by a solar thermal heater

Table 11.8

Hour	6	7	8	9	10	11
Solar irradiance (W/m^2)	30	120	300	500	750	850
Q_{load} (kW)	5	5	5	5	5	5
Ambient temperature	17	19	20	20	20	21

and an auxiliary heater. Table 11.8 shows the solar irradiance, heat load and ambient temperature over morning hours. What is the minimum amount of heat that should be supplied by the auxiliary heater in each hour and what is the resultant water temperature? Assume the efficiency of the solar collector is 100% and that the water temperature is uniform throughout the tank.

Data:

Solar thermal rated power at irradiance of 1000 W/m^2 is 10 kW.
Volume of the tank is 1000 litres.
Loss coefficient of cylinder, UA, is 2.5 W/K.
Specific heat of water = 4190 J/kgK,
Initial water temperature is 45 °C,

[**Answer:** 6am: 50 °C, 10.6 kW; 7am: 50 °C, 3.9 kW; 8am: 50 °C, 2.1 kW; 9am: 50 °C, 75 W; 10am: 52.1 °C, 0 W; 11am: 55 °C, 0 W.]

FURTHER READING

Huggins R.A., *Energy Storage: Fundamentals, Materials and Applications*, 2nd ed., 2016, Springer.
A wide-ranging textbook addressing energy storage.

Jensen J., *Energy Storage*, 1980, Butterworth.
An important early book describing key concepts of energy storage.

Kun Sang Lee, *Underground Thermal Energy Storage*, 2013, Springer-Verlag.
Describes approaches to large-scale, seasonal underground storage of thermal energy.

Plett G.L., *Battery Management Systems Vol. 1, Battery Modelling*, 2015, Artech House.
A comprehensive textbook describing battery modelling and condition monitoring.

ONLINE RESOURCES

Exercise 11.1 Assessment of the degradation of a Li-ion battery: the ageing of lithium batteries is modelled.

Exercise 11.2 Simple dynamic model of a building: this exercise demonstrates a simple dynamic model of the heat stored in the fabric of a small building.

Exercise 11.3 Solar thermal hot water storage: a sequential steady-state spreadsheet model is used to calculate the solar energy stored in a hot water tank.

12 Off-Grid Systems, Microgrids and Community Energy Systems

INTRODUCTION

Renewable energy has an important role in supplying electrical power to off-grid systems, microgrids and community energy systems. Such systems are technically demanding and are usually small with limited budgets. Their design and operation are difficult and often require close interaction with the users of the power. All this can be time-consuming and expensive. The financial and organizational approaches needed to make these systems self-sustaining are only now being demonstrated.

12.1 Off-Grid Systems

Remote communities that live far from grid electricity or on islands are often without access to reliable, affordable energy supply, which constrains their economic development. According to the 'Tracking SDG 7: Energy Progress Report, 2020' jointly published by the IEA, IRENA, UNSD, World Bank and WHO, it is estimated that about 620 million people will remain without access to electricity in 2030 and most of them are in remote communities.

There are three technical approaches to the supply of electricity to remote areas: extending the national grid, standalone individual home energy systems and standalone mini- or microgrids. Extending the national grid can be extremely costly as remote rural communities are often far from the national grid or have difficult terrain that prevents the easy installation of transmission lines. Also, the size of the demand may not justify the costs of the connection. Solar home systems, pico-hydro systems or wind home systems are often the solution when providing energy to isolated households. With a scattered population, the installation of the distribution network covering a large area entails high local connection costs. Thus, for such households, individual home systems can be a good solution. However, the cost of energy from individual household systems tends to be high due to the lack of diversity and economies of scale. Connecting a number of generators into standalone mini- or microgrids can offer an optimal solution for utilizing local renewable energy resources and has the potential to become the preferred technical approach for off-grid electrification.

12.1.1 Standalone Off-Grid Photovoltaic Systems

Photovoltaic (PV) modules are extremely robust, have no moving parts and can operate for more than 20 years without maintenance other than checking the tightness of connections and keeping the module surfaces clean. They are ideal for supplying limited amounts of electrical power to small high-value loads in locations remote from the grid. However, their output is limited by the solar resource, and batteries are needed.

Figure 12.1 shows a schematic diagram of a typical small off-grid PV system consisting of four modules and a lead–acid battery supplying a 24 V dc load. Each module consists of 36 crystalline silicon cells with the performance shown in Table 12.1 at standard test conditions (STC). The modules are connected two in series and then two strings parallel across the battery. A blocking diode is connected at the output of the array. A fuse is connected at the terminal of the battery to protect the wiring and load against short-circuits.

Standalone off-grid PV systems need energy storage to provide power during the night and during days of low solar insolation. A number of new battery types are coming into use (e.g. lithium-ion), although lead–acid batteries are traditionally the most widely used technology for off-grid PV systems.

The life of lead–acid batteries is reduced by both overcharging and excessive discharge. Thus a regulator is connected in the circuit from the solar array to the battery to prevent overcharging, and a low-voltage (LV) disconnect is placed in the circuit from the battery to the load to prevent excessive discharge. For flooded lead–acid batteries, the low-voltage disconnect is the more important device as although overcharging will result in some gassing and loss of the electrolyte, this can be replaced, while the excessive discharge of a lead–acid battery results in a considerably shortened life. For flooded lead–acid batteries, a small amount of occasional gassing may be desirable as this reduces the stratification of the battery electrolyte.

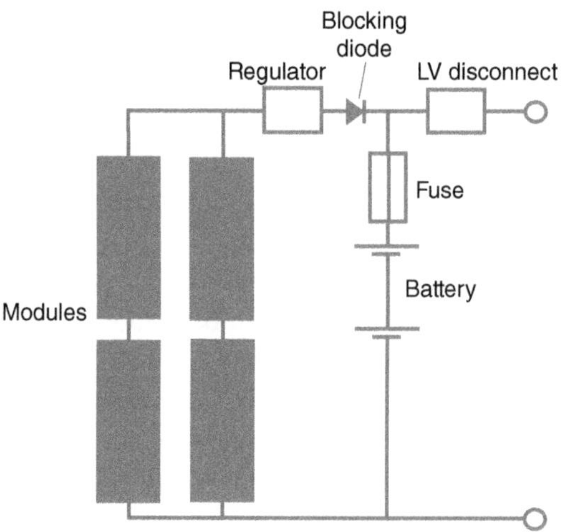

Figure 12.1 Schematic diagram of a 24 V off-grid photovoltaic system.

Table 12.1 Characteristics of the solar modules of Figure 12.1 (at STC)

Open-circuit voltage, V_{OC}	22.3 V
Short-circuit current, I_{SC}	4.15 A
Voltage at maximum power point, V_{mpp}	18.7 V
Current at maximum power point, I_{mpp}	3.75 A

The charge regulator and low-voltage disconnect shown in Figure 12.1 are simple electronic circuits that measure the voltage at the battery terminals to give an indication of the state of charge (SoC) of the battery and control switches. The switch in the regulator opens when the battery voltage (and charge) becomes too high, and a similar relay in the low-voltage disconnect opens when the battery voltage (and charge) becomes too low. In simple controllers the terminal voltage is used as an indicator of the battery state of charge, even though it is influenced by temperature and rate of charge or discharge. More sophisticated controllers apply compensation for the change of battery voltage with the current flow or include a sensor to measure the temperature of the battery and so relate the battery voltage to the state of charge more closely. An alternative technique to assessing the battery state of charge is to integrate the measurement of the current flow into and out of the battery over time.

A large transistor can be used as the series regulator switch but there will always be a voltage drop and hence power loss across any solid-state switch and so mechanical relays are often preferred. Photovoltaic modules have the very benign characteristic that they can be either open- or short-circuited with no ill effects and for small off-grid systems, a regulator that short-circuits the modules is sometimes used. Rather than open circuit the modules when the battery is fully charged, the regulator is placed in parallel with the modules and short-circuits them. The blocking diode must then be positioned to ensure current does not flow from the battery when the modules are short-circuited. A mechanically switched series or parallel regulator can operate in an on–off mode within a hysteresis band around the measured battery voltage at which it switches. If a solid-state switching device (e.g. a power transistor) is used, a pulse width modulated switching pattern can be used to apply a defined voltage to the battery.

Example 12.1 – Operating Point when a Battery is Directly Connected

A 24 V battery is connected in parallel with the photovoltaic modules shown in Figure 12.1. The battery has 12 cells with an open-circuit voltage of 2.1 V/cell. The internal resistance of the battery is 0.35 Ω. Calculate the power that will not be captured due to the direct connection of the battery to the photovoltaic modules.

Solution

The open-circuit voltage of the battery is $(12 \times 2.1)\ \text{V} = 25.2\ \text{V}$.

When the battery is being charged at 8 A, the additional volt drop caused by the internal resistance is (8 × 0.35) V = 2.8 V.

This leads to the straight-line *V–I* characteristic of the battery shown in Figure 12.2. The battery is connected in parallel with the photovoltaic modules and so the operating point of the system is where the battery *V–I* characteristic intersects the *V–I* characteristic of the array of four modules (also shown in Figure 12.2). With an irradiance of 1000 W/m^2, the operating point of the array will be at 28 V and 8 A resulting in power into the battery of 224 W.

Table 12.1 shows that, at the maximum power point of each module, V_{mpp} is 18.7 V and I_{mpp} is 3.75 A, giving a P_{mpp} of 70.1 W. Thus the effect of direct connection of the battery is to reduce the array output to 228.4 W from a possible 280.5 W, i.e. a reduction of 18.6%.

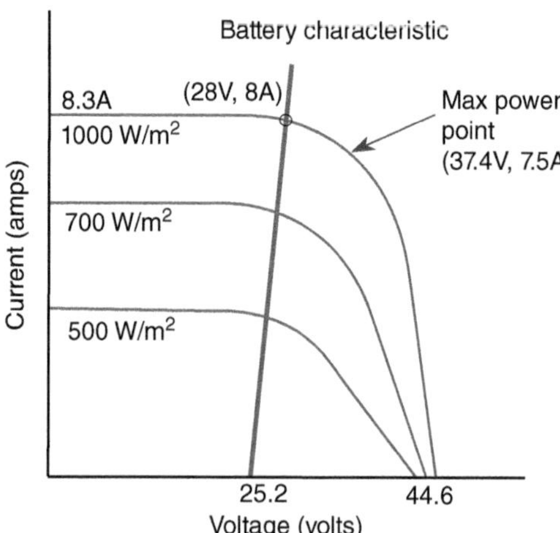

Figure 12.2 *V–I* characteristics of a nominal 24 V lead–acid battery and a four-module photovoltaic array.

In Example 12.1, when directly connected to the battery, the operating point is at a lower voltage than the maximum power point. This is to ensure the operating characteristics intersect, even at high cell temperatures when V_{OC} reduces. Also, an allowance has to be made for voltage drop in the wiring and blocking diode. If the operating characteristics of the modules and battery fail to intersect then no charge is added to the battery. It is possible to use a dc-to-dc converter to ensure the array works always at its maximum power point. However, such devices incur expense as well as electrical losses, and for small installations in remote areas, the simplicity of direct connection has traditionally been preferred.

Example 12.2 – Estimate of the Charge into the Battery of a Standalone PV System

For the standalone system of Figure 12.1 with the module characteristics of Table 12.1, estimate the charge into the battery with a daily solar insolation of 5 kWh/m^2.

Solution

A daily solar insolation of 5 kWh/m^2 is equivalent to 5 Peak Sun Hours, at 1000 W/m^2 (STC). The current rating of the two modules in parallel connected to the battery is 8 A. Hence

$$\text{Ah output} = \text{Peak Sun Hours} \times \text{current rating} = 5 \times 8 = 40\ \text{Ah/day}$$

The charge into the battery will be approximately 40 Ah/day.

Note that for directly connected standalone systems the sizing estimate is made in ampere-hours (Ah) rather than watt-hours (Wh) as the current output of a photovoltaic cell is almost directly proportional to irradiance, and for simple standalone systems the array voltage is held constant by the battery.

The standalone system of Figure 12.1 has 36 cells connected in series in each module to ensure that there is adequate voltage to inject charge into the battery at any credible temperature of the photovoltaic cells. Excessive charging is limited by the regulator. For one or two module systems, e.g. as used for lighting in remote dwellings, a further simplification can be made by reducing the number of photovoltaic cells in series to 32 or 33 per module and discarding the regulator. These are known as self-regulating modules and, as the state of charge of the battery increases, the *V–I* characteristics of the battery and module no longer intersect near the maximum power point and so the current into the battery reduces. Although apparently attractive due to their simplicity, self-regulating modules are used only in very small systems where it is necessary to reduce the cost to a minimum. At high ambient temperatures the *V–I* characteristic of a 32- or 33-cell module may not cross the *V–I* characteristic of a 12 V battery and the solar panel will then not inject charge into the battery.

A standalone off-grid system was installed by a Cardiff University student-led project at a rural health centre in Zambia. The nearest grid power is more than 50 km away and the PV system was installed to provide power for lighting, a fridge and computers for education.

Figure 12.3 shows the twelve 280 W modules that were connected, with three in each series string, and four strings in parallel. Each module has 72 polycrystalline cells connected in series giving a maximum power point voltage of 36 V. The panels were installed facing north at a tilt angle of 12° (the site is at 15.5° south). This tilt angle was chosen by simulation to produce a uniform daily yield throughout the year rather than to maximize total yearly energy production.

Figure 12.4 shows the battery bank of eight 12 V, 220 Ah lead–acid batteries connected, with four in each series module string, and two strings in parallel. The batteries were supplied through a single 60 A maximum power point tracking charge controller.

A 3 kVA continuous, 6 kVA peak inverter generated 230 V ac from the 48 V dc of the battery bank. The 230 V ac was distributed to four buildings around the health

Figure 12.3 Photovoltaic array of a standalone system at a rural health centre in Zambia.
Source: Cardiff University project supported by the Mothers of Africa charity and Cardiff University. System design and photographs provided by D.J. Rogers, L.J. Thomas and J.M. Stevens.
(A black and white version of this figure will appear in some formats. For the colour version, refer to the plate section.)

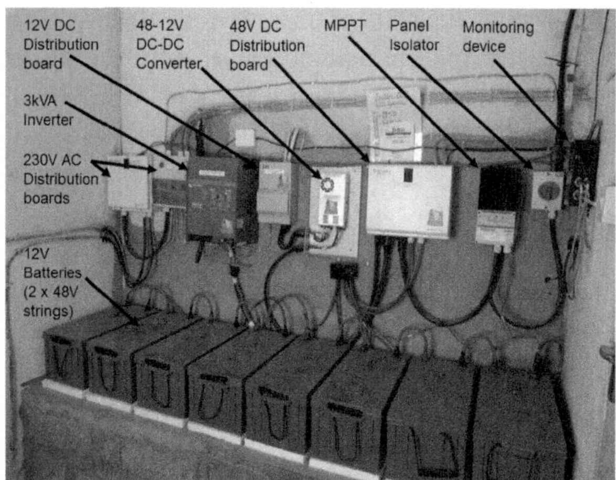

Figure 12.4 Battery bank, inverter and controller of a standalone PV system.
Source: Cardiff University project supported by the Mothers of Africa charity and Cardiff University. System design and photographs provided by D.J. Rogers, L.J. Thomas and J.M. Stevens.
(A black and white version of this figure will appear in some formats. For the colour version, refer to the plate section.)

centre. A 500 W, 48 V dc to 12 V dc converter was used to run 10 low-power PCs. The direct conversion to 12 V dc minimized conversion losses and gave a 48 V dc to 12 V dc efficiency of ~95% rather than only ~65% if a conversion route of 48 V dc to 230 V ac to 12 V dc had been used.

12.1.2 Standalone Off-Grid Microgrids

Off-grid microgrids use more than one energy source together with energy storage. A typical off-grid microgrid is shown in Figure 12.5. This diagram shows that cascaded dc-to-dc and dc-to-ac, or ac-to-dc and dc-to-ac, converters are used for the integration of the different generation sources. For example, photovoltaic systems and energy storage are each connected through a dc-to-dc and dc-to-ac converter to a common ac bus, whereas wind turbines are connected through an ac-to-dc and dc-to-ac converter. Many dc loads such as IT equipment and LED lamps have an internal ac-to-dc and dc-to-dc converter.

Usually off-grid microgrids have a hierarchical management and control scheme using a leader–follower configuration. The microgrid central controller (MGCC) acts as the leader. Local controllers (LCs) at the loads and microsource controllers (MCs) at the generators communicate with the leader and exchange information with it.

One of the inverters, usually the inverter connected to the energy storage battery, acts as the grid-forming inverter. The grid-forming inverter establishes the voltage and frequency of the ac busbar. The MGCC schedules the microgenerators, sends set-points to MCs and LCs, and manages the charging and discharging of the battery to maintain the second-by-second balance of active power. As sun and wind are intermittent and variable, any deficit between the generation and demand must

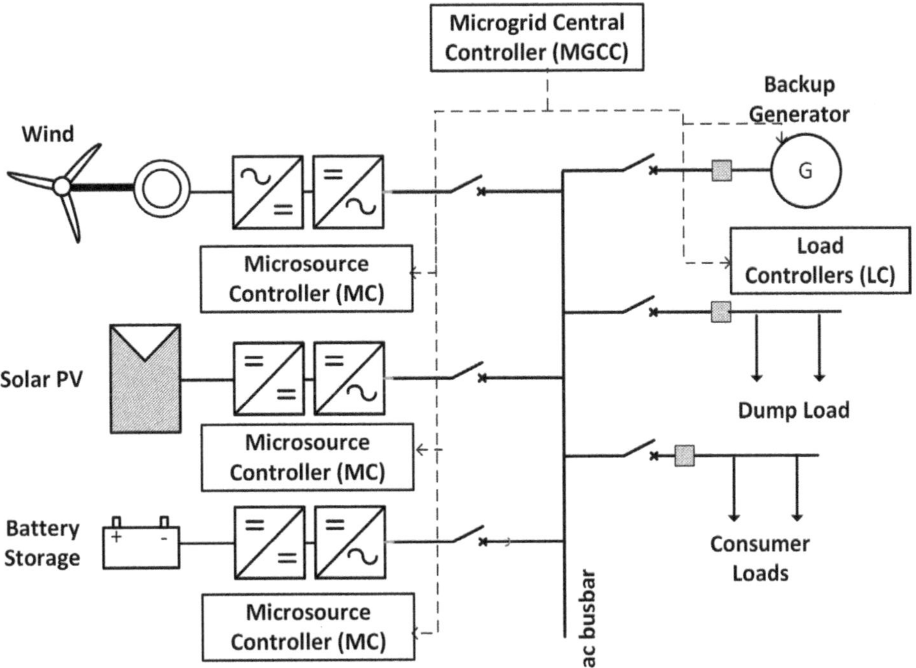

Figure 12.5 Typical architecture of an off-grid microgrid.

be balanced by charging/discharging the battery or by starting the diesel generator. Any excess generation is stored in the battery or dissipated in the dump loads.

For satisfactory operation, it is critical to select the optimum capacities of different energy sources and decide the size of the energy storage during the system design. To demonstrate the steps involved in such design, a study that was undertaken for Eluvaitivu Island is described below. These are the actual steps that were carried out during the design stage of the Eluvaitivu microgrid.

Eluvaitivu Island is situated in the north of Sri Lanka (9° 41′ N, 79° 48′ E) and covers an area of 1.4 km^2. It is 2.9 km from the mainland. The climate is tropical monsoonal with temperatures ranging from 19 °C to 31 °C. The annual average rainfall is 735 mm and the rainy season is from October to December. There are 185 families living in 110 houses with a total population of 787. Most of the population are fishermen; fishing is the island's main income-generating activity. Figures 12.6a and b show the daily load profile and the monthly variation of the community load.

Before installation of the microgrid in April 2016 electricity for the island was provided using a 100 kVA diesel generator. It operated from 4.30am to 6.30am and 6.00pm to 10.30pm, approximately 6½ hours per day; 73 houses were connected to the electricity system and the average monthly consumption of the island was around 3450 kWh. The cost of electricity generation (operational cost) was approximately US$0.5/kWh, while the average tariff charged in the mainland of Sri Lanka was approximately US$0.1/kWh. Given the high cost of electricity and the restricted availability of electricity supply, a hybrid solution was investigated that could deliver a reliable, cost-effective supply to the community.

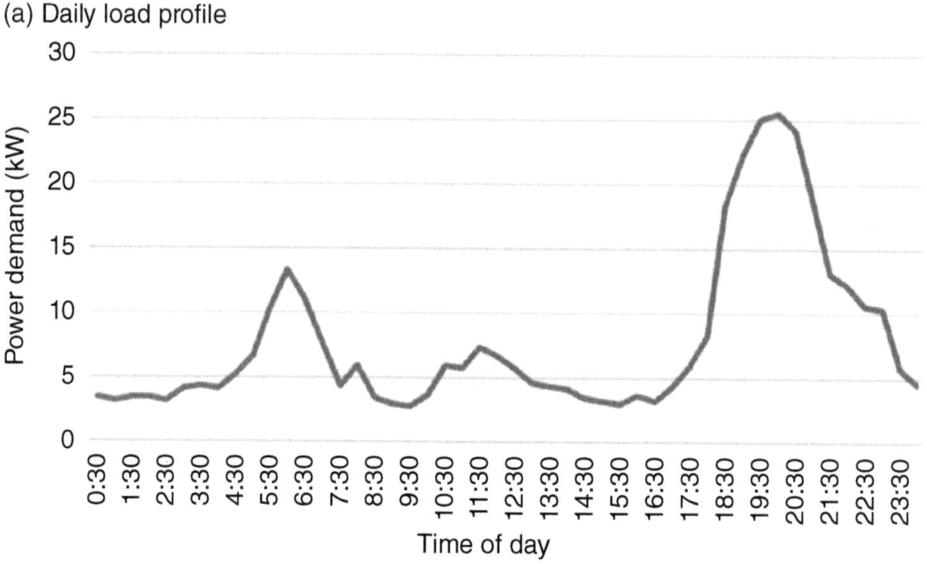

Figure 12.6 Load, solar irradiation and wind speed variations at Eluvaitivu Island in the north of Sri Lanka. (a) Daily load. (b) Monthly load. (c) Solar irradiation. (d) Wind speed.

(b) Monthly load profile

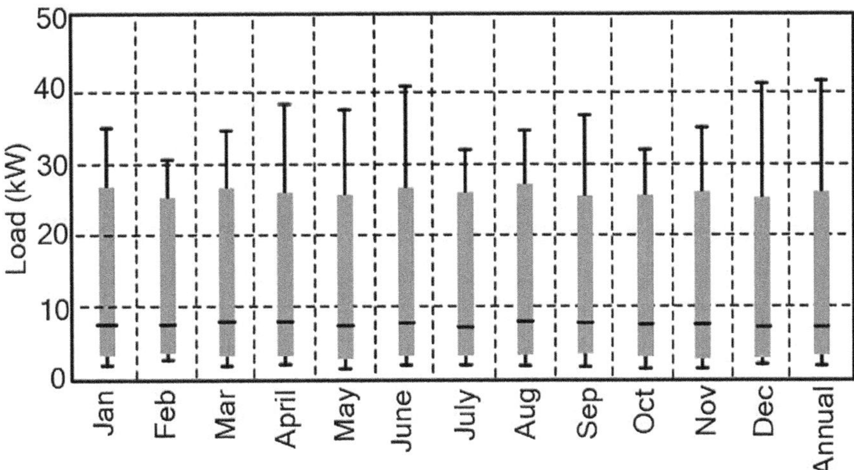

(c) Solar irradiation

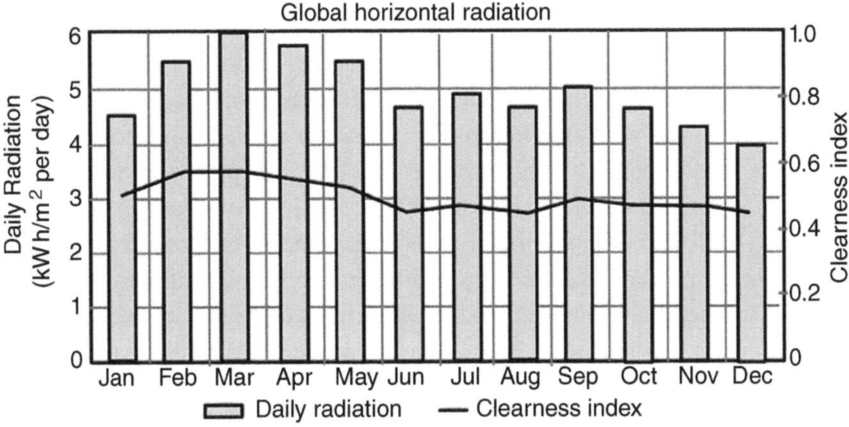

(d) Wind speed

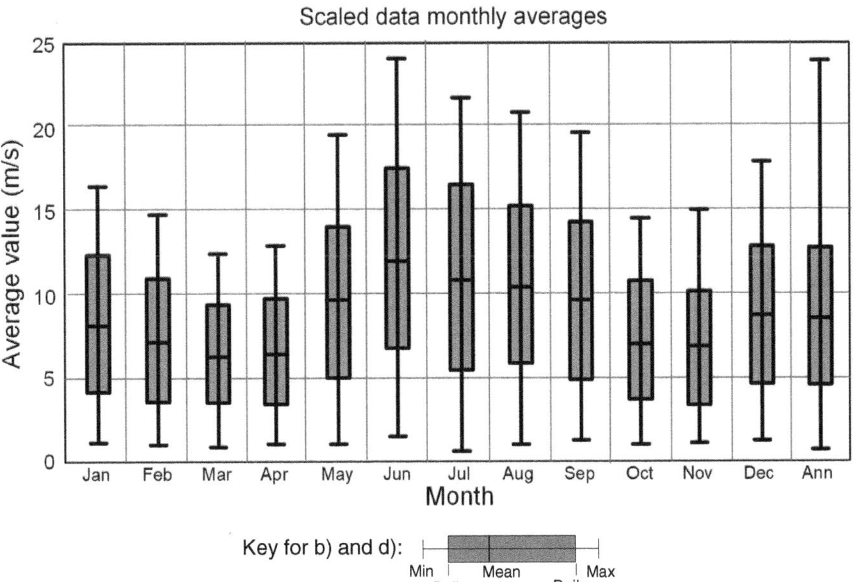

Figure 12.6 (cont.)

Table 12.2 Parameters used for HOMER simulation

	PV plant	Wind	Diesel genset	Li-ion battery
Capital cost including installation cost	Panels \$1800/kWp Inverter \$500/kWp	\$3000/kW	\$800/kW	Battery \$2780/kWh Inverter \$670/kW
Operation and management cost	\$65/kW per year	\$18/kW year	\$0.029/h per kW	\$190 per year per module
Life	25 years	25 years	40 000 hours	5000 cycles for 80% depth of discharge
Derating factor for PV and efficiency for diesel genset and Li-ion battery	80%		35% at full load	95% round trip

Records from the National Renewable Energy Laboratory (NREL) indicate that the solar radiation at Eluvaitivu Island is fairly uniform, varying from 4 kWh/m^2 per day to 6 kWh/m^2 per day. There are two windy seasons: from December to February (average wind speed 7–8 m/s) and from May to September (average wind speed 10 m/s). Figure 12.6c and d show the solar irradiation (insolation) and wind speed.

The HOMER tool, developed by NREL, used hourly optimization to establish the optimal power system configuration for the island community. The optimization calculated the generation needed to minimize the cost, expressed as net present value (NPV) of the system over 20 years, while delivering the required electricity supply to satisfy the projected demand. The cost figures given in Table 12.2 were used. In addition to the data provided in Table 12.2, the following assumptions/parameters were used in the HOMER study.

- The discount rate is 10% per annum.
- A diesel generator major overhaul is undertaken after 8000 running hours.
- Operators' wages are \$2500 per annum.
- The system fixed capital cost is \$50 000 (for powerhouse and associated infrastructure).
- The diesel cost (at site) is \$0.9/litre (this was assumed to be constant throughout the project life).
- No carbon or greenhouse gas emission credits are available.

HOMER calculated the following optimum system capacities:

- solar plant 46 kW,
- wind plant 20 kW,
- diesel generator 30 kW,
- lithium-ion battery 52 kW/110 kWh.

The monthly average generation predicted by HOMER is shown in Figure 12.7. The annual energy generated from the sources was: PV plant 74.4 MWh, wind plant 74 MWh and diesel generator 9.4 MWh.

Table 12.3 Eluvaitivu Island installed capacities

Source	Unit capacity	No. of units	Total capacity
Diesel generator	30 kW	1	30 kW
Wind turbines	3.5 kW	6	21 kW
PV panels	250 Wp	184	46 kWp
Lithium-ion battery modules	13 kW/27.5 kWh	4	52 kW/110 kWh

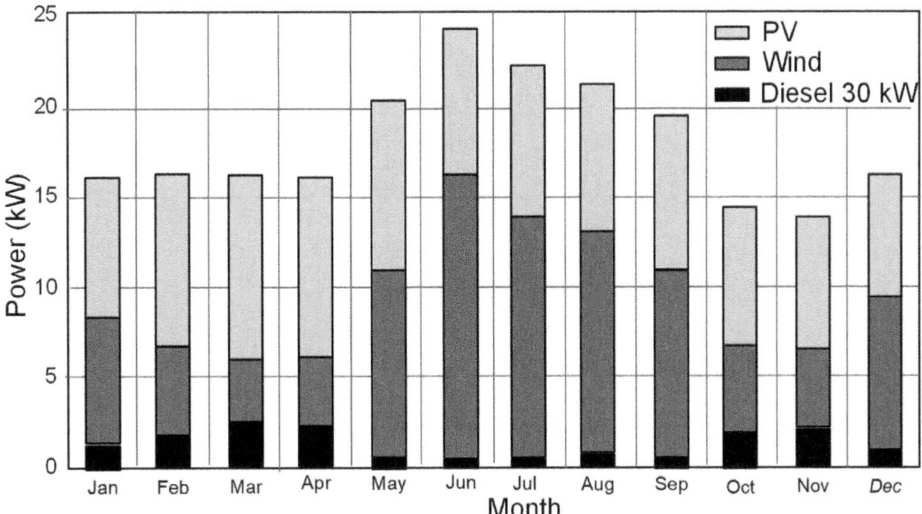

Figure 12.7 Monthly average electricity production.

The optimum generation identified by HOMER is now installed in the island with the rated capacities shown in Table 12.3.

The configuration of the Eluvaitivu Island microgrid is shown in Figure 12.8. The inverter that connects the lithium-ion battery acts as the grid-forming inverter. The solar inverter output is reduced from 100% to 0% as frequency increases from 51 Hz to 52 Hz. Figure 12.9 shows photographs of the installation.

12.2 Grid-Connected Microgrids

In many countries, initiatives to reduce carbon emissions are resulting in many small generating units connecting to the distribution network. These distributed generators are often powered by renewable energy sources. The traditional policy of connecting distributed generators to the distribution network has been based on a 'fit and forget' philosophy. With the increasing deployment of distributed generators, this approach creates difficulties for network operation, reduces the security of supply and increases the cost of operating the electric power system.

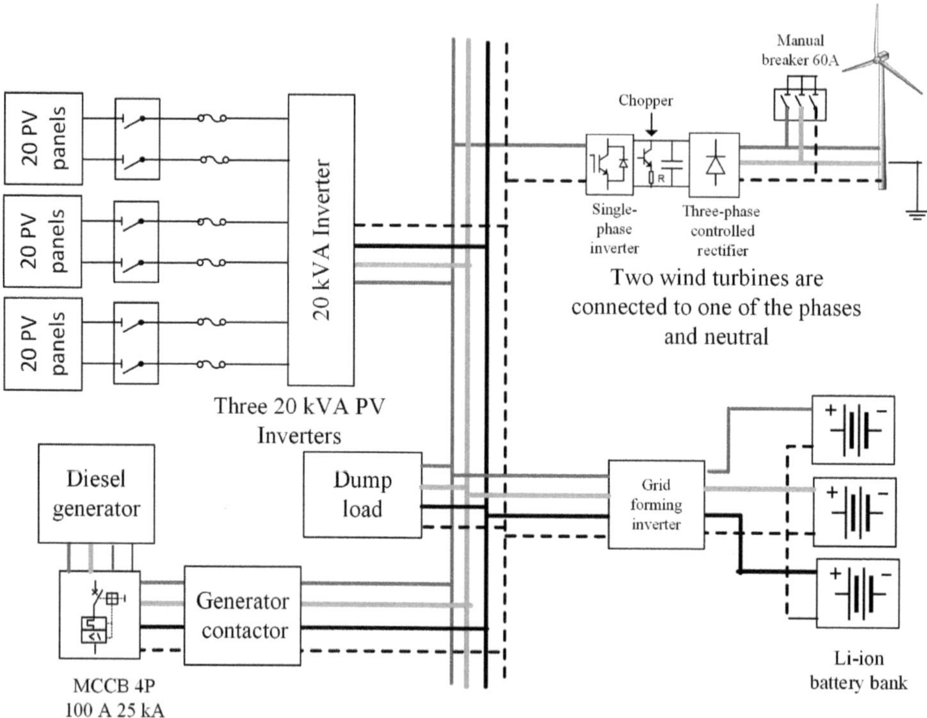

Figure 12.8 Eluvaitivu Island microgrid configuration.

To address these problems, distributed generators must take over responsibility from the large conventional power plants for providing the flexibility and controllability necessary to ensure the secure operation of the power system. Many approaches to providing flexibility and controllability have been discussed, and the two approaches that are currently considered to be promising are (1) microgrids and (2) coordinated control of a large number of distributed energy resources. These approaches shift responsibility for control from large generators to many small distributed generators and controllable loads.

A microgrid can be defined as an electrical network of small modular distributed generating units (whose prime movers may be photovoltaics, fuel cells, microturbines or small wind generators), energy storage devices and controllable loads. A microgrid can operate in grid-connected or islanded mode and increase the reliability of energy supplies by disconnecting from the grid in the case of network faults or reduced power quality.

12.2.1 Microgrid Architectures

Around the world, several microgrid demonstration projects have been developed using different topologies. The most common topology is the ac microgrid. This is similar to that shown in Figure 12.5 but with a grid connection (see Figure 12.10a).

Figure 12.9 Eluvaitivu Island installation. (a) Solar panel and microgrid assembly. (b) Solar inverters. (c) Wind turbines. (d) Li-ion batteries
Photos: Ceylon Electricity Board.
(A black and white version of this figure will appear in some formats. For the colour version, refer to the plate section.)

The alternative microgrid topology is the dc microgrid. Even though the dc microgrid is a relatively new concept, dc power distribution is well established in the telecommunication sector and data centres can be thought of as early dc microgrids. Figure 12.10b shows the configuration of a dc microgrid. Even though each microsource has a controller, as shown in Figure 12.10a, they are not shown in Figure 12.10b, for clarity. There are several advantages of a dc microgrid over its ac counterpart:

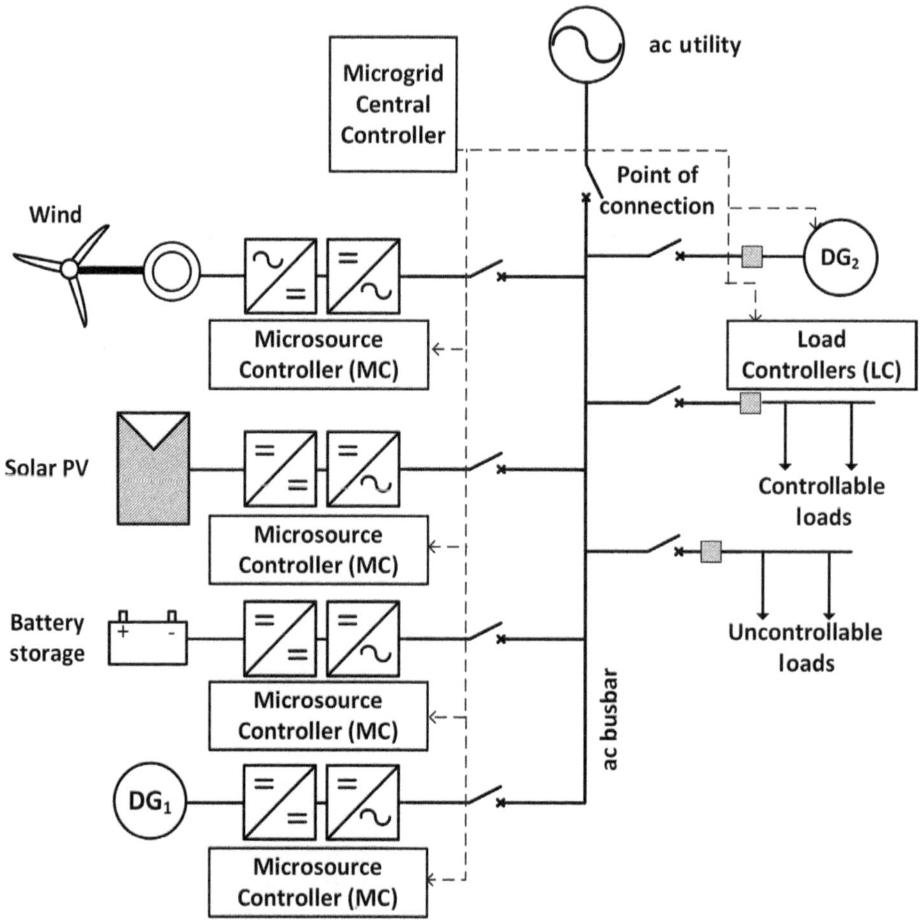

Figure 12.10 (a) ac microgrid.

- They enable the connection of dc operated loads such as IT equipment, LED lamps and renewable energy sources such as solar PV directly or through a reduced number of conversion stages, which enhances overall efficiency.
- They provide a convenient platform to connect energy storage units such as batteries and supercapacitors, which are dc in nature.
- They do not require frequency and reactive power control and so the complexity of the control system of a dc microgrid can be reduced.

A hybrid microgrid is a combination of an ac and a dc sub-grid, which are connected through bidirectional ac–dc converters named ‘interlinking converters’ as shown in Figure 12.10c.

12.2.2 Control of Microgrids

As shown in Figure 12.10, the microgrid has two types of controller: the Microgrid Central Controller and Microsource Controllers. When there are multiple

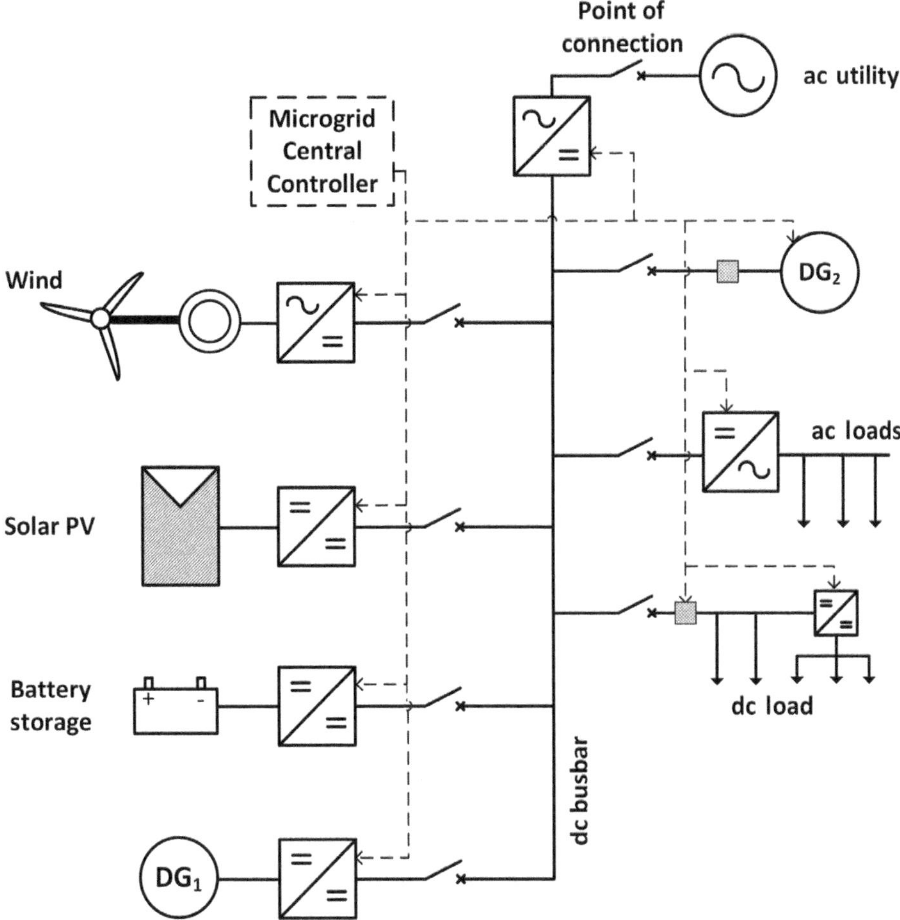

Figure 12.10 (b) dc microgrid (microsource controllers are not shown).

microgrids, an additional energy management level is also employed. These controllers are organized in a hierarchical control architecture (Figure 12.11). The local control of microsources and loads is at a low level with direct communication to the Microgrid Central Controller. The energy management system is at the top and mainly plays a supervisory role.

At the energy management level, conventional approaches to the operation of distributed generators are enhanced with new features. The operation of the microgrid during disconnection and reconnection to the network is of particular importance.

The functions of a microgrid central controller (MGCC) include monitoring the active and reactive power of the generators to optimize the operation of the microgrid by sending control signal settings to the generators and controllable loads. It may perform functions such as load forecasting, demand-side management, economic scheduling of generators and security assessment.

The microsource controllers (MCs) control the power electronic interface of the generators and can be enhanced with various degrees of intelligence. They may use

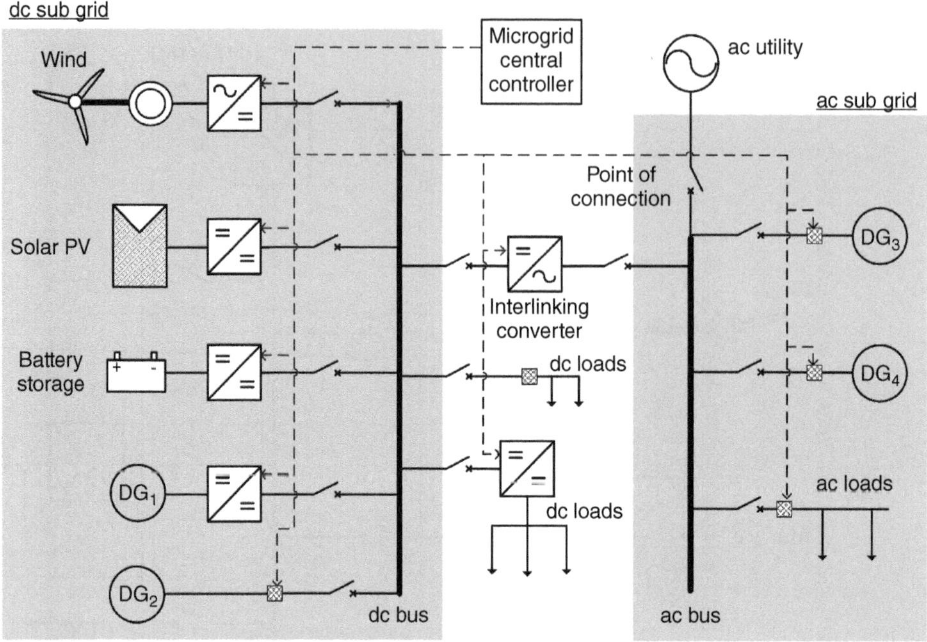

Figure 12.10 (c) Hybrid microgrid (microsource controllers are not shown).

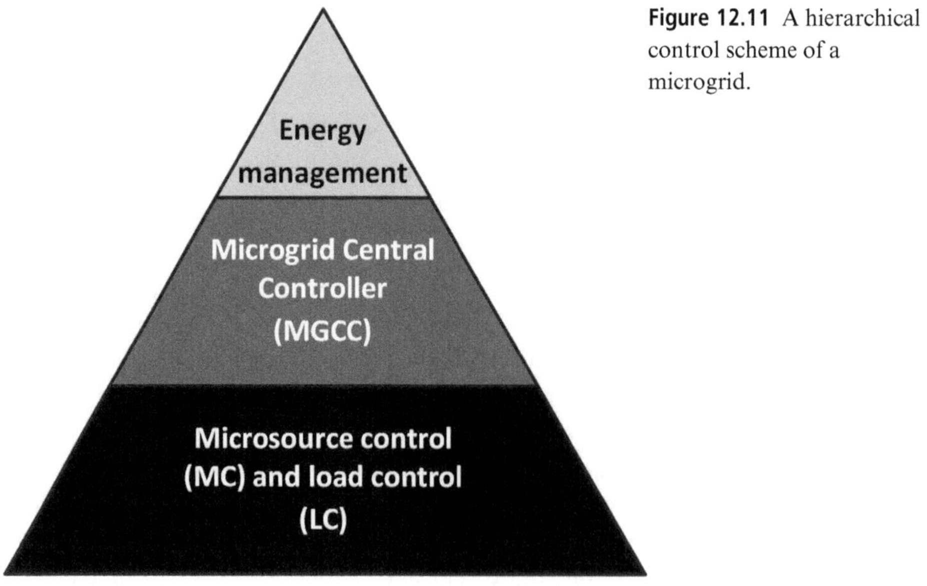

Figure 12.11 A hierarchical control scheme of a microgrid.

local information to control the voltage and the frequency of the microgrid in transient conditions. MCs have to be adapted to each type of microsource (PV, fuel cell, microturbine, etc.). Load controllers (LCs) are installed at the controllable loads to provide load control capabilities.

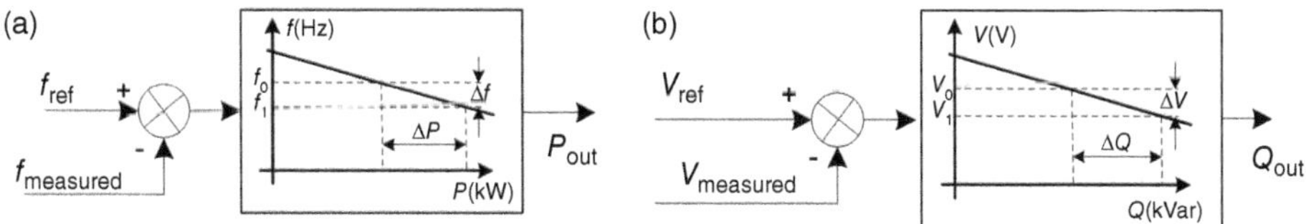

Figure 12.12 Control of the master converter. (a) Active power droop controller. (b) Reactive power droop controller.

The MGCC may use one or more converters as the leader and the other converters as followers. The leader can be operated on a droop controller as in the case of a synchronous generator. Depending on the frequency and voltage magnitude of the microgrid, the output active and reactive power of the leader converter are controlled as in Figure 12.12. The followers are often operated on active and reactive power references set by the MGCC.

Example 12.3 – Microgrid Control

The microgrid shown in Figure 12.13 has three microsources (MS). MS_1 and MS_2 are generators and MS_3 is a battery energy storage system. The combined inertia of the system is 7 kWs/kVA. The system base is chosen to be 50 kVA. The load on the microgrid is 30 kW. MS_1 generates 10 kW, MS_2 generates 15 kW, MS_3 is fully charged and floats. Assume network losses are negligible. If the grid fails, what will happen to the islanded microgrid frequency when

(a) all three sources are operating on fixed power references, i.e. the references for MS_1, MS_2 and MS_3 are 10, 15 and 0 kW,
(b) MS_1 and MS_3 operate on fixed power references of 10 kW and 0 kW, and MS_2 operates on a droop controller with a droop of 0.075 pu.
(c) MS_1 operates on a fixed power reference of 10 kW, MS_2 operates on a droop controller with a droop of 0.075 pu and MS3 operates on a droop controller with a droop of 0.05 pu and a secondary controller with a gain of 3?

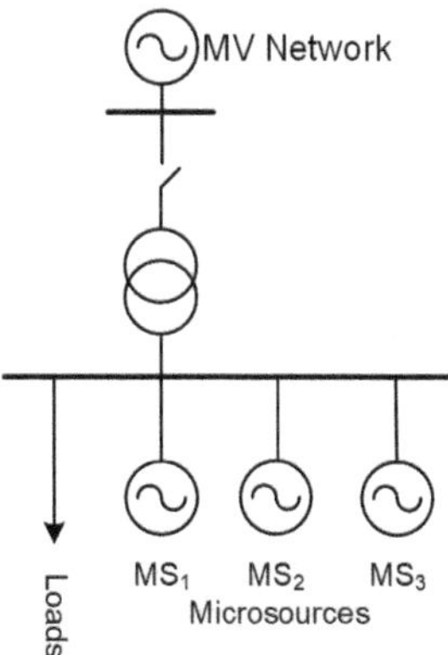

Figure 12.13 Single bus representation of a microgrid.

Solution

(a) When the three generators are operating on a power reference, the power drawn from the grid is 30 – 10 – 15 = 5 kW; when the grid fails and the microgrid is islanded, the microgrid can be represented by the block diagram shown in Figure 12.14a.

Per unit value of total generation $= (10 + 15 + 0)/50 = 0.5$ pu
Per unit value of the load $= 30/50 = 0.6$ pu
The block diagram shown in Figure 12.14a was implemented on MATLAB/Simulink and the frequency response is shown in Figure 12.14b. In the figure, the grid fails at 1 second. As can be seen, the frequency continues to drop until all generators trip on under-frequency and then the microgrid collapses.

(b) When MS_2 operates on a droop and the other two sources on fixed power references, the system can be represented by the block diagram shown in Figure 12.15 a.

Per unit value of fixed generation $= (10 + 15 + 0)/50 = 0.5$ pu
Per unit value of the load $= 30/50 = 0.6$ pu
The block diagram shown in Figure 12.15a was implemented on MATLAB/Simulink, and the power output and frequency, are shown in Figure 12.15b.

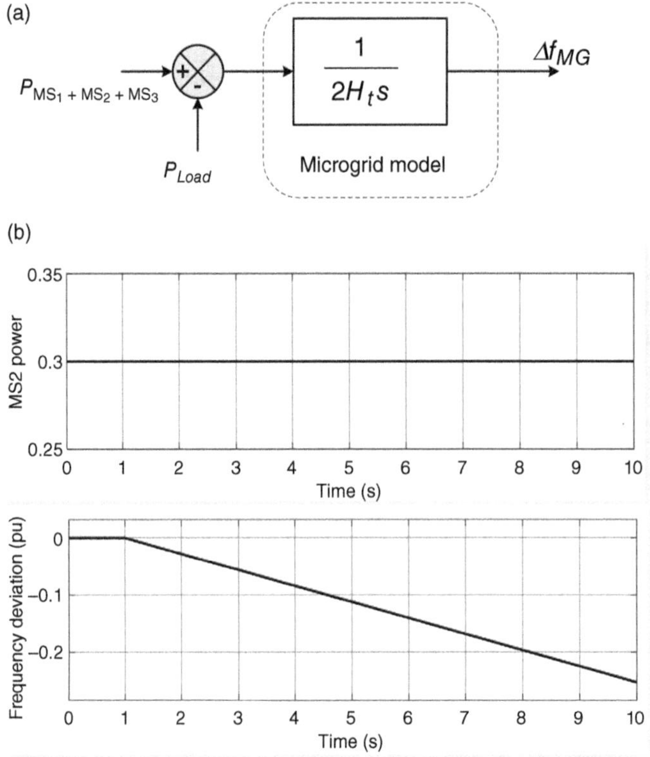

Figure 12.14 Microgrid performance when all three sources are operating on a power reference. The combined inertia of the system (H_t) is 7 kWs/kVA. (a) Block diagram. (b) Power output of MS2 and frequency deviation.

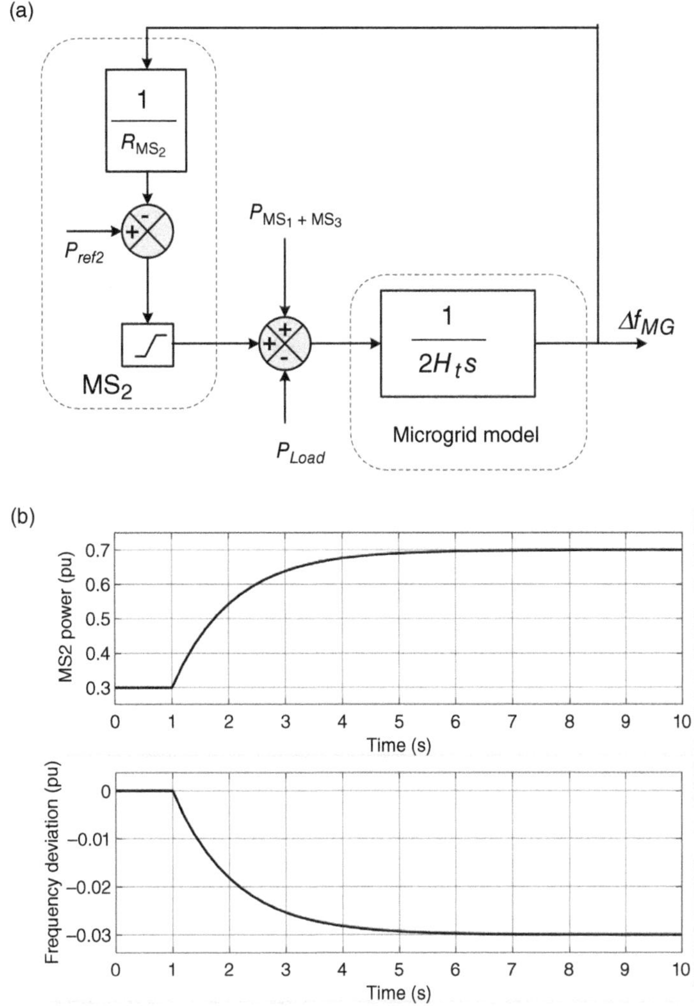

Figure 12.15 Microgrid performance when MS$_2$ operates on a droop. (a) Block diagram. (b) Power output of MS$_2$ and frequency deviation.

As can be seen, the frequency will stabilize at a value lower than the nominal. This can be explained using the droop characteristic shown in Figure 12.16. Initially MS$_2$ operates at P_{ref2} with a frequency equal to the nominal frequency f_0. As the frequency drops, MS$_2$ produces more power and settles at frequency f_p, less than f_0.

(c) When MS$_2$ operates on a droop and MS$_3$ operates on a droop with secondary control the system is represented by the block diagram shown in Figure 12.17a.

Per unit value of fixed generation = 10/50 = 0.2 pu
Per unit value of the load = 30/50 = 0.6 pu

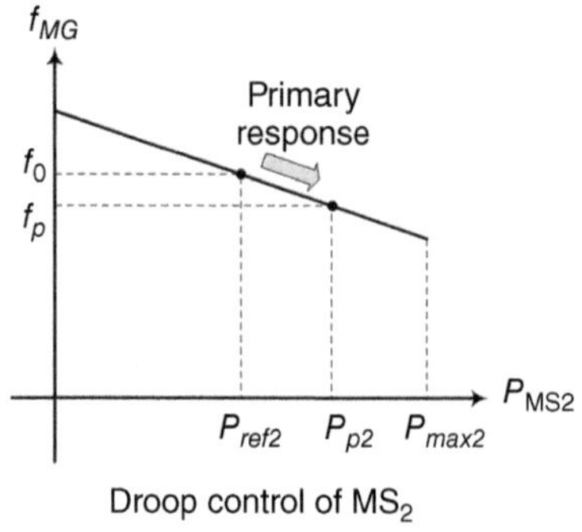

Figure 12.16 Droop characteristic of MS_2.

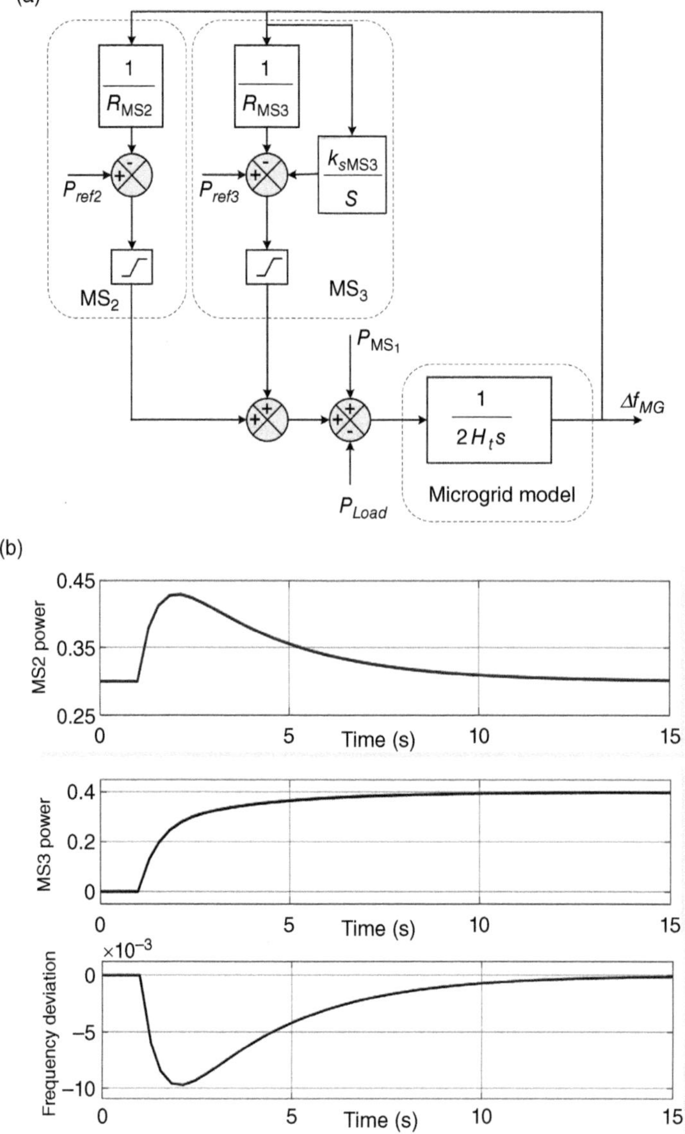

Figure 12.17 Microgrid performance when MS_2 operates on a droop and MS_3 operates on a droop with secondary control. (a) Block diagram. R_{MS2} and R_{MS3} are the droops of MS_2 and MS_3, and k_s is the secondary controller gain. (b) Power output of MS_2 and MS_3 and frequency deviation.

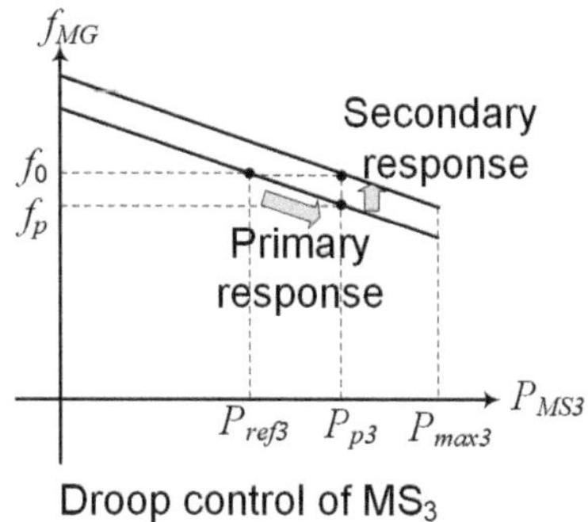

Figure 12.18 Droop and secondary control of MS_3.

The block diagram shown in Figure 12.17a was implemented on MATLAB/ Simulink and the frequency response shown in Figure 12.17b was obtained. As can be seen, the frequency will stabilize at the nominal value. This can be explained using the droop characteristic shown in Figure 12.18. Initially MS_3 operates at P_{ref3} with a frequency equal to the nominal frequency f_0. As the frequency drops, it produces more power and initially settles at a frequency f_p less than f_0. Then the secondary control action restores the frequency to f_0.

12.3 Community Energy

Community Energy (CE) projects are developed and operated by individuals or groups within a local community. Members of the community decide what kind of energy project they want and how they will implement it. CE projects are owned and operated by members of the community and share the benefits of the project with the community as a whole. It is hoped that CE will create a new paradigm of locally controlled energy systems supplied by distributed energy technologies and managed through local trading mechanisms. Figure 12.19 shows the different entities that operate within a CE system and the information flow between these different actors.

As can be seen, a CE system consists of some consumers who occupy dwellings with controllable appliances/devices that can be controlled for demand-side management, and others living in dwellings with traditional passive loads. Some consumers occupy dwellings that have generating equipment such as rooftop solar panels and/or wind turbines and energy storage; these are referred to as prosumers.

The community energy system provides a number of benefits:

- maximum usage of limited resources such as the battery storage and network infrastructure,
- fulfilling the energy demand of consumers within the community while enabling peer-to-peer transactions between prosumers,
- fulfilling the energy and ancillary service requests of the utility distribution grid to which the community power network is connected.

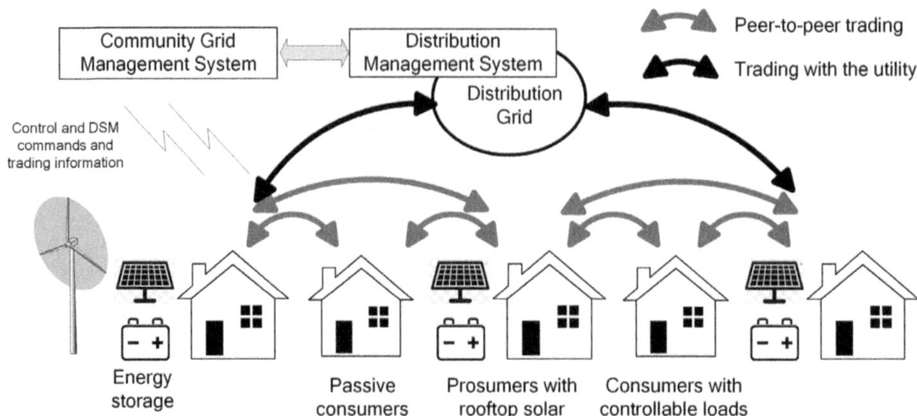

Figure 12.19 Typical community power system.

It is anticipated that community energy systems will result in lower costs and greater engagement of energy users, who will be encouraged to reduce their energy demands. Community energy systems often use peer-to-peer trading and demand-side participation.

12.3.1 Peer-to-Peer Trading

According to an IRENA report published in 2020, 'Peer-to-peer (P2P) electricity trading is a business model, based on a platform, that serves as an online marketplace where consumers and producers "meet" to trade electricity directly, without the need for an intermediary.'

To realize P2P trading, the following are needed:

(a) smart meters that can monitor power production by prosumers within a pre-defined averaging time,
(b) an ICT network to enable communication between participants and facilitate payments,
(c) an Energy Management System that manages energy production and consumption to meet the basic quality of supply, as well as forecast power demand and supply,
(d) a platform for P2P electricity trading that encompasses algorithms for automated execution of P2P transactions,
(e) interoperable protocols for coordination among system/network/market/platform operators, consumers and prosumers.

Several companies are currently creating platforms and developing local ICT systems for community energy systems and some pilots have already emerged.

For some CE systems, P2P trading is automated by Distributed Ledger Technology, an example of which is blockchain. Blockchains connect blocks containing information related to a transaction and arrange them in their chronological order. This ledger contains all the transactions that have taken place until the present state and is shared among all peers. The inherent feature of blockchains is that they facilitate delegation of authority among stakeholders without the supervision of a third-party.

12.3.2 Demand-Side Participation

Demand-side participation, described in Section 10.7, plays a key role in community grids. Appliances with thermal storage such as air conditioning and heating units can easily be controlled to shift the community demand. There is a recent surge of interest to add other appliances such as washing machines, tumble dryers, dishwashers, fridges and ovens to residential demand-side programmes. Smart devices are equipped with communication capabilities and advanced control strategies to respond to external signals. The users can define a time period during which they prefer their smart appliances to be operated.

Example 12.4 – Peer-to-Peer Trading

Three houses have the loads and generation shown in the table. House 2 has an electric car that needs to be charged at 3 kW for one hour.

Loads and generation of three houses

	12:00–13:00			13:00–14:00		
	House 1	House 2	House 3	House 1	House 2	House 3
PV generation	5 kWh			4 kWh		
Domestic load	2 kWh	1 kWh	1 kWh	0.5 kWh	1 kWh	0.5 kWh

The tariffs are:

- House 1 sells electricity to other houses at 7 pence per kWh,
- all houses can draw electricity from the grid at 10 pence per kWh and sell to the grid at 8 pence per kWh.

Obtain the net income to the community under different operating modes and discuss the salient features of each operating mode. What is the best time to charge the electric car?

Solution

From 12:00 to 13:00:

net generation of House 1 is 3 kWh

All possible transactions are given in the table below. √ indicates that the option is active, H1, H2 and H3 are houses and C is the community. The – sign indicates selling and + sign indicates purchasing.

Possible transactions during 12:00–13:00

			Energy to House 1 (kWh)		Energy to House 2 (kWh)		Energy to House 3 (kWh)		Transactions	
Case	Peer-to-peer	EV	Grid	Peer	Grid	Peer	Grid	Peer	With the grid	With the peers
1			−3	0	1	0	1	0	H1: −3×8 H2: 1×10 H3: 1×10	
									C: −4p	
2		√	3	0	4	0	1	0	H1: −3×8 H2: 4×10 H3: 1×10	
									C: 26p	
3	√		1	−2	0	1	0	1	H1: −1×8	H1: −2×7 H2: 1×7 H3: 1×7
									C: −8p	
4	√	√	0	−3	2	2	0	1	H2: 2×10	H1: −3×7 H2: 2×7 H3: 1×7
									C: 20 p	

From 13:00 to 14:00:
net generation of House 1 is 3.5 kWh
All possible transactions are given in the table below.

Possible transactions during 13:00–14:00

			Energy to House 1 (kWh)		Energy to House 2 (kWh)		Energy to House 3 (kWh)		Transactions	
Case	Peer-to-peer	EV	Grid	Peer	Grid	Peer	Grid	Peer	With the grid	With the peers
1			−3.5	0	1.0	0	0.5	0	H1: −3.5×8 H2: 1.0×10 H3: 0.5×10	
									C: −13p	
2		√	−3.5	0	4.0	0	0.5	0	H1: −3.5×8 H2: 4.0×10 H3: 1.0×10	
									C: 17p	
3	√		−2.0	−1.5	0	1.0	0	0.5	H1: −2.0×8	H1: −1.5×7 H2: 1.0×7 H3: 0.5×7
									C: −16p	

Case	Peer-to-peer	EV	Energy to House 1 (kWh) Grid	Peer	Energy to House 2 (kWh) Grid	Peer	Energy to House 3 (kWh) Grid	Peer	Transactions With the grid	With the peers
4	√	√	0	−3.5	1.0	3.0	0	0.5	H2: 1.0×10	H1: −3.5×7 H2: 3.0×7 H3: 0.5×7
									C: 10p	

In both time periods, Cases 1 and 2 are without peer-to-peer trading and the other two cases allow peer-to-peer trading. Comparing Cases 1 and 2 with Cases 3 and 4 shows the benefit of the peer-to-peer trading and the reduction of total cost to all three houses.

However, House 1 could earn a higher income by trading with the grid. From 12:00 to 13:00, House 1 can earn 24 pence for Cases 1 and 2 verses 22 pence for Case 3 and 21 pence for Case 4. From 13:00 to 14:00, it could earn 30.4 pence for Cases 1 and 2 versus 26.5 pence for Case 3 and 24.5 pence for Case 4.

13:00–14:00 is the best time to charge the battery.

PROBLEMS

1. Using suitable diagrams, compare a dc with an ac microgrid for connecting renewable energy sources including solar PV, a variable speed wind turbine, and battery energy storage.
2. A dc motor operates a 10 kW variable speed load. The following efficiencies of power converters are given:

   dc–dc converter 90%

   dc–ac converter 85%

   Calculate the power consumed if the variable speed drive is connected to (a) an ac microgrid and (b) a dc microgrid. Also calculate the percentage reduction of power consumption.

   [**Answer:** (a) 13.07 kW, (b) 11.11 kW; 15%.]
3. Houses with solar photovoltaic (PV) and a Battery Energy Storage System (BESS) are connected to a low-voltage distribution network. In some houses, the PV and BESS are connected at dc and in others they are connected at ac. Using a suitable diagram, discuss how each house could support frequency events on the network.
4. What are the benefits offered by a community energy system? Illustrate each benefit with an example.

Table 12.4 Average power consumption of the house and the irradiance on the PV panel

Time	Average power consumption of the house (W)	Average irradiance (W/m^2)
00:00–06:00	250	0
06:00–07:00	750	100
07:00–08:00	500	200
08:00–09:00	500	300
09:00–10:00	500	400
10:00–11:00	500	600
11:00–12:00	500	800
12:00–13:00	500	1000
13:00–14:00	500	800
14:00–15:00	500	600
15:00–16:00	500	400
16:00–17:00	750	300
17:00–18:00	1000	200
18:00–20:00	1250	0
20:00–24:00	250	0

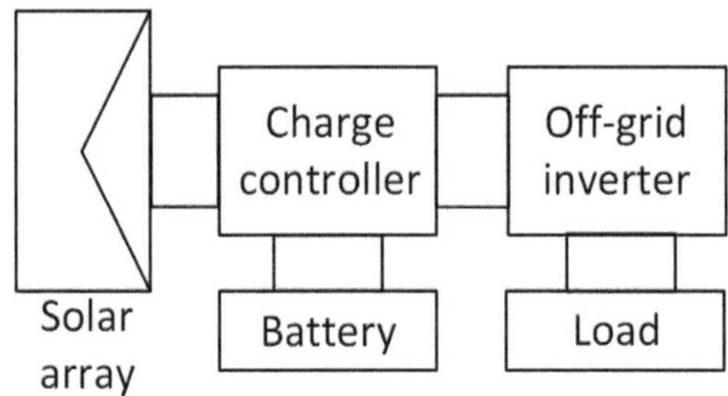

Figure 12.20 A diagram for Table 12.4.

5. A PV module has 60 cells in series. The following data are given for the module.
   - Maximum power at Standard Test Conditions (STC): 250 Wp
   - Open-circuit voltage at STC: 37.6 V
   - Short-circuit current at STC: 8.81 A
   - Voltage at maximum power point: 30.4 V
   - Current at maximum power point: 8.23 A

   An array of 8 modules is connected to a solar home system through a charge controller and an off-grid inverter, as shown in Figure 12.20. The charge controller has an efficiency of 95% and the inverter has an efficiency of 90%. The inverter provides a 230 V, 50 Hz single-phase supply to the house. The battery has an internal resistance of 1 Ω and open circuit voltage of 200 V. Ignore the effect of increased cell temperature.

   The average power consumption of the house and the irradiance on the PV panel on a day of clear sky is given in Table 12.4.

(a) When can the PV array fully supply the house?
(b) Consider operation between 12:00 and 13:00. Using an approximate V–I characteristic of the array, and assuming that the house load is disconnected from the inverter, compare the energy capture (i) if the battery is directly connected to the solar array and (ii) if the battery is connected through the charge controller.
(c) Can the present system supply the daily demand of the house?
(d) How many modules are required to supply the house demand? Assume that the initial and final state of charge of the battery is within limits when balancing the load.

[**Answer:** (a) 08:00–16:00; (b)(i) 1.84 kWh, (b)(ii) 1.9 kWh; (d) 10 modules.]

6. Two options are being considered to supply power to a community.

Option 1:	Option 2:
100 kW solar plant	150 kW solar plant
50 kW diesel generator	50 kW diesel generator
25 kW, 100 kWh battery energy storage	50 kW, 200 kWh battery energy storage
(from SoC_{min} to $SoC_{100\%}$)	(from SoC_{min} to $SoC_{100\%}$)

The costs of generation are:
cost of diesel generation (capital and fuel) = 15p/kWh
cost of solar generation (capital) = 10p/kWh
cost of discharging the energy storage = 12p/kWh

Figure 12.21 shows the power consumption of the community over a day (full line) and irradiance (dotted line).

(a) Compare the daily operating costs of the two options. Assume that the initial state of charge of the battery allows full utilization of its stored energy over the day.

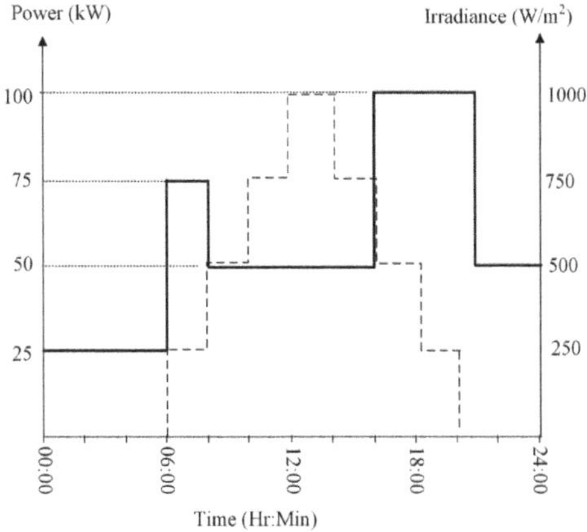

Figure 12.21 Load and irradiance over a day.

(b) Discuss whether the two options meet the demand if the state of charge of the battery is 25% (minimum allowable value) at the beginning of the day.
(c) Calculate the operating cost of the option that can supply the load.

[**Answer:** (a) Option 1 £148.25, Option 2 £129; (c) £209.25.]

7. A microgrid has a total generation of P_G and load of P_L. Show that frequency changes are governed by:

$$\Delta f = \frac{f}{2H}(P_G - P_L)\Delta t$$

where there is only one directly connected microsource with an inertia of H, P_G and P_L are pu quantities and Δt is the time after any dynamic event.
The microgrid has a solar PV system and a diesel generator. At a given time, the following operating conditions hold:
Load on the microgrid = 30 kW
Operating frequency = 50 Hz
Solar PV output = 20 kW
If the load changes to 40 kW, what is the microgrid frequency after 500 ms? Assume network losses are negligible and the system base is 50 kVA. The inertia of the diesel generator is 4 kWs/kVA and solar PV output remains constant at 20 kW.

[**Answer:** 49.375 Hz.]

FURTHER READING

Devine-Wright P., Community versus local energy in a context of climate emerge, *Nature Energy*, **4** (2019), 894–896.
Provides a good insight into community energy and local energy so that one can easily understand what is meant by each term.

Mini Grids for Half a Billion People: Market Outlook and Handbook for Decision Makers, Technical Report 014/19, 2019, International Bank for Reconstruction and Development / The World Bank Group. Available at: https://openknowledge.worldbank.org/bitstream/handle/10986/31926/Mini-Grids-for-Half-a-Billion-People-Market-Outlook-and-Handbook-for-Decision-Makers-Executive-Summary.pdf?sequence=1&isAllowed=y. Last accessed 5 September 2022.
This report provides statistics, definitions and future needs of minigrids and explains where minigrids fit in the electricity sector.

Microgrids 1: Engineering, Economics, and Experience, CIGRE Working Group C6.22, October 2015.
This report covers topics such as definitions, benefits, technologies, business cases and a number of annexures with databases and technologies.

Peer-To-Peer Electricity Trading Innovation Landscape Brief, 2020, International Renewable Energy Agency (IRENA). Available at: https://irena.org/-/media/Files/IRENA/Agency/Publication/2020/Jul/IRENA_Peer-to-peer_trading_2020.pd. Last accessed 5 September 2022.

This brief provides an overview of the peer-to-peer (P2P) electricity trading business model that emerged as a platform-based scheme for increasing the integration of distributed energy resources into power systems.

ONLINE RESOURCES

Exercise 12.1 Island generation optimization: this exercise uses Excel and Solver to determine the optimal mix of generators of a small island power system. It is based on a single annual optimization of the load and generators.

TUTORIAL I

Electrical Engineering

I.1 Direct Current (dc)

The modern electrical power system is an alternating current (ac) system; other than some small, isolated installations, all power systems run on ac (either 50 or 60 Hz). However, for isolated off-grid applications, small dc systems powered by photovoltaic arrays and batteries are often used. Even though these systems produce power at 12, 24 or 48 V, the power generating source, i.e. the photovoltaic (PV) panel or battery bank, consists of a large number of small cells. Each cell produces 0.5–1.5 V and is connected in series to obtain the required voltage. The strings of series connected cells are then connected in parallel to meet the current requirement of the circuit.

As dc is not time varying, dc quantities can be treated as scalars. Figure I.1 shows a 60-cell solar module. Each cell has an open-circuit voltage of 0.6 V. As all 60 cells are connected in series, the voltage at the terminals of the PV module is the addition of the voltage across each cell and is equal to $60 \times 0.6 = 36$ V.

Six modules of the type shown in Figure I.1 are connected to form a PV array as shown in Figure I.2. At 1000 W/m^2 irradiance, each cell has a short-circuit current of 3.6 A.

For the photovoltaic array shown in Figure I.2:

for series connected modules, $V = V_1 + V_2 = 36 + 36 = 2 \times 36 = 72$ V
for parallel connected modules, $I = I_1 + I_2 + I_3 = 3 \times 3.6 = 10.8$ A

I.2 Alternating Current (ac)

Modern electrical power systems mainly follow the arrangement shown in Figure 10.3. Generators operate at medium voltage to supply step-up transformers to create transmission voltages. Power is transported through the high-voltage interconnected transmission network and finally stepped down through a series of distribution transformers to distribution circuits for delivery to the customers. This arrangement is shown in more detail in Figure I.3 with the symbols that are typically used in the one-line representation of power system components. Even though

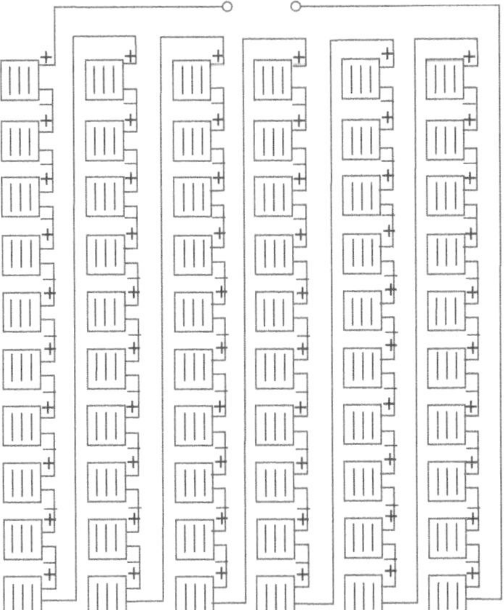

Figure I.1 A photovoltaic module.

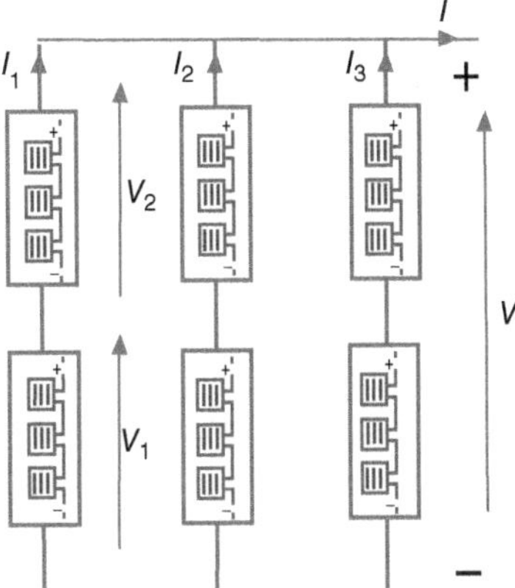

Figure I.2 A photovoltaic array.

single lines are used in this representation, each generator, transformer, overhead line and cable has three phases commonly referred to as the A, B and C phases. The only exception is the connection to the houses that in the UK is usually single phase and consists of a phase and neutral conductor.

The single-phase supply provided to houses is an ac waveform with a shape as shown in Figure I.4. This waveform is characterized by its peak value, V_m, and period T. The period (T) is the time taken for one cycle and is measured in seconds. The positive or negative maximum value is known as the peak value. The peak value is measured in volts (V).

Even though the sinusoidal waveform is characterized by the peak and period, in power systems the ac supply provided to houses is described by frequency and root mean square (RMS) values.

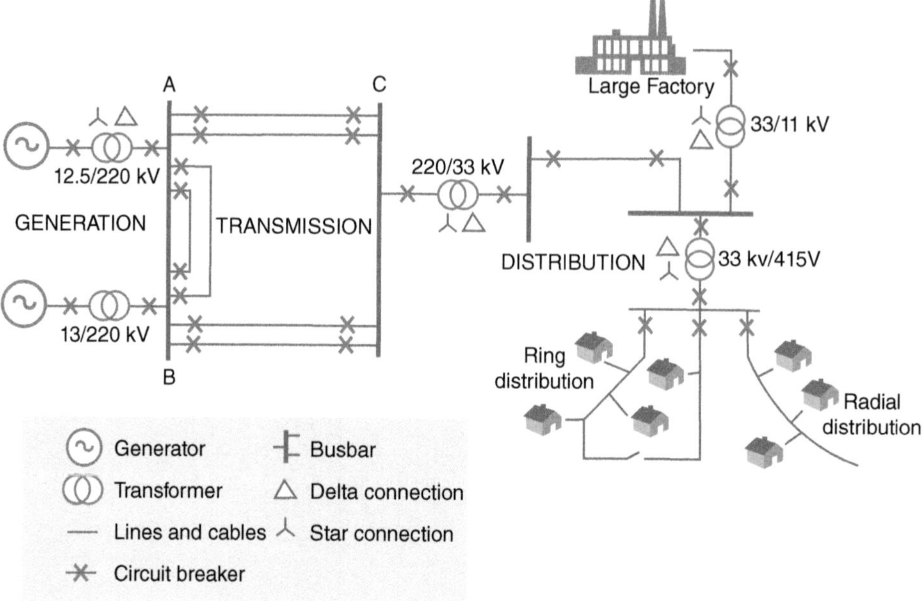

Figure I.3 Modern power system.

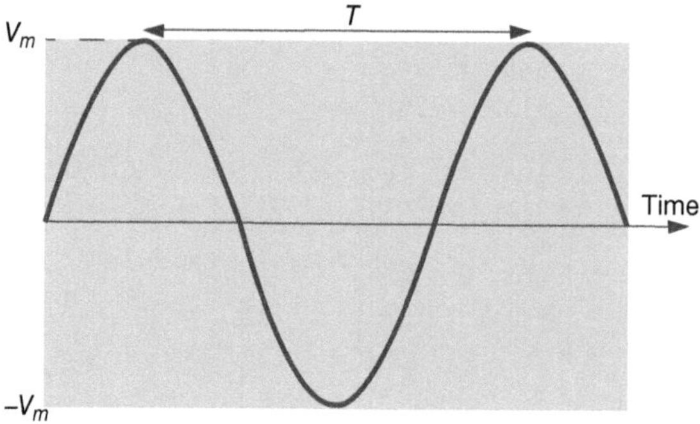

Figure I.4 Sinusoidal voltage waveform.

Frequency (f)

The frequency is the number of cycles that a waveform completes in one second. Frequency has the symbol f and is measured in hertz (Hz). Period and frequency are related by $f = 1/T$; therefore the period, T, is 20 ms in a 50 Hz system and 16.66 ms at 60 Hz.

RMS Value

The RMS current is the value of current alternating as a sine wave that gives the same heating effect as a direct current of the same magnitude. This is given by peak value divided by $\sqrt{2}$. Even though the RMS value is defined for currents, it is also used for the voltage. A house which receives an RMS voltage of 230 V has a peak voltage $V_m = \sqrt{2} \times 230 = 325.3$ V.

Example I.1 – Instantaneous Value of a Sinusoidal Voltage

Write an expression for a sinusoidal voltage with RMS value of 230 V and frequency of 50 Hz. Hence calculate the instantaneous voltage at 2 ms from the zero crossing.

Solution

$$\begin{aligned} v(t) &= 230 \times \sqrt{2}\sin(2\pi \times 50t) \\ &= 325.27\sin(100\pi t) \end{aligned}$$

At 2 ms

$$v(t) = 325.27\sin(100\pi \times 2 \times 10^{-3}) = 191.19 \quad [\text{V}]$$

(Note that the sine term should be calculated using radians.)

I.2.1 Resistors

A heater connected to an ac supply produces heating due to its resistance. Figure I.5b shows the current through a 1 kW heater when it is connected to 230 V mains supply. The voltage supplied is shown in Figure I.5a. As can be seen, the voltage and current waveforms are in-phase.

For a pure resistor, if an ac voltage source $v(t) = V_m \sin(\omega t)$ is connected across it, the current through the resistor is determined by Ohm's law and is given by $i(t) = (V_m/R)\sin(\omega t)$. Further, the instantaneous power is the product of instantaneous voltage and current. Since $v(t) = R \times i(t)$, the instantaneous power, $p(t)$, is given by

$$p(t) = v(t) \times i(t) = \frac{v(t)^2}{R} = i(t)^2 \times R \tag{I.1}$$

The average power dissipation in the resistor is given by:

$$\begin{aligned} P_{\text{ave}} &= \frac{1}{T}\int_0^T p(t)dt = \frac{V_m^2}{R \times T}\int_0^T \sin^2 \omega t \times dt = \frac{V_m^2}{R \times T}\int_0^T \left[\frac{1-\cos(2\omega t)}{2}\right] \times dt \\ &= \frac{V_m^2}{2R} = \frac{V^2}{R} = VI = I^2 R \end{aligned} \tag{I.2}$$

where

V and I are RMS quantities.

The instantaneous and average power dissipation of the heater are shown in Figure I.5c. The instantaneous power varies at double the frequency of the voltage and current with a positive average value. The power of this circuit converts electrical energy into heat and is called active power. The active power is measured in W or kW.

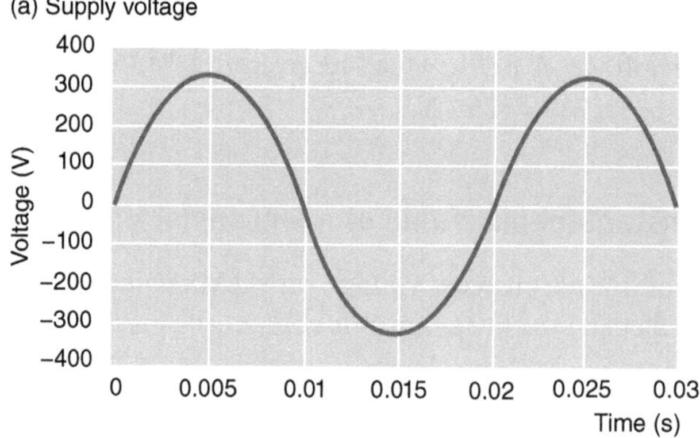

(b) Current through the heater

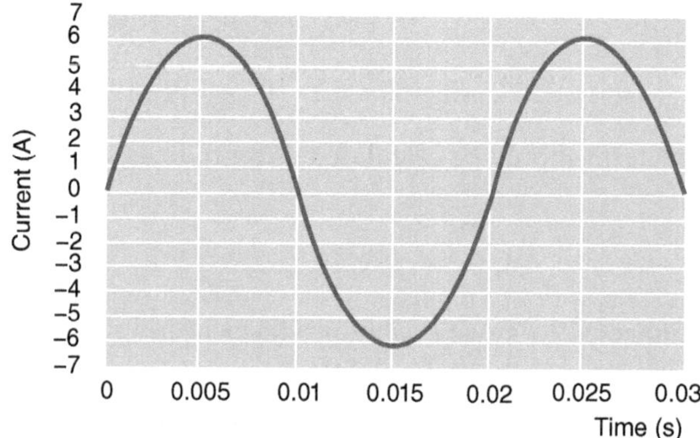

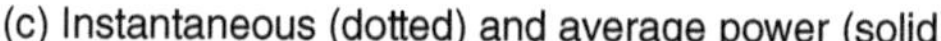

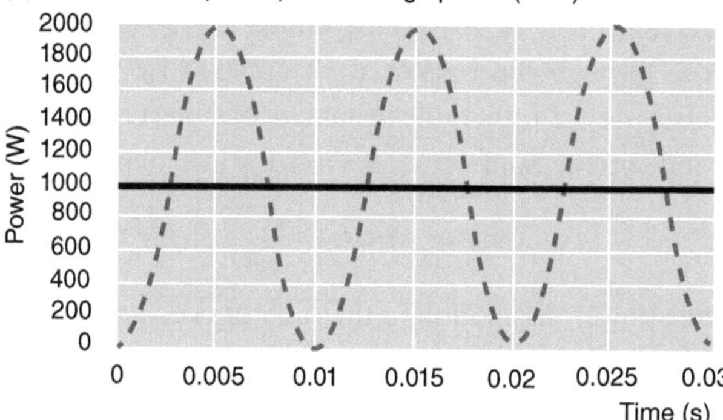

Figure I.5 Voltage across, current through and power consumption of the heater.

For the 1 kW heater

$$V = 230 \text{ V}$$

$$P_{\text{ave}} = 1000 \text{ W}$$

$$I = \frac{1000}{230} = 4.34 \text{ A}$$

The peak value of the current $= 4.34 \times \sqrt{2} = 6.14$ A

Resistance of the heater = V/I = 230/4.34 = 53 Ω

I.2.2 Inductors

An inductor is basically a coil which stores energy temporarily in a magnetic field when a current flows through it. Loads such as fans, pumps and induction heaters have such coils. Even though a coil has a small resistance in addition to its inductance, in the following analysis the resistance is neglected for clarity.

The relationship between instantaneous current and voltage for a pure inductance is $v = L\,{}^{di}\!/_{dt}$. For an applied voltage v as shown in Figure I.6, the current can be obtained by integrating the voltage:

$$\begin{aligned} i(t) &= \frac{1}{L}\int_0^t v\,dt = \frac{1}{L}\int_0^t V_m \sin(\omega t)dt = -\frac{V_m}{\omega L}\cos\omega t \\ &= \frac{V_m}{\omega L}\sin\left(\omega t - \frac{\pi}{2}\right) \end{aligned} \tag{I.3}$$

The voltage, current and instantaneous power waveforms of a 500 mH inductor are shown in Figure I.6. The current through the inductor lags the voltage by 90°.

For a pure inductor, $v(t) = L\dfrac{di(t)}{dt}$ and the instantaneous power is

$$p(t) = v(t) \times i(t) = L \times i(t) \times \frac{d}{dt}i(t) \tag{I.4}$$

If $i(t) = I_m \sin\omega t$, the average power dissipation in the inductor is given by

$$P_{\text{ave}} = \frac{\omega L I_m^2}{T}\int_0^T (\sin\omega t \times \cos\omega t)dt = 0 \tag{I.5}$$

As shown in Figure I.6c, the instantaneous power alternates with a zero average value. The power associated with energy oscillating in and out of an inductor is called reactive power. The reactive power is measured in var or kvar.

I.2.3 Capacitors

The capacitor stores energy electrostatically in an electric field. As discussed later, capacitor banks are often used to compensate for reactive power drawn by loads.

The relationship between current and voltage for a pure capacitance is $i = C\,{}^{dv}\!/_{dt}$. If a sinusoidal voltage v is applied across the capacitor, the current through the capacitor is given by

$$i(t) = C\frac{d}{dt}(V_m \sin\omega t) = \omega C V_m \cos\omega t = \omega C V_m \sin\left(\omega t + \frac{\pi}{2}\right) \tag{I.6}$$

The current through the capacitor leads the voltage by 90°.

I.2.4 Phasor Representation of ac Quantities

Sine waves of a constant frequency are much easier to deal with if they are expressed as phasors. The electrical power system normally operates at a constant frequency (either 50 or 60 Hz) and so phasor representation is used.

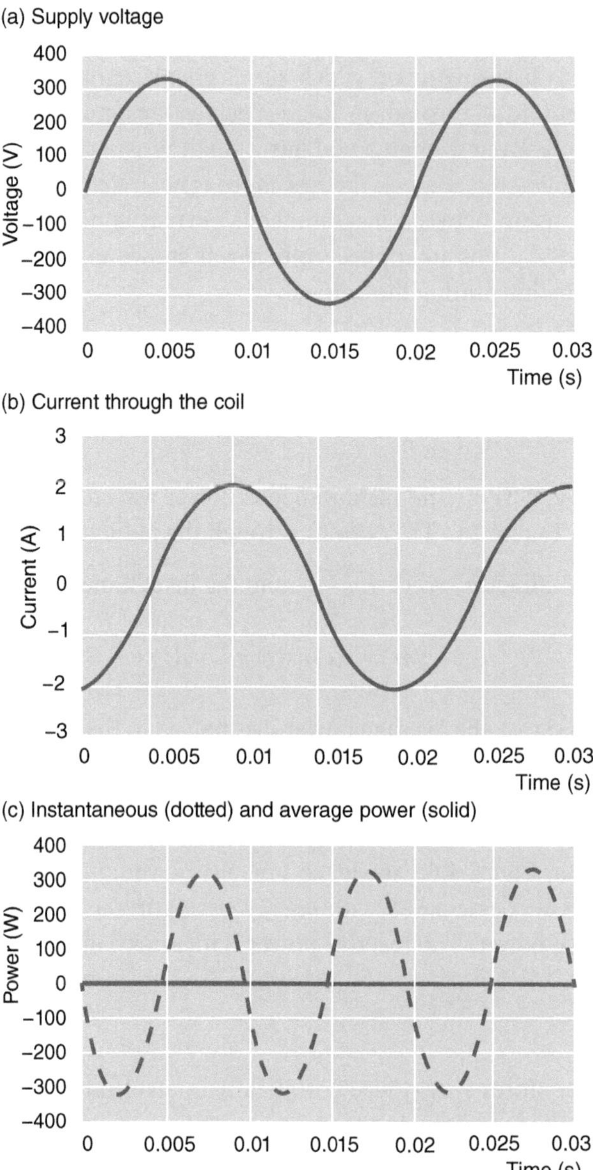

Figure I.6 Voltage across, current through and power consumption of an inductor.

Consider the phasor OA shown in Figure I.7a. Assume that it is rotating anti-clockwise at 100π rad/s. At the position shown, it represents point A on the voltage waveform shown in Figure 1.7b. When the phasor OA rotates by $\pi/2$, the time lapsed is 0.005 s and the projection of the phasor on the waveform corresponds to point B. Similarly, when the phasor OA rotates by π, the time lapsed is 0.01 s and the projection of the phasor on the waveform corresponds to point C. Hence the sinusoidal signal shown can be represented by a rotating phasor with a magnitude equal to the peak value of the ac waveform and angular velocity equal to the angular velocity of the ac waveform $(2 \times \pi \times 50 = 100\pi)$.

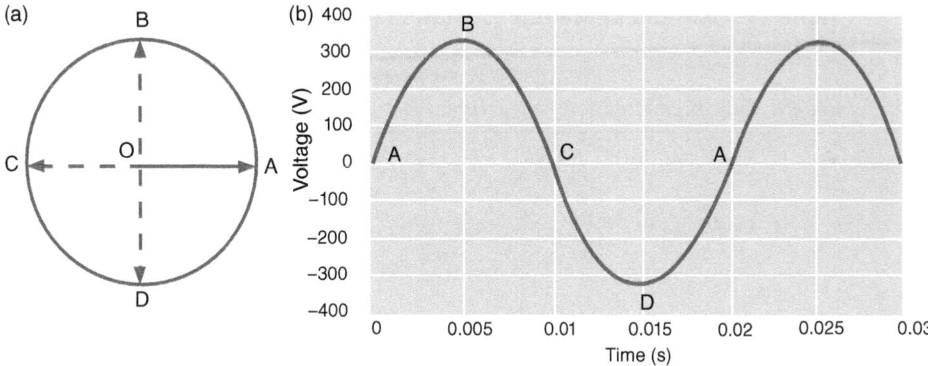

Figure I.7 Phasor representation.

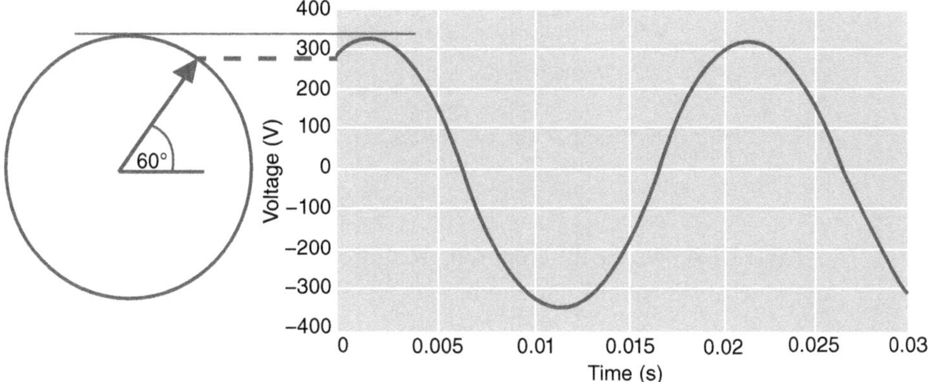

Figure I.8 Phasor representation of a phase-shifted waveform.

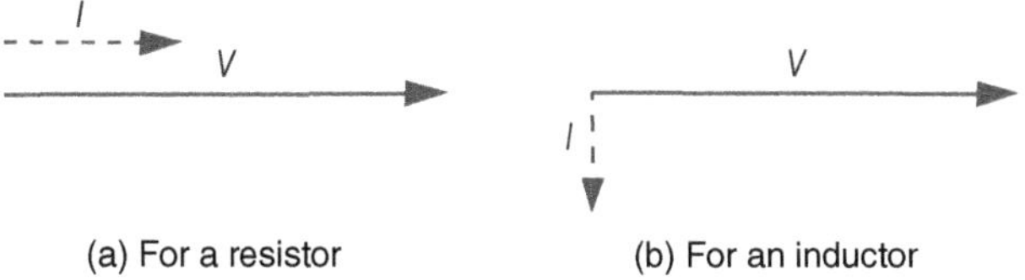

Figure I.9 Phasor representation of a phase-shifted waveform.

Now if the sinusoidal waveform is shifted by 60° as shown in Figure I.8, then the initial position of the rotating phasor is also shifted by the same angle.

In the power system all the waveforms have the same angular velocity (since the frequency is common). Hence it is convenient to represent the sinusoidal quantities by phasors that show only their initial position at time zero. Further, it is common to use the RMS instead of the peak values, and so in phasor representation the magnitude of the phasor is set to the RMS value of the ac waveform.

Figure I.9 shows the phasor representation of current and voltage corresponding to the waveforms shown in Figures I.5 and I.6.

As the phasors have a magnitude and an angle, they can be represented by a vector in the complex plane as shown in Figure I.10.

Table I.1 Time and phase representation of the ac quantities of Figure I.9b

Signal	Time representation	Phasor representation (polar form)	Phasor representation (cartesian form)
V	$\mathrm{V} = V_m \sin(\omega t)$	$\mathrm{V} = \frac{V_m}{\sqrt{2}} \angle 0$	$\mathrm{V} = \frac{V_m}{\sqrt{2}} + j0$
I	$\mathrm{I} = I_m \sin\left(\omega t - \pi/2\right)$	$\mathrm{I} = \frac{I_m}{\sqrt{2}} \angle(-\pi/2)$	$\mathrm{I} = 0 - j\frac{I_m}{\sqrt{2}}$

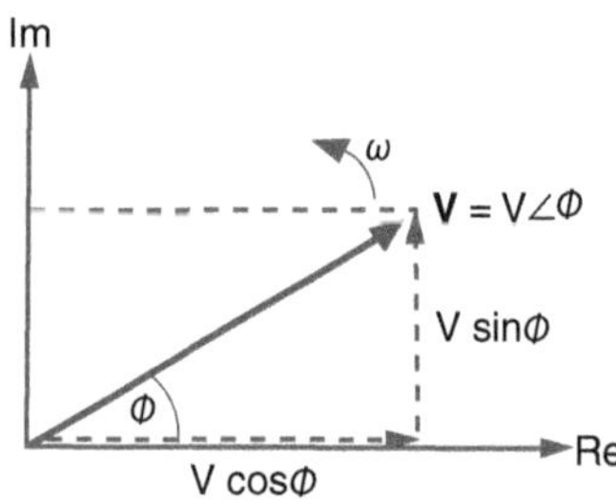

Figure I.10 Phasor diagram of **V** (shown on real and imaginary axes as a vector).

Table I.1 shows the different representation of the voltage and current phasors of an inductor shown in Figure I.9b.

I.2.5 Inductive Loads

Most domestic and commercial electricity consumers have many inductive loads such as fans, air conditioners, fridge-freezer, washing machine, etc. These loads can be represented approximately by a resistor and an inductor connected in series across an ac voltage source as shown in Figure I.11a. The current, **I**, is limited by the impedance of the circuit, given by $\mathbf{Z} = R + j\omega L = R + jX_L$.[1] Then,

$$\mathbf{V} = \mathbf{I}(R + jX_L) = \mathbf{IZ}$$

The phasor diagram is shown in Figure I.11b. The instantaneous current can be divided into two components, a component in phase with the voltage, I_P (which causes the active power), and a component with 90° phase shift, I_Q (which causes the reactive power).

The instantaneous power associated with this circuit and its active and reactive power components are shown in Figure I.12.

From Figure I.11c:

$$\text{Active power, } P = V \times I_P = VI\cos\phi \quad [\mathrm{W}] \tag{I.7}$$

$$\text{Reactive power, } Q = V \times I_Q = VI\sin\phi \quad [\mathrm{var}] \tag{I.8}$$

[1] Complex quantities are shown as **bold** type.

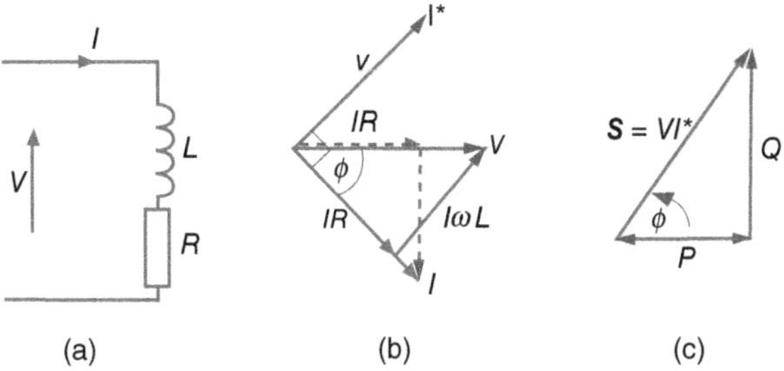

Figure I.11 An R–L circuit (inductive load): (a) circuit, (b) phasor diagram, (c) power triangle.

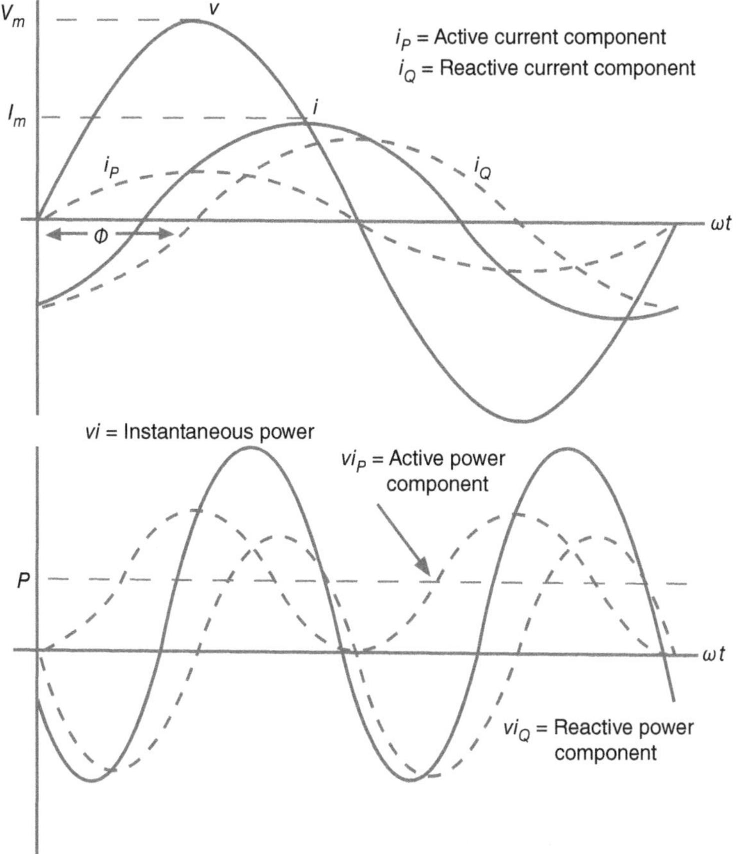

Figure I.12 Active and reactive power associated with an inductive load.

It is conventional to define **S**, the apparent power, as **VI***, where $\mathbf{I}^*$ is the conjugate of the current phasor (see Figure I.11b). In Figure I.11c, $S = \sqrt{P^2 + Q^2} = |\mathbf{VI}^*|$. S is measured in VA or kVA. The cosine of the angle ϕ ($\cos\phi$) is called the power factor of the circuit.

For a given applied voltage and real power load, the current drawn is high if the power factor is low. This in turn increases the size of the cables required both within the consumer premises and on the utility network, the size of the transformers and losses in cables and transformers.

I.2.6 Capacitive Loads

Capacitors are used as a source of reactive power in many power system applications. Impedance of a capacitor, C, is given by

$$\mathbf{Z} = \frac{1}{j\omega C} = \frac{-j}{\omega C} = -\,jX_C$$

Example I.2 – R, L and C Circuit

An inductive load with a 33.86 Ω resistance and an 80.85 mH inductance is connected to a 230 V mains supply. A capacitor of 30 μF is connected in parallel with the load.

Calculate the active and reactive component of the current drawn (a) without the parallel connected capacitor and (b) with the capacitor. Also calculate the active and reactive power drawn by the supply.

Solution

(a) Without the capacitor

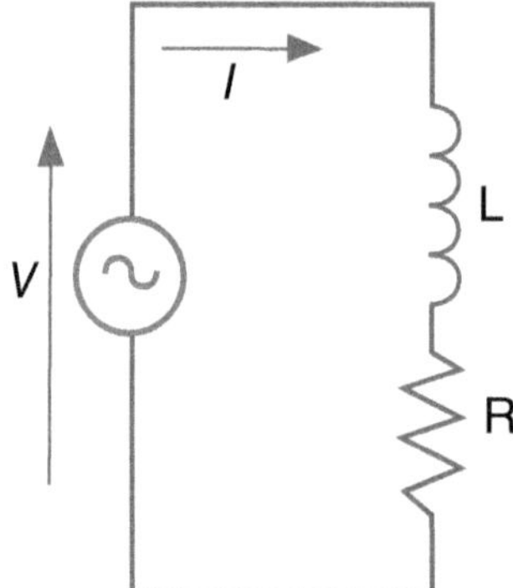

Figure I.13 R–L circuit.

Since the mains frequency is 50 Hz,

$$\omega = 2\pi \times 50 = 314.16 \text{ rad/s}$$

$$\begin{aligned} Z &= R + j\omega L = 33.86 + j \times 314.16 \times 80.85 \times 10^{-3} \\ &= 33.86 + j25.4\,\Omega = 42.33 \angle 36.87°\Omega \end{aligned}$$

$$\begin{aligned} \text{RMS current drawn by the load} &= \frac{230}{42.33\angle 36.87^o} = 5.43\angle -36.87° \text{ A} \\ &= 4.35 - j3.26 \text{ A} \end{aligned}$$

Active component of the current = 4.35 A
Reactive component of the current = 3.26 A
Active power = 4.34 × 230 = 1000 W = 1 kW
Reactive power = 3.26 × 230 = 750 var

(b) With the capacitor

Impedance of the capacitor $= Z = -j/\omega C = -j/\left(314.16 \times 30 \times 10^{-6}\right) = -j106.1\,\Omega$

RMS current through the capacitor $= \dfrac{230}{-j106.1} = -j2.17$ A

As the capacitor supplies part of the reactive current, the new reactive component of the current = 3.26–2.17 = 1.09 A

Active power = 1 kW

Reactive power = 230 × 1.09 = 250 var

I.3 Power System Components

As shown in Figure I.3, the main components in the power system are generators, transformer, overhead lines and cables.

I.3.1 Generator

Consider a single coil moving in a magnetic field as shown in Figure I.14. At the position shown, the two conductors a–a′ and b–b′ are perpendicular to the magnetic field. If this coil rotates in the anticlockwise direction, then according to Fleming's right-hand rule (flux – fore finger, motion – thumb and induced voltage – middle finger), there will be an induced voltage in conductor a–a′ from a′ to a (coming out of the paper and marked as a dot). On the conductor b–b′, the induced voltage is from b to b′, thus into the paper (marked as a cross). Therefore, at the terminals of the coil the induced voltage is the addition of induced voltages on a–a′ and b–b′. When the coil rotates by 45°, the direction of the induced voltages is the same. However, when the coils move outside the pole faces, the flux linking with the coils gradually reduces and when the coil is in a vertical position, the flux linkage and thus the induced voltage becomes zero. When the coil a–a′ is close to the S pole, it takes the original position of b–b′ and the induced voltage reverses. The resultant induced voltage, V, for one rotation of the coil is shown in Figure I.14c. In a real generator, as there are many coils, the shape of the induced voltage converges to a sinusoid.

If three coils, which are physically displaced by 120°, are placed in the stator (Figure I.15a), then the induced voltage in each coil is displaced in time by 120° as shown in Figure I.15b. At the position shown, the voltage across coil A–A′ is zero. When the rotor rotates by 90° in the direction shown, the north pole comes to A and the south pole comes to A′. This induces a positive maximum voltage across coil A–A′. After another 180° rotation, the exact opposite happens and the voltage across coil A–A′ becomes a negative maximum.

In a three-phase system, the voltage in phase B lags the voltage in phase A by 120° ($2\pi/3$ rad) and the voltage in phase C lags the voltage in phase A by 240° ($4\pi/3$ rad).

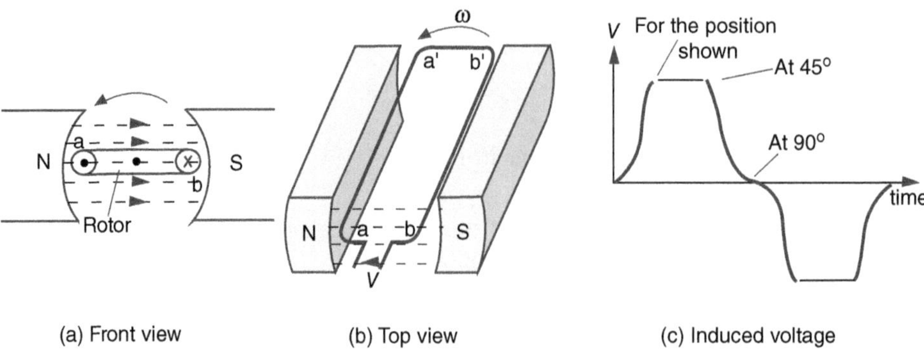

Figure I.14 A single-coil generator.

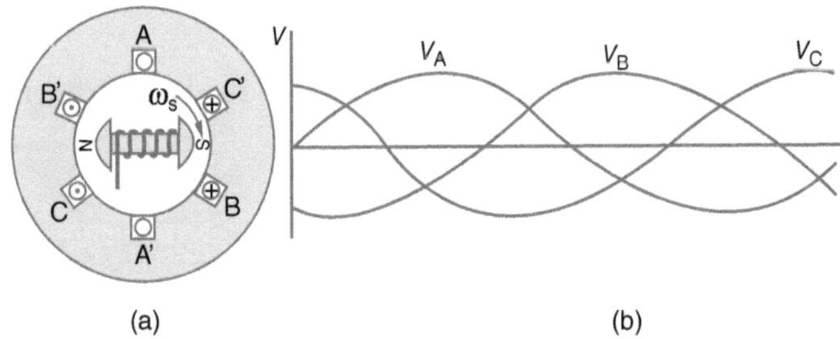

Figure I.15 Three-phase voltages.

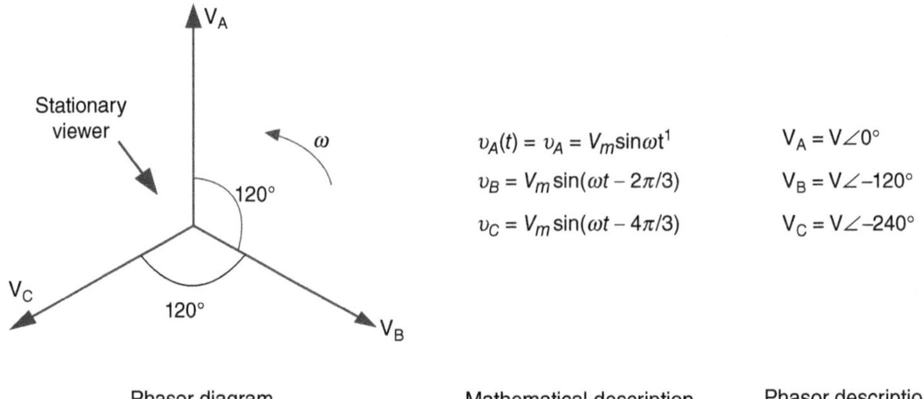

Note: instantaneous values are given in the form of *v*, not as *v(t)* in subsequent sections

Figure I.16 Different descriptions of the three-phase system.

The three-phase system can be represented by the phasor diagram of Figure I.16. The corresponding mathematical description of the three-phase system is also given in that figure.

A stationary viewer will see the phases A, B and C passing him in time in that order. Therefore, the phase sequence of the supply is ABC.

Figure I.17 shows the end windings of a three-phase generator. Three coils A, B and C are placed in slots in the stator and are connected to an ac system (the

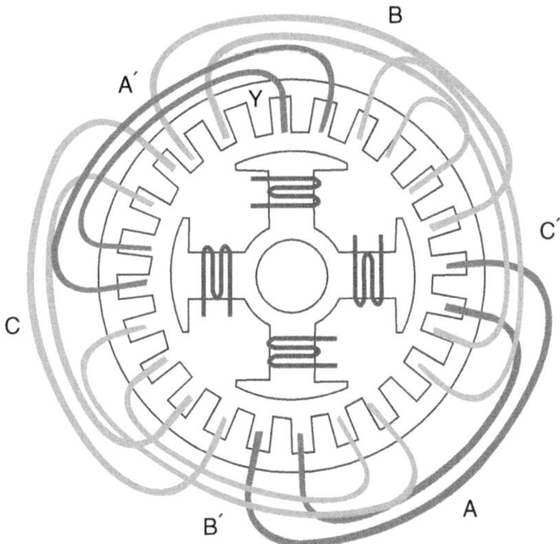

Figure I.17 A schematic of a three-phase generator.

connections are from the remote end of the stator and therefore not shown here). There is more than one coil for each phase thus constituting a winding. The dc field winding of the generator is located on the central rotor.

When the generator is started initially, a dc supply is connected to the field windings that produces a magnetic flux in the rotor poles. When the rotor is turned by a turbine the magnetic flux of the rotor sweeps across the stator conductors. This induces a voltage across the stator coils as shown in Figure I.15. The angular velocity of the voltage induced in the stator is the same as the rotor rotational speed. The voltage induced in the stator is then matched exactly with the voltage of the mains and the generator is connected to the network by closing a synchronizing circuit breaker. The speed of both the rotor and stator voltage is called the synchronous speed and the generator operating on this principle is called a synchronous generator.

Most of the large generators used in hydro power plants are synchronous generators with a similar structure to that shown in Figure I.17. A large thermal or nuclear power plant also uses a synchronous generator, but the rotor construction is cylindrical. The generators used for biomass generation are again synchronous. Some wind power plants use synchronous generators with back-to-back power controllers, as discussed in Chapter 2.

Fixed-speed wind generators and small hydro power plants use asynchronous generators. The stator of an asynchronous generator is very similar to that of a synchronous generator. The main difference is the rotor construction and its operation. In the construction shown in Figure I.18, solid copper or aluminium bars are embedded in a laminated rotor structure and short-circuited by two end rings. Asynchronous generators can use another rotor construction called wound rotor, where the rotor carries three-phase windings. Asynchronous generators are sometimes called 'induction' generators. The terms are synonymous.

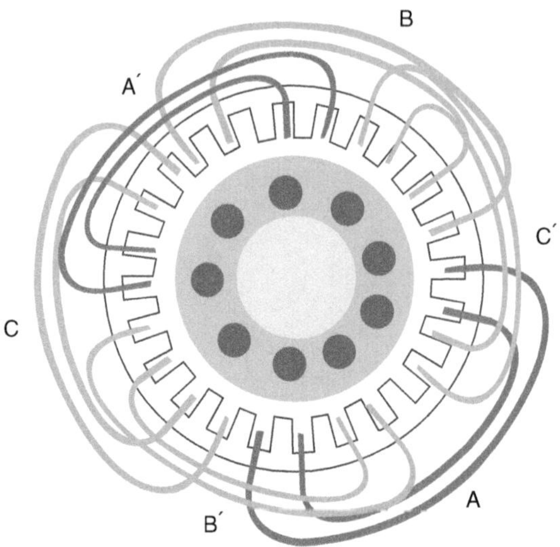

Figure I.18 Squirrel cage induction machine.

In a grid-connected induction (or asynchronous) generator, as used in so-called 'fixed-speed' wind turbines, the rotor is first rotated by the turbine up to a speed slightly less than the synchronous speed. The stator terminals are then connected to the grid through a soft-starter that applies the network voltage to the stator windings gradually. This will create a rotating magnetic flux in the stator windings which rotates at synchronous speed. As the rotor conductors have a relative movement with the stator flux, there will be an induced voltage on the rotor conductors. The rotor consists of a squirrel cage that forms a three-phase winding with the induced voltage in each rotor phase being displaced in space by 120°. All three phases in the rotor are short-circuited and therefore the induced voltage in the rotor produces a circulating current. The three phase currents flowing in the rotor will produce a rotating magnetic field, and power is transmitted from the rotor to the stator by the interaction of the two rotating magnetic fields. After fluxing the generator, the turbine control is changed from initial speed control to power control. This allows the rotor to run slightly faster than the synchronous speed, thus transferring power from the rotor to the stator.

Some variable-speed wind turbines use wound rotor induction generators in which the rotor carries windings as in the stator. In these wind turbines, the rotor is fed with a low-frequency ac voltage and therefore the generators that employ this technology are called Doubly Fed Induction Generators (DFIGs).

I.3.2 Transformer

A simple representation of a transformer is shown in Figure I.19. One coil is connected to a voltage source (directly or through other power system components) and called the primary winding. A mutual magnetic flux produced by the alternating current in the primary winding links with the secondary winding and induces a voltage in it.

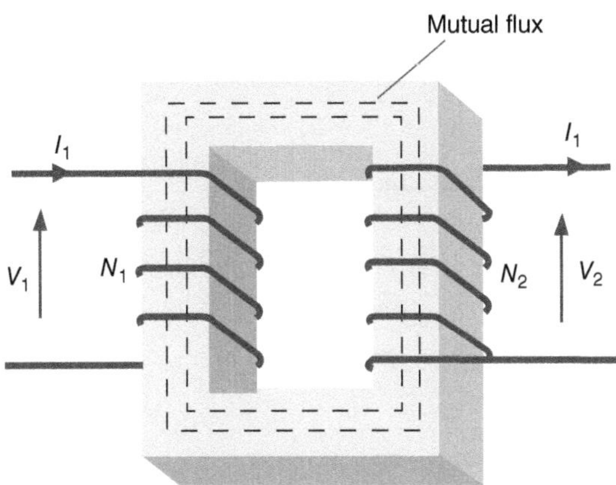

Figure I.19 Ideal transformer.

If the mutual flux in the core is $\phi = \phi_m \sin \omega t$, then from Faraday's law:

$$V_1 = N_1 \frac{d\phi}{dt} = \omega N_1 \phi_m \cos \omega t \tag{I.9}$$

where

V_1 is the voltage of the primary winding,
N_1 is the number of turns.

Assuming that the flux ϕ links completely with the secondary of the transformer, the secondary internal voltage is:

$$V_2 = N_2 \frac{d\phi}{dt} = \omega N_2 \phi_m \cos \omega t \tag{I.10}$$

Dividing Equation I.9 by Equation I.10, the following relationship is obtained for the ratio of the magnitudes of V_1 and V_2:

$$\frac{V_1}{V_2} = \frac{N_1}{N_2} \tag{I.11}$$

Equating the volt-ampere rating of the primary and secondary:

$$V_1 I_1 = V_2 I_2 = \frac{N_2}{N_1} V_1 I_2$$

$$\therefore \frac{I_1}{I_2} = \frac{N_2}{N_1} \tag{I.12}$$

Example I.3 – Ideal Transformer

A transformer has two windings and can be considered as ideal. The primary coil has 210 turns, and the secondary has 630 turns. The primary is supplied by a 230 V ac supply and the secondary is connected to a 5 kVA load. Find the load voltage, the load current and the primary current.

Solution

From Equation I.11,

$$\frac{V_1}{V_2} = \frac{N_1}{N_2}$$

$$\frac{230}{V_2} = \frac{210}{630}$$

$$\therefore V_2 = 230 \times \frac{630}{210} = 690 \text{ V}$$

The load voltage is 690 V.

The load current is 5 × 10^3/690 = 725 A.

From Equation I.12, the primary current $I_1 = \frac{630}{210} \times 7.25 = 21.75$ A.

I.3.3 Connection of Generator and Transformer Windings

The windings in the three phases of Figure I.15a are always connected so that the three phase voltages are carried on three or four wires. Three windings can be connected in star (wye) or delta.

Star Connection

A star connection can be obtained by connecting terminals A′, B′ and C′ (see Figure I.15a) to form a neutral point (N) as shown in Figure I.20. From this connection, three or four wires can be taken. Two voltages can be defined: the phase voltage (the voltage between the neutral and any terminal: V_{AN}, V_{BN} and V_{CN}) and the line voltage (the voltage between any two terminals: V_{AB}, V_{BC}, V_{CA}). In the star connection the phase current and line current are the same and denoted by I_A, I_B and I_C in Figure I.20a.

The phasor diagram of voltages is shown in Figure I.20b. In this diagram, the phase voltages (V_{AN}, V_{BN} and V_{CN}) and the construction of the line voltage V_{AB} are shown. Assuming that the RMS value of the magnitude of phase voltages V_{AN}, V_{BN} and V_{CN} is V_P and the RMS value of the magnitude of line voltages V_{AB}, V_{BC}, V_{CA} is V_L then from the phasor diagram:

$$\begin{aligned} V_{AB} &= |V_{AN}| \times \cos 30^\circ + |V_{BN}| \times \cos 30^\circ \\ V_L &= 2 \times V_P \times \cos 30^\circ \\ V_L &= \sqrt{3} V_P \end{aligned} \quad (I.14)$$

For example, if the phase voltage is 230 V, then the line voltage is $\sqrt{3} \times 230 = 400$ V.

Delta Connection

By connecting terminals A′ to B, B′ to C and C′ to A (see Figure I.15) a delta connection can be formed as shown in Figure I.21. In the delta connection there is no

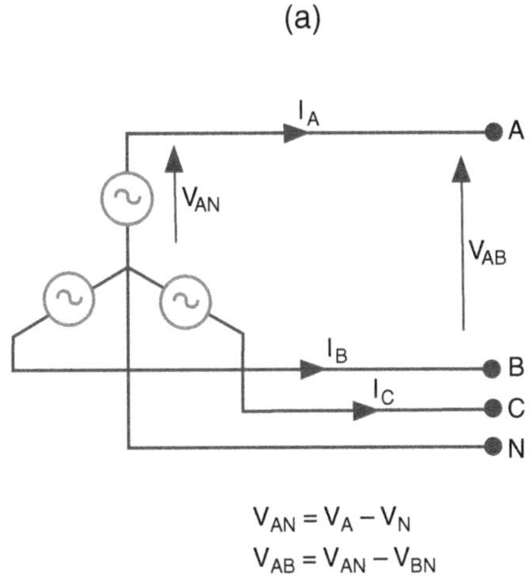

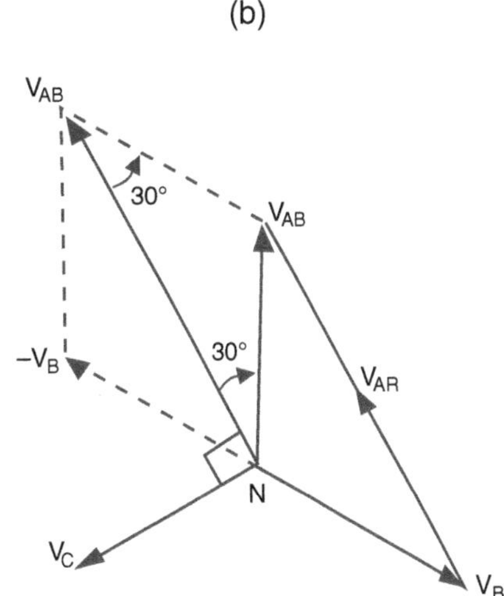

If the three-phase voltages are balanced:

$v_{AB} = V_m \sin \omega t$

$v_{BC} = V_m \sin(\omega t - 2\pi/3)$

$v_{CA} = V_m \sin(\omega t - 4\pi/3)$

Figure I.20 Star-connected three-phase source.

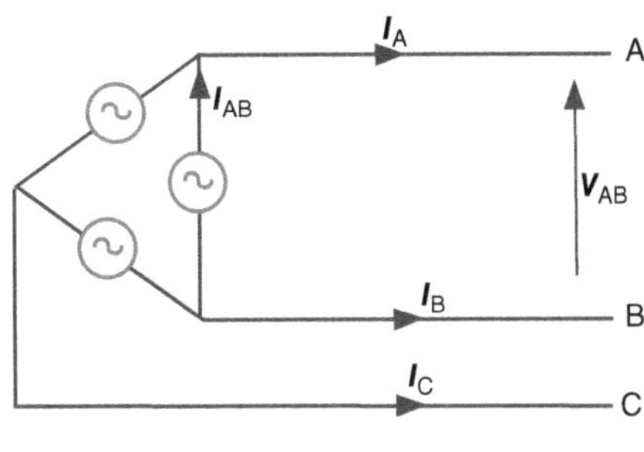

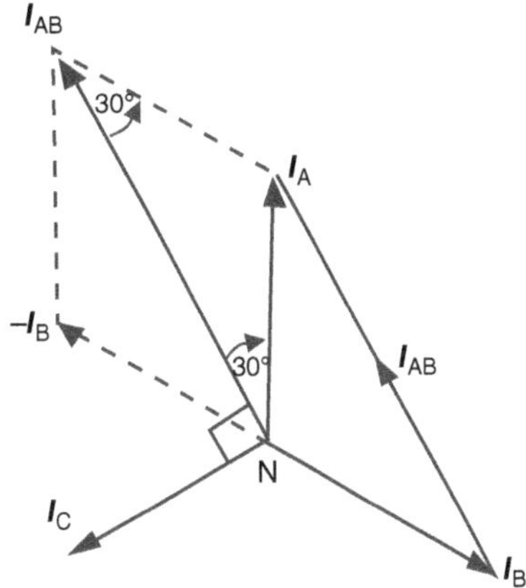

Figure I.21 Delta-connected three-phase system.

neutral, and line voltages and phase voltages are the same. If $V_{AB} + V_{BC} + V_{CA} = 0$, there is no circulating current in the delta loop.

The relationship between the line and phase currents can be obtained by assuming the three phase currents are balanced (i.e. the connected loads are equal). From Figure I.21, the magnitude of the line current is equal to $\sqrt{3}$ times the magnitude of the phase current.

The primary and secondary windings of three-phase transformers are connected in star (Y) and delta (Δ) depending on the voltage level, as shown in Figure I.3. The choice of using either delta or star connection was made when the network was built more than 70 years ago and is now established practice.

I.3.4 Transmission Lines

Part of an overhead transmission line is shown in Figure I.22. Due to the current flowing in the line, there will be a voltage drop along the line and losses created by the resistance of the conductors. The current creates a magnetic field and the voltage creates an electric field as shown in the figure. Therefore the transmission line is represented by a resistor and a reactor in series plus capacitance to ground. The resistor represents the voltage and power loss across the line, and the reactance represents the self and mutual inductance due to the magnetic field. The electric field is represented by the capacitance. In short lines of less than 80 km or so, the capacitance is normally neglected. Then the impedance of the line is represented by resistance and inductance ($R+jX$) per unit length.

I.3.5 Balanced Three-Phase Loads

Most industrial and large commercial electricity consumers have three phase loads. These loads are connected in star or delta. If the loading on each phase is equal, then that load is called a balanced three-phase load. Figures I.23a and b show balanced star-connected and delta-connected loads.

Figure I.24 shows the secondary winding of a three-phase, star-connected transformer connected to a three-phase, star-connected load to form a three-phase, four-wire system.

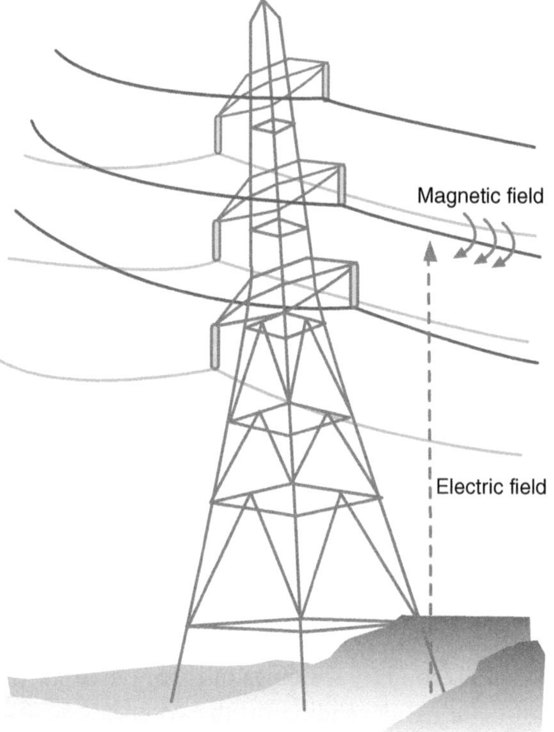

Figure I.22 Overhead line.

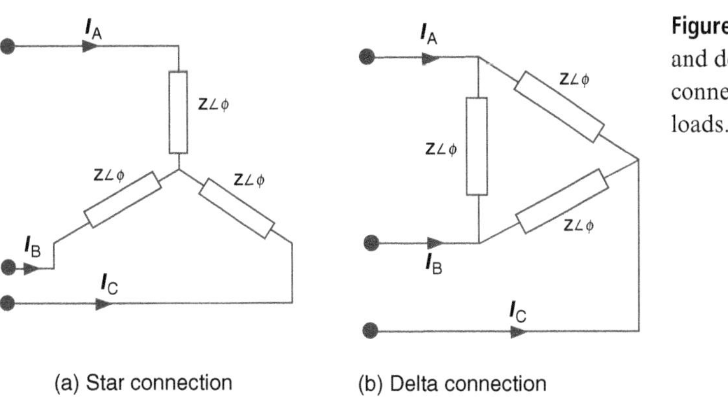

Figure I.23 Star and delta connection of loads.

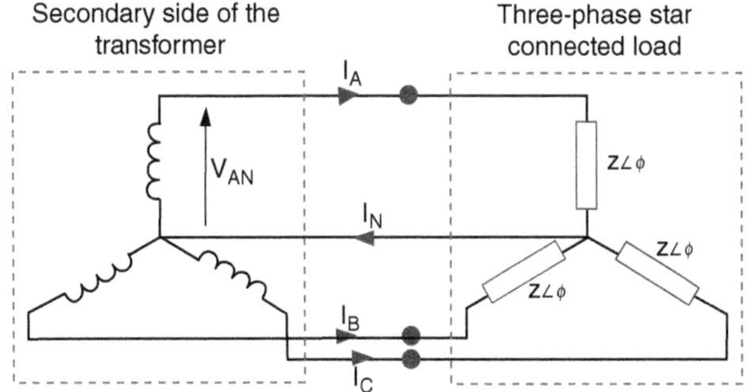

Figure I.24 Configuration of the three-phase, four-wire system.

When the load is balanced, the current in the neutral wire is zero.

$$\begin{aligned}\mathbf{I}_N &= \mathbf{I}_A + \mathbf{I}_B + \mathbf{I}_C \\ &= \frac{\mathbf{V}_{AN}}{Z\angle\phi} + \frac{\mathbf{V}_{BN}}{Z\angle\phi} + \frac{\mathbf{V}_{CN}}{Z\angle\phi} \\ &= \frac{V_m}{Z\angle\phi}\left[\sin\omega t + \sin(\omega t - 2\pi/3) + \sin(\omega t - 4\pi/3)\right] = 0\end{aligned} \tag{I.15}$$

However, when the loads are not balanced (i.e. the three loads are not equal), there will be a current flowing through the neutral wire.

I.3.6 Unbalanced Three-Phase Loads

In the previous section it was assumed that the load connected to each of the phases is equal. This is not often true in the real world. For example, in a low-voltage (LV) distribution network the connected loads are the internal loads of houses and their connection to the LV network depends on the behaviour of the occupants. This is illustrated using Example I.4.

Example I.4 – Unbalanced Loads of Houses

Figure I.25 shows a connection of six houses to a 33/0.4 kV transformer. Each house has some of the appliances given in Table I.2.

Table I.2 Appliances in houses

	Symbol	Active power consumption (W)	Reactive power consumption (var)
Night-time lighting load	LL	200	0
Fridge/freezer	FF	250	100
Television	TV	100	20
Kettle	KE	2500	0
Electric cooker	EC	2000	0
Washing machine	WM	750	300
Microwave	MW	1000	400

Neglecting voltage drop in the LV line, calculate the load current phasors on each phase of the transformer at two time instants (T1 and T2) and hence draw the phasor diagrams. The appliances that are in operation at T1 and T2 are given in Table I.3.

Table I.3 Appliances that are in operation at each time instant

	House 1		House 2		House 3		House 4		House 5		House 6	
	T1	T2	T1	T2	T1	T2	T1	T2	T1	T2	T1	T2
LL	OFF	ON	OFF	ON	OFF	ON	OFF	ON	OFF	ON	OFF	ON
FF	ON	ON	ON	ON	ON	ON	ON	ON	ON	ON	ON	ON
TV	OFF	ON	OFF	OFF	ON	ON	OFF	ON	OFF	OFF	ON	ON
KE	ON	OFF	ON	ON	OFF	OFF	ON	ON	OFF	OFF	ON	OFF
EC	OFF	ON	OFF	ON	OFF	ON	ON	ON	ON	ON	OFF	OFF
WM	OFF	OFF	OFF	OFF	OFF	ON	OFF	OFF	ON	OFF	ON	OFF
MW	OFF	OFF	OFF	ON	ON	OFF	ON	ON	OFF	OFF	OFF	OFF

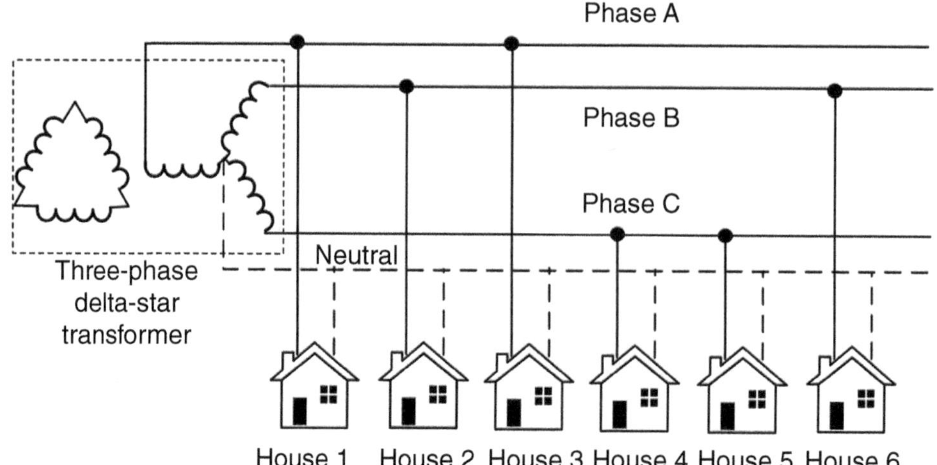

Figure I.25 Connection diagram of six houses.

Solution

Using the connection diagram shown in Figure I.25, and the ON appliances shown in Table I.3, the load connected to each phase at each time instant was calculated and shown in Table I.4. All values in the table are given in W and var. For example, House 1 (H1) and House 3 (H3) are connected to phase A. At time T1, FF and KE of H1 are connected and the load is 2.5 + 0.25 = 2.75 kW, as well as 100 var. Similarly, for H3, FF, TV and MW are connected giving a total of 1.35 kW and 520 var.

Table I.4 Active (P in W) and Reactive (Q in var) connected to each phase at each time instant

		Time T1						Time T2					
		H1	H2	H3	H4	H5	H6	H1	H2	H3	H4	H5	H6
A	P	2750		1350				2550		3300			
	Q	100		520				120		420			
B	P		2750				3600		5950				550
	Q		100				420		500				120
C	P				5750	3000					6050	2450	
	Q				500	400					520	100	

At time T1

The apparent power flowing in phase $A = 4100 - j620$ VA (for inductive loads, Q lags and therefore the j term has a −ve sign)

Corresponding current $= S/230^{*} = 17.83 - j2.7\ A = 18.03\angle -8.6°$

(*Since the transformer secondary voltage is 400 V, the corresponding phasor voltage is $400/\sqrt{3} = 230$ V.)

Apparent power flowing in phase $B = 6350 - j520$ VA

Corresponding current $= S/230^{\#} = 27.61 - j2.26\ A = 27.7\angle -4.7°$

(# Here the current is calculated with respect to the B-phase voltage.)

Apparent power flows in phase $C = 8750 - j900$ VA

Corresponding current $= S/230 = 38.04 - j3.91$ A $= 38.24\angle -5.9°$

The phasor diagram for this instant is shown in Figure I.26.

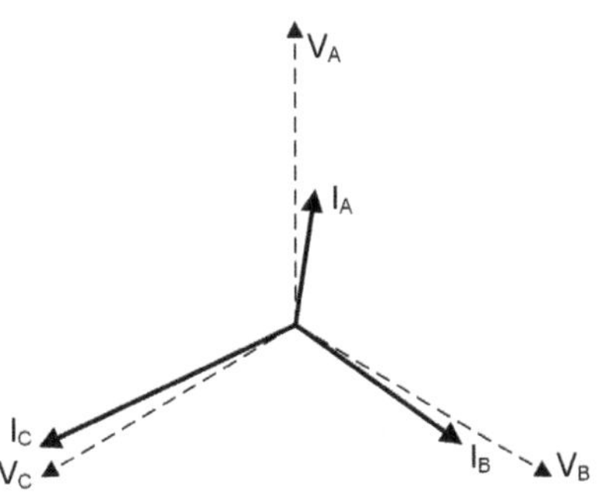

Figure I.26 Phasor diagram for time T1.

At time T2

Apparent power flows in phase A = 5850 – j540 VA
Corresponding current = S/230 = 25.43 – j2.35 A = 25.5∠–5.3°
Apparent power flows in phase B = 6500 – j620 VA
Corresponding current = S/230 = 28.26 – j3.7 A = 28.39∠–5.5°
Apparent power flows in phase C = 8500 – j620 VA
Corresponding current = S/230 = 36.96 – j2.7 A = 37.06∠–4.2°

A phasor diagram for this instant is shown in Figure I.27.

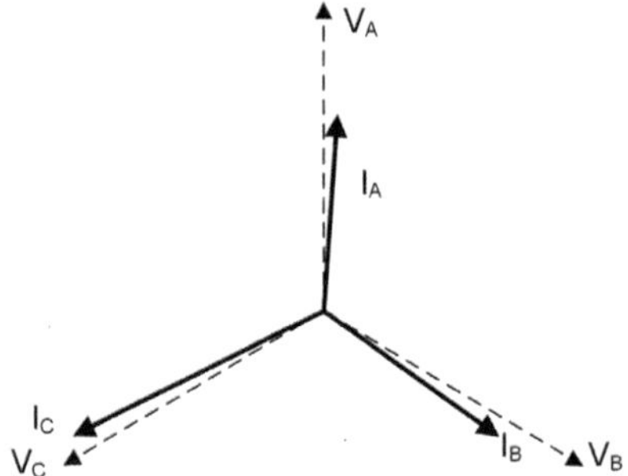

Figure I.27 Phasor diagram for time T2.

I.4 Power in a Three-Phase System

The power in a three-phase load is the sum of the power consumed in each phase. Consider the three-phase unbalanced load of Example I.4. In general form, the voltages and current in each phase can be represented as follows:

Phase	Voltages	Currents
A	$v_{AN} = V_m \sin\omega t$	$i_A = I_{mA}\sin(\omega t - \phi_A)$
B	$v_{BN} = V_m \sin(\omega t - 2\pi/3)$	$i_B = I_{mB}\sin(\omega t - 2\pi/3 - \phi_B)$
C	$v_{CN} = V_m \sin(\omega t - 4\pi/3)$	$i_C = I_{mC}\sin(\omega t - 4\pi/3 - \phi_C)$

Instantaneous power in A-phase
$= v_{AN} \times i_A = V_m \times I_{mA} \times \sin\omega t \times \sin(\omega t - \phi_A)$
Instantaneous power in B-phase
$= v_{BN} \times i_B = V_m \times I_{mB} \times \sin(\omega t - 2\pi/3) \times \sin(\omega t - 2\pi/3 - \phi_B)$
Instantaneous power in C-phase
$= v_{CN} \times i_C = V_m \times I_{mC} \times \sin(\omega t - 4\pi/3) \times \sin(\omega t - 4\pi/3 - \phi_C)$

By adding these equations, the total power consumed by the three-phase load is obtained. If the loads are balanced, $I_{mA} = I_{mB} = I_{mC} = I_m$ and $\phi_A = \phi_B = \phi_C = \phi$. It can be shown trigonometrically that

$$\sin\omega t \times \sin(\omega t - \phi) + \sin(\omega t - 2\pi/3) \times \sin(\omega t - 2\pi/3 - \phi) + \sin(\omega t - 4\pi/3) \times \sin(\omega t - 4\pi/3 - \phi) = \frac{3}{2}\cos\phi$$

Therefore the total three-phase instantaneous power in a balanced three-phase load

$$= \frac{3}{2} V_m I_m \cos\phi = 3\frac{V_m}{\sqrt{2}}\frac{I_m}{\sqrt{2}}\cos\phi = 3V_P I_P \cos\phi \qquad (I.16)$$

When Equation I.16 is compared with Equation I.7 for a single-phase case, it can be seen that the total three-phase power is equal to the addition of the power in the three phases. Therefore, the active power, P, measured in W for a three-phase circuit is given by:

$$\text{Active power} = 3V_P I_P \cos\phi \qquad (I.17)$$

where

$\cos\varphi$ is the power factor of the load.

In a similar manner to that defined for a single-phase circuit, the apparent and reactive power are defined for three-phase circuits.

$$\text{Apparent power} \quad S = 3V_P I_P \qquad (I.18)$$

$$\text{Reactive power} \quad Q = 3V_P I_P \sin\phi \qquad (I.19)$$

Example I.5 – Three-Phase Loads

A three-phase, four-wire 400 V supply is connected to three equal impedances of $10\angle 45°\ \Omega$. Determine the line currents in all three phases and the total active and reactive power consumed by the load.

Solution

$$\text{Phase voltage} = \frac{400}{\sqrt{3}} = 230.94\ \text{V}$$

From Equation I.15, the line current in phase A is

$$\mathbf{I}_A = \frac{\mathbf{V}_{AN}}{Z\angle\phi} = \frac{230.94}{10\angle 45^o} = 23\angle -45°\ \text{A}$$

Since the phase-B current lags the phase-A current by 120°, the line current in phase B is

$$\mathbf{I_B} = 23\angle -45° - 120° = 23\angle -165°$$

Similarly, the phase-C current is

$$\mathbf{I_C} = 23\angle -165° - 120° = 23\angle -285°$$

Active power consumption is

$$P = 3V_P I_P \cos\phi = 3 \times 230.94 \times 23 \times \cos 45^o = 11.3\ \text{kW}$$

Reactive power consumption is

$$Q = 3V_P I_P \sin\phi = 3 \times 230.94 \times 23 \times \sin 45^o = 11.3\ \text{kvar}$$

I.5 Power Electronics

The output of many renewable energy generators, including PV arrays, is dc. In order to connect these sources to the ac system a power electronic interface, known as an inverter, is used. In its simplest form, an inverter is four electronic switches connected to the dc source as shown in Figure I.28a. When the switch pair T_1 and T_4 is ON, then $V_{ac} = V_{dc}$, and when the switch pair T_2 and T_3 is ON, $V_{ac} = -V_{dc}$. If each switch pair is turned ON and OFF for 10 ms in a complementary manner then V_{ac} is a square wave. This is an alternating voltage of 50 Hz, containing a 50 Hz sine wave and an infinite number of sinusoidal components at multiples of 50 Hz. The 50 Hz waveform is called the fundamental, and higher-frequency components are called harmonics.

Utilities impose restrictions on harmonics injected into their networks as these higher frequencies create increased heating losses and other unwanted effects on the network and loads. Therefore the simple square-wave inverter is not employed for utility connections. A commonly used method to generate the ac voltage is Pulse Width Modulation (PWM). The PWM output waveform of the inverter is shown in Figure I.28b. Here the output is obtained by turning ON and OFF switching pairs (T_1 and T_4) and (T_2 and T_3) at a high frequency (often around 1000–2000 Hz). Then the PWM output is sent through a filter before connecting to the utility grid.

PROBLEMS

1. Why is it convenient to use phasors when analysing electric power systems?
2. Compare the starting procedures of a grid connected synchronous and induction generator.

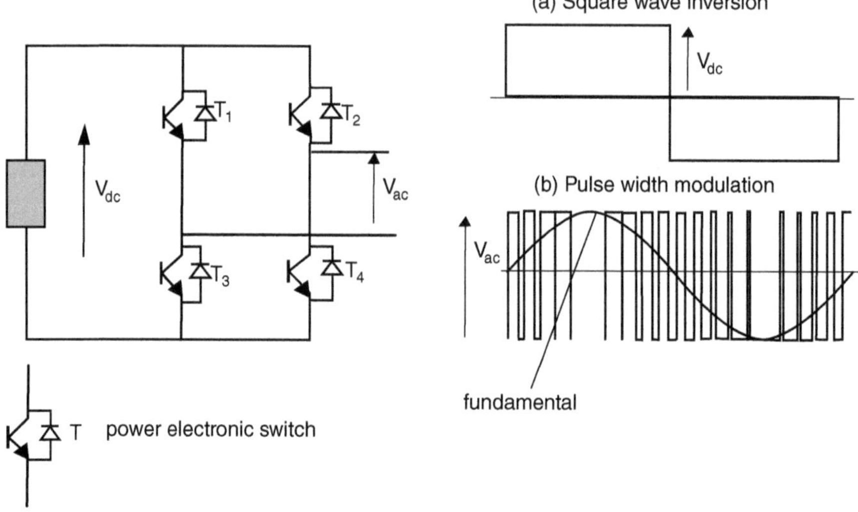

Figure I.28 Single-phase inverter.

3. For a three-phase balanced load, the phase voltage is V and the phase current is I. Compare the phasc and line currents and voltages if the load is connected as star and as delta.
4. A current of $i = 5\sin(200\pi t)$ passes through a series-connected resistor of 10 Ω and an inductor of 20 mH. What is the RMS value of the total voltage applied across both components? Also obtain the angle between the current and voltage.

   [**Answer:** 56.85 V, 51.5°.]
5. A series circuit with a resistance of 5 Ω and an inductance of 25 mH is connected to a 12 V, 50 Hz supply. Calculate the current in the circuit and the active and reactive power drawn from the supply. Then construct the phasor diagram and the power triangle.

   [**Answer:** $1.29\angle -57.5°$A, 8.32 W, 13.06 var.]
6. A 5 kW, single-phase ac motor has an efficiency of 80% when operating at full load. The motor operates at a power factor of 0.8 lagging. The supply is 230 V, 50 Hz.
   (a) What is the active and reactive power drawn by the motor?
   (b) Determine the value of the capacitor bank that should be connected in parallel to the motor to increase the power factor to 0.9.

   [**Answer:** (a) 6.25 kW, 4.69 kvar; (b) 99.9 μF.]
7. The number of turns on the primary side of a 220/110 V transformer is 100.
   (a) What is the number of turns on the secondary side?
   (b) If the transformer secondary side is connected to a load of 100 Ω resistance, what is the current flowing in the primary side?

   [**Answer:** (a) 50 turns; (b) 0.55 A.]
8. The primary winding of a transformer is connected to a 230 V, 50 Hz supply. The number of turns in the primary of the transformer is twice that of the secondary. A load consisting of a resistor of 1.6 Ω in series with an inductor of 3.82 mH is connected to the secondary. Calculate the primary and secondary currents.

   [**Answer:** 28.75 A, 57.5 A.]
9. The following loads are connected in parallel across a 230 V, 50 Hz supply.
   Load 1: 10 kVA at 0.8 power factor lagging
   Load 2: 8 kVA at 0.9 power factor lagging
   (a) Calculate the active and reactive power drawn by both loads.
   (b) What is the line current in the supply?

   [**Answer:** (a) 15.2 kW, 9.49 kvar; (b) $77.9\angle -31.97°$A.]
10. Three star-connected impedances of $Z = 20\angle 45°\Omega$ are connected to a 400 V (line), 50 Hz, three-phase supply. Determine the voltage across each impedance and the line currents.

    [**Answer:** 230.9 V, $11.55\angle -45°$A.]
11. A balanced star-connected load with impedances $\mathbf{Z} = 10 + j5\,\Omega$ is connected to a three-phase generator having an effective line voltage of 400 V. Calculate the active and reactive power consumed by the load.

    [**Answer:** 12.8 kW, 6.4 kvar.]

FURTHER READING

Hughes E., Hiley J., Brown K. and McKenzie-Smith I., *Electrical and Electronic Technology*, 2008, Prentice Hall.

A comprehensive review of electrical and electronic basics, electric machinery and power electronics.

Smith R.J. and Dorf R.C., *Circuits, Devices and Systems*, 1992, Wiley.

Addresses the basic electrical circuits, phasors and transformers.

Chapman S.J., *Electric Machinery Fundamentals*, 2011, McGraw-Hill Education.

Provides basic theory and applications of transformers and electrical machines.

ONLINE RESOURCES

Exercise I.1 Carbon intensity of generation: this exercise uses online resources to investigate the carbon intensity of GB generation.

Exercise I.2 Optimization of generation: this exercise uses Excel and Solver to determine the optimal mix of generators of an island power system. It is based on a single annual optimization of the load and generators.

TUTORIAL II

Heat Transfer

II.1 Heat Transfer

Heat is transferred by conduction, convection and radiation. Conduction is the transfer of heat energy through a solid by the collision and vibration of molecules, without movement of the bulk material. An example is the transfer of heat through the mass of solid thermal insulation. Convection is the transfer of heat energy by the bulk motion of a fluid. An example is the transfer of heat through the movement of air. Convection may be due to natural or forced fluid movement. Radiation is the transfer of heat energy by electromagnetic waves and can take place in a vacuum. Energy from the sun travels across space by radiation.

Heat transfer takes place when there is a temperature difference between a higher-temperature surface or area to a lower-temperature surface or area. Even though heat transfer is a complex process, for simple low-temperature calculations it is usual to assume that heat transferred is proportional to the temperature difference. Equation II.1 gives the general relationship between heat flow and temperature difference:

$$Q = \frac{\Delta T}{R_\phi} \tag{II.1}$$

where

Q is the heat flow in W,
ΔT is the temperature difference in K,
R_ϕ is the thermal resistance of the process in K/W.

Heat transfer in a low-temperature system is analogous to the current flow through a simple electrical resistor from a higher voltage to a lower voltage as described by Ohm's law $\left(I = \frac{\Delta V}{R}\right)$. Therefore, a heat circuit can be considered as an electrical circuit with series and parallel elements of thermal resistance that represent different heat transfer processes.

For example, in a closed heat exchanger, where a tube wall separates hot and cold fluid streams as shown in Figure II.1, the main modes of heat transfer from the hot fluid to the cold fluid are:

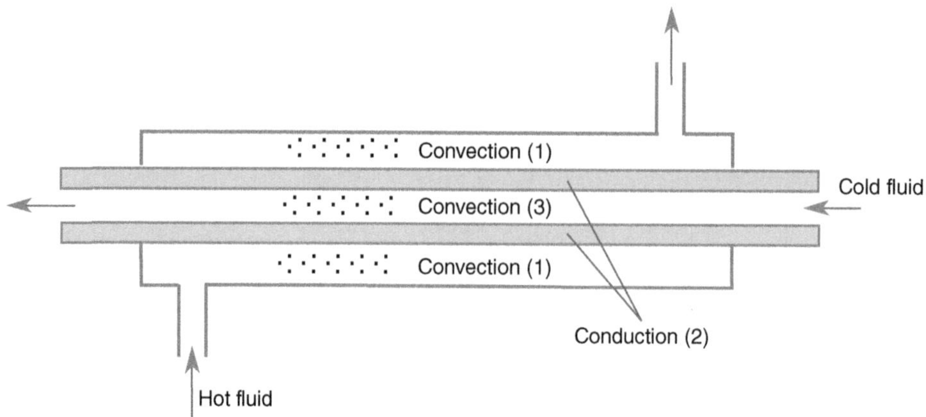

Figure II.1 Heat flow in a heat exchanger.

(1) from the hot fluid to the outer surface of the tube by convection,
(2) through the wall of the tube by conduction,
(3) from the inner surface of the tube to the cold fluid by convection.

Therefore the heat flow from the hot to the cold fluid can be represented by three thermal resistances connected in series: R_{v1} represents the initial convection process, R_c represents the conduction process and the final convection process is represented by R_{v2}. The combined thermal resistance from the hot to the cold fluid is $R_{v1} + R_c + R_{v2}$.

II.2 Conduction

Consider a well-insulated rod shown in Figure II.2 in which the temperature at end G is T_{higher} and at end H is T_{lower}. This will lead to heat transfer from end G to H by conduction. The steady-state heat transfer from G to H is:

$$Q = kA\frac{\left(T_{higher} - T_{lower}\right)}{d} \quad [\mathrm{W}] \tag{II.2}$$

where

Q is the heat flow along the rod by conduction in W,
k is the thermal conductivity of rod material in W/mK,
A is the cross-sectional area of the rod in m^2,
$T_{higher,\,lower}$ is the higher/lower temperature at the ends of the rod in K,
d is the length of the rod in m.

The thermal resistance along the length of the rod is

$$R_c = \frac{1}{A} \times \frac{d}{k} \quad [\mathrm{K/W}] \tag{II.3}$$

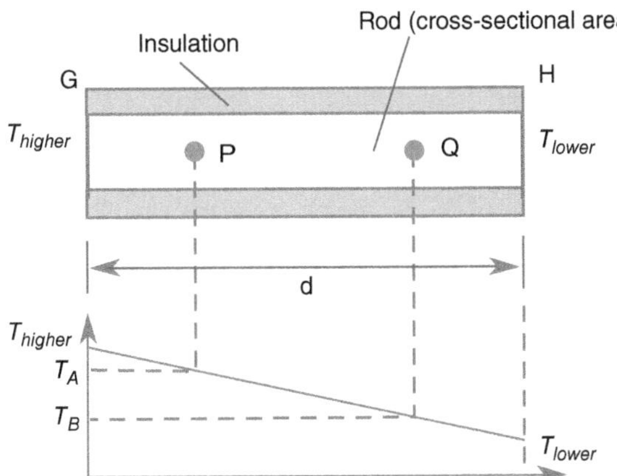

Figure II.2 One-dimensional heat flow in an insulated rod.

where

R_c is the thermal resistance of the rod due to conduction in K/W.

This allows the following simple expression to be written for the heat transfer:

$$Q = \frac{(T_{higher} - T_{lower})}{R_c} = \frac{\Delta T}{R_c} \quad [\text{W}] \tag{II.4}$$

The heat transfer between two points is governed by the temperature gradient between them. For example, the heat transfer between points P and Q of Figure II.2 is proportional to $(T_A - T_B)/\text{PQ}$ where PQ is the length of the rod between points P and Q.

Example II.1 – Thermal Loss by Conduction

The wall of an oven is 100 cm × 60 cm and it is 1 cm thick. It is covered by a layer of insulation 2 cm thick. The thermal conductivity of the oven wall is 15 W/mK and that of the insulation is 0.2 W/mK. The inside wall temperature is 250 °C and the room temperature is 25 °C. These temperatures are constant for a long time. Calculate the temperature at the boundary of the oven wall and insulation. Assume that there is no heat lost at the edges of the wall.

Solution

Assume that the temperature that is common to the oven wall and the insulation is T, as shown in Figure II.3.

Applying Equation II.2 to the oven wall:

$$Q = kA\frac{(T_{higher} - T_{lower})}{d} \quad [\text{W}]$$

$$Q = \frac{15 \times 1 \times 0.6 \times (250 - T)}{0.01} = 900(250 - T)$$

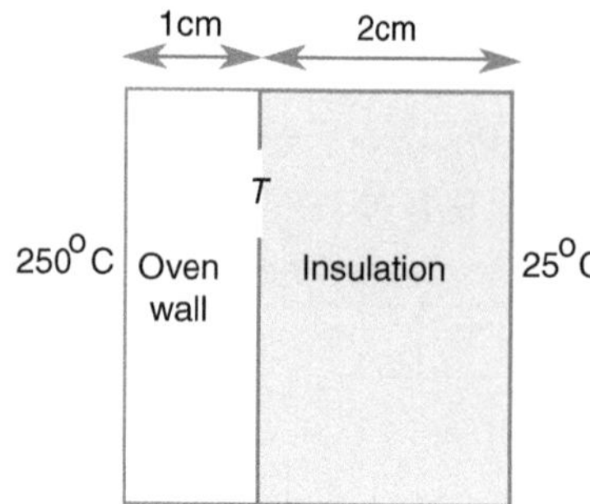

Figure II.3 Side view.

and to the insulation:

$$Q = \frac{0.2 \times 1 \times 0.6 \times (T - 25)}{0.02} = 6(T - 25)$$

Equating the heat flow rate:

$$900[250 - T] = 6[T - 25]$$
$$\therefore T = 248.5\ °\text{C}$$

Example II.2 – Heat Lost Through an Insulated Surface

In the same circumstances as in Example II.1, calculate the heat lost through the insulation in 1 hour using the electrical analogy (Figure II.4).

Solution

Thermal resistance of the oven wall $= R_{\text{oven}} = \dfrac{0.01}{15 \times 1 \times 0.6} = 0.0011\ \text{K/W}$

Thermal resistance of the insulator $= R_{\text{insulation}} = \dfrac{0.02}{0.2 \times 1 \times 0.6} = 0.1666\ \text{K/W}$

Total thermal resistance $= R_{\text{oven}} + R_{\text{insulation}} = 0.1678\ \text{K/W}$

Heat flow $= Q = \dfrac{\Delta T}{R_{\text{total}}} = \dfrac{(250 - 25)}{0.1678} = 1340.9\ \text{W}$

Therefore the heat lost during 1 hour (3600 s) $= 1340.9 \times 3600 = 4.83\ \text{MJ}$

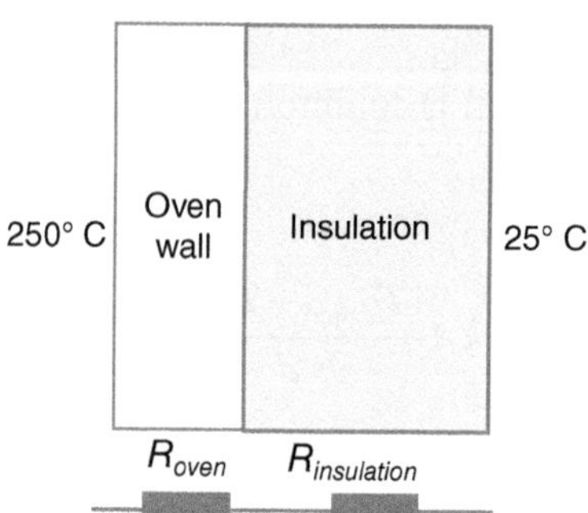

Figure II.4 Resistance analogy.

II.3 Convection

Convection is the transfer of heat by a moving fluid. It is commonly caused by the movement of a fluid that is in contact with a surface at a higher temperature. Free (or natural) convection is the natural movement of a fluid that is warmed by the hot surface, becomes less dense and rises. Forced convection relies on an external force to move the fluid. An example of free convection is the heat transferred by the movement of air inside a closed space that is caused by one side of the space being heated, such as within the cavity of a double-glazed window. The wind blowing on the outer sheet of glass or a fan within the room creates forced convection. Forced convection is a particularly effective form of heat transfer.

Figure II.5 shows a stream of fluid (gas or liquid) moving over a hot surface which is at temperature T_S. At the surface of the solid, the velocity of the fluid is zero and the temperature is equal to T_S. Away from the solid surface, the velocity increases and the temperature decreases. A fictitious boundary is defined above which temperature and velocity become constant and are given by T_f and V_f.

Heat transfer by convection is described by:

$$Q = hA\left(T_s - T_f\right) \quad [\text{W}] \tag{II.5}$$

where

h is the convection heat transfer coefficient in W/m^2 K,
A is the area in m^2,
T_s and T_f are the temperature of the solid and the temperature at the fictitious boundary, respectively, in K.

The heat transfer coefficient depends on the thermal conductivity of the fluid, the shape of the solid object and the conditions in which convection occurs. The shape of the solid object in a given situation is represented by a characteristic length, X. For example, for the flat plate shown in Figure II.5, X is taken as the length of the plate over which the fluid flows. For other surface geometries, the characteristic length is taken as the length perpendicular to the direction of fluid flow. Figure II.6 shows the characteristic length of some surfaces.

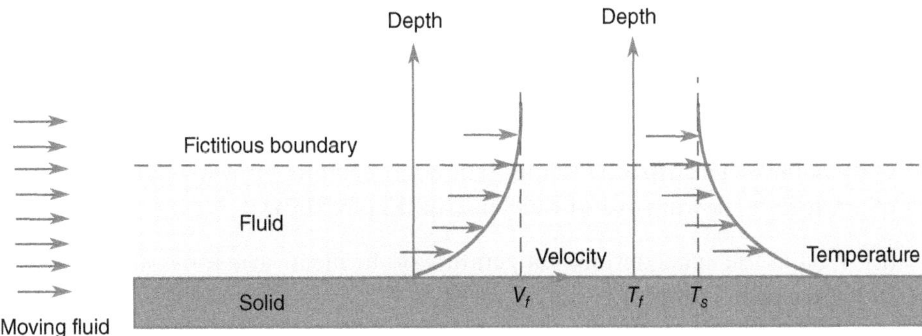

Figure II.5 Simplified representation of convection in a moving fluid.

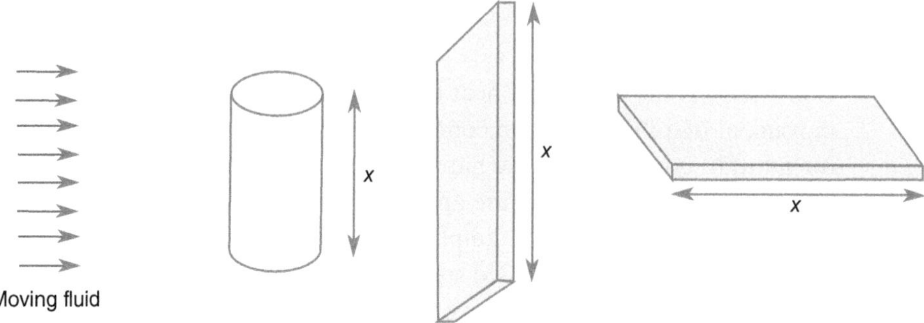

Figure II.6 Characteristic length of some geometric shapes.

The heat transfer coefficient h is defined by

$$h = \frac{\mathbb{N}k}{X} \quad \left[\text{W/m}^2\text{K}\right] \tag{II.6}$$

where

k is the thermal conductivity of the fluid,
$\mathbb{N}$ is the dimensionless Nusselt number,
X is the characteristic length.

From Equations II.5 and II.6, the heat transfer due to convection is obtained as

$$Q = \frac{\mathbb{N}kA\left(T_s - T_f\right)}{X} \quad [\text{W}] \tag{II.7}$$

Then the thermal resistance of the stream of fluid can be defined as

$$R_v = \frac{1}{\mathbb{N}A} \times \frac{X}{k} \quad [\text{K/W}] \tag{II.8}$$

From Equations II.6 and II.8, we have

$$h = \frac{1}{R_v A} \quad \left[\text{W/m}^2\text{K}\right]$$

Example II.3 – Thermal Resistance of Convection

A computer chip of surface area 10 cm^2 and length 5 cm is mounted on an insulator. The top of the chip is covered by a metal strip of thickness 2 cm. The thermal conductivity of the metal strip is 20 W/mK. The chip generates 10 W and is cooled by a stream of air at 25 °C flowing over the metal strip (Figure II.7).

(a) What is the upper surface temperature of the metal strip if the surface temperature of the computer chip is 100 °C?
(b) What is the thermal resistance of the convection process?
(c) What is the Nusselt number if the thermal conductivity of air is 1.0 W/mK?

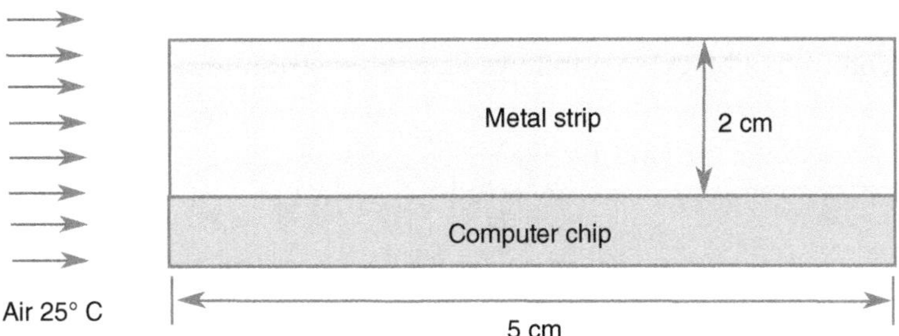

Figure II.7 Computer chip with a metal strip mounted on it.

Solution

(a) The upper surface temperature of the metal strip is obtained from Equation II.2:

$$Q = kA\frac{\left(T_{higher} - T_{lower}\right)}{d} \quad [\text{W}]$$

$$10 = 20 \times 0.001 \times \frac{\left(100 - T\right)}{0.02}$$

Therefore $T = 90$ °C.

(b) The thermal resistance of convection is obtained from Equation II.1:

$$Q = \frac{\left(T_{higher} - T_{lower}\right)}{R_v} \quad [\text{W}]$$

$$10 = \frac{(90 - 25)}{R_v}$$

$$R_v = 6.5 \text{ K/W}$$

(c) The Nusselt number is obtained from Equation II.8:

$$R_v = \frac{1}{\mathbb{N}A} \times \frac{X}{k} \quad [\text{K/W}]$$

$$6.5 = \frac{1}{\mathbb{N} \times 0.001} \times \frac{0.05}{1}$$

$$\mathbb{N} = 7.7$$

II.4 Radiation

Radiation is heat transfer due to electromagnetic waves and does not require any medium for heat exchange, although it can take place through a gas. When radiation meets an object, part of it is absorbed, part is reflected back from the surface and part is transmitted through the material. After absorbing heat the object then re-emits radiation as it warms.

A blackbody is a perfect emitter or absorber of radiation. The spectrum of power emission from a blackbody follows Planck's distribution:

$$E(\lambda,T)=\frac{2\pi hc^2}{\lambda^5\left[\exp\left(\frac{hc}{\lambda kT}\right)-1\right]}=\frac{C_1}{\lambda^5\left[\exp\left(\frac{C_2}{\lambda T}\right)-1\right]}\quad\left[\mathrm{W/m^2\mu m}\right] \tag{II.9}$$

where

$E(\lambda,T)$ is the spectral power density in W/m^{2}μm,
λ is the wavelength in μm,
T is the absolute temperature in K,
k is Boltzmann's constant, 1.38×10^{-23} J/K,
c is the speed of light, 2.998×10^{8} m/s,
h is Planck's constant, 6.626×10^{-34} Js,
C_1 is $2\pi hc^2$, 3.74×10^{8} Wμm^4/m^2,
C_2 is hc/k, 1.44×10^{4} μmK.

The wavelength of maximum radiation (λ_m) is found by differentiating Equation II.9 with respect to wavelength and setting the resulting expression to zero. This leads to Wien's displacement law:

$$\lambda_m T = 2898\ \mu\mathrm{mK} \tag{II.10}$$

Integrating Planck's distribution, Equation II.9, across all wavelengths gives the total power emitted from a blackbody:

$$E_{total}=\sigma T^4\quad\left[\mathrm{W/m^2}\right] \tag{II.11}$$

where

E_{total} is the power density in W/m^2,
σ is the Stefan Boltzmann constant, 5.67×10^{-8} Wm^{-2}K^{-4}.

Figure II.8 (plotted from Equation II.9) shows the blackbody spectra at three temperatures:

6000 K, the approximate surface temperature of the sun,
400 K, 117 °C, the maximum temperature of a flat-plate solar water heater that might be anticipated,
300 K, 27 °C, room temperature.

In Figure II.8 the x-axis of the wavelength uses a linear scale, while in Figure II.9 a logarithmic scale is used. In both plots the y-axis uses a logarithmic scale to accommodate the 6 orders of magnitude difference in the peak emissive power from the low and high temperatures. The sun, a high-temperature source, emits much higher levels of radiation at a shorter wavelength than the blackbodies at the lower temperatures.

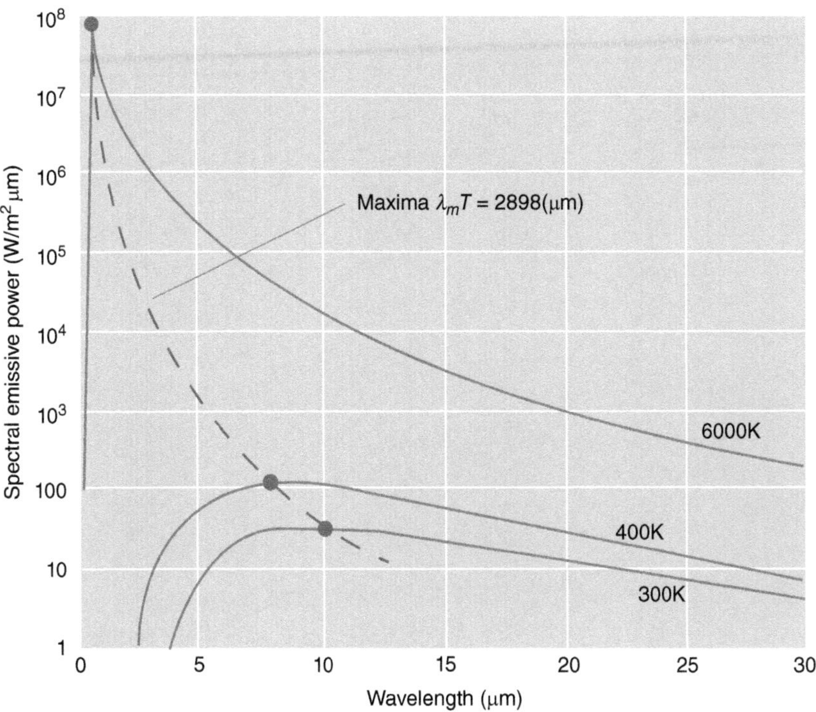

Figure II.8 Spectra of blackbody radiation at 6000 K, 400 K and 300 K (linear–log scale).

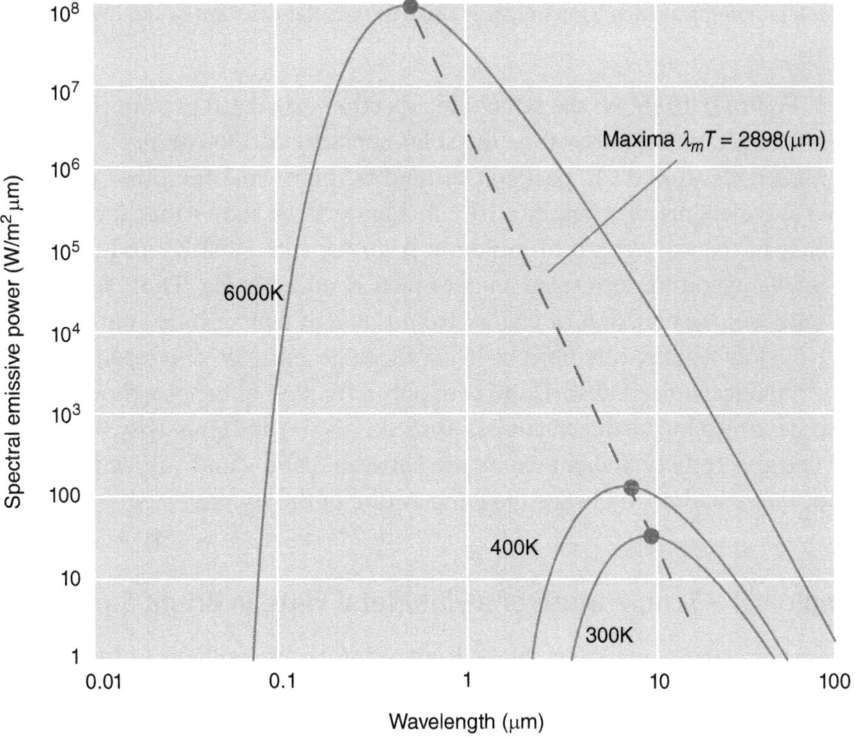

Figure II.9 Spectra of blackbody radiation at 6000 K, 400 K and 300 K (log–log scale).

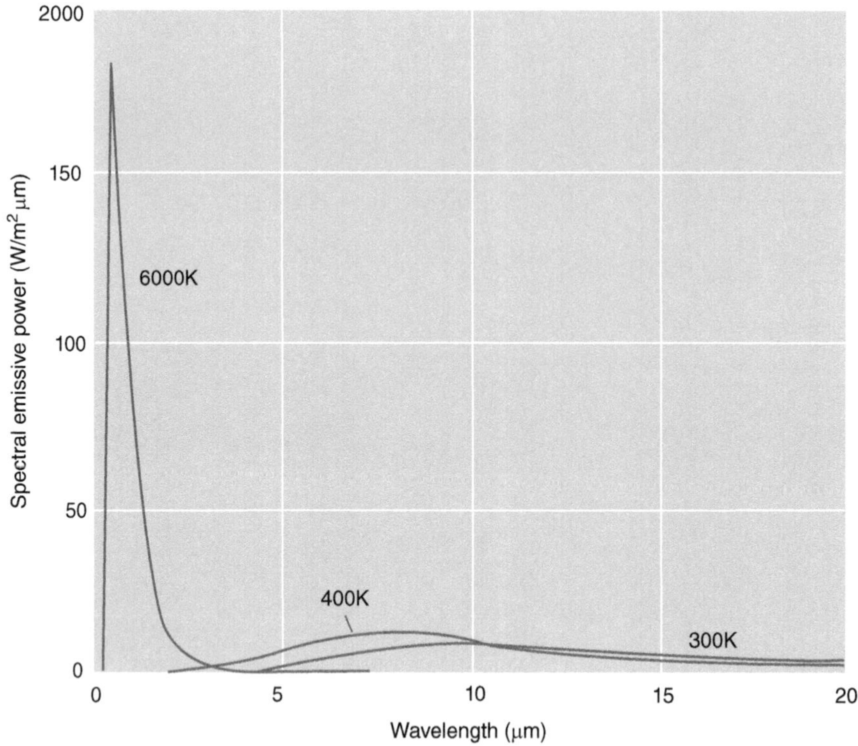

Figure II.10 Spectra of blackbody radiation at 6000 K, 400 K and 300 K, normalized so that total radiation at each temperature is equal to the solar constant of 1360 W/m^2.

Figure II.10 shows the same three spectra normalized to a total radiation at each temperature that is equal to the solar constant of 1360 W/m^2. The spectrum of a blackbody at 6000 K peaks at around 0.5 μm while the other two spectra peak at wavelengths of around 5–10 μm. Figure II.10 shows that there is clear difference in the wavelength of radiation from the sun (6000 K) and that emitted from low-temperature terrestrial sources (300 K and 400 K). Thus, for many practical purposes the emission of energy from the high temperature source of the sun and from low temperature objects on earth can be considered independently.

Practical objects absorb and emit only a fraction of the blackbody radiation. This is described by the dimensionless absorptivity (α) and emissivity (ε) of the material. The absorptivity and emissivity are between 0 and 1 and vary with the wavelength of the radiation and hence the temperature of the source.

Example II.4 – Temperature of a Flat Metal Plate in Bright Sunlight

A flat metal plate painted matt black is placed in bright sunlight of irradiance (G) of 800 W/m^2. Considering only radiative heat transfer, estimate the maximum equilibrium temperature of the plate.

The plate has the following properties:

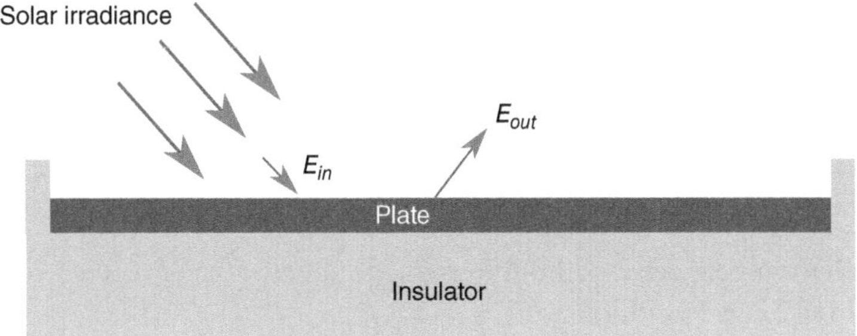

Figure II.11 Solar absorption and emission.

Absorptivity of high temperature radiation (α) is 0.9.
Emissivity of low temperature radiation (ε) is 0.95.
All effects due to convection and conduction are ignored.
σ is the Stefan Boltzmann constant (5.67×10^{-8} Wm^{-2}K^{-4}).

Solution

The plate is shown in Figure II.11.

The irradiance absorbed by the plate is $E_{in} = G\alpha \quad \left[\text{W/m}^2\right]$ (where G is the irradiance):

$$\text{Absorbed irradiance, } E_{in} = 800 \times 0.9 = 720 \text{ W/m}^2$$

The radiation emitted by the plate when it is heated (Equation II.9) is given by

$$E_{out} = \varepsilon\sigma T^4 \quad \left[\text{W/m}^2\right]$$

At thermal equilibrium the absorbed irradiance and emitted radiation balance, so:

$$T = \sqrt[4]{\frac{E_{in}}{\varepsilon \times \sigma}} = \sqrt[4]{\frac{720}{0.95 \times 5.67 \times 10^{-8}}} = 340 \text{ K} = 67\text{ °C}$$

In practice, the temperature would be lower due to heat loss through conduction to the ground and particularly due to convection if there is wind.

Consider two objects, one a radiation object at temperature T_1 and the other an absorbing object at temperature T_2. The radiation exchange between the two objects is given by:

$$E = \varepsilon\sigma\left[T_1^4 - T_2^4\right] \quad \left[\text{W/m}^2\right] \tag{II.12}$$

where

ε is the relative emissivity between the emitting and absorbing surfaces.

With the assumption that all the radiation that is emitted by the radiation surface meets the absorbing surface, the relative emissivity is given by:

$$\frac{1}{\varepsilon} = \frac{1}{\varepsilon_1} + \frac{A_1}{A_2}\left(\frac{1}{\varepsilon_2} - 1\right)$$

where

ε_1 is the emissivity of the radiating surface,
ε_2 is the emissivity of the absorbing surface,
A_1 is the area of the radiating surface,
A_2 is the area of the absorbing surface.

For radiation between two infinite parallel plates, area A_1 is equal to A_2 and then Equation II.12 can be written as:

$$E = \frac{\sigma\left[T_1^4 - T_2^4\right]}{\frac{1}{\varepsilon_1} + \frac{1}{\varepsilon_2} - 1} \quad \left[\text{W/m}^2\right]$$

For a small convex object surrounded by a large enclosure, A_1/A_2 tends to 0 and Equation II.12 reduces to

$$E = \varepsilon_1\sigma\left[T_1^4 - T_2^4\right] \quad \left[\text{W/m}^2\right]$$

where the power density is with respect to area A_1.

The sky can be considered as a blackbody at some equivalent sky temperature T_s so that the net radiation between a horizontal surface and the sky is given by

$$E = \varepsilon_1\sigma\left[T_1^4 - T_s^4\right] \quad \left[\text{W/m}^2\right]$$

Example II.5 – Heat Transfer Through Radiation and Convection

One side of a thin metal plate is placed in bright sunlight of irradiance (G) of 700 W/m^2. The other side of the plate is insulated as shown in Figure II.11. The absorptivity of the exposed surface is 0.7 and the emissivity is 0.7. Assume that radiation takes place to the sky, which is at an equivalent temperature −10 °C. The ambient temperature is 25 °C. The convective heat transfer coefficient is 20 W/m^2 K. Determine the surface temperature of the plate.

Solution

The absorbed irradiance is $E_{in} = G\alpha = 700 \times 0.7 = 490 \text{ W/m}^2$

If the surface temperature of the plate is T [K] then from Equation II.12 the heat transfer per unit area due to radiation is

$$\begin{aligned} E_{out1} &= \varepsilon_1\sigma\left(T^4 - T_s^4\right) = 0.7 \times 5.67 \times 10^{-8} \times \left(T^4 - 263^4\right) \\ &= 3.97 \times 10^{-8} \times T^4 - 190 \quad \left[\text{W/m}^2\right] \end{aligned}$$

The heat transfer per unit area due to convection is

$$E_{out2} = h\left(T - T_a\right) = 20 \times \left(T - 298\right) \quad \left[\text{W/m}^2\right]$$

At thermal equilibrium, $E_{in} = E_{out1} + E_{out2}$:

$$490 = \left(3.97 \times 10^{-8} \times T^4 - 190\right) + \left(20 \times \left(T - 298\right)\right)$$

$$20 \times T = 6640 - 3.97 \times 10^{-8} \times T^4$$

Forming the iterative equation, we have

$$(T)_{n+1} = \frac{\left[6640 - 3.97\times10^{-8}\times(T)_n^4\right]}{20}$$

For the start of the iteration assume $(T)_1 = 300$ K

n	$(T)_n$	$(T)_{n+1}$
1	300	315.9
2	315.9	312.2
3	312.2	313.1
4	313.1	312.9
5	312.9	313.0

$$\therefore T_s = 313 \text{ K}$$

In analogy to conduction and convection, the thermal resistance of radiation is defined as

$$\begin{aligned} R_r &= \frac{[T_1 - T_2]}{\varepsilon\sigma A\left[T_1^4 - T_2^4\right]} \\ &= \frac{[T_1 - T_2]}{\varepsilon\sigma A\left[T_1^2 + T_2^2\right]\left[T_1^2 - T_2^2\right]} \\ &= \frac{[T_1 - T_2]}{\varepsilon\sigma A\left[T_1^2 + T_2^2\right][T_1 + T_2][T_1 - T_2]} \\ &= \frac{1}{\varepsilon\sigma A\left[T_1^2 + T_2^2\right][T_1 + T_2]} \quad [\text{K/W}] \end{aligned} \tag{II.13}$$

where

A is the surface area of the radiating surface.

If $T_1 \approx T_2$, then with $T_{12} = \dfrac{T_1 + T_2}{2}$, Equation II.13 can be approximated to:

$$R_r \approx \frac{1}{4\varepsilon\sigma A T_{12}^3} \quad [\text{K/W}] \tag{II.14}$$

Example II.6 – Thermal Resistance of Radiation

An object with an effective surface area of 0.2 m^2 and temperature 100 °C is radiating to the sky at −10 °C. If the emissivity of the surface is 0.85, calculate the thermal resistance using the exact Equation II.13 and the approximate Equation II.14. What is the percentage difference?

Solution

Using the exact Equation II.13:

$$R_r = \frac{1}{\varepsilon\sigma A\left[T_1^2 + T_2^2\right]\left[T_1 + T_2\right]}$$

$$= \frac{1}{0.85 \times 5.67 \times 10^{-8} \times 0.2 \times \left[373^2 + 263^2\right]\left[373 + 263\right]}$$

$$= 0.78 \text{ K/W}$$

Using the approximate Equation II.14:

$$T_{12} = \frac{373 + 263}{2} = 318 \text{ K}$$

$$R_r \approx \frac{1}{4\varepsilon\sigma A T_{12}^3}$$

$$\approx \frac{1}{4 \times 0.85 \times 5.67 \times 10^{-8} \times 0.2 \times 318^3}$$

$$\approx 0.806 \text{ K/W}$$

The percentage difference is

$$\frac{0.806 - 0.78}{0.78} = 3.3\%$$

II.5 Heat Transfer Through Mass Flow of Fluid

In a flat-plate solar water heater, a collector plate with an array of tubes absorbs heat from solar radiation and passes it to water insides the tubes. The heated water is then transferred to the water tank.

In general, the heat that will be transferred due to mass flow of fluid is described by

$$Q = \dot{m} c_p \Delta T \tag{II.15}$$

where

Q is the heat transferred in W,
$\dot{m}$ is the mass flow rate in kg/s,
c_p is the specific heat of the fluid in J/kgK,
ΔT is the temperature difference in K.

Example II.7 – Heat Transfer in an Unglazed Flat-Plate Solar Water Heater

Solar irradiance (G) of 700 W/m^2 is incident on a 1 m^2 collector plate as shown in Figure II.12. The absorptivity of the exposed surface is 0.9 and the emissivity is 0.9.

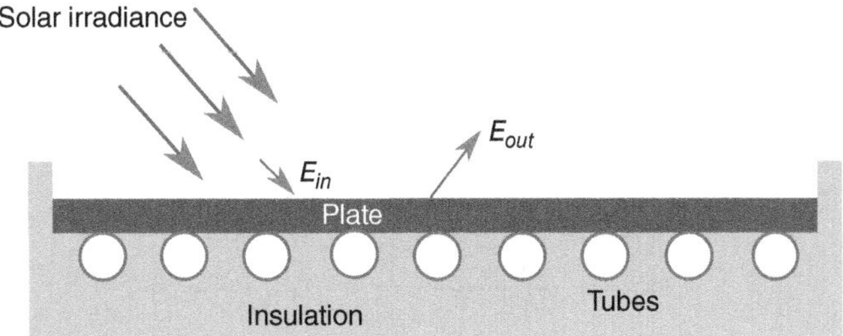

Figure II.12 Solar collector with water tubes.

It can be assumed that radiation takes place to the sky at a temperature −10 °C. The ambient temperature is 20 °C. The convective heat transfer coefficient is 20 W/m^{2}K. The water flow rate through the tubes is 0.009 kg/s and the specific heat capacity of the water is 4.179 kJ/kgK. If the surface temperature of the collector plate is 25 °C, determine the increase in the temperature of the water.

Solution

Absorbed irradiance, $Q_{in} = GA\alpha = 700 \times 1 \times 0.9 = 630$ W (where G is the irradiance).

The heat transfer due to radiation is

$$Q_{out1} = \varepsilon\sigma A\left[T_1^4 - T_2^4\right] \quad [\text{W}]$$

$$\begin{aligned} Q_{out1} &= 0.9 \times 5.67 \times 10^{-8} \times 1 \times \left[298^4 - 263^4\right] \\ &= 158.28 \text{ W} \end{aligned}$$

The heat transfer due to convection is

$$\begin{aligned} Q_{out2} &= hA(T_1 - T_a) \\ &= 20 \times 1 \times (298 - 293) = 100 \text{ W} \end{aligned}$$

The heat transferred due to mass flow of water is

$$Q_{out3} = \dot{m}c_p\Delta T = 0.009 \times 4179 \times \Delta T = 37.61 \times \Delta T \quad [\text{W}]$$

At thermal equilibrium, $Q_{in} = Q_{out1} + Q_{out2} + Q_{out3}$:

$$\begin{aligned} 630 &= 158.28 + 100 + 37.61 \times \Delta T \\ \Delta T &= 9.88 \text{ °C} \end{aligned}$$

II.6 One-Dimensional Heat Transfer

Consider the heat transferred by conduction in the radial direction of the pipe shown in Figure II.13. From Equation II.2

$$Q = \frac{kA\Delta T}{\Delta r} \tag{II.16}$$

where $A = 2\pi rl$ with $\Delta r \to 0$. Equation II.16 is rearranged as:

$$\frac{Qdr}{2\pi rlk} = dT$$

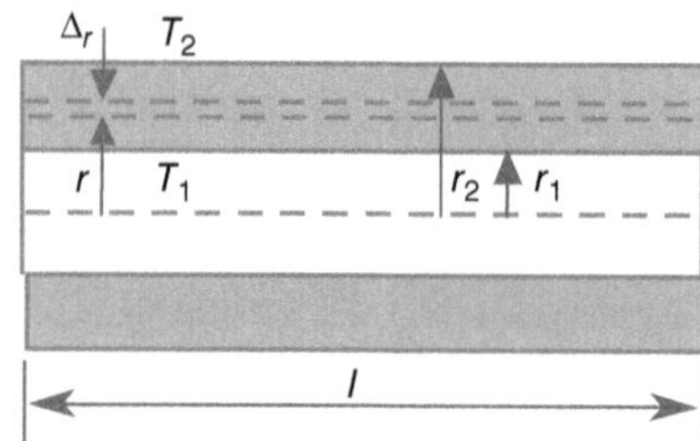

Figure II.13 One-dimensional heat transfer through a pipe.

Integrating from the inner surface to the outer surface, the following equation is obtained:

$$\int_{r_1}^{r_2} \frac{Qdr}{2\pi rlk} = \int_{T_1}^{T_2} dT$$

$$\frac{Q}{2\pi lk}\ln\left(\frac{r_2}{r_1}\right) = T_2 - T_1$$

$$\therefore Q = \frac{2\pi lk\left(T_2 - T_1\right)}{\ln\left(\frac{r_2}{r_1}\right)} \tag{II.17}$$

Therefore the thermal resistance for a cylinder is given by

$$R = \frac{\ln\left(\frac{r_2}{r_1}\right)}{2\pi lk} \tag{II.18}$$

Example II.8 – A Steam Pipe

An insulated steel pipe of thickness 4 mm and radius 10 cm carries steam at 250 °C. The thickness of the insulation layer is 2 cm. The convection heat transfer coefficient inside the pipe is 60 W/m^{2}K and outside is 25 W/m^{2}K. Ambient air is at 25 °C. The thermal conductivity of the steel is 15 W/mK and that of the insulation is 0.2 W/mK. Calculate the rate of heat loss per metre.

Solution

The resistance equivalent is shown in Figure II.14.

A 1 m length of the pipe is considered.

Inside the steam pipe, heat transfer takes place due to convection, and the equivalent resistance is given by

$$R_{\text{Inside}} = \frac{1}{hA} = \frac{1}{60 \times 2\pi \times 0.1 \times 1} = 0.027\ \text{K/W}$$

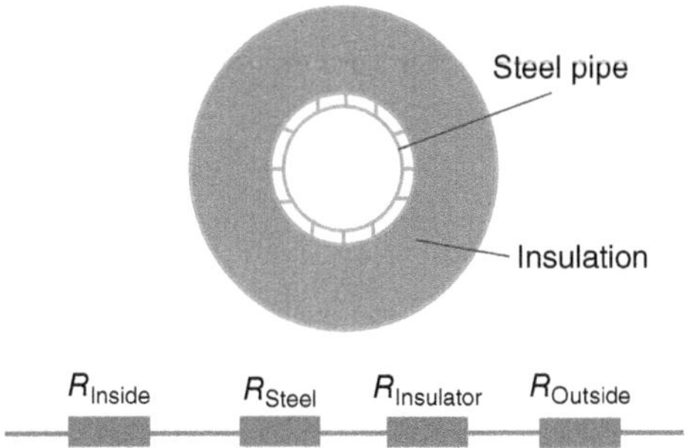

Figure II.14 Resistance equivalent for the steam pipe.

For the steel pipe:

$$R_{Steel} = \frac{\ln\left(\frac{r_2}{r_1}\right)}{2\pi lk} = \frac{\ln\left(\frac{0.104}{0.1}\right)}{2\times\pi\times1\times15} = 0.0004 \text{ K/W}$$

For the insulation:

$$R_{Insulation} = \frac{\ln\left(\frac{r_2}{r_1}\right)}{2\pi lk} = \frac{\ln\left(\frac{0.12}{0.104}\right)}{2\times\pi\times1\times0.2} = 0.114 \text{ K/W}$$

Outside the steam pipe, heat transfer takes place due to convection and the equivalent resistance is given by

$$R_{Outside} = \frac{1}{hA} = \frac{1}{25\times2\pi\times0.12\times1} = 0.053 \text{ K/W}$$

Total equivalent resistance = 0.027+0.0004+0.114+0.053 = 0.1944 K/W

$$\text{The rate of heat loss} = \frac{\Delta T}{R_{Total}} = \frac{250-25}{0.1944} = 1157.4 \text{ W/m}$$

PROBLEMS

1. In a heat exchanger, where a tube wall separates hot and cold fluid streams, discuss the main modes of heat transfer. If the average temperature of hot fluid is T_h and that of cold fluid is T_c, write an equation for the heat transfer by conduction.
2. What is a blackbody? Write an equation for the spectrum of power emission from a blackbody.
3. A glass window of 60 cm × 100 cm has 5 mm thick glass. The thermal conductivity of the glass is 1.0 W/mK. The inner and outer temperatures of the window on a winter's day are 20 °C and −5 °C. Calculate the heat loss through the glass due to conduction.

   [**Answer:** 3 kW.]

4. A cold store has a wall consisting of a 10 cm thick brick wall on the outside, then a 7.5 cm thick concrete wall and then 10 cm of cork. The mean temperature within the store is maintained at −10 °C and the mean temperature of the outside surface of the wall is 20 °C. Calculate the heat gain through the wall per unit area. The thermal conductivities are: for brick 0.69 W/mK, for concrete 0.76 W/mK and for cork 0.04 W/mK.

   [**Answer:** 10.9 W.]

5. The double-glazed window shown in Figure II.15 has two sheets of glass 60 cm × 100 cm separated by an air space of 1 cm. The thickness of each glass sheet is 5 mm. The inner and outer temperatures of the window are 20 °C and −5 °C on a winter's day. If the thermal conductivity of glass is 1.0 W/mK and that of air is 0.5 W/mK, calculate
   (a) the temperature of the glass surfaces A and B,
   (b) the heat loss through the window due to conduction.

   Neglect any heat loss due to convection and radiation.

   [**Answer:** (a) $T_A = 15.83$ °C, $T_B = -0.83$ °C; (b) $Q = 0.5$ kW.]

6. The temperature inside a house is 15 °C and that outside the house is −10 °C. The house has solid brick walls 25 cm thick and the thermal conductivity of the brick is 0.7 W/mK. The convective heat transfer coefficient of the wall surface inside the house is 10 W/m^{2}K and that outside is 20 W/m^{2}K. Calculate the equivalent thermal resistance and hence the heat transfer per unit area.

   [**Answer:** 0.507 K/W, 49.3 W/m^2.]

7. A 2.5 cm radius pipe carries hot water at 100 °C. If the external surface of the pipe has a convective heat transfer coefficient of 5 W/m^{2}K and the surrounding temperature is 20 °C, calculate the heat loss per unit length of the pipe.

   [**Answer:** 62.83 W/m.]

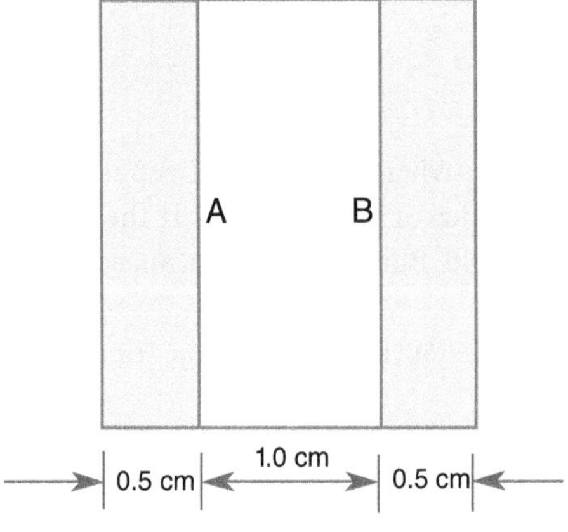

Figure II.15 The double-glazed window.

8. A plate 30 cm × 20 cm with a thickness of 12 mm is made from steel. Its thermal conductivity is 15 W/mK. The bottom side of the plate is maintained at 100 °C by an electric heater. The convective heat transfer coefficient of the top surface with a flow of wind is 25 W/m^{2}K.

(a) If the plate is perfectly insulated on all sides except the top surface, what is the temperature of the top surface of the plate assuming that all heat is lost due to convection? Assume the ambient temperature is 20 °C.

(b) If the emissivity of the plate is 0.85, calculate the temperature of the top surface of the plate assuming that heat is lost due to convection and radiation. Assume that radiation takes place to the sky which is at an equivalent temperature −10 °C.

[**Answer:** (a) 98.4 °C; (b) 90.44 °C.]

FURTHER READING

Incropera F.P., Dewitt D.P., Bergman T.L. and Lavine A.S., *Fundamentals of Heat and Mass Transfer*, 2007, John Wiley.

Comprehensive treatment of convection, conduction and radiation with a large number of worked examples and problems.

Duffie J.A. and Beckmam W.A., *Solar Engineering of Thermal Processes*, 2nd ed., 1991, Wiley Interscience.

Provides fundamentals of solar radiation, solar collectors and their applications.

TUTORIAL III
Simple Behaviour of Fluids

III.1 Types of Flow

III.1.1 Steady Flow

In a steady flow the velocity and pressure at any point do not change with time. The cross-section of the stream of fluid may vary from point to point but again does not change with time. In practice the velocity and pressure may have small instantaneous variations but if the average values over a short time do not vary, then that stream of flow may be considered as steady. All the flows considered in this tutorial are steady.

III.1.2 Compressible and Incompressible Fluids

Fluids are divided into two groups: gases and liquids. The density of a gas changes with temperature and pressure and so a gas is compressible. In a liquid, a very large pressure variation is required to change the density, and for practical purposes all liquids considered in this book are incompressible.

III.1.3 Laminar and Turbulent Flow

A way of visualizing the flow of fluid is to imagine a set of streamlines. A streamline can be defined as an imaginary curve which is tangent to the direction of the instantaneous velocity of fluid. There is no flow of fluid across a streamline. The boundary of a flow is always defined by a streamline as there is no flow across it.

Figure III.1 shows streamlines within a stream-tube: the stream-tube is the volume bounded by the dark lines. The fluid flow within a stream-tube consists of a bundle of streamlines. Fluid can enter or leave the stream-tube only through its ends, as fluid cannot cross a streamline.

If a few drops of ink are inserted into a flow, and parallel streamlines as shown in Figure III.1 are visible, then the flow is considered to be laminar. In practice, laminar flow occurs rarely. Generally fluid moves in an irregular pattern and the streamlines change with time as shown in Figure III.2. This is turbulent flow.

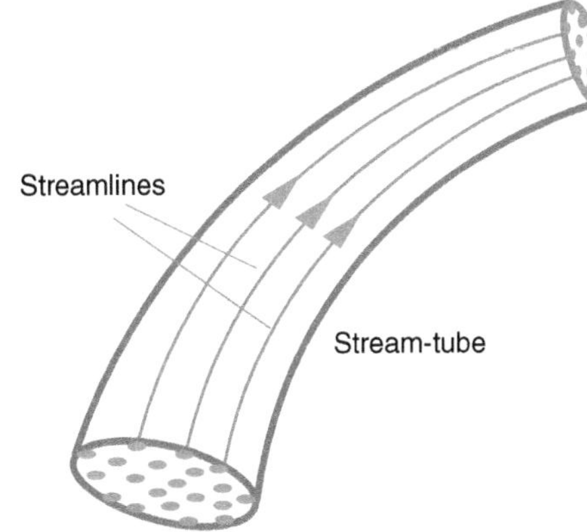

Figure III.1 Streamlines and stream-tube visualization of the laminar flow of fluid.

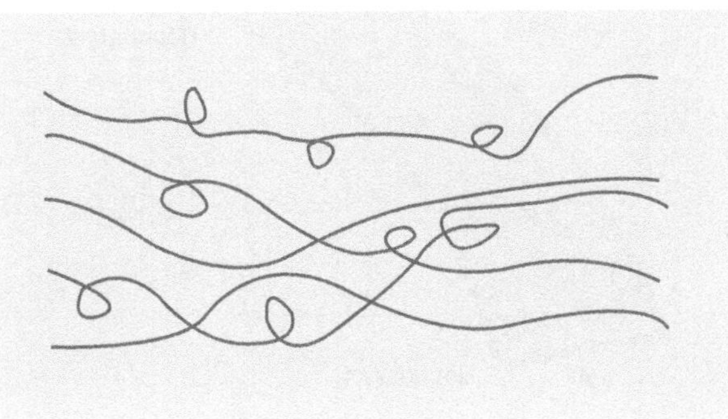

Figure III.2 Streamlines of turbulent flow.

III.2 Viscosity and Ideal Flow

Viscosity is the measure of a fluid's resistance to shear when the fluid is in motion. As a result of viscosity, shear stresses are developed between neighbouring particles in a fluid that are at different velocities. Consider two plates, one stationary and the other moving at velocity V. The velocity v of different layers of fluid will change from zero at the stationary plate to V at the moving plate, as shown in Figure III.3. Then the viscosity is defined as:

$$\eta = \frac{\tau}{\left[{dv}/{dy}\right]} = \tau \frac{d}{V}$$

where

η is the viscosity in Ns/m^2 or Pas,
τ is the shear stress in N/m^2,
V is the velocity of the fluid in m/s.

In a fluid in which the viscosity and shear stress are negligible, the flow is said to be ideal.

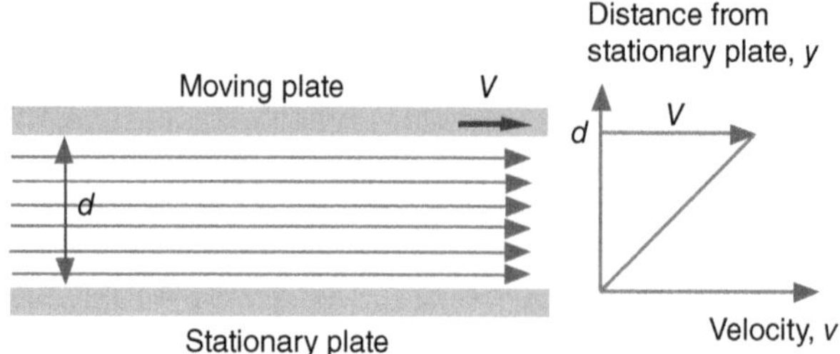

Figure III.3 Flow distribution across two plates: one stationary and the other moving.

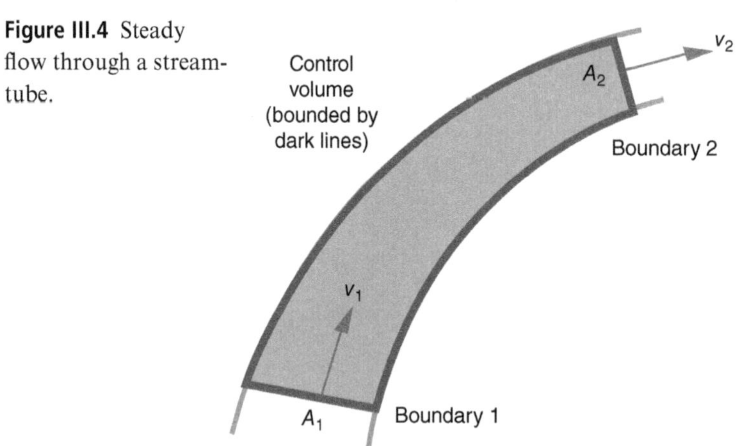

Figure III.4 Steady flow through a stream-tube.

III.3 Mass Continuity Equation

Consider steady flow through the stream-tube shown in Figure III.4. The mass of fluid entering at Boundary 1 leaves from Boundary 2. The volume bounded by Boundary 1, the stream-tube and Boundary 2 is the control volume. The mass flow rate (mass flow per second) is constant at Boundary 1, Boundary 2 and at any cross-section within the control volume.

$$\rho_1 v_1 A_1 = \rho_2 v_2 A_2$$

where

ρ_1 and ρ_2 are the densities of the fluid at Boundaries 1 and 2,
v_1 and v_2 are the average velocities of the fluid at Boundaries 1 and 2.

Example III.1 – Continuity of Mass Flow

At Section 1 of the circular pipe system shown in Figure III.5, the velocity of the water is 1 m/s. Find the velocity at Section 2 and the discharge flow rate.

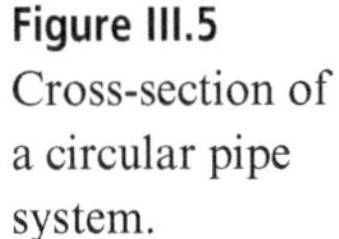

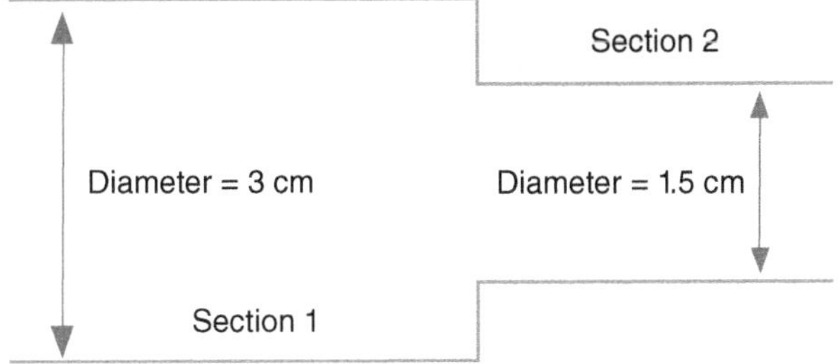

Figure III.5 Cross-section of a circular pipe system.

Solution

Since the density of water is constant, $v_1A_1 = v_2A_2$.

$$\text{Therefore: } 1 \times \pi\left(\frac{3}{2}\right)^2 = v_2 \times \pi\left(\frac{1.5}{2}\right)^2$$

$$v_2 = 4 \text{ m/s}$$

$$\text{Discharge flow rate} = 4 \times \pi\left(\frac{0.015}{2}\right)^2 = 0.0007 \text{ m}^3\text{/s or } 0.7 \text{ l/s}$$

III.4 Energy Balance: Bernoulli's Equation

Energy can enter or leave a control volume in the form of heat transfer, work done or mass flow. In this tutorial, heat transfer is neglected, and the density of the fluid is assumed to be constant.

For the control volume shown in Figure III.6, we use the following nomenclature.

Boundary 1: area A_1, pressure p_1, velocity v_1, height z_1
Boundary 2: area A_2, pressure p_2, velocity v_2, height z_2.

Let W_1 denote the work done by applying a pressure p_1 over an area A_1 that produces a displacement of the fluid by l_1. Then the work done at Boundary 1 by the pressure force (force × distance = pressure × area × distance) is $p_1A_1l_1$ $[\text{J or Nm}]$.

The rate of work done, $W_1 = p_1A_1v_1$ [Nm/s or W]

Since $\dot{m} = \rho v_1 A_1, W_1 = \dot{m} = \frac{p_1}{\rho}$ [Nm/s]

Similarly, the rate of work done at Boundary 2, by applying a pressure p_2 over an area A_2 and producing a displacement of the fluid by l_2, is $W_2 = \dot{m}\frac{p_2}{\rho}$.

The net rate of work done $(\dot{W}_1 - \dot{W}_2)$ is equal to the change in kinetic and potential energies. The addition of these two energies for a unit mass is given by:

$$e = \frac{v^2}{2} + gz \tag{III.1}$$

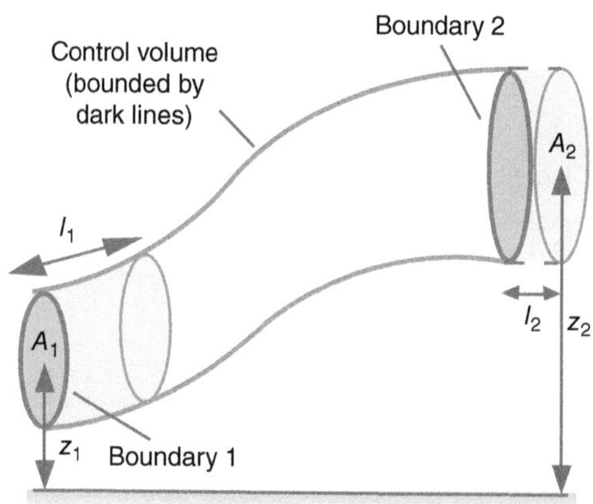

Figure III.6 Control volume for energy balance.

where

$\frac{v^2}{2}$ is the kinetic energy,

gz is the potential energy.

For the control volume shown in Figure III.6, by equating the net rate of work done to the change in kinetic plus potential energy per second, the following equation is obtained:

$$\dot{m}\left[\frac{p_1}{\rho}-\frac{p_2}{\rho}\right]=\dot{m}\left[\frac{v_2^2}{2}+gz_2\right]-\dot{m}\left[\frac{v_1^2}{2}+gz_1\right] \quad \text{(III.2)}$$

Cancelling $\dot{m}$ from both sides and rearranging Equation III.2 gives:

$$\frac{p_1}{\rho}+\frac{v_1^2}{2}+gz_1=\frac{p_2}{\rho}+\frac{v_2^2}{2}+gz_2 \quad \text{(III.3)}$$

For p_1 and p_2 gauge pressures are normally used.

From Equation III.3, $\frac{p}{\rho}+\frac{v^2}{2}+gz$ is a constant. This is one form of Bernoulli's equation, applicable along a stream-tube in a steady, incompressible flow.

Example III.2 – Application of the Bernoulli Equation

Water flows in a pipe of 3 m diameter at a velocity of 5 m/s, as shown in Figure III.7. It then flows down into another pipe of diameter 2 m. The height between the centres of the pipes is 4 m. The density is assumed to be uniform over the cross-sections. A uniform gauge pressure exists across Boundary 1 and Boundary 2. Calculate the velocity inside the smaller pipe.

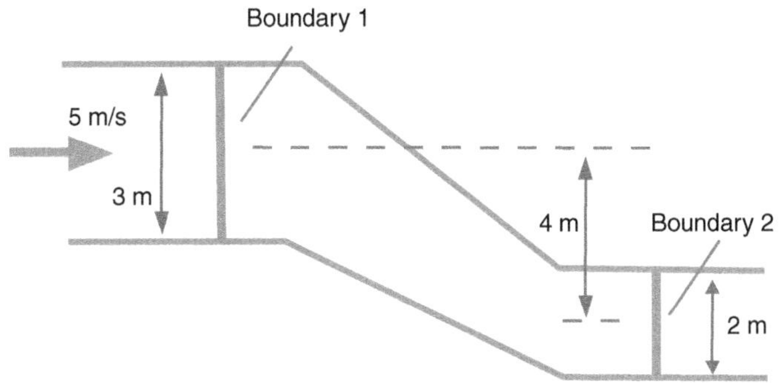

Figure III.7 Flow in two pipes.

Solution

The Bernoulli equation can be written as:

$$\left[\frac{p}{\rho}+\frac{v^2}{2}+gz\right]_1=\left[\frac{p}{\rho}+\frac{v^2}{2}+gz\right]_2$$

As gauge pressure is the same at the two boundaries,

$$p_1 = p_2$$
$$z_1 = 1+4 = 5 \text{ m};\ z_2 = 1 \text{ m}$$
$$v_1 = 5 \text{m/s}$$

Therefore

$$\left[\frac{5^2}{2}+9.81\times 5\right]=\left[\frac{v_2^2}{2}+9.81\times 1\right]$$
$$v_2 = 10.17 \text{ m/s}$$

Example III.3 – A Large Water Tank with Discharge

A large water tank is shown in Figure III.8. The gauge pressure measured at Boundary 1 is 80 kPa and that at Boundary 2 is 100 kPa. Find the velocity at Boundary 2.

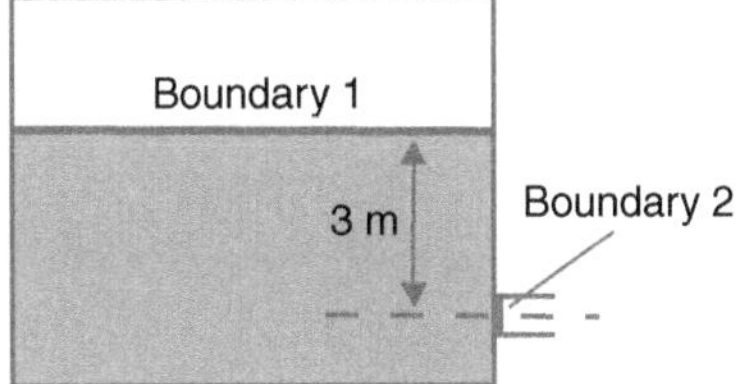

Figure III.8 Discharge from a water tank.

Solution

By selecting the two boundaries as shown in Figure III.8, from the Bernoulli equation, we have

$$\left[\frac{p}{\rho}+\frac{v^2}{2}+gz\right]_1=\left[\frac{p}{\rho}+\frac{v^2}{2}+gz\right]_2$$

$$p_1=80\text{ kPa};\ p_2=100\text{kPa}$$

$$z_1=3\text{ m};\ z_2=0\text{ m}$$

Since the tank is large, it can be assumed that $v_1=0$ m/s.

Therefore

$$\left[\frac{80\times10^3}{1000}+0+9.81\times3\right]=\left[\frac{100\times10^3}{1000}+\frac{v_2^2}{2}+9.81\times0\right]$$

$$v_2=4.34\text{ m/s}$$

In Figure III.6, it was assumed that the flow of fluid does no work within the control volume. However, if the fluid operates a turbine or does other work within the control volume then, from conservation of energy, Equation III.2 is modified as:

$$\dot{W}=\dot{m}\left[\frac{p_1}{\rho}+\frac{v_1^2}{2}+gz_1\right]-\dot{m}\left[\frac{p_2}{\rho}+\frac{v_2^2}{2}+gz_2\right]\ [\text{W}] \qquad \text{(III.4)}$$

where

$\dot{W}$ is the rate of work done inside the control volume.

Example III.4 – Turbine Operation

Figure III.9 shows a simplified representation of a turbine supplied from a large reservoir. If the water is discharged from the pipe at Boundary 2 at atmospheric pressure at a rate of 1 m^3/s, calculate the power generated by the turbine.

Solution

Assume the reservoir is large and $v_1=0$.

As the reservoir and discharge are at atmospheric pressure, $p_1=p_2=0$.

Since the discharge rate is 1 m^3/s, we have

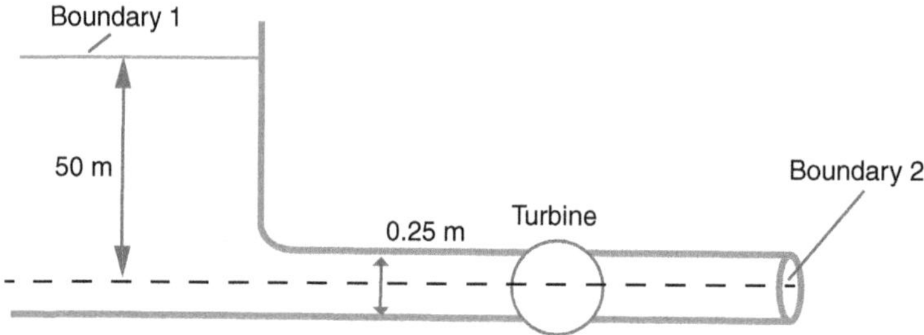

Figure III.9 Simple turbine.

$$v_2 \times \pi \times \left(\frac{0.25}{2}\right)^2 = 1$$

$$v_2 = 20.37 \text{ m/s}$$

$$\text{Mass flow rate} = 1000 \text{ kg/s}$$

Work done per second, or the power produced by the turbine, is

$$\dot{W} = \dot{m}\left[\frac{p_1}{\rho} + \frac{v_1^2}{2} + gz_1\right] - \dot{m}\left[\frac{p_2}{\rho} + \frac{v_2^2}{2} + gz_2\right]$$

$$= 1000[9.81 \times 50] - 1000\left[\frac{20.37^2}{2}\right] = 283 \text{ kW}$$

III.5 Linear and Angular Momentum

Consider a jet of water striking a wall as shown in Figure III.10. The mass flow rate is $\dot{m}$ and the initial velocity of the jet is v_1 in the horizontal direction. The flow hits the wall and is turned through 90°. If we consider motion in the horizontal direction only, velocity v_2 is zero. The fluid exerts a force on the wall as shown.

From Newton's second law,

$$\text{Force} = \text{Rate of Change of Momentum}$$

Thus the force exerted by the water jet on the wall is equal to the difference between the rates of change of momentum of the jet before and after it is diverted through 90°. Since momentum is a vector quantity and equal to mass times velocity, by equating the rate of change of momentum of the jet before and after it is diverted through 90° in the horizontal direction we may write

$$\text{Force} = \dot{m}_1 v_1$$

For a rotating system such as the Francis turbine shown in Figure III.11 (in which the runner of the turbine can only rotate), the net torque acting on the system is equal to the difference in the rates of change of angular momentum of the water jet at the inlet and outlet.

$$T = \dot{M}_{inlet} - \dot{M}_{outlet}$$

where

$\dot{M}$ is the rate of change of angular momentum.

The tangential velocity, radial velocity and radius of the jet of water entering the turbine are $V_{\theta 1}$, V_{R1} and r_1 and those leaving the turbine are $V_{\theta 2}$, V_{R2} and r_2.

Considering velocities and forces in the tangential (circumferential) direction only, the rate of change of angular momentum of the fluid at the inlet is mass flow rate × velocity in the tangential direction × radius of rotation $= m \times r_1 V_{\theta 1}$.

Figure III.10 A jet of fluid hitting a wall.

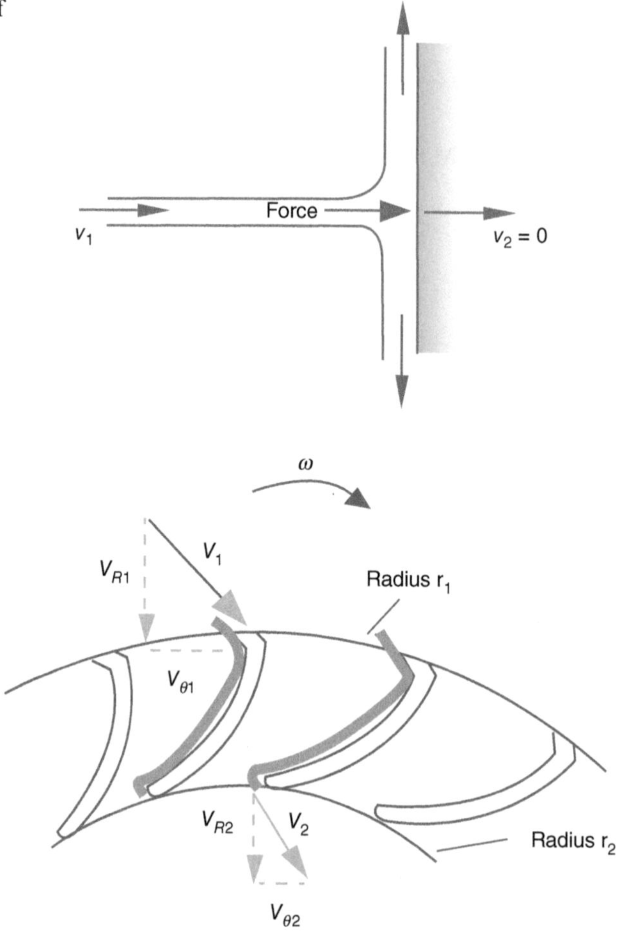

Figure III.11 A part of a Francis turbine.

If the volumetric fluid flow rate through the turbine is Q m^3/s, then the mass flow rate is $\dot{m} = \rho Q$. Therefore

$$\dot{M}_{inlet} = V_{\theta 1} \rho Q r_1$$

Similarly, for the outlet, the rate of change of angular momentum of the fluid is

$$\dot{M}_{outlet} = V_{\theta 2} \rho Q r_2$$

Then the torque, $T = \rho Q r_1 V_{\theta 1} - \rho Q r_2 V_{\theta 2}$.
Since, in a rotating system, power = torque × angular velocity ($P = T\omega$),

$$P = \rho Q r_1 V_{\theta 1} \omega - \rho Q r_2 V_{\theta 2} \omega = \rho Q \left(V_{\theta 1} U_1 - V_{\theta 2} U_2 \right)$$

where

U_1 is the tangential velocity of the runner at radius r_1 $(r_1 \omega)$,
U_2 is the tangential velocity of the runner at radius r_2 $(r_2 \omega)$.

III.6 Flow Through Pipe Systems and the Moody Chart

Fluid flow in pipes can be laminar or turbulent depending on the velocity of flow and viscosity of the fluid. Laminar and turbulent flows are characterized using the Reynolds number.

The Reynolds number of flow in a pipe is defined as:

$$\mathrm{Re} = \frac{vD\rho}{\eta} \tag{III.5}$$

where

Re is the Reynolds number,
v is the velocity of flow in m/s,
D is the diameter of pipe in m,
ρ is the density of fluid in kg/m^3,
η is the viscosity in Pas.

When Re < 2000 the flow is laminar and when Re > 4000 the flow is turbulent. When 2000 < Re < 4000, it is not possible to say whether the flow is certainly laminar or turbulent.

By considering the friction losses associated with flow in a pipe, Equation III.3 can be modified as:

$$\frac{p_1}{\rho} + \frac{v_1^2}{2} + gz_1 - gh_L = \frac{p_2}{\rho} + \frac{v_2^2}{2} + gz_2$$

where

h_L is the head loss due to friction.

The loss due to friction is defined by Darcy's equation, given by

$$h_L = f \times \left(\frac{L}{D}\right) \times \frac{v^2}{2g} \tag{III.6}$$

where

f is the friction factor,
L is the length of the pipe.

For a laminar flow, the friction factor is given by:

$$f = \frac{64}{\mathrm{Re}}$$

Example III.5 – Laminar Flow in a Pipe

A fluid of density 1250 kg/m^3 flows in a 20 m long pipe of diameter 25 cm. If the velocity is 5 m/s and the viscosity of the fluid is 1.0 Pas, determine the Reynolds number and head loss due to friction.

Solution

From Equation III.5:

$$\text{Re} = \frac{vD\rho}{\eta} = \frac{5 \times 0.25 \times 1250}{1.0} = 1562.5$$

As Re < 2000 the flow is laminar.

Since the flow is laminar, the head loss due to friction is given by Equation III.6:

$$f = \frac{64}{\text{Re}} = \frac{64}{1562.5} = 0.041$$

$$h_L = f \times \left(\frac{L}{D}\right) \times \frac{v^2}{2g} = 0.041 \times \left(\frac{20}{0.25}\right) \times \frac{5^2}{2 \times 9.81} = 4.18\text{ m}$$

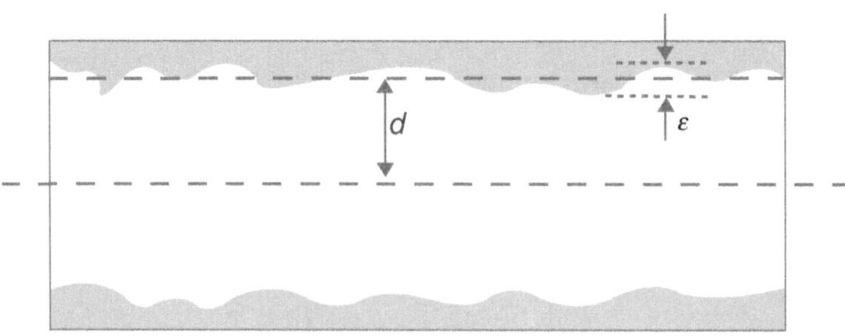

Figure III.12 A cross-section of a rough pipe.

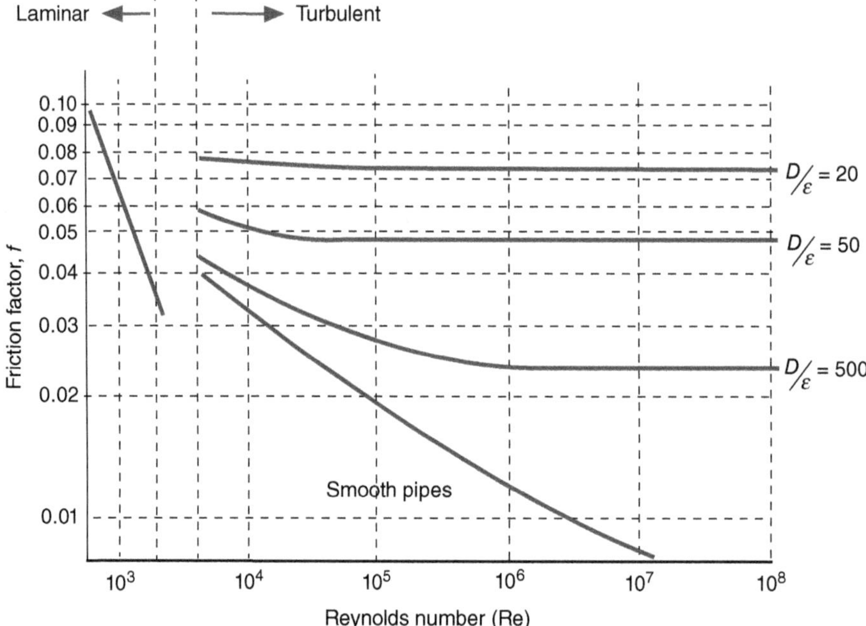

Figure III.13 Moody chart (note that the reciprocal of the relative roughness is shown on the right-hand axis).

For turbulent flow, the friction factor depends on the inside roughness of the pipe and the Reynolds number, so determining it is not straightforward. The roughness is defined by relative roughness, i.e. ε/D. As shown in Figure III.12, D is the mean diameter and ε is the roughness. The Moody chart shown in Figure III.13 gives the friction factor for various values of relative roughness and Reynolds number.

PROBLEMS

1. What is the difference between a laminar and a turbulent flow? Sketch the streamlines for laminar and turbulent flow.
2. What factors determine whether a flow within a pipe is laminar or turbulent? Define the Reynolds number and discuss how the Reynolds number is used to determine whether a flow is laminar or turbulent.
3. The diameter of the inlet of a horizontal pipe is 200 mm and the gauge pressure at the inlet is 100 kPa. Its diameter is reduced at the outlet such that velocity at the outlet is 1.5 times that at the inlet. If the volume flow rate is 0.2 m^3/s, calculate (a) the diameter, (b) the velocity and (c) the gauge pressure at the outlet.

   [**Answer:** (a) 163 mm; (b) 9.58 m/s; (c) 74.5 kPa.]
4. Water flows in a pipe of diameter 5 m at a velocity of 10 m/s. It then flows down into a smaller pipe of diameter 2 m. The height between the centre of pipe sections is 5 m. The density is assumed to be uniform over the cross-sections. The gauge pressure at Boundary 1 is 120 kPa. Calculate the velocity at the smaller pipe section.

   [**Answer:** 20.93 m/s.]
5. Figure III.14 shows a simplified representation of a turbine. If the water is discharged from Boundary 2 at a rate of 0.5 m^3/s, calculate the power generated by the turbine.

   [**Answer:** 243.8 kW.]
6. A fluid of density 1258 kg/m^3 flows through a pipe of 150 mm diameter at 3.6 m/s. Determine the Reynolds number. The viscosity of fluid is 0.96 Pas. Is the fluid flow laminar or turbulent?

   [**Answer:** 707.6, laminar.]

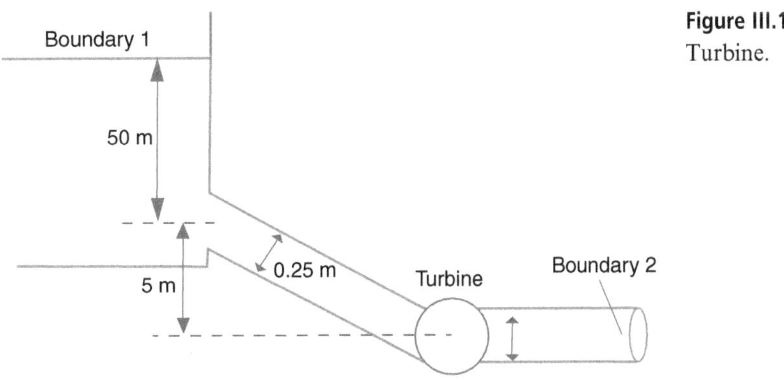

Figure III.14 Turbine.

7. A fluid of density 1250 kg/m^3 flows inside a pipe of diameter 75 mm at 10 m/s. The viscosity of the fluid is 1 mPas. Determine the friction factor for a galvanized iron pipe that has a roughness of 1.5×10^{-4} m.

   [**Answer:** 0.027.]

8. Oil, with density of 1200 kg/m^3 and viscosity of 1 Pas, flows at a volume flow rate of 0.07 m^3/s through a cast-iron pipe of 150 mm diameter. Determine
   (a) the Reynolds number,
   (b) the head losses per metre length of the pipe,
   (c) the pressure drop per metre length of the pipe if the pipe slopes down at 15° in the flow direction.

   [**Answer:** (a) 713; (b) 0.48 m; (c) 2.59 kPa/m.]

FURTHER READING

Massey B.S. and Ward-Smith J. *Mechanics of Fluids*, 9th ed., 2012, Spon Press.
A classic textbook.

Munson B.R., Okiishi T.H., Huebsch W.W. and Rothmayer A.P., *Fundamentals of Fluid Mechanics*, 2013, Wiley.
A comprehensive guide to fluid mechanics with worked examples and problems.

Streeter V.L. and Wylie E.B., *Fluid Mechanics*, 1983, McGraw-Hill.
Advanced book on fluid mechanics that covers fluid statics and dynamics.

Index